WIND ENERGY CONVERSION 1993

Conference Organizing Committee

K.F. Pitcher (Chairman)　　　　　　　J.A. Halliday
B.R. Clayton　　　　　　　　　　　　　P.B. Simpson

Conference Organization

D. Bodger
University of Nottingham

Exhibition Organization

S.R. Powles
Renewable Energy Systems Limited

Conference Papers Committee

K.F. Pitcher (Chairman)

B.R. Clayton　　　　　　　　　　　　J.A. Halliday
G. Elliot　　　　　　　　　　　　　　　D.I. Page
L.L. Freris　　　　　　　　　　　　　　D.C. Quarton
J.M. Graham　　　　　　　　　　　　　P.B. Simpson

Council of the British Wind Energy Association 1992–1993

Chairman　　　　　　　　　　　　　Membership Secretary
J.A. Halliday　　　　　　　　　　　　　C. Tan
Rutherford Appleton Laboratory　　Renewable Energy Systems Limited

Vice Chairman　　　　　　　　　　Editor of Windirections
P.B. Simpson　　　　　　　　　　　　D. Lindley
Wind Energy Group Limited　　　　National Wind Power Limited

Treasurer
M.B. Anderson
Renewable Energy Systems Limited

Council Members

B.R. Clayton　　　　　　　　　　　　N. Jenkins
University of Nottingham　　　　　UMIST, Manchester
L.L. Freris　　　　　　　　　　　　　N.H. Lipman (Co-opted)
Imperial College, London　　　　　Independent Consultant
A.D. Garrad　　　　　　　　　　　　D.J. Milborrow
Garrad Hassan &Partners Limited　Independent Consultant
R. Hunter　　　　　　　　　　　　　D.I. Page
National Engineering Laboratory　Energy Technology Support Unit
D.G. Infield　　　　　　　　　　　　K.F. Pitcher
Rutherford Appleton Laboratory　Yorkshire Water

Scottish Branch Chairman

G. Elliot
National Engineering Laboratory

Scottish Branch Representative

R.M. Morris

Readers wishing to contact Officers of the British Wind Energy Association should write to:
BWEA, 4 Hamilton Place, London W1V 0BQ.

WIND ENERGY CONVERSION 1993

Proceedings of the 15th British Wind Energy Association Conference

York 6–8 October 1993

Edited by

K.F. Pitcher

Yorkshire Water

Published by
Mechanical Engineering Publications Limited
LONDON

Published by
Mechanical Engineering Publications Limited
LONDON

Contents

Energy in tomorrow's world   1
  M. Jefferson

Overview or worldwide wind generation   9
  P. Gipe

Energy generation costs – now and for the year 2000   15
  D.J. Milborrow

The performance and problems of, and the public attitude to, the Delabole windfarm   23
  P.D. Edwards

A windfarm in a mixed industrial and residential area   27
  W. Grainger, D. Still, N. Rogers, and A. Gammidge

Viability of a windfarm in an area of low wind speed   31
  D.C. Corbet

Review of windfarm installations in the UK   37
  D.J. Leivesley and M.L. Hitner

Operating experience from four UK windfarms   41
  D. Lindley, P. Musgrove, J. Warren, and R. Hoskin

The Penrhyddlan and Llidiartywaun windfarms – the first to join the electricity pool   47
  C. Peasley and G. Nicholson

Interaction of Delabole windfarm and South Western Electricity's distribution system   55
  J.M. Haynes and M.J. Birks

Windspeed forecasting and its application to wind power generation   63
  S.J. Watson, L. Landberg, and J.A. Halliday

On the effect of the utility grid characteristics in wind park power output fluctuations   71
  Ana. I.L. Estanqueiro, R.F. Aguiar, J.A. Gil Saraiva, Rui M.G. Castro, and J.M. Ferreira de Jesus

Coastal wind speed prediction   77
  R.J. Barthelmie and J.P. Palutikof

Monitoring at Vindeby: preliminary results   83
  R.J. Barthelmie, M.S. Courtenay, J. Høstrup and P. Sanderhoff

Estimating the wind energy potential   91
  D.M. Hollis

On the turbulent structure of wind turbine wakes over rough terrain   99
  C.G. Helmis, K.H. Papadopoulos, A.T. Soilemes, P.G. Papageorgas, and D.N. Asimakopoulos

A database to aid the commercial development of renewable energy in the North East region
  of England   105
  G. Jenkins, R. Harrison, R. Foster, and N. Johnson

Wind energy resource on the South West of England   113
  H. Scholes and T. Masood

A comparison of physical and statistical methods for estimating the wind resource at a site   119
  L. Landberg and N.G. Mortenson

Determination of areas of similar mean wind velocities in complex terrain using the example of
  the Swabian Alb in Germany   127
  J. Albiger, A. Bohringer, H. Muh, and V. Reich

The influence of transmission system disturbances on the dynamic behaviour of a wind park   133
  Rui M.G. Castro, J.M. Ferreira de Jesus, Ana I.L. Estanqueriro, R. Aguiar, and J.A. Gil Saraiva

Assessing the impact of grid integrated renewable energy sources on the economics of an
  expanding power system   139
  S.J. Watson, A.G. Ter-Gazarian, J.A. Halliday, and S.C. Davis

Power quality measures for isolated power systems   147
  P. Nørgaard, P. Sørensen, and J. O. Tande

The financing of wind energy in the UK compared to other countries and its implications   153
  C. Mitchell

Financial structure for windfarms   161
  J.H. Johns

Realisation of wind park projects by cooperatives   169
  H. Holst

Benefits to the local community: the limits of proper planning gain   171
  G.M. Trinick

Environmental assessment of wind energy projects   175
  P.J. Radmall

Noise from wind turbines: activities in the DTI wind energy programme   181
  M.L. Legerton

Studies of the dynamics of yawed HAWTs in the DTI Wind Energy Programme   189
  J.M. Ward

The free yaw behaviour of upwind HAWTs   197
  J. Noakes, J.T.D. Slater, and M.B. Anderson

Analysis of data from a Howden HWP 330/33 wind turbine   205
  P. Jamieson

Relationship of the controllability of power/torque fluctuations in the drive-train to the wind
  turbine configuration   213
  M.C.M. Rogers and W.E. Leithead

Wind turbine tower structural and dynamic analysis using the finite element method   221
  N. El Chazly

An analytical approach to wind farm design   225
  A.D. Garrad, A.S. Mercer, B.M. Adams, and N.R. Jenkins

A method for the aerodynamically optimal design of wind parks   231
  S.G. Voutsinas and K.G. Rados

Economically optimized design of the electrical layout in wind farms   239
  J. Kristiansen

A wind park linearized model   247
  Rui M.G. Castro and J.M. Ferreiro de Jesus

The implications of fatigue on the cost of HAWTs   253
  N.D.P. Barltrop, I.P. Ward, and D.J. Daw

Methods for the rapid evaluation of fatigue damage on the Howden HWP 330 wind turbine   261
  N.W.M. Bishop, Z. Hu, R. Wang, and D. Quarton

Fatigue testing of wood composites for aerogenerator blades, part IX. Alternative adhesives   269
  C.L. Hacker, I.P. Bond, and M.P. Ansell

Fatigue design curves of fibreglass blade material compared to test data 275
P.A. Joosse and D.R.V. van Delft

Fatigue behaviour of fibreglass wind turbine blade material in the very high cycle range 281
D.R.V. van Delft, P.A. Joosse, and H.D. Rink

Resin transfer moulded aero generator blades 287
V. Middleton, I.A. Jones, and B.R. Clayton

Measurement of stall delay on a model of a stall controlled rotor 293
J.M.R. Graham and C.J. Brown

Pre field test development of the 'Fledge' aerodynamic braking and control device 299
R.S. Hunter, W.A. Derrick, N.I.D. Robertson, J.L. Chapman, and R.A. Court

An attempt to measure dynamic inflow phenomenon on a 1MW wind turbine 305
W.E. Leithead and M.C.M. Rogers

Investigation of fundamental trade-off in tracking the $C_{p\,max}$ curve of a variable speed
wind turbine 313
B. Connor and W.E. Leithead

Some control aspects of a small isolated wind turbine 321
M.T. Iqbal, A.H. Coonick, and L.L. Freris

Prescribed wake modelling of vertical and horizontal axis wind turbines at the University
of Glasgow 327
F.N. Coton, R.A. McD. Galbraith, and D. Jiang

Noise control on the BONUS 300kW wind turbine 335
H. Stiesdal and E. Kristensen

Development and demonstration of an advanced two-bladed HAWT 341
R.S. Haines, R.H. Sauven, and P.B. Simpson

Test results from the AOC 15-50 wind turbine development programme: tip brake life test 347
P. Hughes, S. Childs, and A. Facchetti

300 kW wind turbine erection in Romania 353
A. Gârbacea

Harmonisation of certification approaches 359
J.R. Maguire

Infra-red thermography for condition monitoring of composite wind turbine blades: feasibility
studies using cyclic loading tests 365
G.M. Smith, B.R. Clayton, A.G. Dutton, and A.D. Irving

Thermographic evaluation of static and dynamic damage to composite wind turbine blades 373
M.D. Wakeman, G.M. Smith, and B.R. Clayton

Systematic comparison of prediction and experiement for wind turbine aerodynamic noise 381
M.V. Lowson and J.V. Lowson

Machinery noise investigation and reduction in wind turbine generators 389
P.S. Watkinson and A.R. Clark

Viability of variable speed 395
E.A. Bossanyi

An assessment of the offshore wind potential in the EC 401
A.D. Garrad, B.M. Adams, H. Matthies, M. Scherwent, and T. Siebers

FLOAT – a floating offshore wind turbine system 407
K.C. Tong, D. Quarton, and R. Standing

Delegates to the Conference 415

ACKNOWLEDGEMENTS

The Council of the British Wind Energy Association
gratefully acknowledges the sponsorship given to the
Conference by the following bodies:

NATIONAL WIND POWER

the department for Enterprise

Wind Energy Group

Energy in tomorrow's world

M JEFFERSON
Director, British Energy Association
Deputy Secretary General, World Energy Council

Introduction

I have been asked to report on two recent World Energy Council publications. First, the World Energy Council (WEC) Commission's Report "Energy for Tomorrow's World". Secondly, "Renewable Energy Resources: Opportunities and Constraints 1990-2020", produced by the WEC's Renewable Energy Resources Committee.

These two reports total 700 pages; are each the product of three years work involving over 500 and 80 specialists, respectively; are global in coverage and go into some detail on prospects to 2020 - and in outline to the year 2100. Additionally, I have been asked to mention how the analysis and conclusions in these reports relate to UK wind energy developments.

I shall devote the major part of what follows to the findings of the WEC Commission, which was not merely a top-down exercise. Two features stand out. The multi-disciplinary nature of the Commission's Board - which comprised eminent people from a variety of disciplines and included some leading environmentalists. And the eight Regional Groups (one of which sub-divided to make nine in total), covering the whole world, which ensured the local relevance and practicality of the Commission's bottom-up work.

What I have to say is placed in a global context, noting that between 1900 and 1990 the world's population grew 3.2 times (to 5.3 billion), and world energy consumption grew over 14 times (to 8.8 Gtoe[1]).

Key Analytical Features

The first key feature is the likely evolution of world population. Taking UN and World Bank median projections, one can envisage world population more than doubling over the next century and the world's urbanised population expanding even faster:

Median World Population & Urbanisation Growth Estimates (billions)

	1990	2020	2050	2100
Total	5.3	8.1	10.1	12.0
Urbanised	2.3	4.7	7.0	9.5

---

[1]Gtoe = Gigatonnes oil equivalent

The urbanised total is anticipated to grow some four-fold, by an amount nearly 1 1/2 times the current world population. Given the many problems which confront our cities today, this is a hugely challenging prospect.

There should not be anything too startling in such projections, even if - as many historic examples demonstrate - population forecasting is a hazardous business. Between 1960 and 1990 developing countries accounted for 87% of world population growth, rising from 68% to 76% of the total. In the coming decades they are expected to account for some 90% of world population growth, and to represent about 90% of the world's population by 2100.

Developing countries' demand for the services which energy provides has risen 50% over the past three decades, compared with a 14% rise for the OECD member countries, despite the fact that in over 50 developing countries energy per capita consumption has actually declined in the past 15 to 20 years.

If we look, therefore, to the key upward driving forces on global energy demand they seem clearly to be:

- World Population Growth
- the Drive to Satisfy Basic Needs
- Expanding Material Expectations and Desires

A number of other driving forces can be highlighted:

- local, regional and global environmental concerns
- the lack of fuel wood
- longer-term supply concerns surrounding oil and natural gas (especially by 2050)
- likely energy price rises
- the need for greater efficiency in energy provision and use
- the scope for technological innovation and diffusion
- the more effective pricing of energy to reduce waste and inefficiency, and to discourage further development of inappropriate or inefficient systems
- the accelerated development and diffusion of renewable energy subject to satisfaction of efficiency and environmental criteria.

These, and a number of other factors, were incorporated in four alternative cases developed by the WEC Commission. These cases are not forecasts, and none reflect a Business As Usual evolution. For instance, all assume more rapid improvements in energy efficiency (as rather roughly measured by reductions in energy intensity) than the world has ever achieved in the past - year in and year out for decades. Chart 1 outlines the four Cases.

Energy Demand (or more strictly the demand for the services which energy provides, such as heating, lighting, cooking, cooling, mobility and motive power) is expected to rise in the coming decades in all four Cases. This is principally due to the demands of the developing countries, as Chart 2 shows.

The evolution of demand will provoke different responses in terms of fuel mix dependent upon pace of increase, environmental pressures, and - increasingly over time - availability issues. In all four cases natural gas usage is expected to increase rapidly. The role played by "new" renewables - wind, solar, modern biomass, ocean/tidal, geothermal and small hydro - will expand. However, in the absence of concerted efforts to accelerate the development of "new" renewables they may expand only three-fold by 2020 under Current Policies in the Reference Case; under an Ecologically Driven Case (Case C), a significantly faster evolution is conceivable, as Chart 3 indicates.

One fundamental challenge for "new" renewables is the low base from which they started in 1990 in their contribution to global primary energy supplies. At 164 million tonnes oil equivalent, "new" renewables accounted for less than 2% of the total supplies even of renewable energy. Wind energy itself accounted for only 1 Mtoe. By contrast, traditional biomass accounted for 60% of total renewables, large hydro for 30%, and modern biomass for 8%. By 2020, as Chart 4 indicates, under Current Policies as a "Minimum" evolution "new" renewables might account for 3-4% of total energy supply. With major policy support this proportion might rise to 8-12%.

These figures are significantly less than some others in circulation, such as those in Thomas B. Johansson et al. "Renewable Energy: Sources for Fuels and Electricity", 1993. This is because we are more cautious about what can realistically be achieved in the short space of under 30 years; and more cautious about the contribution of modern biomass (partly due to greater scepticism over the economics of conversion to automotive fuels, partly because of a seeming greater awareness of the competing uses for agricultural land to support a doubled world population (e.g. in India), and fears for further tropical deforestation and loss of biodiversity). It is also what I think is a heightened awareness of the potential adverse local environmental impacts of most forms of renewable energy which caused the WEC Commission and the WEC Renewable Energy Sources Committee to be more cautious about the evolution of renewable energy supply. This is of particular relevance to wind and tidal energy where the UK has high technical potential, as I shall indicate in some closing remarks on UK windfarm development.

Nevertheless, looking much further out in time - to the year 2100 - the WEC reports envisaged the possibility of "new" renewable forms of energy contributing about 25% of global primary energy supply - and perhaps as much as 50% of the much more modest global energy usage (some 20 Gtoe) which could evolve under an Ecologically Driven Case.

But each of the WEC Commission's Cases envisages that CO2 emissions from fossil fuel combustion will continue to rise over the next few decades at a global level. Chart 5 illustrates the outcomes, despite the assumption that the OECD countries will hold their CO2 emissions at or below 1990 levels by the year 2020. Even Case C, with annual energy efficiency improvements some 2 1/2 times the historic best performance, and a creditable OECD improvement, implies a global CO2 emissions rise of nearly 7%. And even this gives a misleadingly optimistic impression, because for at least 50 years after 2020 global CO2 emissions are likely to continue to rise due to developing countries' demand.

More importantly, atmospheric CO2 concentrations are expected to rise significantly. Over the past 200 years or so, atmospheric CO2 concentrations have risen from 280 ppmv (parts per million by volume) to 355 ppmv. Under the Commission's Case A concentration is expected to rise 80 ppmv by 2020, more than the entire rise over the past 200 years. Even under the Ecologically Driven Case C the rise to 2020 represents two-thirds of the rise of the past 200 years.

As Chart 6 shows, even the very testing assumptions of the Ecologically Driven Case - testing particularly for energy consumers and energy policy-makers - imply that atmospheric CO2 concentrations are unlikely to start falling until after the year 2070. Only this case, if pursued, would avoid the significant global mean temperature, sea level and climatic changes which some fear will be disastrous for much of the world.

None of these comments should be taken as implying that we are dealing with certainties regarding enhanced global warming and climate change. The WEC Commission has been at pains to point out the uncertainties, and has quoted at length and with approval the Intergovernmental Panel on Climate Change; "Climate Change 1992: The Supplementary Report to the IPCC Scientific Assessment", 1992. Nor are we unaware that carbon dioxide is only one of the greenhouse gases, and fossil fuel combustion only one of several significant sources of anthropogenic greenhouse gas emissions. Indeed, we continue to work closely with the IPCC in their scientific and technical assessments.

Key Messages

The key messages of the WEC Commission "Energy for Tomorrow's World" (and which, in respect of renewable forms of energy, closely mirrored those of the WEC report on renewable energy resources) were:

1. The Developing Countries will be the key to the future: accounting for 90% of world population growth and at least 85% of the increasing demand for energy services. Huge numbers are at present unable to satisfy basic needs.

2. The need for greater efficiency in energy provision and use, as the immediate priority world-wide. Against the theoretical optimum,

   we in Western Europe waste about 95% of our energy. We need a more considered cost/benefit assessment of priorities and opportunities and more effective pricing.

   Even the developing countries could probably increase their energy efficiency 20-30% using existing plant and equipment more efficiently. Electricity prices in developing countries commonly cover only 40% of costs, which must encourage waste and inefficiency. Non-OECD energy subsidies more generally average 30-50%. There is a worldwide need to incorporate "externalities" into energy costs and pricing as rapidly as politically and economically feasible, to provide "a level playing field" between different forms of energy and their use - to cover everything from coal and gasoline emissions, through the impact of cooling towers and traffic congestion, to destruction of estuarine habitats by tidal barrages or visual intrusion of windfarms.

3. Fossil fuel supply availability is satisfactory to 2020 and somewhat beyond, but higher energy demand trajectories would mean growing uncertainty over oil and natural gas availability towards 2050. If nothing significant was done to extend the quantity and range of energy forms available long before then, the idea that people could simply switch to coal (with in excess of 200 years availability on current Reserves/Production ratios) would be found severely wanting. By 2050 there would on higher global energy demand trajectories be well under 100 years availability, and a major switch to coal would imply constrained availability by 2100.

   These are the more complex realities behind the apparent dominance of fossil fuels in the global energy mix of the next few decades.

4. A further complication on the supply side is that import dependency will increase, supply lines will lengthen, and demand will grow for all the fossil fuels (only under environmentally driven policy constraints might this not be the case for coal). This implies impacts on price levels and supply stability casting their shadow well forward from the period when physical shortages might be expected to occur.

One indicator of the cause and scale of problems in this area can be seen in Chart 7, which gives our estimate of the numbers of people dependent on energy imports in 1990 (2.9 billion) and the numbers likely to be import dependent in 2020 (7.2 billion). The rise of 4.3 billion people is equivalent to the world's total population as recently as 1978.

5. Almost everything that I have said so far underscores the need to expand renewable energy provision, provided the economics and the local environmental impacts can be justified. Under most current policies, renewable energy - especially "new" renewables - is likely to grow only steadily and make only a modest contribution. Concentrated governmental support will be required if accelerated development is to occur.

   Technology development and diffusion will play a crucial role here, as in other energy fields. Heavy responsibilities will be placed on industrial countries; there will need to be greater emphasis on local suitability; institutions and markets will need to be developed (including local capital markets, joint ventures, etc.); and some technology and financial transfers will be inevitable. After all, developed country signatories to the UN Framework Convention on Climate Change will be required to make such transfers.

6. The Commission found that in 8 out of its 9 Regions potential climate change is not a priority. Expectations of curbing global greenhouse gas emissions and atmospheric concentrations (except for specific cases such as CFCs) over the next few decades are unrealistic.

   Nevertheless, the Commission found the potential impact of climate change is sufficiently great to justify immediate, cost-effective, precautionary action. The Commission believed that such a response was one of "minimum regret", recognising that costs would have to be borne before benefits - of a size which would not be determined accurately in advance - became available.

7. Investments in the energy sector will increase substantially. The Commission believed cumulative global energy sector investments 1990 - 2020 would be at least US$ 30 trillion at 1992 prices, and for "new" renewables would be at least US$ 2.5 trillion. For wind energy alone the figures are likely to fall within the range US$ 180 billion (minimum) to US$ 400 billion (maximum).

8. Much is made in both WEC reports of the need to get the energy/environment balance right in the future, however grave and whatever the sources of past errors.

The Commission stated: "In pushing ahead with "new" renewable forms of energy it is imperative that consistent, sensitive environmental criteria are applied throughout the field of energy provision and use. The sort of problems that have already loomed large with many large-scale hydro developments need to be avoided."

"Energy for Tomorrow's World", p.96

The Renewables Report stated: "There will be a need to apply strict environmental standards to renewable energy developments, as part of a consistent process toward total energy provision and use."

"Renewable Energy Resources: Opportunities and Constraints 1990-2020", p.1.1

The latter report specifically warned of the need for large-scale applications of hydro, biomass, wind and tidal to be applied with sensitivity to their possible environmental impacts.

UK Windfarm Developments

How do wind energy developments in the UK measure up against the foregoing analysis, key messages and criteria?

I have recently reviewed renewable energy developments more generally, up to and including Mr. Tim Eggar's Ministerial Statement of 21 July, 1993 on the third Order under the NFFO.[1] I have noted the issue of three relevant Planning Policy Guidance Notes recently, especially PPG22 on Renewable Energy (February, 1993) which stated that wind energy is on the verge of widespread commercial exploitation as a source of electricity.

With 11p per Kilowatt/hour being offered to successful windfarm developers under the NFFO against costs of 6p to 8p KW/hr, there are clear attractions for developers. Most observers would probably claim that truly commercial exploitation, that is on a non-subsidised basis, must await the early years of the next century. The present incentives are also geared to encouraging large-scale windfarm development in upland areas of high average windspeed (over 7.5 m/sec). This is a logical objective, but one which may nevertheless risk in some cases unacceptable visual intrusion in areas of great landscape value, and discourage technical developments which could make larger areas of flatter landscape and more modest but steadier wind velocity attractive for windfarm development.

I believe there are already a couple of UK windfarms in existence which should be regarded as involving unacceptable visual intrusion. So far few public objections have been raised, but there are signs that a few schemes may be fuelling a groundswell of wider opposition. The bigger scheme (Penrhyddlan and Llidiartywaum windfarms near Llandinam in mid-Wales, comprising 103 relatively noisy

Mitsubishi turbines) is only briefly visible from the nearest main road. A deliberate approach, on foot, is required to gauge the scale and locus of this development. Not far away, near Cemmaes, a much smaller scheme has greater - though distant - visibility from the main road. Again, a specific detour up a winding lane is required to gain full view. A third, even smaller and more recent, scheme above Kirkby-in-Furness overlooking the Duddon Estuary in Cumbria is also visually intrusive. Considerable effort has been made to minimise visual intrusion at Kirkby and Cemmaes. The same could also be claimed for Llandinam, but the sheer scale means the effort would only meet with qualified success.

The central point is that despite the best endeavours and skills of developers these windfarms are located on sites which were too sensitive and unspoilt in visual landscape terms.

Two further schemes arouse my concern. The first is at St. Breock Downs, at the head of the Camel Estuary. I have not visited this site, but I am advised by windfarm and planning specialists in Cornwall that recent approval of this scheme is highly unfortunate on grounds of reduced visual amenity and disturbance of Bronze Age remains.

The second is the most worrying of all, the proposal that up to 267 300 KW wind turbines might be placed on and around Humble Hill, Tynedale on the edge of the Kielder Forest. The site is a mere 13 kms north of the finest stretch of Hadrian's Wall (from Greenhead eastwards to Housesteads), and 1.5 kms outside the Northumberland National Park. I find it astonishing that this ancient northward vista of great cultural significance (only marginally intruded upon by the monoculture of the Kielder Forest) is now threatened with the apparent approval in principle of Northumberland County Council, and the Northumberland National Park authorities. Any windfarm of any size visible from Hadrian's Wall should be regarded as unacceptable[1]. However, I have some sympathy for the developers, Trigen (composed of EcoGen Ltd, Sea West (US) and Tomen (Japan)). They put on a public exhibition about the proposed scheme at the Wentworth Centre in Hexham, Northumberland and EcoGen have done a lot of preparatory work - and still the level of local public awareness of the scheme is extraordinarily low. Having walked stretches of the wall many times over the past 50 years, and known Professor Sir Ian Richmond during the last few years of his life (Professor Richmond completed the revised 12th Edition of J. Collingwood Bruce's "Handbook to the Roman Wall", 1965, just before his death), I think many of our forebears will turn in their graves if this proposal becomes a reality.

---

[1]"Renewable Forms of Energy in the UK during 1992", in British Annual Energy Review of 1992/93, September, 1993. Available from the British Energy Association.

[1]It is a sad commentary that David Breeze and Brian Dobson remark in "Hadrian's Wall", Allen Lane, 1976 that the tower on Walltown Crags (turret 45a) which has the "best" view of Humble Hill "occupies a commanding position in advance of the Stanegate, which, designed as a road, rarely offered a good view to the north." (p.24)

That this possibility exists suggests that neither our planning criteria nor our valuation methods are satisfactory, and while I recognise the difficulties in valuing unique landscapes and historic/cultural assets there may - as the WEC Commission advocates - be a case for experimenting with the sort of valuation techniques being explored by Professor David Pearce.

In advance of such steps, I believe it is possible to rank UK windfarms according to level of acceptability. Chart 8 sets out my ranking, which accords closely with the views of a number of wind energy specialists and windfarm developers. It follows from this ranking that I am delighted to note that later in this Conference we will hear about Delabole windfarm from Mr. Peter Edwards himself, and about Blyth Pier windfarm from Mr. Grainger, two excellent examples of windfarm development. It should be noted in passing that both windfarms are highly visible to passers-by, yet neither are intrusive in a negative way. As Chart 8 suggests, there are numerous other very acceptable UK windfarms to place in the balance with my earlier remarks. [1]

It will also be noted that I make no reference to noise (except for an earlier comment about Llandinam), though there have been a few temporary and one or two more ongoing complaints on this score. In general this does not seem a significant problem. Nor do I refer to bird mortality arising from windfarms, simply because to date UK experience does not suggest this is a significant problem either. Experience in Denmark, The Netherlands and Sweden tends to confirm that of the UK. Tidal barrages pose much more of a threat to birds than windfarms.

My underlying worry is that if accelerated development of wind energy is to be maintained, and in a manner consistent with broad environmental principles, then the industry and the nation cannot afford to have two or three apples which threaten the larger crop. There is a risk that a few unwise schemes may provoke a backlash and undermine the pace of sounder developments.

Conclusion

The main conclusion of the WEC Reports is that we cannot carry on providing and using energy as we now do and avoid a disastrous outcome. There is therefore a need for change. Paradoxically, because substantial change takes such a long time - we need to think in terms of a century and several generations of technology as well as people - we must start that process of change now.

The years 1995-2020 represent the opportunity for transition. But we should not seek change for change's sake. What is required is quality change, the "right sort" of changes while heeding costs and benefits. The firm and consistent application of environmental criteria to renewable forms of energy as well as other human activities to ensure a sustainable future growing in quality.

The UK's experience with windfarms, recent though it is, highlights both the opportunities and the pitfalls. But for the WEC Commission it was clear that the word change signified opportunity rather than threat. Change implies improved economic and social development; expanded business opportunities; and a better local and global environment.

It was, I think, with some justice that one reviewer of the WEC Commission Report "Energy for Tomorrow's World" concluded:

"This report is a major statement which not only signals a broadening of perspectives of the global energy community, which the WEC effectively represents, but also a landmark in addressing issues of sustainable development."

---

[1]There was also a technical paper at the Conference on the excellent Vindeby windfarm, off the west coast of Denmark's island of Lolland.

WEC Commission's 4 Cases

Case	A	B₁	B	C
Name	High Growth	Modified Reference	Reference	Ecologically Driven
Population	UN/World Bank Median			
Economic Growth % p.a.	DCs 5.6 World 3.8	DCs 4.6 Others 2.4 World 3.3		
Energy Intensity Reduction % p.a.	Quite High	Moderate	High	Very High
OECD	1.8	1.9	1.9	2.8
CIS/CEE	1.7	1.2	2.1	2.7
DCs	1.3	0.8	1.7	2.1
World	1.6	1.3	1.9	2.4
Energy Demand (Gtoe)	17.2	16.0	13.4	11.3
CO_2 Emissions (GtC)	11.5	10.2	8.4	6.3

NOTING:

- Not forecasts
- None are "Business As Usual" evolution
- In terms of energy efficiency improvements
  - globally - up to 2 1/2 times previous performance, year in year out

Chart 1

Energy Demand to 2020

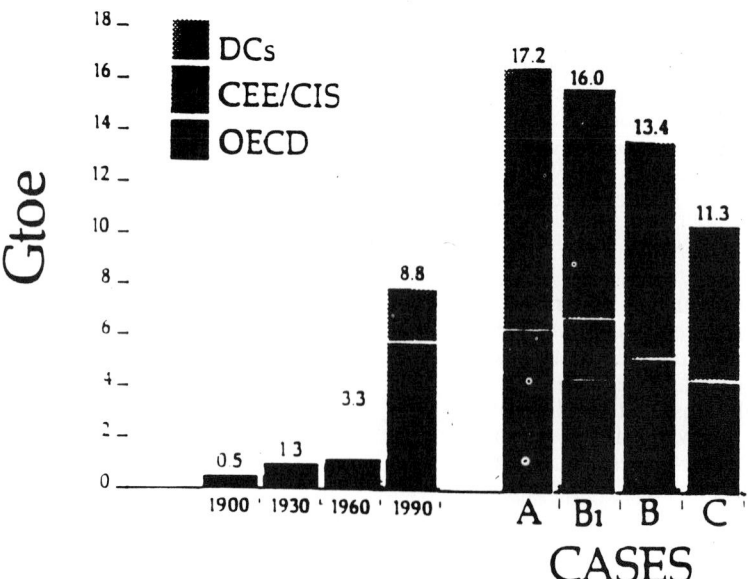

Chart 2

Fuel Mix

	1960	1990	in 2020			
			A	B_1	B	C
Coal	1.4	2.3	4.9	3.8	3.0	2.1
Oil	1.0	2.8	4.6	4.5	3.8	2.9
Natural Gas	0.4	1.7	3.6	3.6	3.0	2.5
Nuclear	—	0.4	1.0	1.0	0.8	0.7
Large Hydro	0.15	0.5	1.0	1.0	0.9	0.7
"Traditional"	0.5	0.9	1.3	1.3	1.3	1.1
"New" Renewables	—	0.2	0.8	0.8	0.6	1.3
Total	3.3	8.8	17.2	16.0	13.4	11.3

Energy Mix - Annual Past And Future Global Fuel Use (Gtoe)

Chart 3

Estimated Cumulative Total Renewable Investment (Billion U.S. Dollars, 1990)

	Current Policies			Ecologically Driven		
	2000	2010	2020	2000	2010	2020
Solar	52	134	313	65	265	1205
Wind	14	62	181	16	98	412
Geothermal	15	20	30	20	50	80
Biomass	50	100	150	66	140	260
Ocean	1	10	55	2	50	150
Small Hydro	21	50	100	36	88	150
Subtotal	153	376	829	205	691	2257
Transmission	10	23	55	15	47	141
Total	163	399	884	220	738	2398

Chart 4

CO$_2$ Emissions

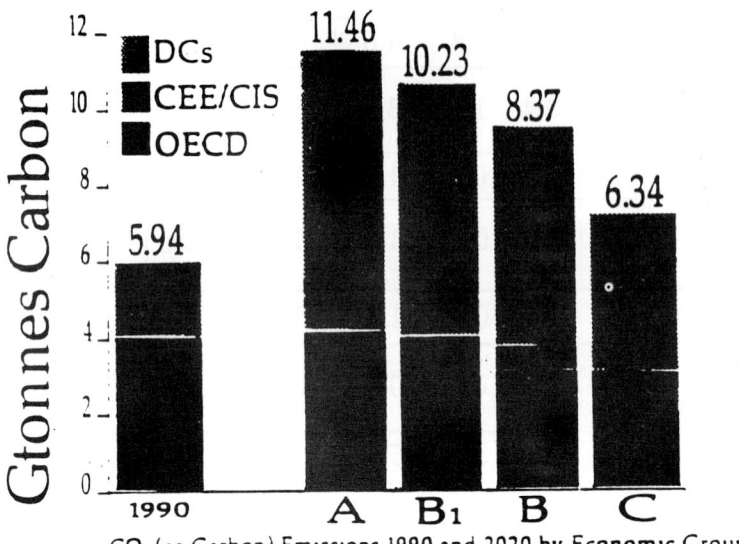

CO$_2$ (as Carbon) Emissions 1990 and 2020 by Economic Group

Chart 5

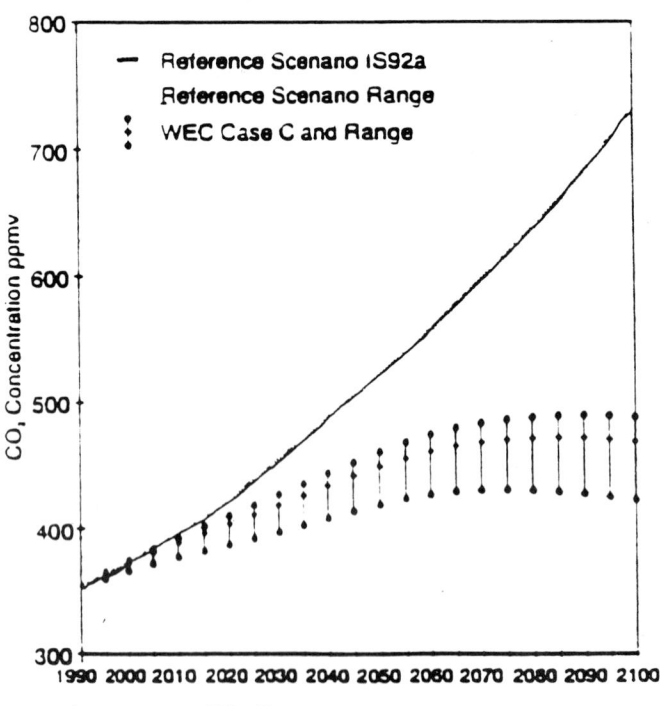

Atmospheric CO$_2$ Concentrations :

The outcome implied by the Commission,s
ecologically-driven case

Chart 6

Overview of worldwide wind generation

PAUL GIPE
Paul Gipe and Associates, Tehachapi, CA, USA

Synopsis: An overview of activities in the United States and Europe including a look at installed capacity, generation, specific yield, price, and environmental impact.

Portions of this paper have appeared in past issues of *Independent Energy Magazine*. Data herein has been compiled from numerous sources including Finn Godtfredsen, Risoe National Laboratory; Birger Madsen, BTM Consult; and Andrew Garrad, Garrad Hassan.

The 1992-1993 period saw the wind industry reach several significant milestones. Worldwide wind generation exceeded 4 TWh during 1992 for the first time and will surpass 5 TWh in 1993. See Table 1. By 1995, total generation will top 6 TWh. And during 1993, Europe's total installed wind generating capacity will reach 1,000 MW for the first time: the Netherlands and the United Kingdom will each top 100 MW and Germany will exceed 250 MW.

Table 1. North American and European Wind Development

	North America		Europe		World	
	MW	TWh	MW	TWh	MW	TWh
1992	1,700	2.7	850	1.6	2,650	4.4
1993	1,750	2.8	1,200	2.1	3,000	5.1
1994	1,800	2.9	1,450	2.6	3,350	5.7
1995	1,850	2.9	1,850	3.3	3,800	6.3

Europe's rapid growth will push wind energy beyond another milestone this year when sales of wind turbines and wind-generated electricity exceed $1 billion (thousand million), an increase of nearly $200 million over 1992 sales. 1993 turnover includes $600 million in project development, and sales of wind-generated electricity worth an estimated $450 million. The industry has not seen turnover in excess of $1 billion (thousand million) since 1985 during the height of California's *wind rush*. At that time, revenues were due almost entirely to sales of new wind turbines.

Europe's feverish development of wind energy shows little sign of abating. Altogether, Europe will install 350 MW of capacity in 1993, up from nearly 250 MW last year. European development is far outpacing the 15 MW installed in North America during 1992, and the 25 MW expected this year. By the end of 1993, Europeans will be operating 1,200 MW of wind generation, and if the present pace continues, will outstrip the United States in total installed capacity by the end of 1995.

Denmark currently accounts for half the capacity installed in Europe and continues to add 40-50 MW per year in one of Europe's most stable domestic markets. Germany accounts for another fifth, with new capacity additions in Germany now rivaling that of any other country in the world, including Denmark. See Figure 1.

EUROPE TO SURPASS NORTH AMERICA IN 1995

North America's share of world wind generation fell to 66 percent in 1992, the lowest level in a decade. As recently as 1990, California alone accounted for 78 percent of worldwide wind generation. The slide in America's leading role in the world's wind industry began with a boom in European wind development while growth stagnated in the United States. A lack of new installations and a poor wind year in 1992 contributed to the first fall in California production since wind development began in the early 1980s. If present trends continue, European generation will exceed that of North America by the end of 1995. See Figure 2.

TEHACHAPI LEADS IN GENERATION

California electric utilities report that the Tehachapi Pass led the world in wind-electric production during

1992. See Figure 3. For the first time, production in the Tehachapi Pass exceeded that of the Altamont Pass east of San Francisco. Southern California Edison Co. estimates that wind turbines near Tehachapi, Calif. generated 1.158 TWh during 1992, while Pacific Gas & Electric Co. estimates that total Altamont production fell from a high in 1991 of 1.079 TWh to approximately 1.010 TWh in 1992. Despite poor winds during 1992, Tehachapi's generation increased over that of the 1.064 billion kWh produced during 1991. Altamont contains about one-sixth more wind turbines than Tehachapi, but Tehachapi's more powerful wind resource, and greater number of state-of-the-art turbines has led to continuing improvements in production. The gap between the two sites will widen further this year. Production by mid-1993 is down 20-30% in the Altamont Pass, while above average winds in Tehachapi have led SCE to project record-breaking production approaching 1.4 TWh. By the end of 1993, Denmark should surpass 1 TWh placing it alongside the Altamont Pass in annual generation.

CURRENT ACTIVITY IN NORTH AMERICA

During 1993 and early 1994, wind developers will erect 100 MW of capacity in North America: 25 MW in California, 18 MW in Alberta, 26 MW in Iowa, and 25 MW in Minnesota. Between 1995 and 1997, wind companies will install 300-350 MW in North America outside of California: 10 MW in Saskatchewan, 5 MW in Quebec, 100 MW in Washington, 60 MW in Wyoming, 50 MW or more in Texas, 75 MW in Minnesota, 10 MW in Wisconsin, and 65 MW in Maine. Another 200 MW or more may be added in California during this period. However, this may be insufficient to offset retirements of existing capacity in the state. Several firms are also active in Mexico, where the first small project should be completed in 1993.

PROJECT PRICE DECLINES

The market price paid for wind power plants has decreased dramatically since the first projects were built in California. However, the price of installed wind plants bears little resemblance to the much-touted cost of $1,000/kW of installed capacity with the exception of projects installed by Danish utilities.(1) See Figure 4. California projects reached a low of $1,250/kW in 1987. Subsequent projects were more costly because of the construction of a 45 mile high voltage transmission line.(2) These projects also included the developer's up front profit. Similarly, non-utility projects in England and Wales are costing more than $2,000/kW, among the most costly in Europe. Only Danish utility projects are showing a steady decline in installed costs, resulting from competitive procurement and low-interest financing. Danish utilities also take their profit out over the life of the project.

SPECIFIC YIELD STILL INCREASING

According to data from the CEC's Performance Reporting System, BTM Consult and Denmark's Risoe National Laboratory, the specific yield of individual wind turbines and wind power plants in California and Denmark have increased steadily since the early 1980s. See Figure 5. The exception is average yield in California wind plants and the yield of post-1985 wind plants in California. The specific yield in California reached a peak in 1990 and has declined since. There are several possible explanations. Wind speeds have suffered during the late 1980s and early 1990s due to *El Nino* and from a volcanic eruption in the Philippines. The turbine stock may also be showing signs of age.

Of note is the steady improvement in specific yield with succeeding design iterations, both in California and Denmark. Medium-sized wind turbines have steadily increased in size since introduction of the 55 kW Danish wind turbine (15-16 meters in diameter) in the early 1980s. Machine designs have increased in modest increments from 55 kW to 100 kW, to 200 kW, to 400 kW, and to today's state-of-the-art machines in the 500 kW range. See Figure 6.

Though the specific yield (the kWh generated per year per square meter of rotor area) increases for each Danish wind turbine model in succeeding years, the most dramatic improvement is from one wind turbine model to the next.(3) The average specific yield of the 450 kW model of 850-900 kWh/m2/yr is twice that of the early 55 kW model. Even the later 55 kW model shows a marked performance improvement over the previous version. The same results can be seen in California, when the specific yield of all wind turbines is contrasted with those installed since 1985. The improved performance is attributable to greater reliability, improved airfoils, and taller towers. Next to greater reliability, the most important contributor to higher specific yields is taller towers. Tower heights have nearly doubled from 18 meters to 35 meters since the early 1980s. Doubling the tower height alone will increase the power available more than 30%. For sites yielding 600 kWh/m2/yr the taller towers will add nearly 200 kWh/m2/yr, bringing total yield to 800 kWh/m2/yr at a good site.

At exceptional sites, such as on the west coast of the Jutland peninsula (7 m/s), or atop Whitewater Hill near Palm Springs, contemporary turbines should yield 1,000-1,250 kWh/m2/yr. See Figure 7. San Gorgonio Farms, which operates the world's most productive wind plant, consistently produces within this range. There are numerous sites in Northern Europe where specific yields exceed 1,000 kWh/m2/yr, including coastal Germany, and the Netherlands.

REPOWERING CALIFORNIA

As mentioned, California's turbine stock is comprised largely of early, less cost-effective designs. Some of the turbines are ten years old. The aging stock may have partially contributed to the decline of average specific yield during 1991 and 1992. Nearly 3,100 turbines comprising 230 MW of capacity are of first generation design. See Table 2. These turbines are costly to maintain, often sited poorly, installed on short towers, and unreliable. Many of them are unsalvageable. The bulk of the capacity in the state is provided by second generation designs, including 100-150 kW Danish machines and U.S. Windpower's 56-100. There are 1,300 state-of-the-art turbines representing 300 MW of capacity.

Table 2. Repowering California Wind Plants
Current Fleet

	Units	MW	TWh/yr
Junk	3078	233	
Second Generation	12509	1211	
State-of-the Art	1286	317	
	16873	1761	2.8

After Repowering

Junk	0	0
Second Generation	0	0
State-of-the Art	7012	1761
		3.6

The American Wind Energy Association's west coast office, in a study for the California Energy Commission, estimated that repowering California's wind plants by replacing first and second generation designs with contemporary machines could lower operations and maintenance costs, reduce the density of turbines on the landscape by more than 50%, and increase annual generation by 30% to 3.6 TWh. The AWEA study concluded that repowering with modern turbines would make the California industry more competitive with other resources, preserve jobs, and reduce the industry's aesthetic impact, thus improving its public acceptance.

The AWEA study took the first-ever truly comprehensive employment survey of California wind plant operators and their service providers. See Table 3. AWEA found that there are 1,250 people working directly with wind energy in California for 460 jobs/TWh/yr. This compares well with estimates by BTM Consult of the number of people employed in Denmark. BTM Consult estimates there are nearly 600 manufacturing jobs in Denmark for 100 jobs/MW of manufacturing and another 400 people are employed in the service sector for 440 jobs/TWh.(4)

Table 3. Wind Industry Jobs in California and Denmark

	California		Denmark	
	Jobs	Jobs/TWh	Jobs	Jobs/TWh
Manufacturing	0	0	600	(100/MW)
O&M and support	1250	460	400	440
Indirect	4350	1500	?	?

ENVIRONMENTAL IMPACT

During 1992 and 1993, the wind industry has made considerable strides towards quieting wind turbine noise. Bonus, Vestas, and WEG have demonstrated that noise emissions can be reduced significantly by focusing attention on tip design, trailing edge thickness, drive-train compliance, noise insulation, and nacelle isolation. Keeping tip speeds to 60 m/s or less is an important means for reducing aerodynamic noise. These wind turbines are 5-7 dB(A) quieter in their noise emissions than competitive machines designed during the mid-1980s operating at higher tip speeds.

Despite the industry's best efforts, wind turbines will introduce noise into many rural environments previously noted for their solitude. This intrusion creates concern, fear, and objections until the public has had ample time to become familiar with the technology. See Figure 8. It has taken the community of Tehachapi a decade to become comfortable with its wind industry. It was Tehachapi's experience that prior to development there was broad support of wind energy in the abstract. Yet when wind projects were first developed, there was a loud outcry and consternation that the wind turbines would devour the small town of 5,000. Several years after installation qualitative acceptance has resumed to near pre-project levels. Energy Connection, a Dutch wind developer, has observed the same effect in the Netherlands.(5)

Wind energy's most significant impact remains its use of the visual amenity. Wind energy will only reach its potential when the industry addresses aesthetic impact. However, public opinion surveys and architectural studies in the United States and Europe can provide guidelines for minimizing wind's aesthetic impact.

The single most important measure is aesthetic uniformity. All wind turbines and towers within a wind plant must look similar. They need not be identical, but they must appear similar. This is less of a problem in Europe than in the United States. In California, for example, it is common for a developer to install a wind plant containing hundreds of different wind turbines in a seemingly incoherent mix.

To maintain visual uniformity within a wind plant that comprises one visual unit, all wind turbines must spin in the same direction, have the same number of blades, use the same tower, and use the same color scheme. If another wind turbine and tower combination will be installed nearby, there must be sufficient visual separation to make the projects distinct from one another.

Turbines should all be of the same height unless they are part of a coherent *wind wall*, with alternating groups of turbines on towers of different heights. If turbines are of different heights, all towers should appear similar if not identical. This provision alone would eliminate much of the jumble and visual clutter typical of many California wind plants.

Further, developers must minimize roads or eliminate them altogether. This prevents unsightly cut and fill slopes in steep terrain, and minimizes the amount of land used by the wind plant.

If two-bladed turbines are used, all turbines must park their rotors in the same position. This "synchronized stop" provides visual balance for what many consider an ungainly design choice. Three-bladed rotors, which appear more visually symmetrical to observers than two-bladed rotors, can be parked in any position.

No wind turbine should ever go out undressed: none should operate without a nacelle cover. Nose cones and nacelle covers serve a valuable purpose. They smooth the angular lines of the drive train and aid in making the nacelle a part of a visual whole with the tower. Nacelle covers should be replaced immediately if they blow off.

Designers should strive toward visual unity between rotor, nacelle, and tower. A wind turbine need not be a *box on a stick*. Lattice towers need not appear cluttered with cross braces and angular lines. Lattice towers can be designed with graceful curves and a sparing use of cross braces.

The industry must address the question of aesthetics or the publics' general support will be lost. Consider the reversal of public attitudes towards nuclear power during the 1960s. Wind is currently a preferred technology even when accounting for wind's aesthetic impact.(6) See Figure 9. This support is tenuous and can be squandered by inappropriate development.

COMMUNITY ASSIMILATION

Community acceptance of wind energy can be aided by addressing community concerns, by providing information about the operation of nearby wind turbines and the companies involved, and by low-key, long-term participation of the wind industry in community events.

In Tehachapi the Kern Wind Energy Association is the focal point for inquiries about wind energy and the local wind industry. KWEA and member companies offer speakers for local service clubs, participate in local festivals, and provide simple, inexpensive brochures and postcards to local merchants for distribution to tourists and residents alike. KWEA also operates a low-power radio transmitter for broadcasting information about the wind industry to motorists on Highway 58, a major artery crossing the Tehachapi Mountains. More recently KWEA has sponsored a new local event: the Tehachapi Wind Fair. The success of this weekend event, which drew 14,000 the first year and 12,000 the next, confirms that the Tehachapi wind industry has become an accepted part of the community.

CONCLUSION

Wind generation now meets 1% of California's electrical supply and 3% of Denmark's electrical consumption. Worldwide wind generation will exceed 6 TWh in 1995, when Europe will surpass North America in total generation and installed capacity. At good sites, medium-sized wind turbines today should produce specific yields in excess of 800 kWh/m2/yr. Wind energy can become an accepted part of the community if designers and developers keep community interests in mind, especially those of wind's aesthetic impact on the landscape.

REFERENCES

1. Jens Vesterdal, ELSAM, "Experience with Wind Farms in Denmark," European Wind Energy Association special topic conference on "The Potential of Wind Farms," Herning, Denmark, September 1992.
2. California Energy Commission, "Wind Project Performance Reporting System," Sacramento, Calif., 1985-1991.
3. Finn Godtfredsen, Risoe National Laboratory, "Wind Energy in Denmark: Development in Wind Turbine Technology and Economics Since 1980," paper presented at Windpower 93, American Wind Energy Association annual conference, San Francisco, Calif., July 1993.
4. BTM Consult, personal communication, April 1993.
5. C. Westra and L. Arkesteijn, "Physical Planning, Incentives, and Constraints in Denmark, Germany, and the Netherlands," European Wind Energy Association special topic conference on "The Potential of Wind Farms," Herning, Denmark, September 1992.
6. Robert Thayer and Heather Hansen, "Consumer Attitude and Choice in Local Energy Development," Center for Design Research, Department of Environmental Design, University of California, Davis, Calif., May 1989.

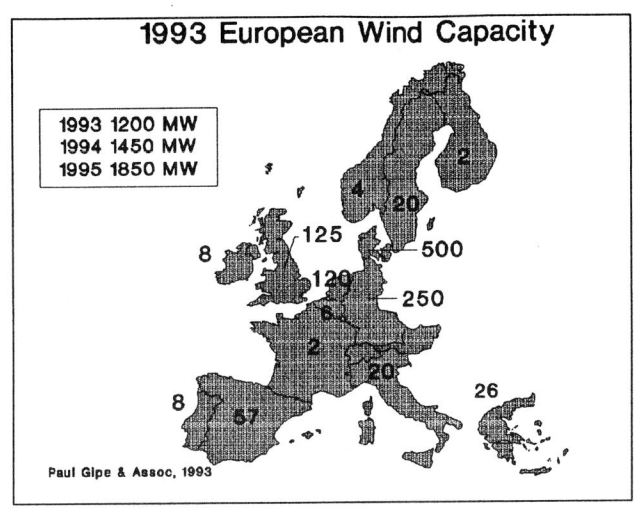

Figure 1. Installed capacity in Western Europe by year end 1993.

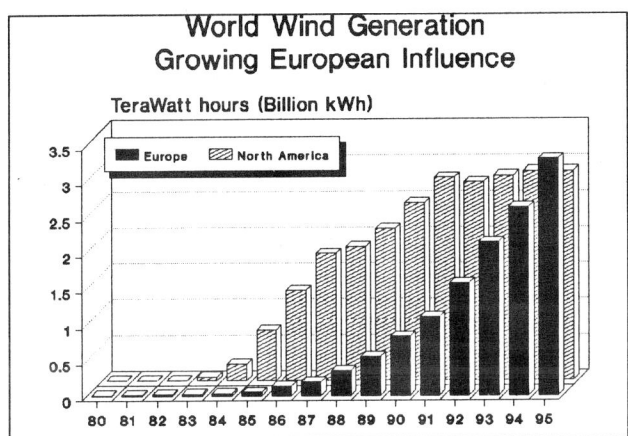

Figure 2. North American and European wind generation.

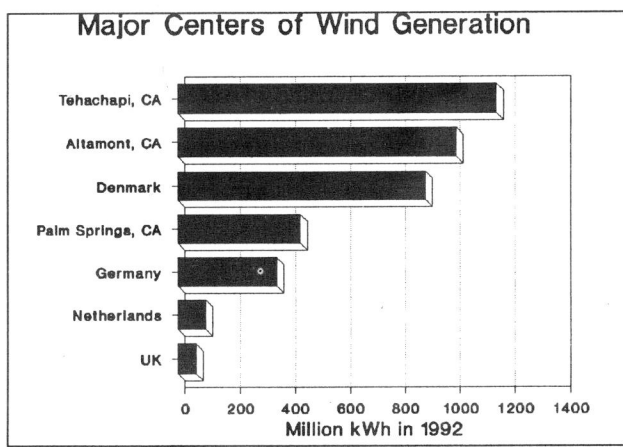

Figure 3. Ranking of major producing centers.

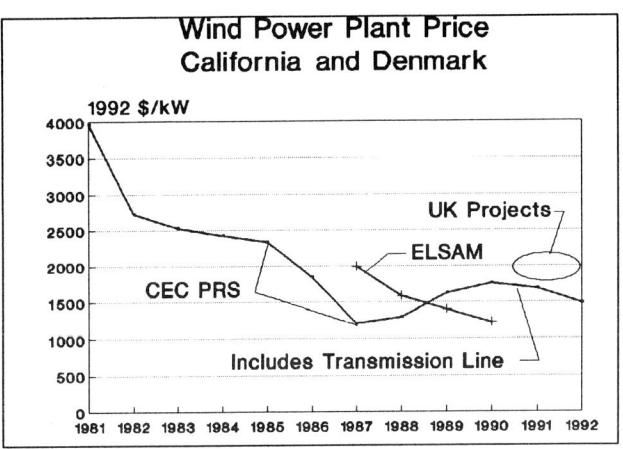

Figure 4. Market price of installed capacity.

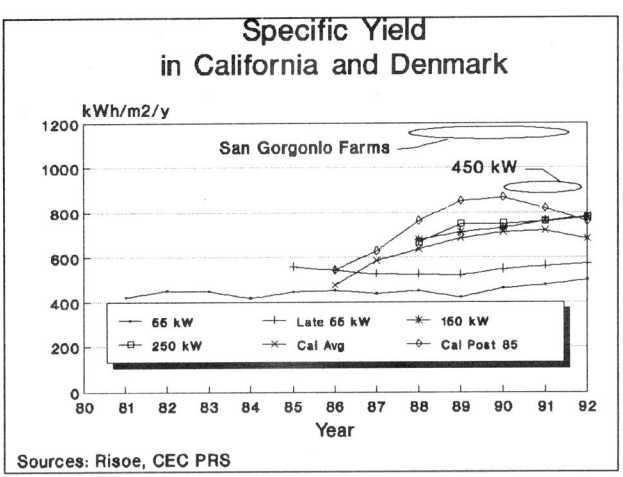

Figure 5. Average actual specific yield.

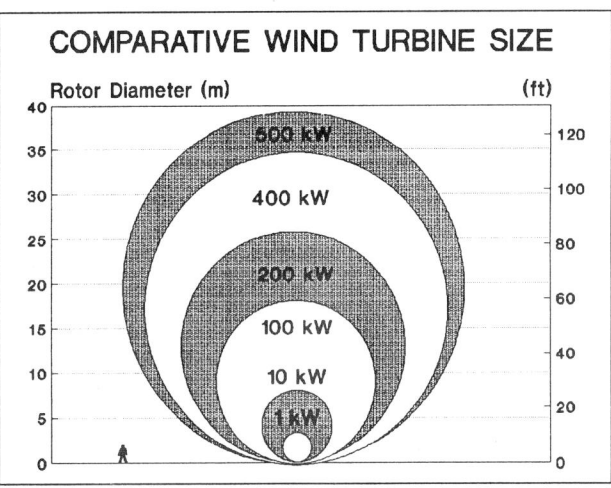

Figure 6. Increasing rotor diameter in successive design iterations.

13

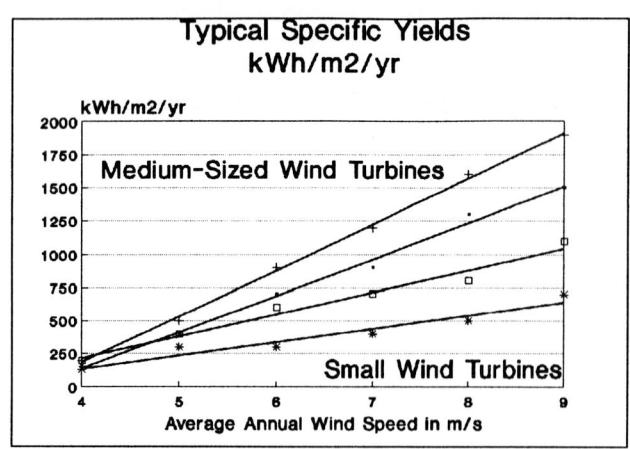

Figure 7. Typical specific yield.

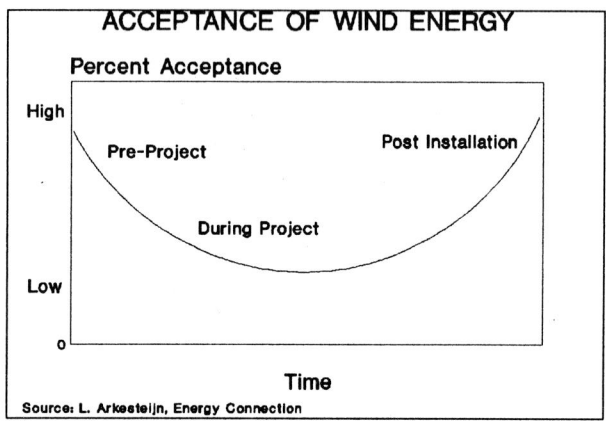

Figure 8. Acceptance after installation.

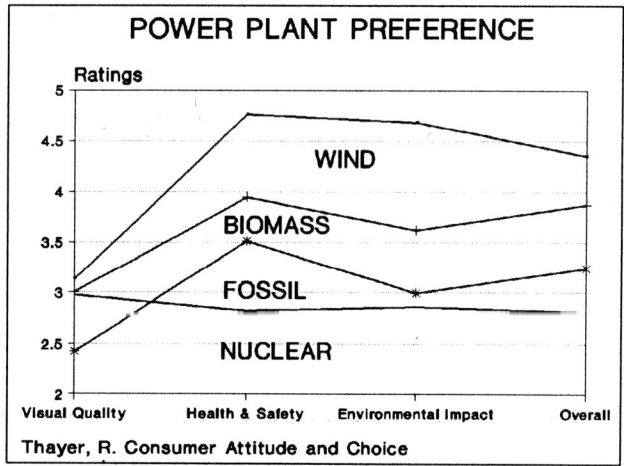

Figure 9. Preference of generating technologies.

Energy generation costs - now and for the year 2000

D. J. MILBORROW, BSc, CEng, MIMechE
Consultant, Horsham, UK

SYNOPSIS It is expected that support for renewable energies such as wind will gradually be withdrawn - possibly by the year 2000. However, a change in the relative costs of the thermal generation sources may stem from higher fuel costs and the introduction of carbon tax. A review of possible energy cost scenarios for 2000 shows that wind energy may be competitive with the most of the thermal sources

1 INTRODUCTION

The precise date at which the renewables are expected to compete directly with the conventional sources of generation is uncertain but may well be around the turn of the century. When he announced that a third renewable order would be made as part of the Non-Fossil Fuel Obligation, the Minister for Energy expressed the hope that "in the not too distant future the most promising renewables can compete without financial support". This analysis looks forward to the year 2000 and assesses the data available on current and future generation costs from the three principal thermal sources - gas, coal and nuclear. The implications of the introduction of a carbon tax - which might dramatically influence cost comparisons - is also considered.

2 METHODOLOGY

The calculation of energy generation costs follows procedures which are reasonably standardised. For wind energy plant, guidelines have been issued by the International Energy Agency (1) and can be applied to other plant.

The following items are included in the generation cost calculations:-

- planning costs)
- capital cost of plant) Capital
- construction costs)
- land costs (either as part of the capital or as annual leasing payments)
- fuel costs
- operating costs (O & M) (fixed and variable) including labour, materials, rents, taxes and insurance
- decommissioning

The following items are, however, not included:-

- company overheads
- research and development costs, except those chargeable directly to the specific installation

For each technology, three sets of generation costs are examined -if the data are available - so as to build up a complete picture:-

- present day levels from operational plant
- costs from new plant built today, i.e.. in 1993
- costs from plant to be built in the year 2000

Wherever possible "mid-range" estimates are specific data sets, which are representative of the technology; "high" and "low" values are also quoted. These are not necessarily the extreme values and it may be noted that low plant costs will not always be linked to low fuel costs. This point is discussed later.

This paper quotes levelised costs over plant (or contract) lifetimes; the cost base is January 1993 and the test discount rate is 10%. The latter has been chosen as representing the minimum level likely to be used in private sector electricity generation (2). The use of higher levels of discount rate raises overall price levels but the relative positions of the various technologies generally does not change.

3 PLANT PERFORMANCE

As costs depend on plant performance, construction time and life, a standard set of assumptions has been used. These are set out in Table 1 and are generally in line with a review carried out by the IEA (3), modified to allow for current UK trends. Where two entries are made in any column, this reflects improvements expected by the year 2000. The efficiency of coal plant has been taken as 37% (with FGD) and of combined cycle gas plant, 42%, at present, rising to 48% by 2000 (4).

Table 1 Plant Performance Details

PLANT TYPE	BUILD TIME	LOAD FACTOR	LIFE
	Years	%	Years
Coal			
- New plant	5	75	25
- CEGB plant, as sold	-	75	10
Gas	2.5-2	80	15
Wind	0.8-0.5	30-33	20
Nuclear	7-6	72	20-30

4 GENERATION COSTS FROM GAS

4.1 Existing Plant

Only a few Combined Cycle Gas Turbine (CCGT) power stations have been commissioned and there are few data on operating costs. However the "dash for gas" took place in the light of experience from abroad and costs are known with reasonable accuracy.

4.2 Costs for Plant built in 1993

The costs from plant recently commissioned, or now being built, have been debated with some thoroughness (5). Capital costs range from £380 to £560/kW, with a central value of £490/kW (6). In the early days of the "dash for gas" some generators obtained supplies at below 16p/therm, but most obtained them at this level and this is the basis of the fuel cost quoted in Table 2. The spread in construction prices is responsible for a range of generation costs between 2.4 and 2.8 p/kWh.

4.3 Future Costs

The "dash for gas" ended when British Gas raised its prices to 21p/therm (which pushes the mid-range cost up to around 3.2p/kWh) and dampened enthusiasm for further gas fired stations.

Interest centres on the behaviour of gas prices between now and the end of the century. Demand in Europe is now increasing more rapidly than production (7) and some observers predict a gap between supply and demand from about 1997 (5). This could have the effect of pushing prices upwards. Gas prices for power stations in the year 2000 are therefore subject to uncertainty. Estimates range between 186 p/GJ (7) and 261p/GJ (8) by that time. In the absence of a clear consensus, the mean of these, i.e. 233p/GJ has been taken as a central

estimate. Assuming a modest decrease in the price of plant the estimated costs for gas generation in 2000 lie in the range from 2.5 to 3.2p/kWh, made up as shown in Table 2.

Table 2 Generation Costs from CCGT plant

Date		Plant Cost	Generation Cost, p/kWh			
		£/kW	Capital	Fuel	O & M	Total
	Low	380	0.8	1.3	0.3	2.4
1993	**Central**	**490**	**1.0**	**1.3**	**0.3**	**2.6**
	High	560	1.2	1.3	0.3	2.8
	Low	380	0.8	1.4	0.3	2.5
2000	**Central**	**425**	**0.9**	**1.7**	**0.3**	**2.9**
	High	470	1.0	2.0	0.3	3.2

5 GENERATION COSTS FROM COAL

5.1 Existing Plant

No coal plant is currently being built in the U.K. and an analysis of costs needs to take into account three complications:-

♦ The plant now owned by the two English generators (National Power and PowerGen) was sold at privatisation at a price which reflected its age and the need to ensure a successful flotation of the industry. The price was £85/kW (9) and it is assumed that the most efficient plant has a life of 10 years.
♦ The recent changes in contract prices between the generators and British Coal. Until March of this year the BCC pit-head price was 186p/GJ; from April onwards it was reduced to 151p/GJ and will fall in future years.
♦ Flue gas desulphurisation. All future plant is likely to be fitted with FGD, but experience in the UK is limited.

Generation costs from former CEGB plant before and after March 1993 are estimated in Table 3. It may be noted that these (2.8 and 2.5p/kWh) straddle the price for gas (2.6p/kWh) quoted in Table 2. This illustrates, in a nutshell, the arguments at the core of the "coal crisis" debate. (The generators' selling prices are slightly higher than these levels, probably reflecting the use of a higher discount rate and the inclusion of company overheads).

5.2 New Plant

The generation costs from new plant were discussed at length during the recent debate and capital costs for FGD plant, together with the corresponding running costs, have been drawn from the Coal Market Review (5) and elsewhere. The April 1993 BCC coal price has been used.

5.3 Future Costs

Between now and 1997 BCC coal prices will fall to world levels (currently around 130p/GJ). Beyond 1997, there is a range of estimates for the year 2000. Whereas it is demand in industrialised Western Europe and United States which will influence the price of gas, it is the growing demand for coal from the third world - especially India and China, which may raise the world coal price between now and the year 2000. Estimates range from 103p/GJ (8) to 192p/GJ (3) and the corresponding generation costs, set out in Table 3, range between 3.5 and 5.1p/kWh. The central estimate is based on a delivered coal cost of 153p/GJ (10) and the IEA capital cost estimate (3).

Table 3 Generating Costs for Coal

Description	Capital Cost £/kW	Generation Cost p/kWh			
		Capital	Fuel	O & M	Total
NP/PG Plant					
March 1993	85	0.2	2.0	0.6	2.8
April 1993	85	0.2	1.7	0.6	2.5
New Plant					
Low	850	1.9	1.7	1.0	4.6
Central	**970**	**2.2**	**1.7**	**1.0**	**4.9**
High	1090	2.4	1.7	1.0	5.1
2000					
Low	850	1.9	1.0	0.6	3.5
Central	**1040**	**2.3**	**1.6**	**0.8**	**4.7**
High	1040	2.3	1.9	1.0	5.2

5.4 Advanced Coal and other Technologies

There are a number of "clean coal" cycles which may evolve to commercial fruition around the turn of the century - particularly the Integrated Gasification Combined Cycle (IGCC) and the Topping Cycle. Some authorities consider these will yield cheaper costs (4), others the reverse (19). British Coal does not consider the short-term prospects to be bright (2).

If such plant should become available the relevant costs will lie within the ranges quoted above.

The economics of other renewable sources were examined in detail by the Renewable Energy Advisory Group (18).

6 NUCLEAR POWER

Although the costs of nuclear power are the subject of debate it is possible to correlate data from a number of sources and obtain a reasonably consistent picture of price levels both now and in the future. The situation in Great Britain is slightly complicated by the fact that no data are available for the costs from current construction. (The generation costs from Sizewell 'B' are not regarded as representative since it is now a "one off" project)

6.1 Current Cost Levels

Scottish Nuclear (SN) provided detailed generation costs to the most recent Select Committee Enquiry (11). Their average cost (for magnox and AGR), updated for inflation, as 6.5p/kWh. This used an 8% discount rate and using a 10% rate gives a generation cost of 7.6p/kWh, as shown in Table 4. Nuclear Electric (NE) declined to give comparable data to the enquiry but operational costs are quoted in their annual report (12). Although NE quotes a figure for depreciation this does not appear to include interest. When allowance is made for this, and the fact that the implicit capital cost of the NE plant is just under half that of the SN plant, this brings the two generation cost levels roughly in line, with the NE figure about 10% higher. Other analysts have reached a similar conclusion (13). Although AGR costs are lower than Magnox, this has little relevance in the context of future costs, since interest is now focused on the PWR.

6.2 Future Costs for PWR

Much has been made of the difference between the 2.2p/kWh price level quoted by the CEGB at the Hinkley Point C Public Enquiry and the "privatisation" cost of 6.25p/kWh quoted by Lord Marshall (15) in November 1989. However this was simply a reflection of the "change in the rules" imposed by the privatisation programme and by the cancellation of the complete PWR programme. Using Lord Marshall's figures, reinstating interest during construction (which was subtracted out), eliminating company overheads and correcting for inflation yields a generation cost for Hinkley Point C of 6.6p/kWh. This corresponds reasonably with data given by National Power to the Select Committee Enquiry on nuclear cost in 1989/90 (16); a levelised cost was not quoted, but a figure of 5.9p/kWh can be derived from the data. This is probably the best estimate that can be made for the year 2000 and, even so, may be optimistic. Figures compiled for the Nuclear Energy Agency and published by the IEA (3) are included in Table 4 but these were compiled in 1987 and are now out of date.

6.3 Sizewell B

Although generation costs from Sizewell B are not really relevant to this discussion, the final construction cost is included in the Table 4. (The National Power estimate of costs for a 4-PWR programme was about 30% less than this). Looking further ahead, Nuclear Electric are forecasting generation costs of about 3.5p/kWh for Sizewell C (5). A breakdown of the costs is included in the last entry in the table which reflects the application of the 10% discount rate and raises the generation cost to 4.2p/kWh.

6.4 A Note on Decommissioning Costs

There has been considerable discussion of decommissioning costs in recent months following a report by the National Audit Office (14). The debate centres more on finding the very large sums needed for decommissioning and does not, in practice, influence generation costs to a large extent. Estimates of decommissioning costs vary between about 200 and 2000 £/kW (17) and there is also considerable discussion as to the best method of funding this cost.
There appear to be three methods:-

1. The decommissioning cost is included as a final sum in the last year of operation. This is then discounted backwards (which considerably reduced its value) and added to the cost of construction. This method does not address the origin of the money or how it is to be repaid and is now rarely used.
2. A second method is to pay for the decommissioning costs of the first generation of nuclear stations out of the revenue from the second. As the Daily Telegraph has remarked (3.6.93), this "sounds like the alcoholic's justification for another drink"
3. The method adopted in this analysis is to assume that a sinking fund is set up during the operating life of a power station which accumulates to fund the decommissioning. (Assumed to take 100 years). In line with the recommendation of the National Audit Office (14), a real interest rate of 2% is used for this fund, during its accumulation and use.

Decommissioning costs have been taken as £1000/kW - although some estimates are higher (17). This method results in a "running charge" of £10/kW/yr , or an additional generation cost of about 0.2p/kWh. .

Table 4 Nuclear Power Costs

Description	Capital Cost £/kW	Generation Cost, p/kWh			
		Capital	Fuel	O & M	Total
Current, SN	1950	4.8	1.4	1.4	7.6
NE	850	2.1	1.7	1.8	5.6
Sizewell B	2380	6.5	0.7	0.8	7.9
Production PWR					
Lord Marshall	1890	5.1	0.7	0.7	6.6
National Power	1680	4.4	0.7	0.8	5.9
IEA	1684	4.1	0.5	0.6	5.2
Sizewell C (2005?)	1450	3.3	0.4	0.5	4.2

7 CARBON TAX

Generation of electricity from gas, coal and nuclear will attract carbon tax if the European Commission's proposals are eventually adopted by the British Government. At present, they appear opposed but have decided to impose VAT on domestic fuel - which will raise the cost of electricity for most consumers by more than the proposed carbon tax. VAT will be phased in over two years; carbon tax in seven.

Carbon tax is a way of reflecting the external costs of the various energy sources and although it may be imperfect, it avoids the necessity for protracted discussions on the precise levels of external costs, some of which are very difficult to quantify.

Estimates of the effect of carbon tax have been made using the plant efficiency data quoted in Table 1. This results in the following estimates for the final level of carbon tax - scheduled for 2000. (The proposals envisage it being introduced at 3/10ths of this level, rising in equal steps over 7 years).

Coal..............1.2p/kWh
Gas...............0.8p/kWh
Nuclear.........0.5p/kWh

8 WIND ENERGY

Although the primary purpose of this Paper is to examine the target prices wind energy needs to meet by the year 2000, it is instructive to examine the costs projections for wind energy. There are complications however,:-

- The broad range of wind speeds across the British Isles
- The very wide range of future cost projections
- Only limited data are available on actual operating costs in the UK

8.1 Current Costs

The report from the Renewable Energy Advisory Group (18) quoted costs for a site with a mean wind speed of 7.5m/s and this is a convenient reference windspeed. Reported costs for the first UK windfarms generally lie in a range from £850/kW to around £1200. An average level of £1000/kW has been used, which yields a "current" energy cost which is consistent with the REAG estimate. The "NFFO" entry in Table 5 assumes a six-year contract period and illustrates the distortions introduced by the current arrangements.

8.2 Future Costs

Future projections for capital costs go as low as $500/kW but it is conservatively assumed that UK installed costs will fall to £750/kW - a level slightly higher than predicted by an authoritative analysis by the US Solar Energy Research Institute (4). The "high" capital estimate conservatively assumes that costs only fall to the current UK minimum and the "low" capital cost is slightly lower than <u>current</u> Danish levels (22).

Modest improvements in energy productivity and O&M costs (20) yield a central estimate for energy cost of 4.1 p/kWh, a 30% reduction from current levels. This reduction is in line with an analysis carried out for West Germany (21). The trends predicted by that study are compared with those of the American study (4) in Fig 1.

The range of generation costs for 2000 in Table 5 does not explicitly take windspeed variations into account but allowances can easily be made. Alternative combinations of windspeed and cost are set out below:-

Windspeed	7 m/s	7.5 m/s	8m/s
Low estimate	£568/kW	£650/kW	£732/kW
Central estimate	**£655/kW**	**£750/kW**	**£845/kW**
High estimate	£743/kW	£850/kW	£958/kW

Table 4 Wind Energy Costs

Description	Capital Cost £/kW	CF %	Generation Cost p/kWh		
			Capital	O&M	Total
NFFO	1000	30	9.1	1.3	10.4
Current	1000	30	4.6	1.3	5.9
2000 Low	650	34	2.6	0.9	3.5
Central	750	33	3.1	1.0	4.1
High	850	32	3.6	1.1	4.7

Notes: CF - Capacity Factor

8.3 Value of Wind Energy

The foregoing discussion implies that investment decisions are made solely on a basis of cost comparisons. This is not necessarily be the case but the virtually fixed nature of wind energy prices once the capital investment has been made is a favourable consideration. Previous discussions have drawn the distinction between the cost and the value of wind energy but, it may be argued, there is no longer such a distinction in the UK

Following privatisation there is no longer a central body (such as the CEGB) to determine the value of wind energy; in theory the pool price determines the market value. However, the latter is working rather imperfectly at present and it may be noted that:-

- capacity payments are low and irregular

- most electricity by-passes the pool

This situation seems unlikely to change in the short term. It follows that energy contracts will continue to be negotiated between generators and Regional Electricity Companies on the basis of total costs (which is what happened during the coal price negotiations) As capacity payments are uncertain this removes at a stroke the contentious issue of the "capacity value of wind energy" and it is therefore reasonable to make comparisons on the basis of levelised costs.

Even if capacity payments are re-established, the "pool" system would automatically - over a period of time - pay wind plant a capacity credit roughly in line with its capacity factor. Studies have shown that the nature of wind in the UK is such that, on average, it is available at times of peak demand, in proportion to its capacity factor (23).

9 CONCLUSIONS

When all costs are put on a common footing, wind energy in the UK is presently cheaper than nuclear, but dearer than coal and gas. No allowances are presently made for external costs.

In the absence of carbon tax, the central estimate for wind in the year 2000 is a little over 4p/kWh, which is markedly cheaper than nuclear (at least 5p- even on favourable assumptions) and coal (4.7p), but dearer than the central estimate for gas (2.9p). The likely trends, must, however, be considered in more detail.

9.2 Costs in 2000 - "Environmental" Scenario

Looking ahead the most likely trends in energy supply and demand are (24) :-
   * Continuing, or increased concern over environmental impacts, leading to -
   * Recognition of external costs, and -
   * Greater energy efficiency, but
   * A strong demand for gas and oil

Under this scenario:-
   * Wind plant costs would probably fall due to reductions in the cost of steel, concrete and other components. (The CEC directive allows for relief from the effects of carbon tax in energy-intensive industries).
   * The price of coal may stay low, due to the reduced demand - but not plant costs.

The introduction of carbon tax would put wind energy below the "low" estimate for coal (4.7 p). Gas would remain the cheapest option (3.7p). This scenario is shown in Fig 2. However, there is a strong possibility that gas prices would rise significantly in response to increased demand, i.e. to the "high" fuel cost. If this happens, and/or wind costs fall by more than about 30%, then wind would undercut gas. These possibilities are shown in Fig 3.

9.2 Costs in 2000 - less Environmental Concern

If concern for the environment wanes fuel prices will probably move towards the "high" levels, due to the higher demand, although plant costs would be low. Wind costs would probably be "high". Using central O&M estimates, wind (4.6p) would still be slightly cheaper than coal (4.8p), as shown in Fig 4.

Under most scenarios, therefore, the costs of wind energy move towards parity with those of thermal generation. Beyond the year 2000, wind costs are expected to fall further. The prospects for wind, even with conservative estimates of windspeed and plant costs, therefore appear bright.

10 REFERENCES

(1) Nitteberg, J (Ed), 1983. Estimation of cost of energy from wind energy conversion systems. Expert Group Study submitted to IEA WECS Executive

(2) House of Commons Energy Committee, Second Report, Session 1991-92. Consequences of Electricity Privatisation. HMSO, London

(3) International Energy Agency, 1990. Projected costs of generating electricity from power stations for commissioning in the period 1995-2000. OECD, Paris

(4) Johansson, T B, Kelly, H, Reddy, A K N and Williams, R H, 1993. Renewable Energy. Earthscan, London

(5) House of Commons Trade and Industry Committee, 1993. British energy policy and the market for coal. Memoranda of evidence. HMSO, London

(6) Harlow, I. Nuclear Power in the OECD: is there life after dearth? WEC Journal, July 1992

(7) Special Report. *Energy in Europe*, 9.1992

(8) Caminus Energy, 1993. Markets for coal. DTI/HMSO, London

(9) British Wind Energy Association, 1993. Wind energy economics. Factsheet 11

(10) House of Commons Trade and Industry Committee, First Report, Session 1992-93. British energy policy and the market for coal. HMSO, London

(11) House of Commons Energy Committee, Third Report, Session 1991-92. Information on nuclear costs. HMSO, London

(12) Nuclear Electric plc, Report and Accounts, 1991/92

(13) MacKerron, G, 1992. UK Nuclear: giving nothing away. *Power in Europe* No 123

(14) National Audit Office, 1993. The cost of decommissioning nuclear facilities. HMSO, London

(15) Marshall, Lord, 1990. The future for nuclear power. *Atom*, 2.1990

(16) House of Commons Energy Committee, Fourth Report, Session 1989-90. The cost of nuclear power. HMSO, London

(17) Large, J H, 1992. Decommissioning of nuclear reactor systems. Journal of Power and Energy, 206, A4, 273-78

(18) Renewable Energy Advisory Group: Report, November 1992. DTI, Energy Paper 60, HMSO, London

(19) (Anon) USCEA finds nuclear can be competitive. *Atom*, 7/8. 1992

(20) Bossanyi, E A and Strowbridge, A G, 1992. Assessing windfarm operation and maintenance requirements. Proc 14th BWEA Conf, MEP, London

(21) Jochem, E and Hohmeyer, O, 1992. The economics of near-term reductions in greenhouse gases. In: Confronting climate change. Cambridge University Press

(22) Vesterdal, J K, 1992. Experience with wind farms in Denmark. EWEA Special Conf. Assoc of Danish Windmill Manufacturers, Herning

(23) Holt, J S and Milborrow, D J, 1992. Wind Power Penetration Study in the case of the United Kingdom. CEC, Brussels

(24) World Energy Council, 1993. Energy for Tomorrow's World. Kogan Page, London

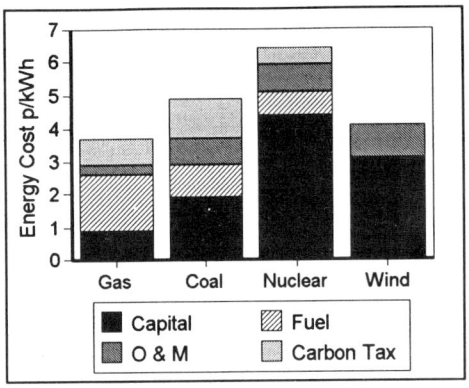

Figure 2 Generation Costs for 2000 (Central estimates - except coal)

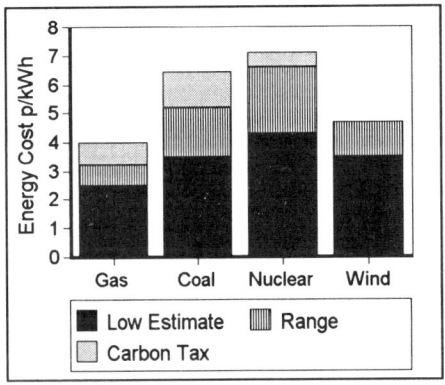

Figure 3 Range of Generation Cost Estimates for 2000

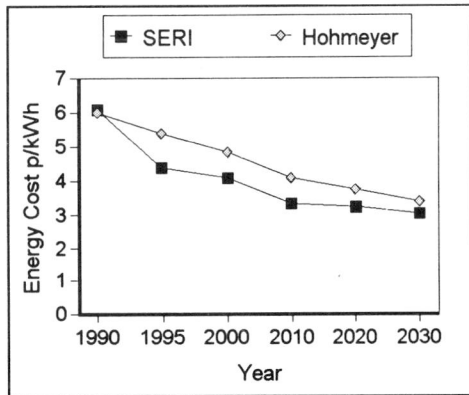

Figure 1 Forecasts of Wind Energy Costs

Note: The data from SERI (4) have been recalculated for 10% tdr and 7.5 m/s, but not the data from Hohmeyer (21)

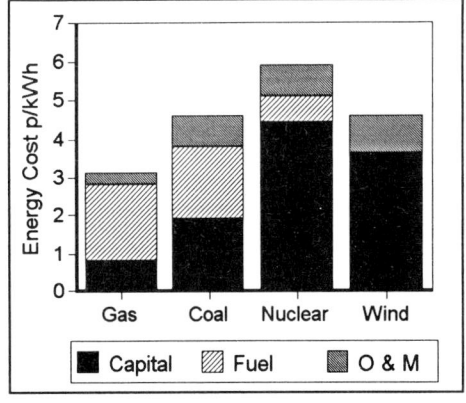

Figure 4 Estimate of Cost Levels in 2000 - reduced environmental concern

The performance and problems of, and the public attitude to, the Delabole windfarm

P D EDWARDS, BSc
Windelectric Ltd. UK

SYNOPSIS: The Delabole Windfarm was commissioned in December 1991. This Paper describes briefly the history of its development but in the main reports on its performance during the first year of operation and relates actual windspeeds to those predicted. Also described are problems encountered. The Paper concludes with a summarised result of a Public Attitude Survey conducted before the Windfarm was constructed and again after its operation.

1. HISTORY OF THE DELABOLE WINDFARM

The idea of Wind Energy at Deli Farm, Delabole was conceived as early as 1980 as an alternative to a nuclear power station that was proposed for Cornwall.

A study tour of Denmark was made in 1983 but it was not until 1988 that the 'green light' appeared for a commercially viable operation to be put into action.

The three main factors being:
1. Greatly improved technology.
2. Privatisation of the Electricity Industry and the N.F.F.O. (Non-Fossil Fuel Obligation).
3. Public concern for the Environment.

1.1 Planning

The Planning authorities were first approached in March 1989 and written approval received two years later in August 1991 with strict conditions applying particularly to noise.

These were:
Less than 45dBA at 350m at a windspeed of 9m/s.
Less than 39 dBA " " " " "6m/s.
5 dBA (L50) over background at 420 m.

1.2 Finance

The machines chosen were the Vestas Windane 34 a 400 Kw. turbine therefore ten of these giving a capacity of 4Mw.

The total cost of the project was £3.4m or £850/kw installed. This included all set up, professional and administrative costs.

The project was funded by a combination of equity and loan capital through a Company called Windelectric Ltd.

The loan element required considerable negotiating for the Bank had three fears:
- Construction would take longer than planned.
- There would be cost overruns.
- The income stream would be less than projected.

Equity partners are the Edwards Family, National Power and S.W. Electricity with the family maintaining control. A loan of £2.1 million was provided by County Nat West.

1.3 Construction

Foundations started on the 30th. August 1991. The first turbine arrived on 13th. November by road having been sent by sea from Denmark and landed at Dartford Docks.

The turbines were transported around the site without roads. This was achieved by dropping the trailer section on a hard standing and reconnecting using a dolly pulled by a bulldozer. Cranes used for erection were either all terrain or tracklaying.

Erection was completed by 17th. December and all ten were connected to the grid via the Delabole sub-station which was conveniently already situated on the property.

The first two major fears of the Bank were now allayed. The project had been completed on time and on cost.

2. PERFORMANCE:

The third fear of the Bank had now had to be faced. Was the performance to be as predicted? Were the wind speed predictions correct and the sum of these, was the financial model with the interest payments and capital repayments correct? As for the Shareholders were they going to receive the returns on their investment?

2.1 Windspeed.

Since the wind is the key variable factor lets look at the predicted wind speed for the site, the actual for the year and whether 1992 was an average year. As Figure 1. shows - the original prediction using only a 8.5m mast was 7.6m/s. The computer prediction based on NOABL was 7.2 to 8.0m/s. The measured figure at hub height at Delabole in 1992 was 6.7m/s. Mean wind speed at St Mawgan in 1992 was 5.7 which is 8.6% below the long term mean of 6.2. The wind speeds measured at Deli in 1992 are of course with the turbines in situ so takes into account array loss.

PREDICTED WIND SPEED (25m)	
	m/s
RAF St Mawgan 20 yr av.	6.2
Predicted at Deli	7.6
Computer Prediction (NOABL)	7.2 - 8.0

RECORDED WIND SPEEDS 1992	
RAF St Mawgan	5.7 (92% of average)
Deli (32m)	6.7 (88% of Prediction)

Fig 1.

Fig. 2 shows the mean wind speed at St Mawgan for the last 20 years. 1992 was therefore one of the three lowest windspeed years in that recorded period.

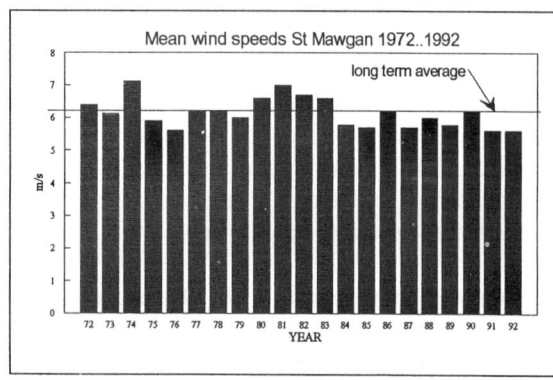

Fig 2.

2.2 Production:

How did the turbines perform with such low windspeeds and how was the budget affected?

Fig 3. shows actual production against that predicted.

	mwhrs
Power Curve Prediction at 7.6 m/s	1270
Budget Prediction (85%)	1080
Actual Production 6.7 m/s	1008
Power Curve Prediction at 6.7 m/s	950

fig 3.

The actual production during the first twelve months was 1008 MWhrs for the ten turbines. This was only 72 MWhrs short of the budget prediction of 1080 MWhrs. The 6% shortfall was well within the financial tolerance allowed so the Bank's third fear was allayed. The windfarm was not only on time and on cost, but also within budget.

The power curve for the Windane 34 shows an annual production of 1270 mwhrs per turbine at 7.6 m/s, but only 85% of this was used for budget purposes to allow for array loss, line loss and availability. The same power curve shows a production of 950 mwhrs at the recorded windspeed of 6.7 m/s. The performance thus being 106% for the whole windfarm. This agrees with individual machine testing. Availability, even allowing for commissioning and early teething problems, averaged 97%. The best individual turbine being 98.97%. The worst 94.77% which was the one struck by lightning.

Availability is being defined as the hours a turbine is available to run over the total hours the grid is able to receive. Hours taken out for servicing or for any other reason the turbine is manually set to "pause" are not included in the available hours.

The percentage of time the windfarm is actually generating is 77%.
The overall capacity or load factor of the 4mw windfarm is 29%.

The monthly variations (Fig. 4) are of interest for cash flow purposes. Individual months show considerable variation but they average out into a seasonal pattern. Fig 5. shows Accumulative Production against that predicted. The predictions are based on 60% production October to March and 40% April to September.

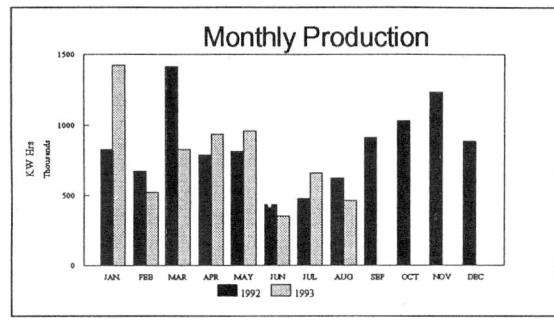

fig 4

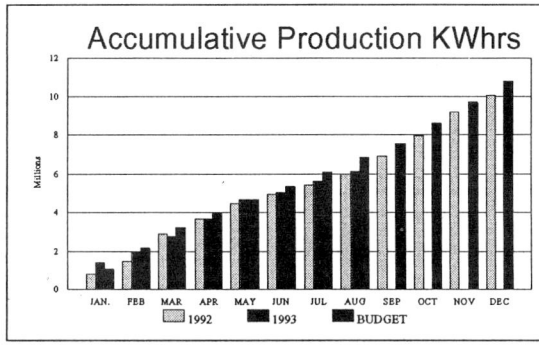

fig 5.

3. **PROBLEMS:**

What problems were encountered?

1. Lightning:
On the 20th. December 1992, only three days after the Danish erectors departed, one machine was severely hit by lightning causing a high pressure inside the nose cone and blades forcing off the outer resin layer. No damage was done to the generator or the computer. There was also no possibility of harm to third parties since only the smallest of fragments blew more than 50m from the turbine.

A second lightning strike occurred in June 1993 but this time not a direct strike. A strike on the SWEB system caused an exceptional voltage surge to jump the protection on one of the transformers causing damage to the windings.

2. A grid phase imbalance occurred usually around 1300 - 1400hrs when the economy 7 tariff cut in causing the turbines to stop if production was under 80kw. Over this the windfarm smoothed the imbalance and no trip occurred. The problem was solved by reducing the sensitivity of the turbines.

3. Two generators needed replacing. This was done the same time as the damaged blades were changed, thus requiring only a single crane visit.

4. A faulty batch of Yaw Brake castings. These were replaced within minimal production loss.

5. An aerodynamic squeak from the blades in certain wind conditions. This caused several complaints from neighbours and in particular from the Camp Site. This was cured by fixing fluted tape on a small section of the trailing edge of the blade. Since that was done no more complaints have been received.

6. A resonant ring from the generator in higher windspeeds which breached our planning conditions. This was considerably dampened by fitting a sand jacket to the top 2 metres of the tower.

7. Large numbers of sightseers blocked the roads. This was countered by constructing a large car park and viewing area and subsequently a Visitors or Information Centre. The Visitors' Centre attracted over 35,000 visitors in the first 9 months.

4. **PUBLIC ATTITUDE:**

A Public Attitude survey was carried out by Exeter University and funded by E.T.S.U. The survey was in two parts:
> Before construction.
> After operation.

The first survey showed that only 17% approved of the concept of a Windfarm, 32% disapproved and 51% were not sure.

After operation however, this had changed to 85% approved, 4% disapproving and 11% not sure.

Acknowledgements:
The Author is greatly indebted to E.T.S.U. for help and assistance given during planning stages as well as during operation. He is particularly grateful for the funding of surveys and of the Monitoring Programme.

Also acknowledged for their valuable assistance are:
Garrad Hassan
W.E.G.

REFERENCES:

1. Attitude Towards Windpower - ETSU
   W/13/00354/038/REP.
2. Edwards P.D. - Considerations to be taken into Account when Planning a Windfarm.
   Wind Energy Conversion 1990.
3. Edwards P D - The Legislative Planning and Commercial Challenge encountered in Developing the U.K.'s First Commercial Windfarm - CRES Seminar - Rhodes 1992.

A wind farm in a mixed industrial and residential area

W. GRAINGER, D. STILL, N. ROGERS, and **A. GAMMIDGE**
Blyth Harbour Wind Farm Co. Ltd, UK

SYNOPSIS

The wind farm at Blyth is unique in the UK being the only one at the moment in an industrial area. It is also very close to housing. The design and installation of the wind farm were very different from most of the other wind farms in the UK.

Some energy strategists are suggesting that future wind farm activity should be concentrated in industrial areas like Blyth rather than in upland areas, to reduce the environmental impact of 'the move to wind'. The details of the Blyth scheme may be useful in the debate on this idea.

1 INTRODUCTION

Blyth is on the Northumbrian Coast 10 miles north of Newcastle. It has a long history in the energy field, being a major coal exporting port until the seventies and having an aluminium smelter fuelled by the local coal.

The harbour is protected by a one and half kilometre long breakwater to the east. This breakwater is well exposed to the predominant winds from the south west and is the site of the wind farm. The breakwater is over one hundred years old. It is 6.5 m wide, regularly washed by waves and only has vehicle access to the north end.

The harbour is operated by Blyth Harbour Commission who were involved in the project from the outset.

2 WIND TURBINES

There are nine Windmaster 300 kW wind turbines on the site spaced 200 m apart in a line along the breakwater. Windmaster turbines were selected because of their track record and recommendations form existing owners. The technical support was excellent and helped produce a good proposal for the second round of the NFFO in a short time.

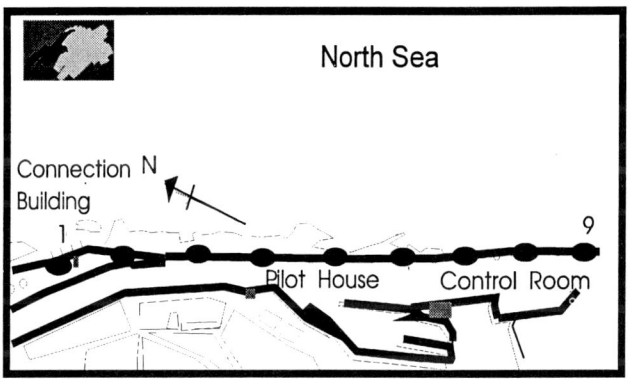

The turbines are on 30 m towers and have 25 m diameter blades. The three turbines nearest dwellings are dual speed with lower source noise levels than the single speed machines used in the rest of the site.

3 PLANNING

A district boundary runs through the harbour, with the breakwater in Wansbeck District and the rest of harbour and the town in Blyth Valley Council. All planning information was supplied to both councils.

3.1 Noise

As with most wind farm planning applications, noise was an important aspect of the environmental impact assessment. Probably uniquely in the last round of the NFFO, a British Noise Standard was directly relevant, BS4142:1990, *Method for Rating industrial noise affecting mixed residential and industrial areas.* A row of houses backs onto the port, parallel to the line of the turbines, 300 m away. The river is 100 m wide at this point and there is an area of derelict land between the river and the houses. The houses do not have gardens to the rear.

Due to the nature of the port, ships come and go at any time and more importantly, unload at any time. However ships are not being unloaded every night. Daytime noise levels can be as high as 60 dB(A) L_{90}. At night the noise can fall to 43 dB(A) L_{90}. The environmental health officers measured the background level on the river side of the houses to be 43 dB(A) in the middle of the night. The sources of the noise at night were trees in a nearby park and distant traffic. The noise from the sea is screened by the breakwater except during storms or when heavy seas are running.

When this project was being considered there were no wind farms in the UK. The planners and council officials were cautious in their approach to the noise issue. This was reflected in the planning condition on the noise - the noise from the wind farm was not to exceed 43 dB(A) ie no allowance above background. This was strict compared to other sites.

3.2 Visual Impact

As mentioned before, the site is in an industrial area. There are cranes, electricity pylons and storage silos for alumina around the site. The sandy beach to the south is popular locally, but to the north are old coal loading facilities, a number of large factories and a large, elderly coal fired power station. The visual impact did not feature significantly in the planning review once photomontages were submitted.

3.3 Wildlife

The wildlife on the site did feature prominently in the planning review. The developers were concerned about the impact on birds and commissioned an expert report on the likely impact on local birds and more importantly on migrant birds, which would not know the area. A bird monitoring program was started before the planning application was considered, to obtain a good 'before' picture to assess the impact if the scheme went ahead.

The expert report concluded that there would be little effect on locals or migrants with the possible

exception of the relatively rare, winter visitor the purple sandpiper which roosts on the breakwater. The monitoring program paid special attention to these purple sandpipers and as a result the construction was planned for the early autumn before the most of the sandpipers arrived. The installation teams did not work near the two roosts sites at the same time to minimise the disturbance to the birds.

4 FOUNDATIONS

The most demanding aspect of the wind farm was the foundations for the seven turbines installed on the breakwater. The construction of the breakwater varies along its length, but generally consists of mass concrete with a masonry or rubble-filled core.

The ground conditions beneath the breakwater were determined by core drilling through the concrete and into the bedrock beneath at each of the wind turbine locations. This established that while in some cases the breakwater was founded on bedrock, at other locations a layer of boulder clay, gravel or silt existed up to about 1m thick.

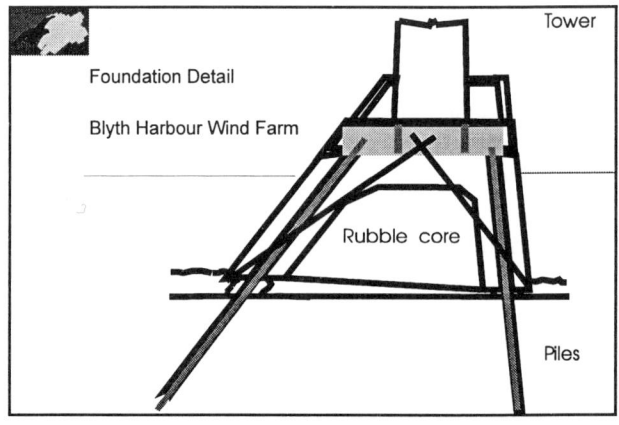

The foundation design adopted consisted of a series of underpinning piles (up to 22 per foundation, each up to 15m long), installed by drilling through the breakwater and into the bedrock beneath. Steel tubes were then installed and grouted up afterwards. Extending the steel tubes above the top of the breakwater enabled the reinforced concrete block foundation containing the anchor bolts for the wind turbines to be cast on top of the breakwater and firmly attached to it.

Concrete for two of the foundations was pumped and for the remaining five on the pier, it was ferried across the river in skips. This required careful timetabling to allow for tides and other shipping.

5 GRID CONNECTION

The grid connection is into the local Northern Electric plc grid, 4 km from the site. Unlike rural wind farms the grid in the area is very strong, with two coal fired power stations and the east coast, National Grid link to Scotland within sight of the wind farm.

6 INSTALLATION

The nacelles, blades and towers arrived by sea and were unloaded onto the quayside on the other side of the harbour. The turbines on the breakwater were erected by floating crane moored in the harbour.

The floating crane was expensive to hire, so the installation was carefully planned to minimise the hire period. When shipping permitted the crane was in use on both high tides each day. A couple of days were lost to bad weather. Three lifts were cancelled due to high winds and three lifts were abandoned due to waves or swell. At the peak fifty people were working during this period.

There were two hitches in the installation. One blade was blown off a trailer by a freak gust, but luckily the blade manufacturer had a spare available. Some of the power cables in one tower in storage were stolen under the nose of the security guard. The tower was erected and the power cables replaced later without too much difficulty.

7 COSTS AND FUNDING

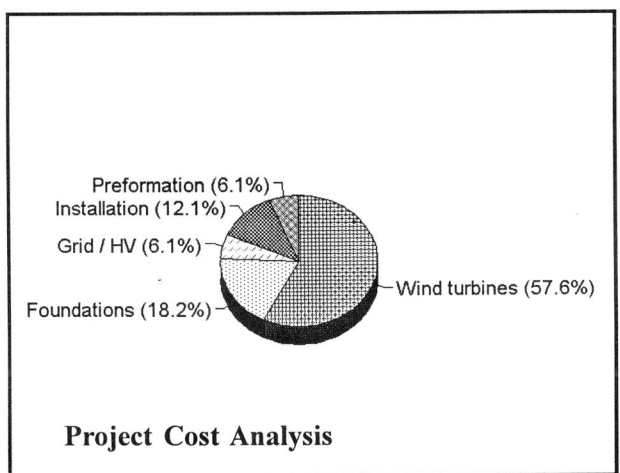

Project Cost Analysis

Preformation (6.1%)
Installation (12.1%)
Grid / HV (6.1%)
Foundations (18.2%)
Wind turbines (57.6%)

The project cost £3.3M. The project received fundingof £0.9M from the EC under the Thermie scheme. The equity was provided by Blyth Harbour Commission, Windmaster and Border Wind. The balance was funded by a loan from Hambros Bank.

8 PERFORMANCE

The wind farm started generation at the beginning of January 1993. The performance of the machines is on target for 6 GWh a year, as predicted. Independent power performance tests are under way as a check.

The turbines are linked to a central computer in the Harbour Master's Office by radio links. These initially proved unreliable and more error checking had to be introduced in the software. The turbine performance was not effected just the record keeping, as the turbine controllers in each machine perform all control functions.

The noise from the turbines has turned out to be slightly quieter than predicted with the dual speed turbines having a source noise level of 94 and 98 dB(A) in low and high speed modes respectively, and the single speed machines having a value of 101 dB(A). The environmental health officers of both councils are happy with the scheme.

Visually the turbines have been well received. They have rapidly become part of the background, only being really noticeable on a sunny day from the beach. The linear arrangement is generally thought to be attractive.

The purple sandpipers were unaffected by the installation, using the roost beside one turbine even when the crane was alongside. They have taken advantage of the shelter provided by the high voltage cable duct. As a result of this, additional wooden shelters have been provided for the birds when the wind is from the opposite direction. Given the baseline data available, and given that there are over 100 species of birds seen in the vicinity of the wind farm, it was decided to expand the scope of the bird study and continue it with the support from ETSU.

9 CONCLUSION

This project may well demonstrate that a wind farm can be built in an industrial area close to houses without an adverse effect on the local environment. The foundation designs and installation techniques developed are relevant to offshore wind farms in shallow water.

The performance to date of the wind farm is in line with predictions of 6 GWh per year.

Viability of a wind farm in an area of low wind speed

D. C. CORBET, MA, CEng, MIMechE
Carter Wind Turbines Ltd, UK

SYNOPSIS This paper reviews how the economics of wind energy influences siting decisions and how an improvement in the current economics can be achieved. An example of a wind farm in an area of relatively low wind speed is discussed.

Introduction

Wind energy is beginning to be taken seriously as a source of electricity in many parts of the world. Many countries have targets of up to 20% of electricity consumption to be produced from wind. The declared aim of the European Wind Energy Association (EWEA) is to have 10% of Europe's electricity generated by wind power by the year 2030. This will be equivalent to 100GW, or around 250,000 wind turbines, each rated at 400kW, a typical size in today's wind power market.

In contrast, there is currently only about 2GW of wind-power-generating capacity installed world-wide, of which little more than a quarter is installed in Europe. Most wind power plant is still to be found in California.

However, even with wind power's relatively low penetration into the energy market there is already a split in public opinion over the benefits versus the disadvantages of wind power. The main objections revolve around the relatively obvious impact of wind farms on the landscape versus the seemingly small and expensive contribution that they make to our energy needs. This paper reviews how the economics of wind energy influences siting decisions and how an improvement can be achieved. An example of a wind farm constructed in an area of relatively low wind speed is discussed.

The Economics of Wind Power

Current methods of calculating electricity generation costs show that in many cases it is more expensive to harness the wind than to burn fossil fuels. The precise difference depends on fuel subsidies, financing arrangements and how external costs are accounted for. Wind speed has a dramatic effect on the energy production from a wind site and hence its competitiveness. Wind energy is generally only considered cost effective on higher wind speed sites.

In the UK that means sites with an annual mean wind speed of greater than 7.5 m/s

In general, most comparisons of costs take no account of the downstream consequences to the environment of burning fossil fuels. These have been variously estimated at up to 3p/kWh (4.5 c/kWhr) which, if they were included, would make wind generation look a lot more attractive.

For every 1% of fossil fuel generation displaced by wind energy, there will be a 0.3% reduction in CO_2 emissions. This contribution has been acknowledged to be significant in helping the UK meets its commitments made at the Rio Earth summit.

Furthermore, within its first year of operation, a wind-turbine-generator will convert more energy to electricity then went into its construction. The environmental benefits of wind energy cannot be ignored.

Site Selection

Until the environmental benefits of wind energy are fully realised the development of wind energy, particularly in Europe, will depend on government support. Furthermore, developers will continue to take advantage of the windier, generally upland, sites in order to improve the economics of their projects.

It is the use of such upland sites which appears to causing much of the public concern about the large scale development of wind energy. A further problem is that concentration of wind development on only the windiest sites will eventually limit the extent to which wind can contribute to our energy needs. In the UK, it is generally recognised that wind energy projects in areas with annual mean wind speeds of less than 7.5m/s are not economically attractive. Yet if this figure could be reduced by just

15% to 6.5m/s, then the land area suitable for exploitation would be trebled. Furthermore, the total wind resource available would be more than doubled. However, such a decrease in wind speed would need a one third increase in the cost of wind energy production in order to stay viable.

Reducing this "viability threshold" not only opens up new areas and resources, it also takes the pressure off the highest wind speed sites. In the UK at least, these sites tend to be in upland areas which are not readily accessible, expensive to develop and, more often than not, in areas perceived to have outstanding natural beauty or great amenity value.

When it comes to assessing the environmental impact of wind power projects throughout Europe, beauty is definitely in the eye of the beholder.

The visual impact of wind farms is a highly subjective issue. Noise, the other main environmental concern, is much easier to quantify. While many people agree with exploiting wind energy in principal, they feel differently when the wind farm is going to be built in their "backyard".

Studies have shown that even in some countries where little attention is has been given to visual impact, the type of landscape has a strong influence on the acceptance of a wind farm. Wind turbines are much more acceptable in industrial or large-scale agricultural areas. Conversely, they are much less welcome in areas of natural beauty or in smaller-scale landscapes, where the machines can look intrusive.

This may seem an obvious point, but it is very important for the general acceptability and long-term development of wind energy. Wind farms must not be perceived as intrusive "eyesores".

Again, if the viability threshold of sites could be lowered by improvements in wind turbine technology, then wind farms could be developed in less attractive areas of lower wind speed, which at the present time would not be considered viable.

Improving the economics

Generating electricity from wind power will need to become more cost-effective, both in terms of reducing unit electricity costs generated and increasing the geographical areas which can be developed.

There are four major elements affecting the viability of a wind energy project: the unit price paid for electricity produced; the financing terms for the project;; the site wind resource and the the lifetime costs and performance of the wind farm.

The price paid for the electricity depends on the level of government support and must eventually fall to become directly competitive with other sources of energy if the wind energy industry is to develop. The financing terms of the project will largely depend on the perceived risk which should fall as wind projects become more familiar to the financial markets. Given a particular site wind resource the only factors left which we can influence are the lifetime cost and performance of the wind farm.

Wind Farm Performance

It was said earlier that the amount of energy which can be captured by a wind turbine relates to the average wind speed on a site. In fact, the energy content of the wind is directly proportional to the cube of its wind speed. A doubling of the wind speed produces an eight-fold increase in the amount of energy it carries.

The ability of a wind turbine to capture this energy is not perfect. In practice, on low wind speed sites, the relationship between energy capture and wind speed is nearer a square than a cube. Nevertheless, a doubling of the wind speed will produce a four-fold increase in energy capture by the wind turbine.

The problem for low wind speed sites, to reverse the calculation, is that halving the wind speed decreases the energy captured to a quarter. And, since the annualised costs of a wind farm are independent of wind speed, the unit cost of electricity will quadruple with a halving of wind speed.

The full equation for energy capture also includes the area swept by the turbine blades. As this area increases, so the energy captured increases by the same proportion - double the area; double the energy.

The full equation is of the form:

$$E = k_1 k_2 k_3 V^n A \text{ kWh/year},$$

where

E =	Energy production	
V =	Annual mean wind speed	
A =	Area swept by the turbine blades	
n =	Constant, between 2 and 3	
k_1, k_2, k_3 =	Constants, dependent on machine efficiency, reliability, wind farm array losses.	

The cost of the energy is simply the total annualised cost of the wind farm divided by annual energy production.

From the above equation, there are only a limited number of areas where effort can be concentrated to improve energy production from a wind turbine: increase the swept area; increase the wind speed at the rotor; increase machine efficiency and reliability or decrease the wind farm array losses.

Increasing machine efficiency will yield marginal benefits, but it will never achieve a breakthrough. The efficiency of the machines currently installed in Europe could be improved by fractions of a percent, perhaps even by one or two percent, but this route will never achieve the one third increase in energy capture mentioned earlier in this article. Similarly, machine reliabilities are already relatively high with availability in the range 95% to 98% being typical. There has been much research into the micro-siting of wind turbines and it is unlikely that wind farm array losses can be reduced significantly.

Increasing Swept Area

Increasing the swept area at a site can be achieved simply by increasing the number of machines on the wind farm. This approach brings little gain since most of the costs will increase nearly in proportion to the number of machines.

However, by increasing the swept areas of an individual rotor while keeping all other aspects of the design constant, including its peak power output, more wind energy will be captured more cost-effectively. This strategy will increase the cost of the rotor, but as it is only about 5 or 10% of the total cost of the installed machine, there will be a net reduction in unit energy cost.

The problem with this approach is keeping all other aspects of the design constant. The stresses and strains on a wind turbine are directly related to the rotor swept area: the bigger the swept area; the higher the thrust loading on the tower; gearbox and other mechanical components, and the larger the costs to relieve the additional loads.

Furthermore, the tower's natural frequency has to be well away from the rotational frequency of the rotor. The tower's natural frequency depends on its stiffness and on the weight of the rotor and nacelle assembly. Increasing the swept area increases the nacelle mass, which in turn requires the tower to be stiffer and more massive, and hence more costly.

The maximum power developed by the wind turbine can be limited by a control system, such as pitch-regulation. However, even with an active control system, it is difficult to avoid increasing the transient loads on the machine.

In addition, the control systems themselves introduce more complexity, more cost and more components to go wrong. There is also a limit to how much the swept area can be increased without increasing the height of the tower and having further to increase its stiffness, mass and overall cost.

The alternative to pitch-regulation is stall-regulation, which limits the peak power of the machine by aerodynamic means. No additional components are required as the rotor blades are designed to stall naturally at peak power. However, for a given rotor diameter, maximum stall power depends on the speed of the rotor, the solidity of the blades and their airfoil characteristics. Speed and solidity are dictated by noise and structural requirements. Therefore, the stall power is mainly dependent on the airfoil used and cannot easily be kept fixed if the rotor diameter is increased.

Much work has been undertaken on the design of airfoils specifically for use with wind turbines. The thrust of the work has been to design airfoils which are less sensitive to dirt and bugs and which have a lower peak lift coefficient. New airfoils have been developed by the NREL in the US which have a greater rotor swept area for a given peak power. However, these are still relatively uncommon on commercial wind turbines.

Reducing rotational speed only results in an increase in mechanical torque and increased gearbox costs. It is thus difficult to increase the swept area of a stall-regulated machine and keep its maximum power output constant, without affecting its performance.

Increasing Tower Height

The remaining option to improve the energy production of a wind turbine generator is to increase the wind speed at the rotor. The most practical way of achieving this is to make the tower taller.

Wind is slowed down as it passes over the ground by an amount dependent on the roughness of the terrain. The effect is known as "wind shear". On a flat, open site, a doubling of tower height can lead to an increase in the wind speed at the rotor of up to 10%.

Raising the height of a rotor increases its energy production potentially by the cube of the increased wind speed at the greater height. This energy improvement can either be achieved by taking existing rotors and mounting them on higher towers or by scaling up existing designs.

For the former option to be cost-effective, the rotor would have to be extremely light. Further elevating most of the rotor designs currently available would require stiffer and more massive towers and would result in a commensurate increase in costs. Any energy production benefit would be matched by cost increases. As a rule of thumb, as the height of a tower is increased, its weight, and by association, its price, is cubed.

Scaling up by doubling the size of a particular design increases its weight eight-fold. Furthermore, the swept area is only doubled, so the energy captured only doubles. The result is that simply scaling up a design decreases its cost effectiveness. However, other factors bring economies of scale.

Every wind turbine design has an optimum cost-effective size. Broadly speaking, as the design is scaled up, the material cost per kilowatt-hour produced increases. However, the unit labour cost in producing a larger version of the same thing decreases. Furthermore, the wind farm balance of plant costs tend to fall as machines become larger . In addition, generally the total energy yield from a given site can be improved by using larger machines. At some point, there is a balance.

Reducing Wind Farm Lifecycle Costs

The final option is to decrease the cost of the wind farm. Since the lifecycle cost of a wind farm is dominated by paying off the capital costs and the capital cost is dominated by the machine cost, this means lowering the lifetime costs of the machines. The prospects for doing this depend on a breakthrough in technology being achieved at today's prices. The present trend is to refine existing technology using standard components. This strategy is unlikely to achieve either a breakthrough or any great cost reductions.

It is much more likely that a breakthrough will come from an the use of an integrated design approach, resulting in a flexible, lightweight turbine which can be mass produced to give economies of scale.

The Carter Approach

There are no doubt many approaches which could achieve the dual aims of improved energy capture combined with low cost. However the CWT model 300 already exhibits many of the features discussed above. In particular the machine uses an integrated design which eliminates many of the components

required on other wind turbines. In addition, the structure is very flexible so that component weights and costs are kept to a minimum. These features in turn allow the rotor to be put on a very tall tower for a machine of its diameter. The tower is typically twice the height of machines of a similar size. The machine therefore exhibits a very low installed price and a high energy capture.

The lightweight approach also offers advantages in scaling up to larger designs. Most designs of machine currently installed in Europe, show decreasing cost effectiveness above about 750kW-1MW. However, a design currently under development by Carter Wind Turbines shows increasing cost-effectiveness even up to 1.5MW.

This two-times scale-up of the standard CWT Model 300 machine is intended to prove that wind energy can be produced for under 5 US cents/kWh on sites with wind speeds of under 6m/s. This compares with current unit wind energy costs int he US of 7-8 cents/kWh for similar sites.

Meanwhile, ten CWT model 300 wind turbines have been installed on a wind farm at Great Orton in Cumbria, UK. The site has an annual mean wind speed of only 5.8m/s at 30m height. The wind speed improves to 6.3m/s at the hub height of 50m. This is still considerably less than the wind speeds typical of other wind farm sites in the UK. Even so, the wind farm is profitable, providing an acceptable commercial return to its owners.

Carter Wind Turbines strongly believes Great Orton will represent a breakthrough in wind turbine electricity generation in Europe, the first step over the "viability threshold" for which we are all waiting.

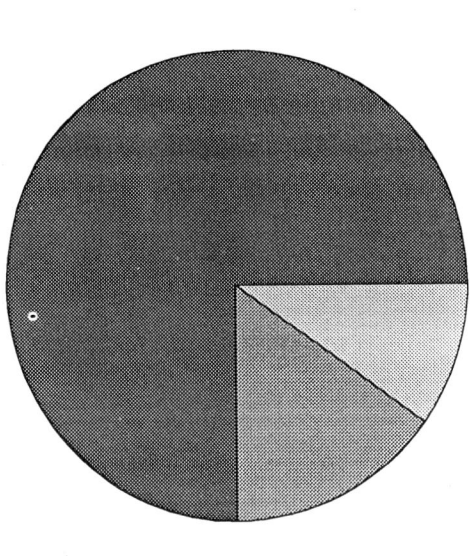

Debt & Equity ◼ O&M ◻ Other

Figure 1 Lifecycle Cost of a Wind Farm

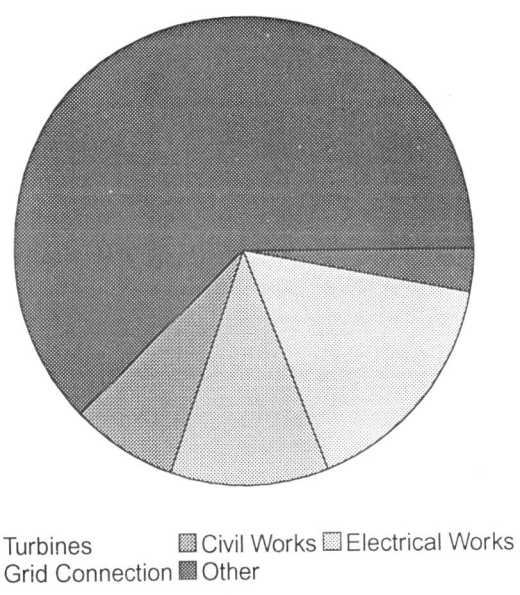

◼ Turbines ◻ Civil Works ◻ Electrical Works
◻ Grid Connection ◼ Other

Figure 2 Capital Cost of a Wind Farm

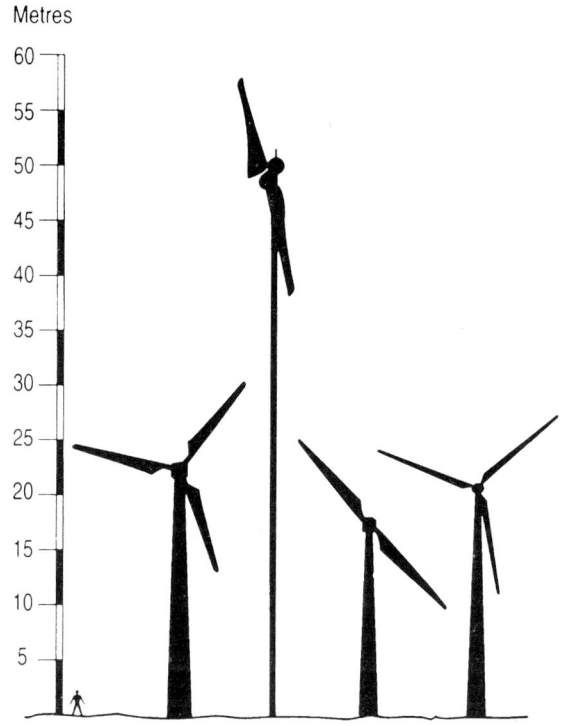

Figure 3 Relative hub heights of various wind turbine designs, scaled to the same rotor diameter.

Figure 4 Great Orton Wind Farm

Review of wind farm installations in the UK

D. J. LEIVESLEY and **M. L. HITNER**
Wind Energy Group Ltd, UK

The end of the summer of 1993, saw the completion of four wind farms supplied and installed by the Wind Energy Group, following orders under NFFO awarded power contracts. A total of 70 machines were installed in England and Wales.

Prior to these four wind farms, WEG had historically supplied small numbers of machines of varying sizes for specific customer requirements, in addition to the twenty MS-2 machines supplied and erected in Altamont Pass, California.

1 MANUFACTURE

For the manufacture of larger numbers of machines, it was essential that an appropriate Quality Assurance programme was in place. WEG had already obtained BS 5750 for design and blade manufacture, therefore similar certification would be required of our component suppliers. WEG engineers carried out extensive auditing, inspecting and evaluation of component manufacturers to ensure conformity with WEG designs. Even more attention was given to the supplier selection process as WEG adopted a policy of dual sourcing, for reasons of security as well as competition.

Blade and cladding production posed a less complex problem as WEG-designed blades and nacelles were already manufactured in Southampton for the existing machines. In addition rotor blade design and development had continued at Southampton as part of WEG's objective to achieve lighter and more cost effective machines.

The expansion of WEG's production capacity was a smooth operation, with the acquisition of the adjacent factory unit in Southampton and the purchase of new equipment for laying up GFRP claddings, which led to a significant increase in production.

WEG subcontracted an Assembly Plant located in South Wales, with a company that had excellent experience in mechanical fittings, good machine shop back-up, suitable cranage facilities and an assembly hall set apart from the main works. In addition, South Wales was readily accessible and close to the majority of our wind farms.

2 INSTALLATION

In eighteen months, WEG installed a total of 70 machines in four wind farms. 24 machines at Cemmaes, Mid Wales, 21 machines at Cold Northcott, Cornwall, 4 machines at Chelker Reservoir in Yorkshire and 21 machines at Llangwyryfon, Mid Wales. Each wind farm was built on varying terrain from hilltops to flat and rolling fields, which gave WEG engineers a variety of challenges to overcome. However, now successfully completed, WEG has gathered a wealth of knowledge and experience from what are recognised to be some of the toughest sites in Europe.

Although each wind farm was located on different terrain, the challenges encountered on each site were essentially the same, it was the solutions that were different.

- Landowners
- Terrain
- Land usage

Landowners

Each wind farm occupied land owned by more than one landowner, Cemmaes and Llangwyryfon both involved five landowners, Cold Northcott had nine landowners and Chelker had two tenant farmers. In collaboration with our client, this meant there was a need to closely involve and discuss all activities with each landowner at the beginning of the project as well as constant communication throughout the windfarm construction. As most of the landowners were farmers, their agricultural activities had to be taken into consideration when planning the construction phases. WEG employed a number of methods: personal visits, monthly newsletters and open evenings that were also open to the surrounding residents.

Terrain Difficulties

Cemmaes, the first wind farm to be installed was located on a hill top site, 1400 ft above sea level and was essentially bleak and barren with many marshy areas. The initial problem was access to the base of the hill. At that time access was via a narrow, winding 'C' road, totally unsuitable for heavy traffic of the type needed to transport concrete, tower and nacelles to the site. Whilst this road had to be widened, and passing bays constructed to allow satisfactory access, considerable effort was employed to retain the existing environment. There would be little point installing an environmentally friendly wind farm, if a motorway had to be constructed as well.

On getting to the base of the site, the next challenge was obtaining access to the hill top, which existed in the form of a narrow track, totally inadequate in width, strength and gradient for heavy vehicles. A new road was constructed, with a total of five hairpin bends to incorporate the correct gradient. It was fortunate that a number of small quarries were located on the site from which stone could be extracted and used for all the roads, including the access road on the top of the ridge between turbines, that would later be used by the maintenance team. The stone quarries were subsequently returned to their original state.

Torrential rain during the construction period, forced a change of plan, as the heavy traffic could not obtain access up the hill. As a result, all materials and equipment had to be unloaded at the base and reloaded on to purpose-built vehicles suited to all weather conditions. For example, concrete for the foundations was transferred onto 'Moxy' vehicles fitted with special concrete drums.

Cemmaes was the first Welsh wind farm to be connected to the grid in October 1992.

Cold Northcott, unlike Cemmaes, was located in rolling countryside, the wind farm was sited either side of the A395, the main link to Launceston (A30). The problem of access was different as there could be no central gateway, and instead several points along the A395 were used for access as well as some of the side lanes.

One major concern at Cold Northcott was the heavy flow of traffic on this busy road, therefore adequate signage was essential to direct and warn traffic of site accesses. As with Cemmaes and Llangwyryfon, the WEG teams adopted a 'park-up and escort' policy to ensure that the heavy vehicles delivering the nacelles and towers did not miss the accesses, and to minimise traffic conjestion.

As some of the land was for arable farming, many of the site roads had to be of a temporary nature and special top soil storage areas were established to ensure that the land would not deteriorate when re-laid. Cold Northcott was initially connected to the grid in November 1992.

Chelker wind farm was located at a reservoir site on the A65, north of Ilkley in Yorkshire. This was an upland site and was generally fairly level, which enabled relatively smooth access for construction. In addition, there was already road access on site to the pumping station operated by our client Yorkshire Water. Other permanent tracks had to be laid for the remaining turbine locations, one major consideration was to avoid damage to the two main water outflows to Wakefield and Bradford. This was achieved by bridging the outflows to allow access to the windfarm site by heavy construction traffic.

Chelker was connected to the grid in December 1992.

Llangwyryfon was the fourth wind farm to be completed. The site is situated south east of Aberystwith. The turbines are located on a ridge 950 ft above sea level. Although the terrain combined both hills and rolling countryside, the installation was similar to Cemmaes, so similar methods of construction were employed. Access was once again, via a 2 mile long narrow 'C' rated road which had to be widened and major components had to be off-loaded and on-loaded on to all weather vehicles before climbing to the site.

Llangwyryfon was completed and providing electricity to the grid by June 1993.

Land Use

On all four of the sites, the land was used predominately for sheep grazing and in some cases, cattle. It was essential to isolate construction works to ensure the safety of the livestock, and to allow normal farming activities to continue during windfarm construction. Seasonal activities such as lambing and calving etc, had to be considered as well, but consultation prior to commencing work, prevented any problems.

In addition, parts of the land at Cold Northcott, were arable, so it was necessary to take into account farming activities such as ploughing, planting and harvesting. Again, all potential problems were avoided, by discussing and employing a flexible programme approach.

As well as observing specific land usage, preliminary discussions with Archeological Societies, Wild Life and Nature Conservation groups together with the local Authorities enabled WEG to map out any Rights of Way, bridleways, areas of historic and ecological significance, to ensure these sites were protected during construction.

In conclusion, all four wind farms were completed ahead of schedule and within budget. This was achieved by adhering to, throughout the turnkey project, the following methods;

• Thorough Investigation

Prior to commencement of construction works, detailed studies need to be undertaken to identify any potential obstacles that could effect the overall construction programme.

Such investigations should be constantly updated and if necessary adjustments made to the agreed programmes, whilst aiming to achieve the agreed completion programme.

• Constant Communication

In collaboration with the customer, communicate with everyone at the earliest possible time: landowners, residents, local authorities, and any environmental groups, and throughout all stages of construction. Ensure everyone is well informed of the construction activities.

• Planning

At the offset of the wind farm project planning in collaboration with the customer and landowners must be undertaken to make sure that everyone's needs are known and if possible can be accommodated.

• Continual Assessment

To ensure that any unforeseen situations that may arise as the project proceeds can be effectively dealt with as it is necessary to have the whole project continually monitored and assessed.

• Conclusion

The valuable experience gained in undertaking the turnkey construction of the four windfarms has provided a wealth of practical knowledge and skills in windfarm construction, such knowledge will ensure that future windfarms are designed and installed to meet all our customers needs.

Its clear that wind turbine designers, manufacturers and wind farm constructors have two major obligations. The first is a duty of care to the landowner, the countryside and the local community. The second is a commitment to customers to provide wind turbines that are fit for purpose. It is the extent to which we are able to achieve these twin goals which will largely shape the success of our industry.

Operating experience from four UK windfarms

D. LINDLEY, BSc, PhD, FEng, FIMechE, **P. MUSGROVE,** BSc, PhD, CEng
J. WARREN, BSc, PhD and **R. HOSKIN,** BSc, PhD
National Wind Power Ltd, UK

SYNOPSIS. National Wind Power now has four operational windfarms in the U.K. and a fifth under construction. They comprise 99 turbines purchased from three different manufacturers and have a combined rating of 34.6MW. For the three windfarms where operational data is now available, availability of upto 99% is being achieved and at the Cemmaes windfarm, where the longest operational record is available, the energy output is close to that originally budgeted. The different terrain at the five sites allows some comparisons to be made between wind speed variability and individual turbine performance in an early attempt to determine the influence of site characteristics on windfarm design and operation.

1 INTRODUCTION

National Wind Power Ltd (NWP) has four operational windfarms and a fifth under construction. Details are given in Table 1.

Table 1 National Wind Power Windfarms

Location	No.of Tur-bines	Manu-factur-er	Turbine Dia.(m)	Rating (MW)	Opera-tional
Cemmaes (Wales)	24	WEG	33	7.2	Oct.'92
Cold Northcott (Cornwall)	22	WEG	33	6.7	Jun.'93
Llangwyryfon (Wales)	20	WEG	33	6.0	Jun.'93
Kirkby Moor (Cumbria)	12	Vestas	34.8	4.8	Sep.'93
Bryn Titli (Wales)	22	Bonus	37	9.9	Jun.'94

In addition, National Power PLC, National Wind Power's majority shareholder has made an investment in the Delabole Windfarm comprising ten Vestas 400 kW turbines and operated by Wind Electric Ltd which makes National Power the largest investor in windfarms in the U.K.

The performance of the first three windfarms listed in Table 1 is being monitored in considerable detail with a major wind farm Research and Development programme co-funded by National Wind Power and the Department of Trade and Industry and undertaken by National Wind Power in consultation with the Energy Technology Support Unit (ETSU). This programme embraces detailed individual turbine monitoring, environmental studies and comprehensive anemometry investigations; it also includes the construction and testing of the first production WEG400 turbine at Cold Northcott.

The following represents a selection of the early performance results obtained from this programme.

2 ENERGY OUTPUT

The total energy output from the Cemmaes windfarm from January 1st to end of August 1993 was 13,790 MWh compared with the original budget figure of 14,183 MWh. The way in which this output was distributed on a month by month basis is shown in Figure 1 and compared with budgeted output. The latter is based on an original assessment of site wind speed data and correlations with long term meteorological data. The budgeted monthly distribution assumes an annual mean wind speed of 8 m/s and not 8.3 m/s which is the actual estimate based on measured data.

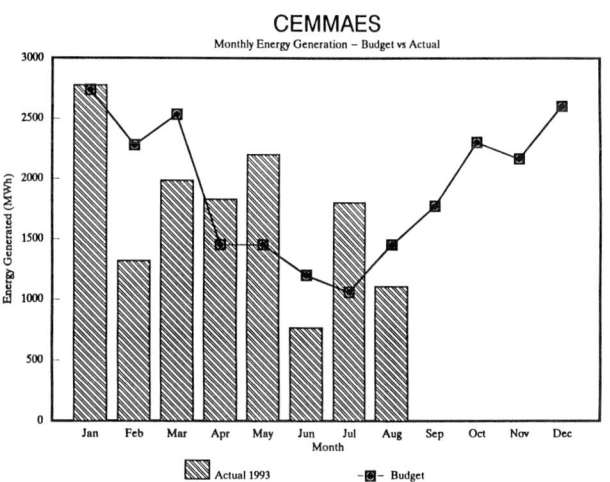

Figure 1. Monthly output versus budget at Cemmaes

It is important to recognise at this point that actual outputs in any year will differ from those predicted on the basis of long term mean statistics. What is important is whether the economic viability of the windfarm is likely to be impacted by a 'bad' wind year in its first year of operations compared to a 'bad' year towards the end of the NFFO contract period. NWP uses Monte Carlo simulations to assess this effect which show, for example, that for a 10 year forecast (e.g. a 10 year NFFO contract) at a discount rate of 10%, there is a 95% probability that the Net Present Value of the energy outputs in the 10 year period is at worst 94.3% of the average and could be 105.7% of the average.

The Cold Northcott and Llangwyryfon Windfarms have only been fully operational since July 1st so that only limited data is available. Figures 2 and 3 provide operational and budget data where it can be seen that at all three sites, July output exceeded budget and August fell below budget.

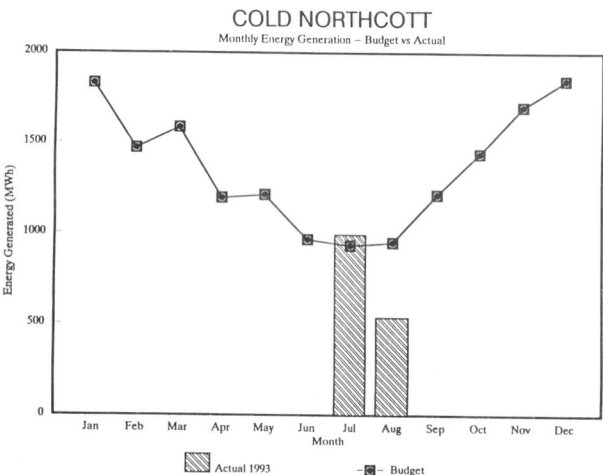

Figure 2. Monthly output versus budget at Cold Northcott

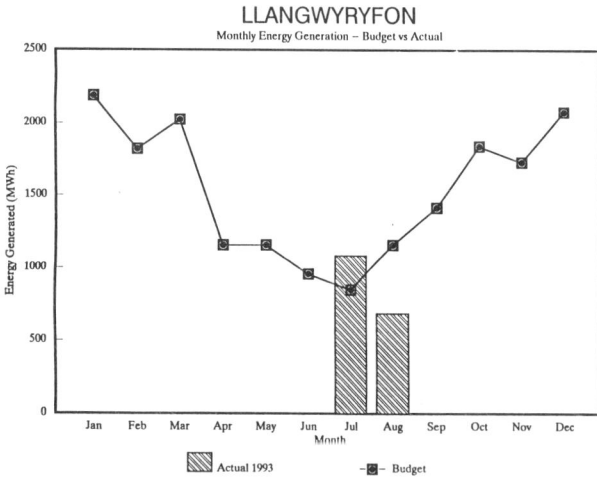

Figure 3. Monthly output versus budget at Llangwyryfon

3   SITING EFFECTS

It is of considerable interest to the designer of a windfarm to know how turbines might be sited to maximise output and minimise costs. The NWP windfarms present a considerable opportunity to investigate the effects of topography and local meteorological conditions. Figure 4, for example, gives the energy generated per month (averaged over the period January 1st to the end of August 1993) for each of the 24 turbines at Cemmaes. The mean wind speed measured by an anemometer located on the nacelle of each turbine is given in the same figure. This data raises more questions than we have answers for at this stage of our investigations. It shows, for example, that turbine 2

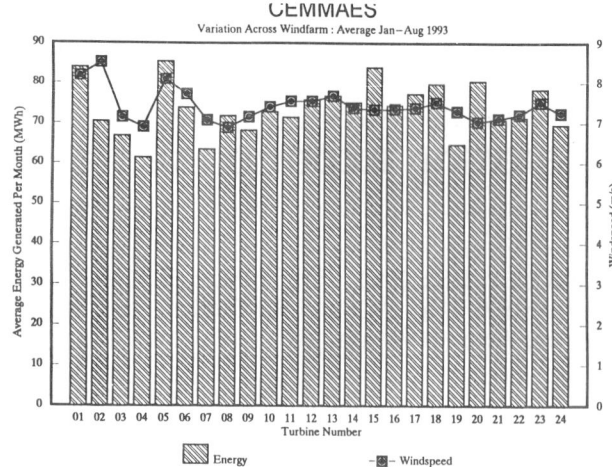

Figure 4. Output and mean wind speed for each turbine at Cemmaes (Jan. to Aug. 1993)

has a lower output than turbine 1 even though a higher wind speed has been recorded. Turbine 19 also has a higher windspeed but lower output than Turbine 20. The lower outputs are not the result of lower availability (which is almost the same for all turbines). The wind speed versus output anomalies could be the result of different anemometer calibrations (which are being checked now) but are also likely to be the result of directional and turbulence effects resulting from the inter-relationship between the complex topography and the wind rose. Figure 5 shows the location of the turbines in relation to the topography of the site (note that the contours shown are at 10m intervals and the turbines are generally located at an elevation of 400m above sea level). It can be seen that turbines 3, 4, 7 and 9 are generally in the lee of turbines 1, 2, 5, 6 and 8 for

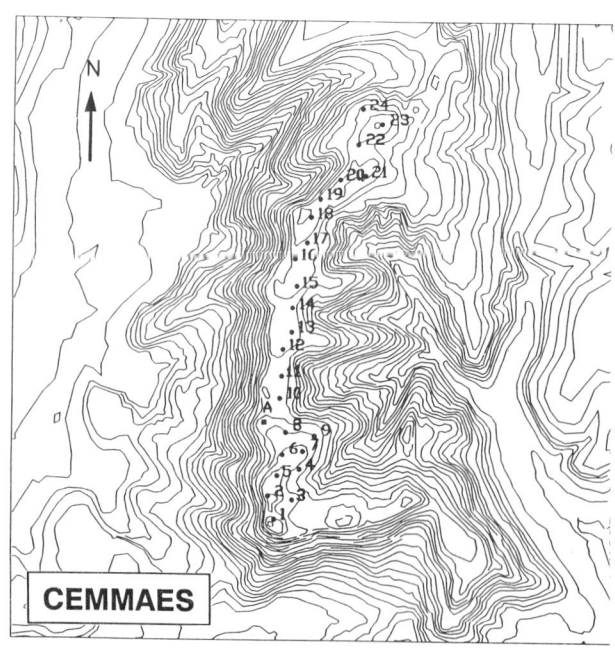

Figure 5. Turbine locations at the Cemmaes Windfarm (contours at 10m intervals)

winds coming from the westerly sector. Table 2 shows that winds from the westerly sector are particularly dominant at this site. It can also be seen that turbine 2 is close to the escarpment (closer than any other turbine)

Table 2. Wind Distribution Table for Cemmaes, Cold Northcott and Llangwyryfon

WINDSPEED DISTRIBUTION TABLE

Direction Sector	CEMMAES	COLD NORTHCOTT	LLANGW-YRYFON
345° - 015°	3.0 %	14.0 %	4.0 %
015° - 045°	4.5 %	13.3 %	3.0 %
045° - 075°	8.9 %	10.0 %	2.8 %
075° - 105°	12.2 %	5.5 %	6.3 %
105° - 135°	5.0 %	11.1 %	10.6 %
135° - 165°	2.9 %	2.7 %	7.6 %
165° - 195°	2.2 %	0.8 %	10.0 %
195° - 225°	6.8 %	3.0 %	10.2 %
225° - 255°	34.9 %	11.9 %	11.0 %
255° - 285°	13.0 %	15.8 %	15.2 %
285° - 315°	4.4 %	2.0 %	10.2 %
315° - 345°	2.4 %	10.0 %	9.2 %

and is therefore likely to be operating in a more turbulent flow field with some strong vertical velocity component in the approaching wind vector. To enable comparison with Cold Northcott and Llangwyryfon, Cemmaes data is presented for the months of July and August in Figure 6. The Cold Northcott and Llangwyryfon data is presented on the same basis in Figures 7 and 8.

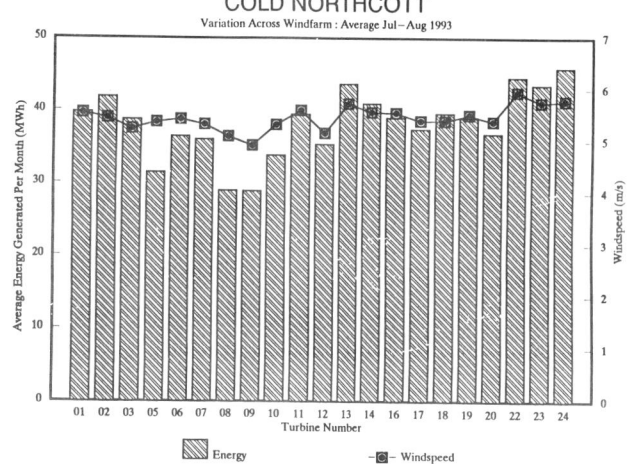

Figure 7. Output and mean wind speed for each turbine at Cold Northcott (July/August 1993)

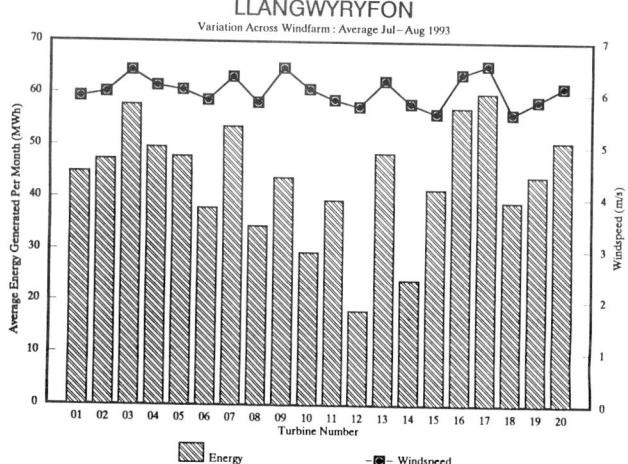

Figure 8. Output and mean wind speed for each turbine at Llangwyryfon (July/August 1993)

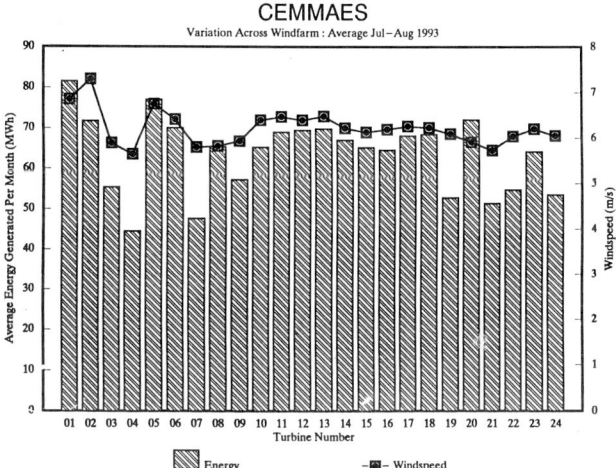

Figure 6. Output and mean wind speed for each turbine at Cemmaes (July/August 1993)

Though Cold Northcott is at a lower elevation than the other two sites and is generally flatter and less complex, the turbines with the larger outputs (Nos.13, 22 and 24) are generally on higher and more exposed locations. Whilst turbines 8, 9 and 12 are in the least advantageous locations.

Llangwyryfon is again a site with a complex topography and wind regime. Contours are given (at 10m intervals) in Figure 9 and wind statistics in Table 2. One of the factors, which probably has a significant effect on

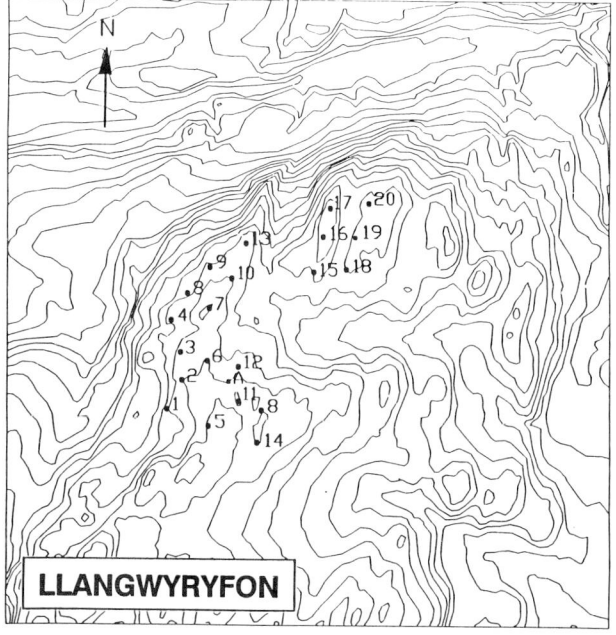

Figure 9. Turbine locations at the Llangwyryfon Windfarm (contours at 10m intervals)

43

the data given in Figures 6, 7 and 8, is that results are for the lowest wind speed months of the year, when mean wind speeds (see Table 3) are only marginally above the cut-in wind speed of the turbines. The turbulence of the wind has a greater impact on individual turbine outputs in these circumstances.

Additional data is now being acquired at all three sites. Further and more detailed analysis will then be carried out to obtain a better understanding of the relationship between site topography, wind speed and direction, turbine location and operational characteristics.

4   OTHER OPERATIONAL STATISTICS

Table 3 presents some operational statistics for the windfarms at Cemmaes, Cold Northcott and Llangwyryfon. It can be seen that availability is very high and that Cemmaes has the greatest annual mean wind speed. It is not easy to compare statistics on a rigorous basis at this stage as there are only two months of operational data (obtained in the lowest wind speed months) at Cold Northcott and Llangwyryfon to compare with the longer operating record at Cemmaes.

Table 3.   Operational Statistics for Cemmaes Cold Northcott and Llangwyryfon Windfarms.

OPERATIONAL STATISTICS

SITE	CEMMAES	COLD NORTHCOTT	LLANGWYRYFON
Availability (August)	99.10 %	96.03 %	99.08 %
Operation Period (1993)	January - August	July - August	July - August
Hours of Operation (1993)	76,000 hrs	15,500 hrs	16,500 hrs
Operating/Available Time (1993)	54 %	44 %	46 %
Energy Generated (1993)	13,400,000 kWh	1,500,000 kWh	1,700,000 kWh
Mean Power Level (1993)	2300 kW	1010 kW	1160 kW
Number of Turbines	24	22	20
Mean Power per WTG (1993)	96 kW	42 kW	48 kW
Mean Operating Power (1993)	176 kW	97 kW	103 kW
Mean Windspeed (1993)	8.0 m/s	5.6 m/s	5.4m/s
Annual Mean Windspeed	8.3 m/s	7.7 m/s	8.1 m/s

The mean power level for each turbine (and therefore the mean power level of the windfarm) will of course increase as they are exposed to the higher winds of September to December. At Cemmaes, the table shows that in 8 months, 13,400,000 kWh have been generated. This amounts to an output of approximately 560,000 kWh per turbine or 650 kWh/m^2 of rotor swept area. In a full year we expect the turbines to produce not less than 1120 kWh/m^2 of swept area.

5   WIND FARM VARIABILITY

Table 4 presents data for each turbine at Cemmaes, where it can be seen that the energy output, average power, mean nacelle windspeed, number of starts, number of speed changes (the WEG turbines have two-speed operation depending on wind speed), and number of yaw operations all vary across the windfarm. These variations are almost certainly the result of the impact of topography and its interaction with wind speed and direction and the spatial variation of wind speed, direction, turbulence and gustiness across the site.

Table 4.   Variation in individual turbine operations at Cemmaes windfarm

WINDFARM VARIABILITY

CEMMAES : AUGUST 1993

WTG	Operating Time/ Available Time (%)	Avail- ability (%)	Energy (% total)	Average Power (kW)	Capacity Factor (%)	Mean Nacelle Windspd (m/s)	Starts (% total)	Speed Changes (% total)	Yaw Opns (% total)
1	53	98.60	5.40	160	28.34	5.95	5.1	4.1	3.7
2	59	98.41	4.83	130	25.32	6.36	3.5	4.0	2.5
3	45	99.91	3.69	128	19.35	5.24	4.6	5.9	4.3
4	39	99.49	2.89	118	15.15	4.91	5.4	4.5	4.0
5	57	99.91	5.05	139	26.46	5.90	4.5	3.3	2.9
6	55	99.61	4.62	133	24.24	5.62	4.6	3.8	3.9
7	42	99.31	3.13	118	16.43	5.17	5.8	6.4	3.1
8	43	99.46	4.20	153	22.04	5.07	4.3	4.2	3.5
9	45	99.45	3.72	130	19.50	5.19	4.6	4.6	4.3
10	45	99.87	3.91	137	20.49	5.57	3.4	3.1	8.3
11	45	97.73	4.37	151	22.93	5.56	3.1	3.2	4.0
12	47	99.64	4.49	152	23.56	5.53	3.9	3.5	3.0
13	47	99.34	4.47	149	23.42	5.57	3.5	3.5	5.1
14	44	99.61	4.27	151	22.42	5.35	4.0	3.6	4.1
15	46	99.88	4.31	148	22.61	5.34	4.7	5.4	4.4
16	46	98.52	4.24	146	22.25	5.39	3.8	3.8	3.5
17	47	99.30	4.55	151	23.84	5.43	3.4	3.7	2.7
18	48	97.88	4.56	148	23.90	5.46	3.8	4.1	2.4
19	45	99.75	3.72	130	19.52	5.39	4.4	4.4	4.2
20	40	96.56	4.64	162	24.34	5.24	4.1	2.8	2.9
21	43	99.36	3.39	124	17.77	5.09	5.1	4.5	3.1
22	43	98.60	3.55	131	18.64	5.34	3.2	3.6	7.1
23	46	99.68	4.39	149	23.03	5.51	4.1	5.3	2.5
24	45	98.60	3.61	128	18.93	5.34	3.0	4.8	10.4

6   POWER CURVE

Figure 10 presents power curve measurements taken according to the IEA standard method (Ref:1) for one of the turbines at Cemmaes. These are compared with the guaranteed power curve for the turbine. Other measurements are under way to determine effect of location (if any) on power curve measurements.

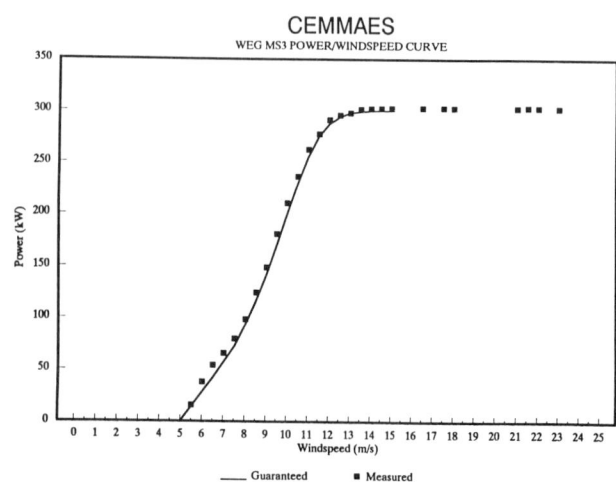

CEMMAES
WEG MS3 POWER/WINDSPEED CURVE

—— Guaranteed   ■ Measured

Figure 10.   WEG MS-3 measured power curve at Cemmaes

7 GRID CONNECTIONS

Arguably one of the highest risk aspects of building a windfarm is associated with the risks involved in obtaining wayleaves for and building the connection to the local distribution system in a timely way. NWP has had connections built by MANWEB, SWEB, SWALEC and NORWEB. Details are given in Table 5. Construction in each case has been accomplished in 9 months or less and connections up to 6 km in length have been required. All NWP windfarms have turbines connected at 11 kV with 11kV/33kV substations in four out of five cases. The Bryn Titli windfarm has an 11kV/66kV substation.

Table 5. Grid Connection Statistics for five NWP Windfarms

GRID CONNECTION STATISTICS					
SITE	CEMMAES	COLD NORTHCOTT	LLANGW-YRYFON	KIRKBY MOOR	BRYN TITLI
REC	MANWEB	SWEB	MANWEB	NORWEB	SWALEC
Connection Voltage (kV)	33	33	33	11	66
Description	6km 33kV line from existing Manweb substation	200m 33kV line Teed to existing 33kV line	6km 33kV line from existing Manweb substation	New Norweb substation built. 5km 33kV line. 11kV connection	5km 66kV line Teed to existing 66kV line
Construction Time (months)	9	7	8	9	9 expected

8 OTHER WORK

In addition to windfarm performance monitoring, NWP are currently undertaking, or planning, a wide range of related studies within the Windfarm Research and Development Programme. Topics include:

- Environmental investigations involving ecological and ornithological studies before, during and after windfarm construction.

- Individual turbine and windfarm noise investigations including propogation and tonal assessment studies.

- Anemometry investigations including airflow modelling and monitoring, and resource assessment techniques.

- Individual turbine performance investigations relating to power quality, micro-siting, turbine control optimisation and health monitoring.

- Electrical integration investigations.

9 CONCLUSIONS

The early results of the programme to monitor the performance of windfarms developed by National Wind Power are raising many interesting questions which will lead to a greater understanding of site assessment and windfarm design and operations.

10 REFERENCES

i) IEA recommended practices for wind turbine testing and evaluation.
1. Power Performance Testing.   Edition 2, 1990.

11 ACKNOWLEDGEMENTS

To the Wind Energy Programme Management team at the Energy Technology Support Unit for their contribution to the joint NWP/DTI Research and Development programme and to the shareholders (National Power PLC and Taylor Woodrow Construction Holdings Ltd) in the windfarms at Cemmaes, Llangwyryfon and Cold Northcott for permission to publish these results.

The Penrhyddlan and Llidiartywaun windfarms - the first to join the electricity pool

C. PEASLEY and **G. NICHOLSON**
Ecogen Ltd, UK

SYNOPSIS All power stations above 10MW in size must belong to the Pooling and Settlements system, which is the market for trading electricity in England and Wales. The P&L windfarm of 30.9MW near Newtown in Wales is currently the largest outside California and the first windfarm to join the Pool.

1 INTRODUCTION

Ecogen was formed in 1990, and has offices in Machynlleth, Wales; Truro, Bradford, and Northumberland, England; and Glasgow, Scotland. The company formed a joint venture partnership in order to bid wind power contracts into NFFO2 with Seawest of California and Tomen of Japan. This company, Adaprojects Ltd., has achieved contracts and planning permission for five wind projects totalling 47.55 MW capacity. The same partners have formed a joint venture company, Trigen Windpower Ltd., to develop wind power sites for NFFO3 contracts.

2 DEFINITIONS AND ABBREVIATIONS

Company	Ecogen Ltd., Adaprojects Ltd., Trigen Windpower Ltd. and its successors .
P&L	Penrhyddlan and Llidiartywaun.
NFFO	Non fossil fuel obligation - the power purchase contract system for renewable energy in England and Wales.
NFPA	Non Fossil Purchasing Agency - administers NFFO contracts and payments.
OFFER	Office of Electricity Regulation - the government appointed watchdog and regulator of the electricity industry.
REC	Regional Electricity Company - the owner/operator of the electricity distribution system.
MANWEB	The local REC for the windfarm.
Generator	A company or body which owns and operates a power station (including a windfarm).
NGC	National Grid Company.
Embedded	Connected to a REC system and not directly to the NGC Grid.
GSP	Grid supply point - substation with metering on the national grid system at 275 or 400kV.
UOS	Use of System - a basis for charging for the transfer of power through an electricity system.
Pool	The term for the market of electricity in England and Wales.

3 DEVELOPMENT OF THE P&L WINDFARMS

Adaprojects obtained contracts under NFFO2 in November 1991 to build the windfarms at Penrhyddlan (P.) and Llidiartywaun (L.) near Llandinam, Newtown, Powys, Wales. Planning permission was obtained in principle on 21st February 1992 subject to an agreement with the BBC over possible television interference to a re-broadcast link which was achieved on 2nd June 1992. Following several iterations on siting and grid connection the substation received planning permission on 3rd August 1992. The two 33kV overhead lines linking the substation to the two windfarms received approval from the Department of Trade and Industry under Section 37 of the Electricity Act 1989 on 2nd September 1992. The layout of the windfarm and the overhead lines from the substation is shown in Figure 1.

Construction commenced on 8th June 1992. The P windfarm comprises of 43 Mitsubishi 300kW pitch controlled turbines, capacity 12.9MW and the L windfarm 60 turbines capacity 18MW.

The P. site was connected to MANWEB's system on 23rd November 1992. The L. site was energised on 29th December 1993. The whole windfarm was fully operational at the end of January 1993.

4 THE ELECTRICITY SYSTEM IN ENGLAND AND WALES

The principle transmission system operates at 275 and 400kV and is owned and operated by the NGC. Large power stations, customers and suppliers are connected to this system. Most customers and smaller generators are connected to the systems of the 12 regional electricity companies. They own and operate distribution networks at 132kV and below, the principle voltages are 66, 33, 20, 11, etc). The electrical system design and operation vary between the 12 RECs and within RECs territories, although many common practices, standards and recommendations apply.

4.1 Connection of windfarm to system.

MANWEB is the REC that operates in the Newtown area of Wales where the P&L windfarms are located. Supply to Newtown is via a 65km long single circuit 132kV line from the grid supply point (GSP) at Legacy substation on the 400kV grid. A single transformer at Newtown feeds 33kV circuits which are either radial or looped back to Newtown. Substations with isolation, circuit breakers, protection and 33/11kV transformers are located at various points on the 33kV system. Most customers are fed from 11000/415-240V transformers on the 11kV network. The location of the windfarm and connecting grid system is shown in Figure 2.

Connection of the windfarm was made at 33kV and a new substation was established at Lower Penrhyddlan Farm, Llandinam which is close to the windfarm and within 50m of the existing MANWEB 33kV ring system emanating from Newtown. The substation contains seven circuit breakers with associated busbars, protection, isolation and earthing. One circuit breaker protects each of MANWEB's incoming circuits, one circuit breaker acts as a bus switch between the two circuits, and each windfarm feeder has a metered circuit breaker owned by MANWEB in series with a protection circuit breaker owned and operated by the Company.

In line with normal MANWEB practice, the substation is operated with the bus switch closed, maintaining a 33kV ring network. However as windfarm output increases, the current flowing to Newtown in the low impedance leg of the ring rises beyond the design threshold. At this point the bus switch is opened and the output of the smaller P

windfarm is forced round the long way, on the higher impedance leg back to Newtown. When conditions permit the bus switch is closed again.

In the event of a fault on the 33kV system only one circuit would be available from Llandinam to Newtown and the maximum windfarm output has to be curtailed to 20MVA.

In the event of a fault on the 132kV system or transformer at Newtown the windfarm output is forced to travel approx. 22km further on the 33kV system to Welshpool, which is the next adjacent 132kV transformer. The windfarm can continue to operate in such conditions provided that the 33kV connection voltage remains within statutory limits, which itself depends on the local load, windfarm output and power factor.

From the substation the Company constructed two 33kV overhead lines 2km and 4km to connect the P & L windfarms to the substation. From the terminal poles of these lines the distribution on both windfarm sites is by underground cables to padmount transformers at the base of each turbine.

5 INTRODUCTION TO POOLING AND SETTLEMENT

On March 1990 a new market was established for the trading of electricity between those who produce it - the Generators - and those who use it themselves or buy, transmit and resell to smaller users - the Suppliers.

Prior to privatisation the Electricity Supply Industry was effectively a state owned monopoly. Nowadays the generators: National Power; PowerGen; Nuclear Electric (still state owned); supplies from France and Scotland via the Interconnectors; and the new independent generators all provide electricity to the 'Pool' a concept conjured up to deal with a new market for electricity that cannot have a surplus or a shortage at any time.

The opportunity afforded by this new free market does not always appear 'golden'. The method of trading has been set up with the large experienced generators in mind and with all necessary safeguards for security of supply. Small independent generators must wade through acres of paper to find the relevant rules and comments. In the early stages of the P&L windfarm ways were considered in which joining the Pool could be avoided (see below). On the positive side for NFFO2 contracts the Pool guarantees a market place for electricity beyond the end of 1998, but at an uncertain price.

5.1 How does the Pool Work.

Put very simply, the generators bid into the Pool how much power they can supply and at what cost. This is done a day in advance and in half hour periods. The expected demand is calculated a day ahead and again in half hour periods. Bids are selected to meet this demand starting with the lowest for each half hour. The highest bid needed to satisfy the demand sets the Pool Purchase Price. It is not clear how this system would work for a windfarm! Fortunately this complex bidding system only applies to generating stations with a capacity of more than 100MW known as Centrally Despatched. Generating stations under 100MW but over 10MW are considered large enough to be accounted for in the control and management of the whole system, but not significant enough to join the complex price setting bid process.

5.2 Joining the Pool

To join the Pool a small independent generator must agree to the rules and procedures of the Pooling and Settlement Agreement. A Grid Supply Point (GSP) must be identified as the notional point of connection to the National Grid System, and a Use of System Agreement must be made with the National Grid Company.

The process of application involves considerable paperwork. Agreed Procedures relating to metering, payment and keeping of data must be satisfied. These must be considered at an early stage as some require a six week lead time. The applicant must demonstrate that a Connection Agreement and Generating Licence are either in place or under way and an independent engineer must certify the expected date of first generation.

There are two stages of Pool Membership: becoming a Party to the Pooling and Settlement Agreement; and becoming a Pool Member. These applications can run contemporaneously. A Party is entitled to certain information related to the Pool but cannot trade and does not have voting rights. Each Pool member has voting rights in proportion to the amount of Energy traded in the system.

6 THE PROCESS OF POOLING THE WINDFARM

At an early stage of the process various options were considered to avoid having to join the Pool. These included:

(a)   limiting P. to 10MW by curtailing (from 12.9 MW);

(b)   splitting the 30.9 MW into 3x9.9 (29.7)MW and rejecting 1.2 MW;

(c)   downsizing each project P&L to 9.9MW and rejecting 11.1MW.

Option (a) still left all the overhead and learning curve of an application for L. Option (b) meant an additional point of supply, metering, larger substation, and an additional overhead line to the L. site. All options were rejected on economic and timescale grounds.

6.1 The affect on other projects

As a result of the company being a Pool Member the affects on its other projects were assessed. It was decided not to Pool the site at Rhyd-y-Groes Anglesey (7.2MW). There was some debate as to whether this was possible, or whether all projects of a Pooled Generator must be licenced (and Pooled). The question was answered under Clause 8.5.1 of the Pooling and Settlement Agreement, which indicates that just because a generator is Pooled by virtue of one site does not mean that all its sites must be Pooled.

6.2 Other licenses and agreements

Anyone who generates electricity is required under Section 6 of the 1989 Electricity Act to hold a licence unless exempted under Section 5 of the Act. The current OFFER exemptions include any generator of less than a maximum of 10 MW from any one generation station.

A generation licence was applied for and granted to Adaprojects (P&L) Ltd. , the specific site owning company. The Company became a Party to the Pooling and Settlement Agreement on 23rd July 1992 and a Pool Member on 16th November 1992.

A non Pooled generator would normally have a Connection Agreement and a Supply Agreement with the REC for electricity consumed. A Pooled generator has a Connection and Use of System Agreement.

6.3 Metering

Metering had to comply with Pooling and Settlement Agreed Procedures. The metering was sealed by the NGC in the presence of the Company and the REC and breaking of the seals is only permitted with the agreement of all parties. The metering is read by MANWEB (the meter Operator) using their Second Tier Data Collection System. The Scaling Factor (see Section 7) is applied to each half hour reading before data is sent on to the Pooling and Settlement System. Pulse signals from the metering are available to the Company for checking purposes at the substation.

In theory any of the RECs can be the meter operator and the charges for the service vary considerably.

6.4 Power Purchases

Power is purchased from the system when the windfarm is not generating for the following purposes: substation supplies (heat lights, protection equipment); losses (transformer iron losses); turbines (lights, dehumidifiers, hydraulics, yawing). A Pooled generator purchases power from the Pool at Pool selling price, on to which are added Use of System charges by the REC. A non Pooled generator will purchase power from the REC as a franchise customer.

7 CALCULATION OF SCALING FACTOR

The commercial boundary of the Pool is normally the GSP, in our case Legacy substation on the 400kV system. However, an embedded generator is connected to the REC system some distance away. In the absence of multiple metering systems a Scaling Factor is required to reconcile the network losses. The scaling factor can be positive or negative depending on whether the generator is reducing or increasing system losses. A high local load and a small generator will tend to reduce losses. A large generating station and a small local load will tend to increase losses. The system impedance (primarily the distance) between the generator and the GSP will finally determine the magnitude of the Scaling Factor, positive or negative.

The scaling factor is calculated under 8 different time periods representing different system loads conditions and shown in Table 1. (1).

Table 1 Time periods for scaling factor calculation

Summer and Winter	Time periods
Working day Peaks	1600-1900
Working days	0700-1600
Nights	0000-0700
Rest inc. weekends	All other times

The winter months are defined as November to March inclusive. MANWEB calculated the system loads during the specified periods and the expected output from the windfarm was calculated on the basis of wind data collected. As expected this showed some diurnal variation in the summer but was hardly discernible in the winter. Variations at weekends or bank holidays were not looked for in the wind data!

A network analysis package was run simulating the system with the eight different average demands without wind generation and the same with wind generation. The losses under both scenarios were obtained and the difference attributed to the windfarm. A Scaling Factor for each time period was calculated to credit the reduction in losses to the windfarm. The scaling factor is applied to the half hour meter readings from the windfarm before being transmitted to Settlement. Any gain or loss is also reflected in the payments from settlement and the NFPA.

8 NETTING OFF NGC CAPACITY CHARGES TO REC

Changes were being made to the NGC Use of System (UOS) charges during our application and negotiations. These changes were intended to make the charges for the grid system more directly related to the actual costs. The changes included negative UOS charges for generators (i.e. payments) in areas short of generation. The idea was to encourage generation where it is needed and encourage demand in areas of surplus generation. The crucial change related to NGC UOS charges for capacity for embedded generators.

Suppliers of electricity taking supplies from the NGC system at a GSP must pay capacity related NGC UOS charges based on their demand at the 3 highest half hour demand points in a year, known as the system peak demand Triad. The presence of embedded generation however can reduce the demand from the GSP if it is supplied locally from the generator. Rather than charging the embedded generator for this capacity, the generator should be credited for reducing load on the NGC system. This change actually took place on 17th March 1993.

A contract was made with a supplier taking power from the Legacy GSP to share the cost savings resulting from this change. The value of this credit varies regionally as shown in figure 3. The demand charge in region 3 (N. Wales and W. Lancs) is 9.54£/kW.

The capacity charges are based on the three maximum demand 1/2 hour periods in the months November to February. In 1992-3 these occurred during the half hour periods ending:

17.30 on Tuesday 17th Nov. 1993
17.00 on Wednesday 9th Dec. 1993
17.30 on Monday 4th Jan. 1993

If the amounts being generated in each half hour period were $G1$, $G2$, $G3$ kW then the average Triad peak generation

$$GT = (G1 + G2 + G3)/3.$$

To calculate the total saving multiply by the NGC UOS demand charge for that region (in £/kW) and by the percentage share agreed with the supplier under the netting off agreement.

9 OTHER PAYMENTS AND CHARGES

The Company is paid the Pool purchase price for units of electricity sold to the Pool through the Pooling and Settlements System. A contract for differences exists with the NFPA so that the net payment for energy is the agreed NFFO price.

(NFFO contract price) - (Pool purchase price)
$$= (NFPA \; payment \; price)$$

Occasionally the Pool purchase price has exceeded the NFFO price and the company must reimburse the NAPA under these circumstances.

Table 2 Summary of payments

From	For
NGC	Units generated;
NFPA	Units generated;
REC	Netting off NGC UOS demand charges.

Table 3 Summary of charges

To	For
Offer	Generation Licence & Annual Fee;
NGC	UOS Application Fee;
NGC	Party and Membership to Pooling and Settlement;
REC	Meter Operator Charges;
NGC	Membership of Settlement System - data fee;
NGC	Pool Funds Administration Charges - energy costs, data costs & administration;
NGC	UOS Demand charges;
REC	UOS Demand charges;
NGC	Meter registration fee;
NGC	Generator registration fee;
NGC	Energy Purchase for imports

Energy imported by the windfarm (during no wind periods) is purchased from the Pool at the Pool Selling Price and wheeled through the RECs system incurring REC's Use of System Charges.

A summary of charges associated with Pooling and Settlement is presented in Tables 2&3.

10 CONCLUSIONS

The Company has become a Pool member and operates the P&L windfarm through the Pool. This has brought both additional costs and benefits. Each power station or windfarm project must be assessed on its merits to see whether there is an overall financial gain or loss in crossing the Pool threshold of 10MW. This assessment needs to be one of the early factors affecting windfarm capacity, particularly in regard to projects around 10MW in size. The paperwork and process of joining the Pool is daunting for first timers but not insurmountable.

ACKNOWLEDGEMENTS

There have been far too many people and even organisations involved with this project to mention by name but the authors would like to thank all those involved, landowners, planners, councillors, surveyors, lawyers, consultants, financiers, suppliers, engineers, contractors, our colleagues and joint venture partners, who made this project successful.

REFERENCES

(1) SETTLEMENT SUBCOMMITTEE OPERATIONS Guidance note for calculation of loss factors for embedded generators in Settlement, SSC(OP)1390(Revised), March 1992.

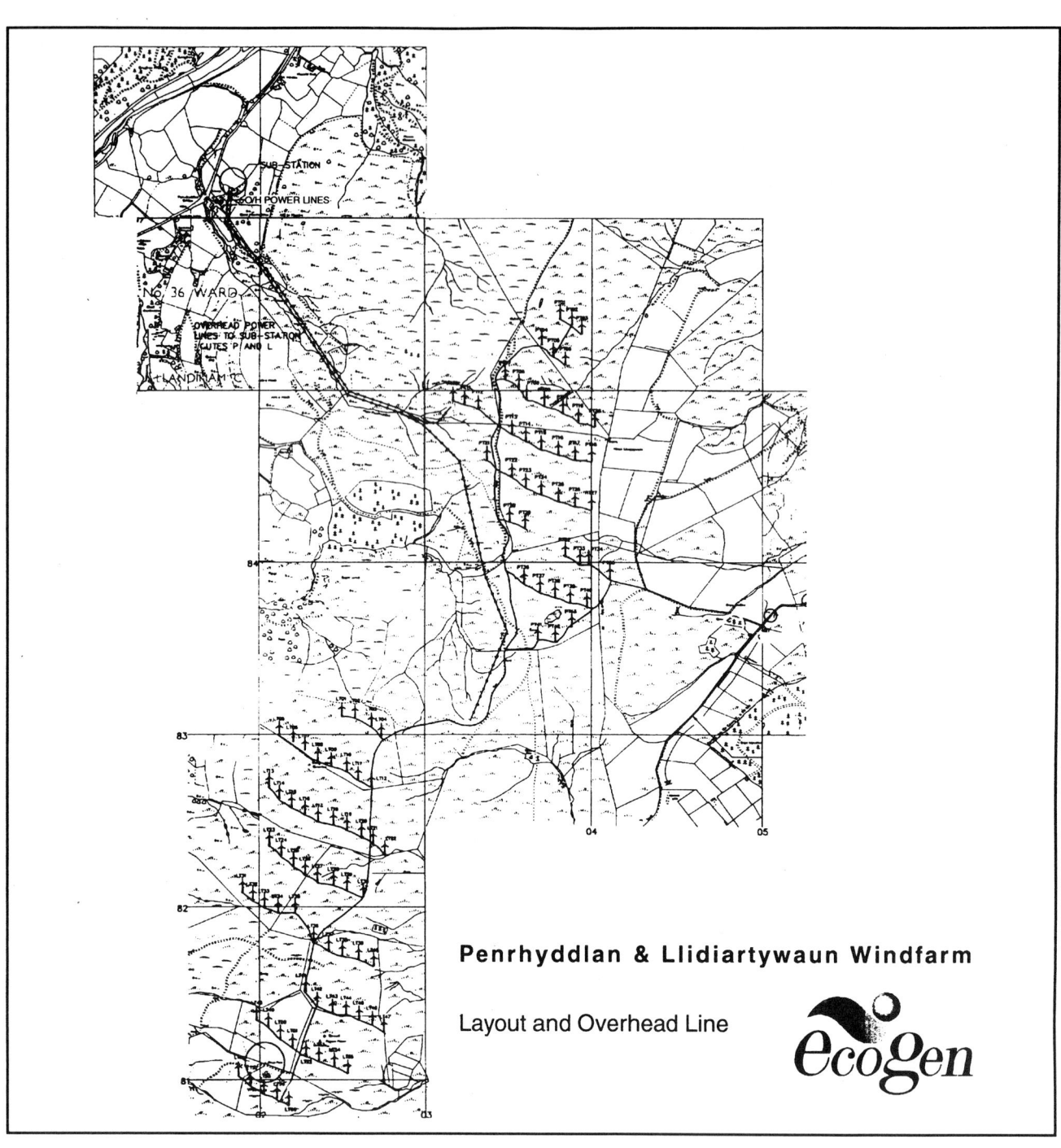

Penrhyddlan & Llidiartywaun Windfarm

Layout and Overhead Line

ecogen

Fig 1 Location of turbines, roads, overhead lines and substation

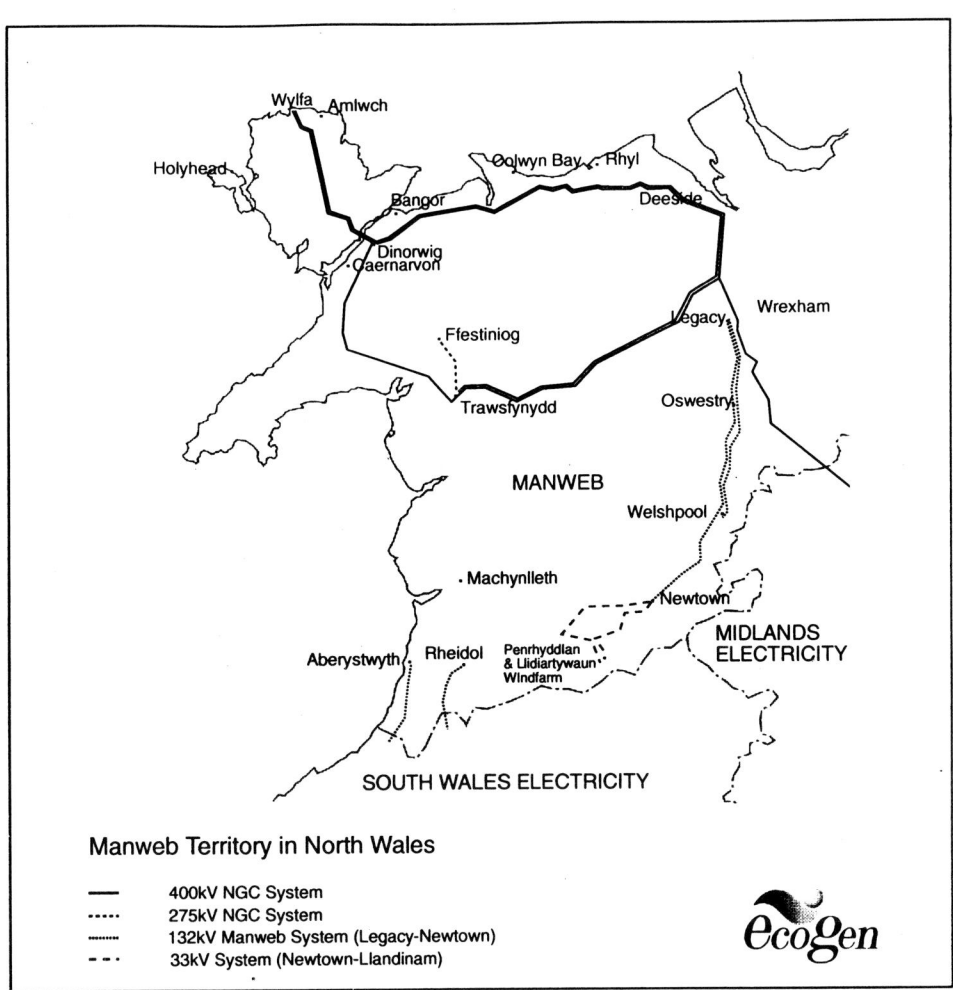

Fig 2 Location of the windfarm and connecting grid

Manweb Territory in North Wales

——	400kV NGC System
·····	275kV NGC System
▪▪▪▪	132kV Manweb System (Legacy-Newtown)
– ▪ –	33kV System (Newtown-Llandinam)

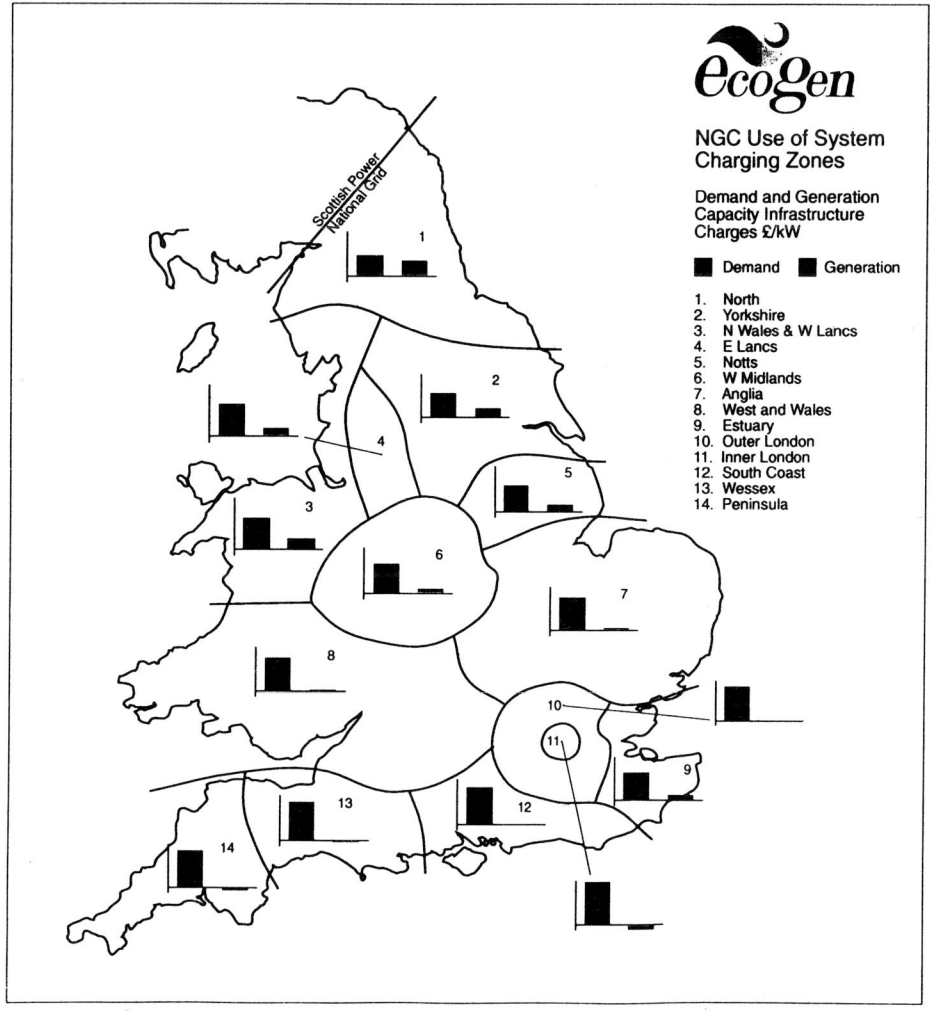

Fig 3 National Grid Company use of system charges scales and zones

NGC Use of System Charging Zones

Demand and Generation Capacity Infrastructure Charges £/kW

▪ Demand ▪ Generation

1. North
2. Yorkshire
3. N Wales & W Lancs
4. E Lancs
5. Notts
6. W Midlands
7. Anglia
8. West and Wales
9. Estuary
10. Outer London
11. Inner London
12. South Coast
13. Wessex
14. Peninsula

Interaction of Delabole windfarm and South Western Electricity's distribution system

J. M. HAYNES
South Western Electricity Board, UK
M. J. BIRKS
Energy Technology Support Unit, UK

SYNOPSIS: This paper describes some aspects of the work being carried out by SouthWestern Electricity to monitor the electricity distribution system associated with Delabole windfarm Cornwall.

INTRODUCTION

South Western Electricity are carrying out electrical monitoring studies of Delabole wind farm in north Cornwall. This paper discusses some of the early results and conclusions, when the work is completed guidelines will be suggested for the connection of future windfarms to 11kV systems. The work is being sponsored by the Department of Trade and Industry through the Energy Technology Support Unit.

PROJECT OBJECTIVES

The work has the following objectives:

- To monitor the electrical output from the windfarm.

- To demonstrate, using measured data, that the windfarms in the UK will not cause any noticeable effects on the electricity distribution system.

- To demonstrate that disturbances on the distribution system will not cause any noticeable effects on the windfarm.

- To produce guidelines for the efficient design of future windfarm sites.

DESCRIPTION OF THE DELABOLE ELECTRICAL SYSTEM AND ASSOCIATED NETWORK

The windfarm consists of ten asynchronous 400kW Vestas Windane 34 machines having a total maximum output of 4MW. On starting, each generator is temporarily connected to the electrical network by means of 'soft start' thyristor equipment limiting the output current from each generator to its full output value. Figure 1 shows a schematic of the distribution system associated with the windfarm.

The output voltage from each wind turbine generator of 690V is stepped up locally to 11kV by means of dedicated 500kVA transformers interconnected by underground 11kV cable to Delabole Primary 33/11kV Substation. Fed by two 33/11kV, 15MVA transformer oil-circuit breakers (OCB's), the 8 panel 11kV switchboard also feeds Stannon Clayworks, Delabole Quarry and 3,250 customers. Maximum demand is 11.2MVA (0230 hours winter night peak).

The 33kV supply is via a wood pole overhead line ring circuit from St Tudy 132/33kV Substation (Bulk Supply Point). System fault levels are 195MVA (33kV) and 90MVA (11kV), under normal 33kV network feeding arrangements.

Electrical system disturbances were monitored at the windfarm intake substation using the following two instruments: 1) Dranetz Power Quality Analyser; capable of monitoring and recording events triggered by impulses or excursions on the voltage and current wave forms. 2) A Flickermeter; designed to accurately model the response of the eye and brain to voltage fluctuations based on the flicker produced by a 60 watt tungsten lamp. Both instruments are capable of recording voltage harmonics. Results from the disturbances recorded and specific tests carried out are discussed in more detail later in this paper.

WINDFARM OUTPUT COMPARED TO LOCAL 11KV DEMAND

The installation of high accuracy import and export power measurement instrumentation in combination with SWEB's Data Acquisition and Telecontrol system (SWEDAT) allowed day to day charts of MWh supplied both by the Delabole wind farm and by SWEB's 33/11kV transformers to be compiled.

On three occasions in March and April 1992 the windfarm output reached a maximum on the same days that the substation demand also peaked. It is suggested that the strong winds providing the high windfarm outputs also caused increased power usage by local households in the area. The majority of housing in the area is pre-1939 and it is well known that cold, strong winds cause the 'chill factor' to increase, encouraging electricity users to switch on extra heating.

Also during March 1992 the windfarm generated over one third of the total units provided to the Delabole network. This was a particularly windy period where the windfarm output exceeded 40 MWhs on 15 days of the month. During some spring/summer days the windfarm has supplied the complete local 11kV load and occasionally back fed the 33kV network.

COMPARISON OF MEASURED POWER OUTPUT WITH THEORETICAL VALUES

This work did not set out to monitor the performance of the windfarm, however comparisons have been made between measured total power output from the windfarm and the theoretical curve. By plotting half hour averages of total windfarm output against the wind speed at hub height for the months of Oct./Nov. 1992, reasonable correlation has been confirmed. Any deviation from the theoretical performance curve can be explained by timescale of the plotted data, wind gusting over the windfarm site and reduced high wind speed data.

WINDFARM OUTPUT CURRENT FLUCTUATIONS

From the time that the Windfarm began operating in December 1991 large and rapid fluctuations have been observed in it's generated current. Fluctuations as rapid as 102 per minute with a swing of 30A (at the full output of 200A) have been noted on numerous occasions.

Two possible causes of the above phenomenon were explored.

1. Effect of the tower on each of the three blades in turn as they rotate at 35 rpm.

2. The possibility that the fluctuations originate on the SWEB distribution network and are amplified by the windfarm eg. from a quarry and clayworks.

However, when two other windfarms with identical wind turbines began to operate in Cornwall approximately one year after Delabole windfarm, very similar current fluctuations and swings were recorded.

It is concluded that these fluctuations are caused by the effect of the windturbine tower on the 3-bladed machines. At the standard rotational speed of 35rpm, one blade passes the tower at a rate of 105 per minute. These continuous fluctuations in current have only had a minimal effect on the voltage, with no apparent ill-effect to consumers supplied from the 11kV busbars.

WIND TURBINE START-UP TESTS AT REDUCED FAULT LEVELS

The fault level at the Delabole Primary Substation 11kV switchboard, under normal 33kV network feeding arrangements, is 90MVA. By carrying out a sequence of switching on the 33kV and 11kV networks, the fault level at the 11kV switchboard can be progressively reduced. On Thursday 27 August 1992, a series of tests was arranged during which the fault level was lowered in 8 steps from 90MVA to 14MVA.

During each of the tests the wind turbines were stopped and then released to generate at intervals of 10 seconds.

The effect on the substation voltage and current of the wind turbines starting up was monitored using a Dranetz Disturbance Analyser. In addition, harmonic distortion was recorded by the analyser.

As the fault level was reduced, the effect of starting the windturbines became more and more pronounced. By Tests 7 and 8 (16 and 14 MVA), the recorded current and voltage became very interesting. In both cases, as the wind turbines start to generate the output current rises from zero to a maximum in a series of well defined steps. As several of the wind turbines cut in, a dip in the current occurs momentarily. This is due to those particular turbines drawing power from the mains network immediately before reaching the speed required to generate.

The voltage recording for Test 7 showed a dip of 1.87% as the first windturbine started. The voltage dip recorded became progressively less as more turbines started up.

The voltage recording for Test 8 (at 14 MVA the lowest fault-level achieved), showed a dip of 2.6%. In this case it was when the second windturbine started.

EFFECT OF ONE WINDTURBINE STARTING (AT 14MVA)

Test 8 took place between the times of 15.12 hours and 15.14 hours. At 15.12.16, both the current waveform and the corresponding voltage waveform were recorded on the Dranetz instrument.

The current waveform shows the first wind turbine starting and causing high frequency oscillations to the current wave over half a cycle (10 milliseconds). The HF oscillation has a frequency of 533 Hz.
In the case of the voltage waveform, a small distortion occurred on the waveform on the negative half-cycle, but only persisted for 5 milliseconds.

These are typical of waveform distortions which have been recorded for windturbine starts on numerous other occasions during the course of the Delabole project. The fault level has no effect on these waveform distortions.

FLICKERMETER PRINTOUT FOR START-UP TESTS

The effect of the network switching and the wind turbine 'start-up' tests on the amount of flicker introduced into the system, was analysed using the Digital Flickermeter.

Ten minute severity values (PST) were recorded between 11.29 hours and 16.09 hours on the day of the tests. The PC printout was as Figure 2 which shows a peak value of 0.69 and three lesser peaks.

The numbers shown in brackets on the graph refer to the corresponding short-circuit level in MVA.

These values indicate that although increased flicker resulted from the wind turbine start-ups, the highest values of PST do not necessarily correlate with the lowest fault levels.

As a comparison, a previous Flickermeter recording made over a 6 hour period on 28th May 1992 reached a peak PST value of 0.29, but with the majority of values being below 0.1.

Notes
a) 1.0 per unit represents marginal flicker visibility to 50% of observers.
b) PST is based on 10 minutes of time series data output.

VOLTAGE HARMONICS RECORDED AT DELABOLE SUBSTATION

In an attempt to quantify the effect of the Windfarm on voltage harmonics, two complete 24 hour periods were analysed.
1. With the windfarm producing a very low output. The day selected was 24th June 1992 when the maximum output was only

730kW at 15.30 hours and the remainder of the 24 hour period was very low.

2. With the windfarm producing approximately 3.8 MW for the whole of the 24 hour period. The day chosen was 2nd December 1992 which was also the day of the reduced fault-level tests.

The barcharts produced for these two contrasting days are shown in Figure 4. The two barcharts are similar, showing a reduced level of harmonics between midnight and noon, which then increase to peak at approximately 2000 hours (GMT), falling away again towards midnight.

As only the 3rd and 5th harmonic were of significantly high values, these two have been plotted, with all other harmonic numbers being excluded. It can be seen that the 5th Harmonic was predominant. This harmonic number is associated with capacitance in the electrical network. The windfarm has ten capacitors each of 180 kVAr and 54 kVAr of 11 kV cable capacitance linking the windturbines to the primary substation. It is also known that the 5th harmonic is TV linked, which accounts for the increase in the afternoon and evening from the 'background level'.

By inspection of the barcharts it appears there is a constant background of voltage harmonics present for the whole 24 period, of which the 5th harmonic is by far the highest. This 5th harmonic value, excluding the effect of TV evening peak, is consistent at a level of approximately 1.3% regardless of the windfarm output.

CHANGES IN WINDFARM POWER OUTPUT DUE TO RAPID VARIATIONS IN WINDSPEED

During the early hours of 23rd March 1993 the weather at Delabole consisted of squalls and heavy rain showers. The large variation in windspeed over a relatively short time period caused a very erratic power output from the windfarm. Over a 27 minute period the windfarm output varied from 50A (1MW) at

0139 hours to over 200A (4 MW). The output then fell to 140A (2.7 MW) at 0210 hours, ie 4 minutes later.

Initial concern that with very strong wind gusts all ten windturbines would generate instantaneously to produce step increases in current and voltage, has not materialised in the field. Even under severe gusting conditions, the windfarm output has been found to ramp up over a period of 2 or 3 minutes.

VOLTAGE-DIP DUE TO WINDFARM START-UPS

From the reduced fault-level tests described earlier in this paper, it was seen that as the system short-circuit level was reduced, the voltage dip caused by each windturbine start-up increased. In order to investigate this further, follow-up tests were carried out during November 1992 and April 1993. From all the test results the percentage voltage-dip, measured at the time of the first windturbine start-up, was plotted against the corresponding fault-level. Figure 3 clearly shows the effect of reducing the fault-level. The Engineering Recommendation G59 limit of 1% voltage dip is exceeded when the fault-level falls below 40 MVA.

Engineering Recommendation P28 describes a method, using the short-circuit level, of calculating the voltage dip resulting from load changes. Assuming the change at Delabole is 400 KVA per windturbine start, the method was used to calculate and plot points on the same axes as the test results. The two plots show close agreement. It should be noted that the graph relates to Vestas 400 kW, three-bladed, soft-start machines and that other makes and design of windturbine will cause a different electrical disturbance on start-up.

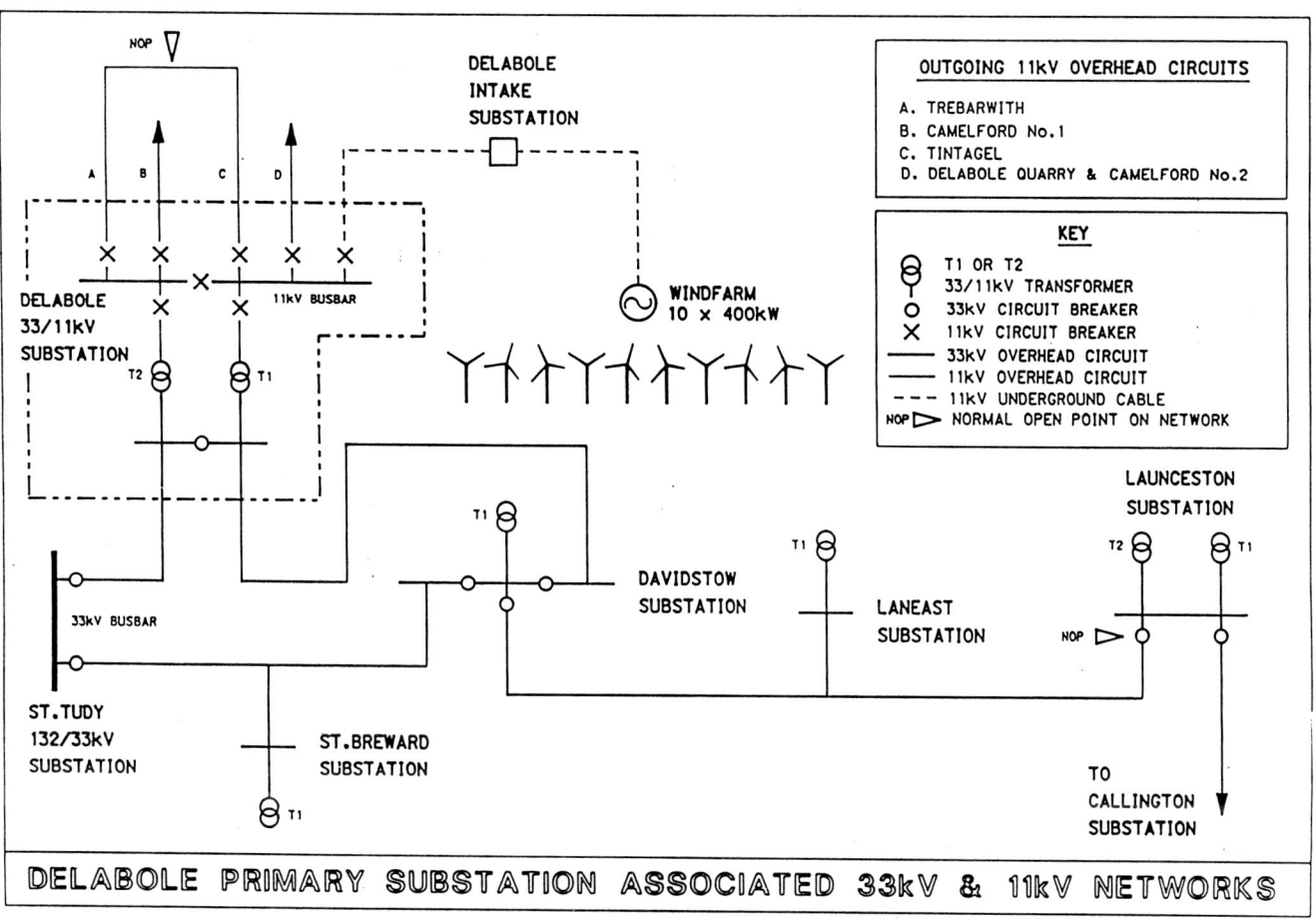

FIG 1

59

Delabole Flicker Severity – 27 Aug 1992

Start–Up Tests (Various Short Circuit Levels)

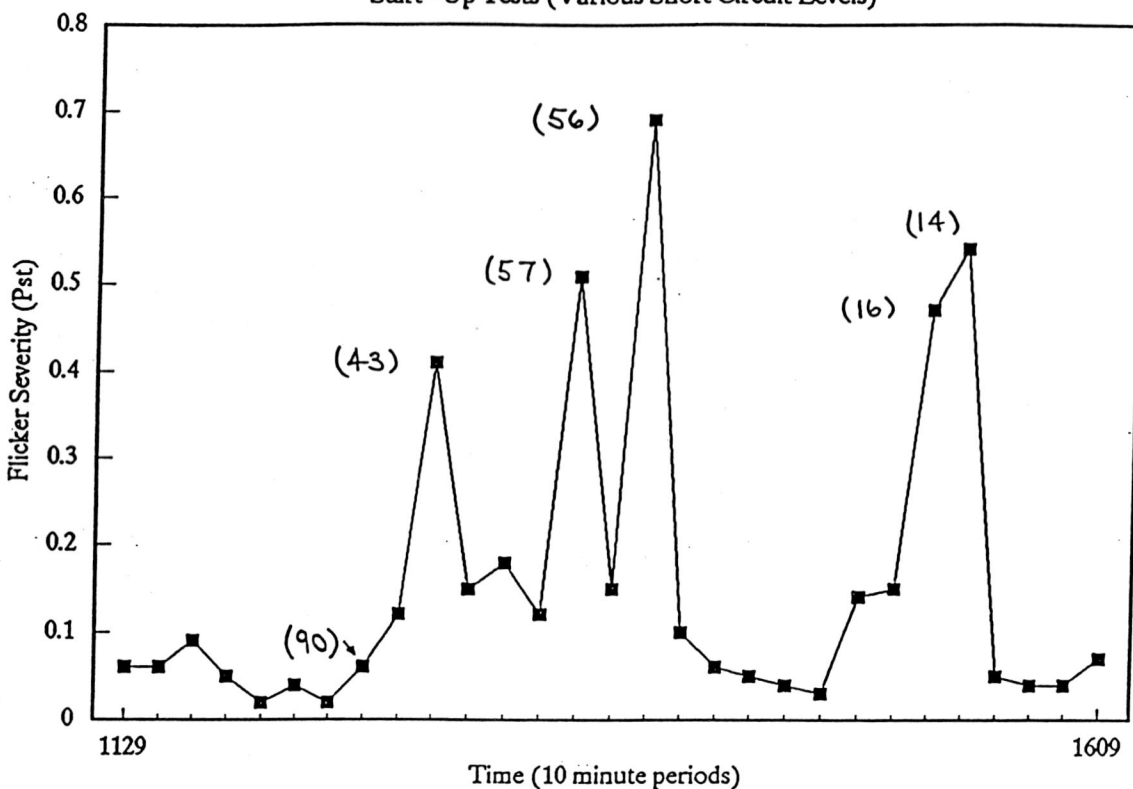

FIG 2

Delabole Windfarm

Disturbance v Fault Level

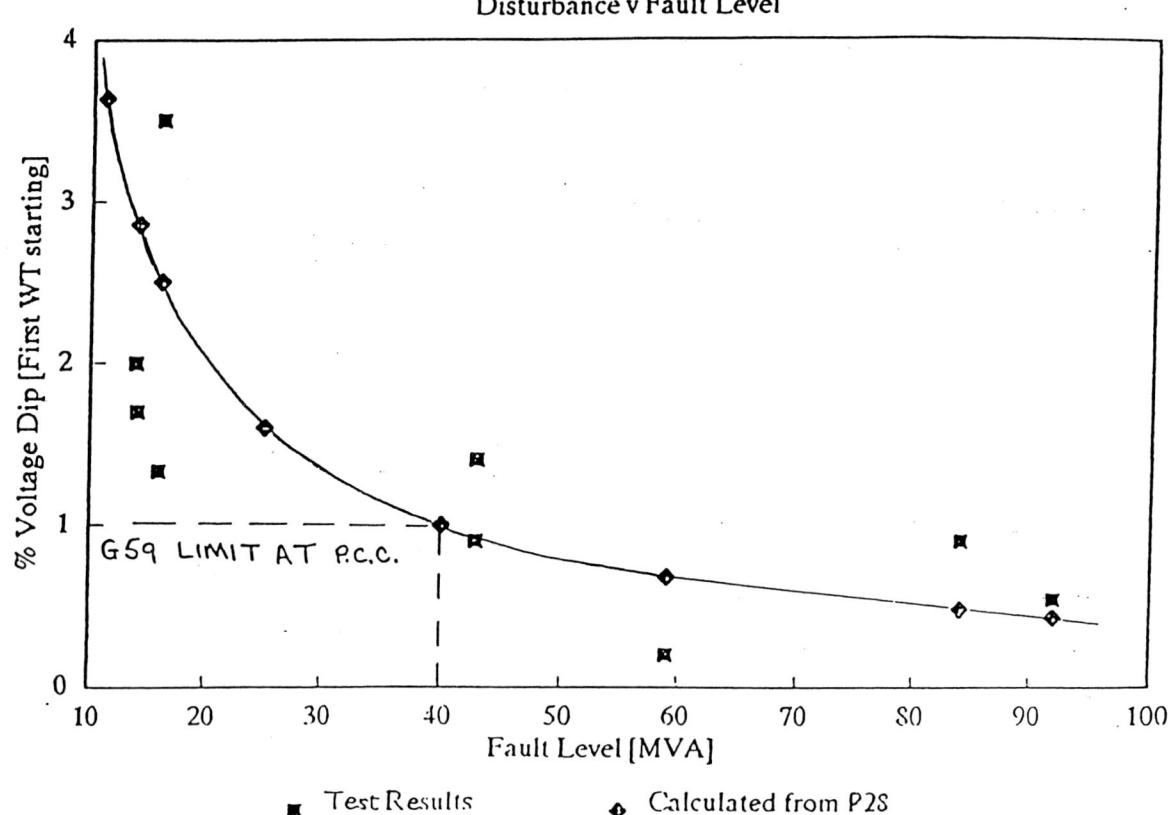

FIG 3

Delabole Windfarm

Background Voltage Harmonics 24th June 92

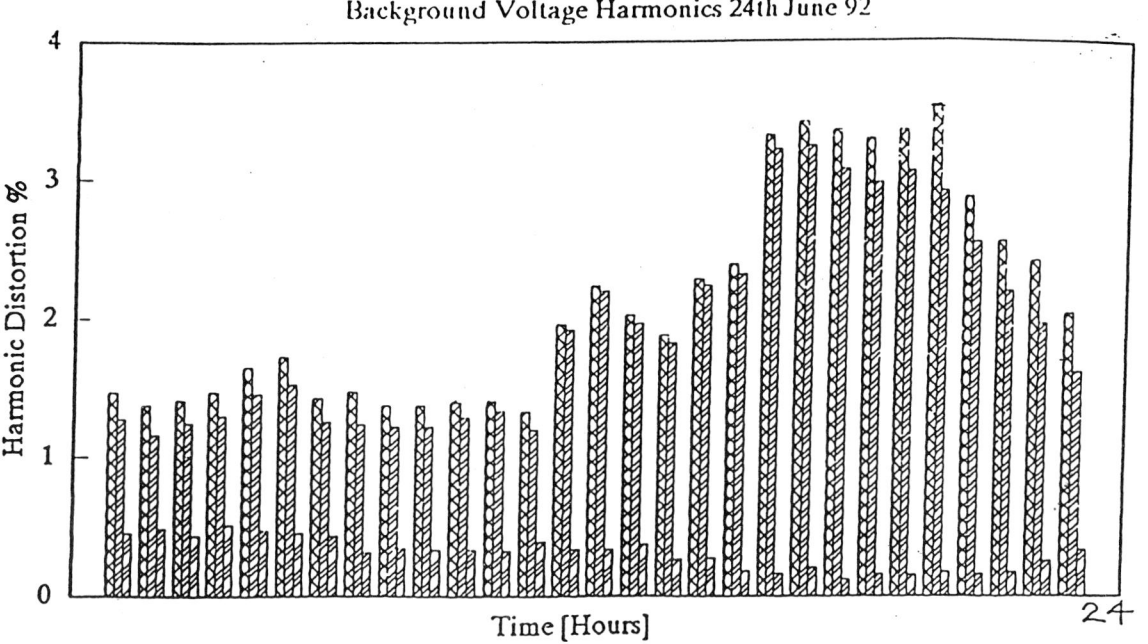

Background Voltage Harmonics 2nd December 92

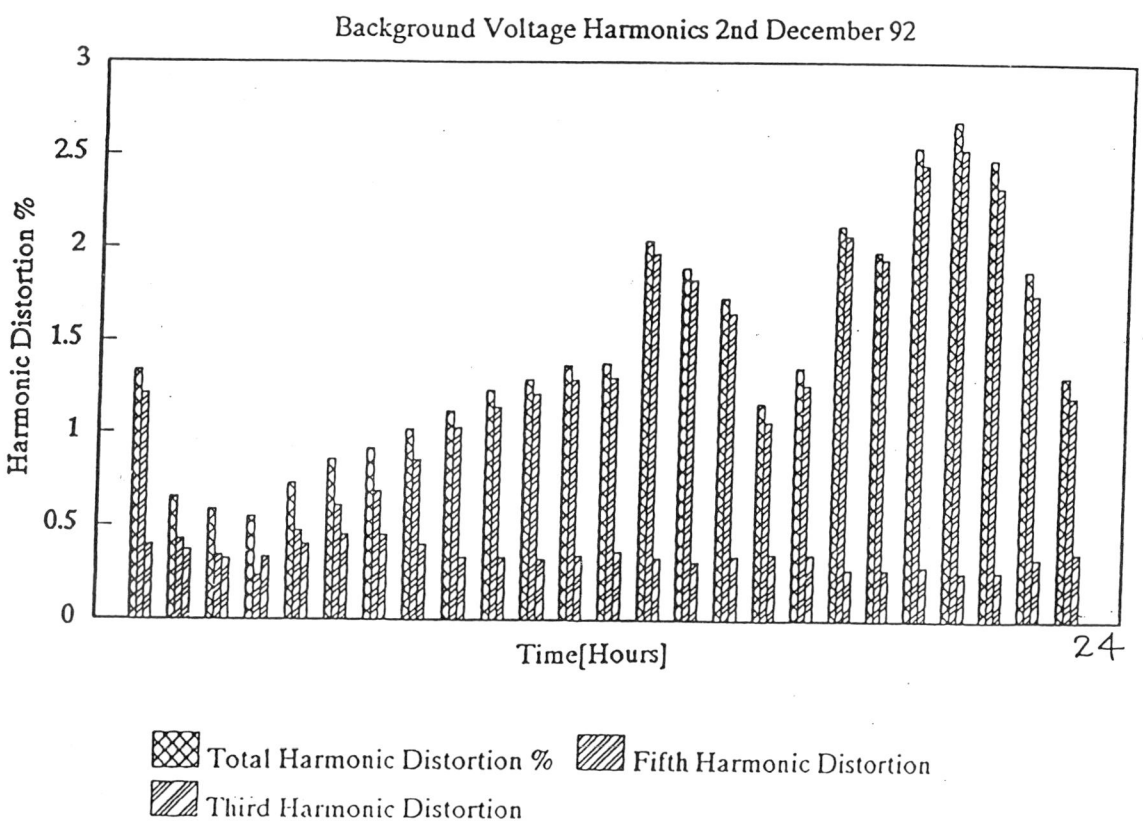

Total Harmonic Distortion %   Fifth Harmonic Distortion

Third Harmonic Distortion

FIG 4

Wind speed forecasting and its application to wind power integration

S. J. WATSON BSc, ARCS, PhD, **L. LANDBERG** BSc, MSc and
J. A. HALLIDAY BSc, PhD, CEng, MBCS, FRMetS, MIEnvSc
Rutherford Appleton Laboratory, UK and Risø National Laboratory, Denmark

Synopsis: An assessment is made of several different methods to forecast wind speed a number of hours ahead. These methods are then quantified in terms of fossil fuel savings to the England and Wales electricity grid if a large amount of wind power plant were to be installed into this grid. The results suggest that fossil fuel savings significantly greater than using persistence could be achieved by using 'intelligent' forecasts of wind speed but that knowing the accuracy of each forecast is at least as important as improving this accuracy.

1 Introduction

With ever increasing numbers of wind turbines becoming grid connected throughout Europe, it is becoming increasingly important to look at the integration aspects of wind energy both from a practical and an economic point of view. This paper will summarise the use and accuracy of various wind speed forecasting methods as a tool to improve the scheduling of conventional power plant to accommodate a large amount of wind power connected to a large scale electricity grid. The economic benefits will be defined in terms of fossil fuel savings to the England and Wales grid system. The paper will conclude by examining the problems caused by an uncharacteristically poor wind power forecast for one hour of a year and how the way in which one deals with this can affect fossil fuel savings from wind power.

2 Forecasting Methods

This paper uses basic methods to forecast the wind speed a number of hours ahead:

1. Persistence

2. The combination of an NWP (Numerical Weather Prediction) model and a physical flow model (HIRLAM/WASP).

3. The combination of an NWP and a statistical method (MESO/MOS).

These three methods will be described briefly here.

2.1 Persistence

This simplest form of forecasting uses the assumption that a parameter at time $t + x$ is the same as it was at time t, where x is a period of time into the future. This method is useful for small values of x (say three hours in the case of wind speed) but rapidly becomes inaccurate for a long period of time ahead.

2.2 HIRLAM/WASP

The second forecasting method to predict wind speed uses the Danish Meteorological Institute's HIRLAM (HIgh Resolution Limited Area Model) NWP model [1]. This model gives output of several meteorological parameters including wind speed and direction at several vertical levels with a horizontal grid point separation of 55 km covering a major part of the North-Atlantic Ocean, Greenland and Europe. Forecasts are made twice a day at 00Z and 12Z (Z=GMT) up to 36 hours ahead in increments of three hours. HIRLAM forecasts of the high level geostrophic wind ($\sim$700 m above the site of interest) which is esssentially undisturbed by the terrain below, are then transformed to 10 m above ground level (agl) by the use of the wind flow model WASP [2, 3] which 'folds-in' the physical effect of the topography, surface roughness and surrounding buildings close to the site. Terrain, roughness and building data are obtained from digitised and printed maps of the area surrounding the site.

2.3 MESO/MOS

The third forecasting method makes use of the U.K. Meteorological Office mesoscale model (MESO) [4].

This model produces output at several vertical levels and has a horizontal grid point spacing of 15 km and covers the U.K. plus some of mainland Europe. Forecasts are made twice a day at 00Z and 12Z up to 18 hours ahead in increments of one hour. Forecasts of the wind speed at 10 m agl cannot take account of terrain features on a scale of less than ~30 km so in this case a weighting is taken of the three nearest grid points to the site at the 10 m height and this is compared with the concurrent observed wind speed at the site. Model Output Statistics (MOS) [5, 6, 7] are then use to assess the correlation between a year of hourly forecast data and a year of concurrent hourly observed data. A mean regression of the MESO forecast wind onto the observed wind is then formulated where the constant of regression is dependent on the level of a class parameter (or combination of class parameters). The class parameters deduced as factors which significantly influenced the ratio of the MESO predicted to observed wind speed were the time of day, forecast wind speed direction, season and whether precipitation was forecast by the MESO model. Subsequent MESO model forecasts are then rescaled using this class parameterised mean regression to produce what will be referred to as MESO/MOS forecasts.

3 Relative Accuracy of Forecasting methods

Fig. 1 [8] shows a comparison of the mean (first group of three bars per station) and root mean squared (second group of three bars per station) residual error for a selection of six European met. stations for HIRLAM/WASP forecasts, HIRLAM forecasts without WASP corrections and persistence at 18 hours ahead $(t + 18)$ averaged over the period June 1991 to November 1991. The stations in this figure are coded by an arbitrary station number. The first three stations, (19) Abbeville, (56) Manchester and (57) Birmingham, are considered 'good' stations i.e. WASP appears to perform well at these stations; the latter three, (29) Salamanca, (35) Braganą and (41) Munich are classified as 'bad' stations and WASP appears to perform badly at these stations due to large directional errors in HIRLAM forecasts. In the case of the 'good' stations, WASP mean errors are significantly less than the 'raw' HIRLAM forecasts. You would expect, on average, the mean persistence error to be zero. The WASP rms error is slightly less in two of the three

Table 1: Mean and standard deviation of observed minus forecast wind speed residuals in m/s for the period of data April 1989 to March 1990 comparing MESO and MESO/MOS forecasts.

| Site | Forecast Method | | | |
| | MESO | | MESO/MOS | |
	Mean	*S.D.*	*Mean*	*S.D.*
Aberporth	1.78	2.64	-0.02	2.56
Belfast	-1.37	1.99	-1.54	2.13
Benbecula	-1.07	2.33	-0.28	2.22
Birmingham	-0.89	1.79	-0.01	1.59
Blackpool	-0.76	2.22	0.27	1.99
Cairngorm	8.80	6.22	0.47	7.20
Coltishall	-1.50	2.02	-0.26	1.87
Ronaldsway	-1.33	2.51	0.13	2.25
St. Mawgan	0.13	2.24	0.15	2.21
Wick	-1.52	2.34	0.14	2.13

'good' cases. Both HIRLAM and HIRLAM/WASP forecasts perform significantly better than persistence at $(t + 18)$.

Table 1 shows a comparison of mean and standard deviation wind speed residual errors for 'raw' MESO forecasts, MESO/MOS forecasts and persistence at ten U.K. met. stations. These statistics are calculated over all forecast lead times up to $t + 18$ for the period of data April 1989 to March 1990, MOS regressions having been obtained from April 1988 to March 1989 data. In eight out of ten cases the absolute value of the mean residual error is reduced by using the MOS regressions. Similarly, in eight out of ten cases the standard deviation of the residual error is reduced by the application of MOS. The standard deviation of residual errors from persistence are, in all but one case, significantly larger than either MESO or MESO/MOS forecasts. The one exception is Cairngorm which is very high mean wind speed site on the side of a mountain and the wind speed is poorly predicted by the MESO model.

Fig. 2 shows a comparison of rms residual wind speed errors over the nine U.K. sites: Benbecula, Birmingham, Blackpool, Bournemouth, Eskdalemuir, London, Manchester, Valley and Wick. Statistics are calculated over a year of data (December 1990 to November 1991) are plotted as a function of forecast lead-time for MESO/MOS, HIRLAM/WASP and persistence forecasts. The first thing to note is that persistence performs better than both MESO/MOS and HIRLAM/WASP for lead-times of two hours or less. Up until five hours ahead, persistence performs better than

HIRLAM/WASP though worse than MESO/MOS. Above five hours ahead, both 'enhanced' NWP forecasting methods perform significantly better than persistence, both with a fairly constant rms residual error over all lead-times up to $t +$ 18. It is clear that MESO/MOS out-performs HIRLAM/WASP at all lead-times up to $t+18$ which is presumably due to the higher spatial resolution of the MESO NWP model.

4   The National Grid Model

4.1   Description

To assess the usefulness of these forecasts in terms of wind power integration, use was made of a time domain simulation model, the Reading University/Rutherford Appleton Laboratory National Grid Model (NGM) [9, 10, 11, 6, 7]. This model simulates, on an hourly basis, the scheduling of conventional power plant, (in the case of steam plant up to eight hours ahead) to meet the predicted load demand on the England and Wales electricity grid. The model is capable of including a hypothetical installed capacity of wind power. The scheduling includes an amount of on-line 'spinning reserve' consisting of part-loaded steam turbine plant which can be used to meet any unforeseen short-falls in wind generated power or surges in load demand which could not otherwise be met by fast response plant such as open cycle gas turbines and water pump storage, either because of insufficient installed capacity or high running cost. This spinning reserve is calculated as the sum of a fixed fraction (SR1) of the predicted load and a fixed fraction (SR2) of the predicted wind power. The model optimises the values of SR1 and SR2 to give the minimum fossil fuel costs for conventional power plant subject to the constraint that there are *no loss-of-load events* during the year long simulation period. Load demand data is taken from actual system data. These data form the *predicted* load data used by the model. *Actual* load data used by the model is assumed to be actual system data multiplied by a random number with mean 1 and standard deviation 0.015 to reflect the uncertainty on the load prediction. Wind power is calculated from observed wind speed at a number of met. station sites (ten sites were used in this case) spread around the U.K. to minimise average wind speed variability [11] , assuming a wind turbine power curve similar to a Vestas V27 [12]. Wind power predictions are calculated from either persistence,

MESO, MESO/MOS, HIRLAM/WASP or perfect forecasts. Similarly, these forecasts are turned into power forecasts using the V27 power curve. In addition, an extension was made to the MESO/MOS forecasts to include assigned forecast errors. These errors were calculated using the class parameters as described in § 2.3. The standard deviation residual error for each combination of class levels (*e.g.* day-time, for a westerly predicted wind, during the summer when no rain is forecast) was calculated for the year of data used to formulate the MOS classified regression. This was then used as the assigned predicted error for the subsequent MESO forecasts to which this MOS regression was applied. This was then used to plan the wind power uncertainty part of the spinning reserve instead of a fixed fraction of the predicted wind power. This set of forecasts will be referred to as MESO/MOS(E).

4.2   Results

Firstly, a comparison was made between persistence, MESO/MOS, MESO/MOS(E) and perfect forecasting for the financial year 1989/90 (FY 89/90). MOS regressions were calculated using the previous year's data. The NGM was used to assess the fossil fuel costs for a range of *hypothetical* installed nameplate capacity of wind power in the U.K. (all of which is assumed to feed into the England and Wales National Grid) between 0 GW and 40 GW. Table 2 shows a summary of the results. Note that the parameter SR2a is is used instead of SR2 for MESO/MOS(E) forecasts because of the different way spinning reserve is planned in this case. Fig. 3 shows the fossil fuel *savings* calculated by subtracting the fossil fuel cost at each penetration of wind power from the fossil fuel cost when there is no wind power installed in the grid. In addition, the 'ideal' fossil fuel savings were calculated by multiplying the unit cost for power generated by conventional power sources when there is no wind power installed in the grid by the number of units of wind power that could have been generated for each penetration were conventional power displaced. The 'ideal' savings are those which would result if wind power behaved as a conventional power source, *i.e.* it could be switched on and off at will (no 'operating' penalty) and all available wind power could be absorbed by the grid (no 'discarding' penalty).

The main points to note about Table 2 and Fig. 3 are:

- Fossil fuel costs decrease (savings increase) for

Table 2: SR1, SR2, SR2a, average spinning reserve and fossil fuels costs for different forecasting methods for NGM optimised for FY 89/90.

Wind Power (GW)	Forecast Method	SR1	SR2	SR2a	Average Spinning Reserve (MW)	Fossil Fuel Costs (£M)
0	-	-0.013	-	-	315	4434
5	Persistence	0.000	0.15	-	1042	3939
5	MESO/MOS	0.001	0.01	-	646	3932
5	MESO/MOS(E)	-0.016	-	0.92	659	3930
5	Perfect	-0.019	0.04	-	270	3919
10	Persistence	0.000	0.45	-	3185	3491
10	MESO/MOS	-0.022	0.35	-	1957	3457
10	MESO/MOS(E)	-0.046	-	2.21	1730	3454
10	Perfect	-0.019	0.02	-	316	3416
20	Persistence	-0.004	0.67	-	7527	2783
20	MESO/MOS	-0.008	0.45	-	5211	2645
20	MESO/MOS(E)	-0.010	-	2.30	5205	2636
20	Perfect	-0.007	-0.03	-	410	2447
30	Persistence	-0.012	0.78	-	9951	2539
30	MESO/MOS	-0.004	0.56	-	7934	2292
30	MESO/MOS(E)	-0.001	-	2.40	7192	2173
30	Perfect	-0.007	-0.04	-	365	1604
40	Persistence	-0.015	0.80	-	10689	2442
40	MESO/MOS	-0.004	0.64	-	9442	2240
40	MESO/MOS(E)	-0.001	-	2.30	7712	1944
40	Perfect	-0.001	-0.06	-	313	1017

increasing penetrations of wind power.

- Fossil fuel costs are reduced (savings increased) most of all for perfect wind speed forecasting and then, in order, MESO/MOS(E), MESO and persistence. The average spinning reserve requirement is similarly reduced.

- The importance of 'good' forecasts of wind power increases as the penetration of wind power increases.

- Perfect forecasting cannot achieve ideal savings at high penetration due to the effect of the operating and discarding penalties.

- Using MESO/MOS(E) forecasts instead of persistence can increase fossil fuel savings by up to 25%.

Fig. 4 shows fossil fuel savings for the year from December 1990 to November 1991, this time also including HIRLAM/WA^{S}P forecasts. It is clear from this graph that the enhanced NWP forecasts perform only slightly better than persistence and do not show the large increase in savings seen in Fig. 3. To investigate this further, the constraint which ensured that there were no losses of load in a given period was relaxed to allow up to three loss-of-load events. Fig. 5 shows the fossil fuel savings for FY 89/90 at 40 GW penetration comparing persistence, MESO/MOS and MES/MOS(E) forecasts. It can be seen that the fossil fuel savings are a fairly flat function of loss-of-load events. Fig. 6 shows a similar plot for the period December 1990 to November 1991 comparing persistence, HIRLAM/WA^{S}P, MESO/MOS and MESO/MOS(E) forecasts. This time there is a sharp rise in fossil fuel savings between zero and two losses of load, particularly for the MES/MOS and MESO/MOS(E) forecasts. Closer inspection of the raw wind speed data showed that there was a large increase in wind speed for an hour in the year resulting in the furling of a large number of turbines thus dramatically reducing the available power. This was poorly forecast particularly by the MESO NWP model. This resulted in a disproportionately large amount of spinning reserve being carried throughout the year to compensate for this one large shortfall. Such spinning reserve is required due to insufficient installed capacity of fast response plant. By relaxing the zero loss-of-load requirement this spinning reserve can be substantially reduced. This illustrates the significance of one 'poor' forecast.

5 Conclusions

The use of site-enhancement of NWP forecasts has been shown to reduce forecasting errors both in the case of the physical model (WA^{S}P) and more clearly in the case of the statistical model (MOS). The advantage of the WA^{S}P model is that it does not require a large amount of observed and forecast wind statistics. The advantage of the MOS model is that it does not require explicit knowledge of the terrain surrounding the site. MESO/MOS forecast errors have been shown to be less than HIRLAM/WA^{S}P forecast errors but this reflects the higher spatial resolution of the HIRLAM model. The HIRLAM model covers a much larger area than the MESO model and predictions can be made further ahead.

When used to predict wind power installed in the England and Wales National Grid, the MESO/MOS and MESO/MOS(E) forecasts have given fossil savings significantly greater than using persistence. Savings can be up to 25% greater than persistence. HIRLAM/WA^{S}P forecasts have given greater saving than persistence but to a lesser extent than the MESO/MOS and MESO/MOS(E) forecasts. Once again this reflects the greater resolution of the MESO model. The MESO model covers the U.K. with relatively high resolution and thus is the better choice in this case, bearing in mind that only predictions up to eight hours ahead are required.

The results have also indicated that one poor forecast of wind power can adversely reduce fossil fuel savings due to the increase in the required spinning reserve. The attractiveness of wind power as an economic means to reduce the burning of fossil fuel could be increased by:

- A better knowledge of when one forecast is likely to be poor.

- Improved NWP forecasting of wind power.

- A larger installed capacity of fast response plant to meet large shortfalls in wind power generation.

- By accepting the risk of one or more loss-of-load events.

The results in this paper have concentrated on the fossil fuel savings from wind power. Further work is required to determine the sensitivity to other factors such as capital costs and limits on power flows within the grid itself.

6 Acknowledgements

The authors would like to thank Mr. Gil Ross of the UK Met. Office for guidance in the use of Model Output Statistics to apply wind speed forecasts to the U.K. sites. The project was funded by the Commission of the European Communities under the DG XII (Directorate General for Science Research and Development) JOULE programme, contract number JOUR-0091-C(MB).

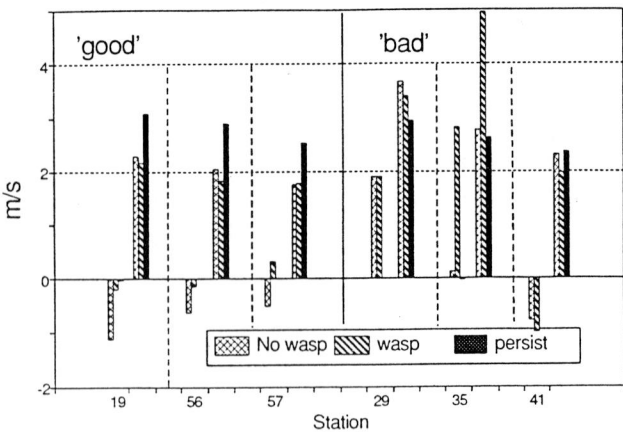

Figure 1: Comparison of mean and rms residual errors for HIRLAM, HIRLAM/WASP and persistence for six European sites.

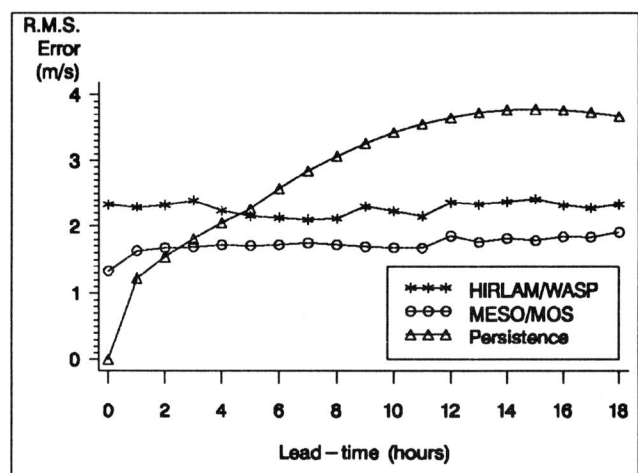

Figure 2: Comparison of rms residual errors for MESO/MOS, HIRLAM/WASP and persistence for nine U.K. sites.

References

[1] Machenauer, B. (editor), (1989) The HIRLAM Final Report. HIRLAM Technical Report No. 5, Danish Meteorological Institute, Denmark.

[2] Troen I. and E.L. Petersen, 1989 *The European Wind Atlas.*

[3] Mortensen, N.G., L. Landberg, I. Troen and E.L. Petersen, (1993) Wind Atlas Analysis and Application Program (WASP), User's Guide. Risø National Laboratory, Denmark.

[4] Golding B.W., 1990 *Met. Mag.* **119**, 81–96.

[5] Glahn H.R. and D.A. Lowry, 1972 *J. Appl. Met.* **11** 1203–1211

[6] Watson, S.J., J.A. Halliday and L. Landberg, (1992) Proc. 14th British Wind Energy Association Conference, Nottingham, pp 291–298. Mechanical Engineering Publications Ltd, London.

[7] Landberg, L., S.J. Watson and J.A. Halliday, (1993) Proc. European Community Wind Energy Conference, Lübeck-Travemünde, Germany, pp 677–680. H.S. Stephens and Associates, Bedford, UK,

[8] Landberg. L., S.J. Watson, J.A. Halliday, J.U. Jørgensen, A. Hilden (1993). *Short-term Prediction of Local Wind Conditions*. Report to the Commission of the European Communities on contract no. JOUR-0091-C(MB).

[9] Bossanyi, E.A., 1983 *Wind Engineering* **7**, 233–246.

[10] Bossanyi, E.A. and J.A. Halliday, 1983 Proc. 5th British Wind Energy Association Conference, Cambridge University Press, 62–74.

[11] Halliday, J.A., 1988 PhD. thesis, University of Strathclyde. Published as Rutherford Appleton Laboratory internal report RAL T 075.

[12] Risø Measurement Summary No. 6.1. 1989.

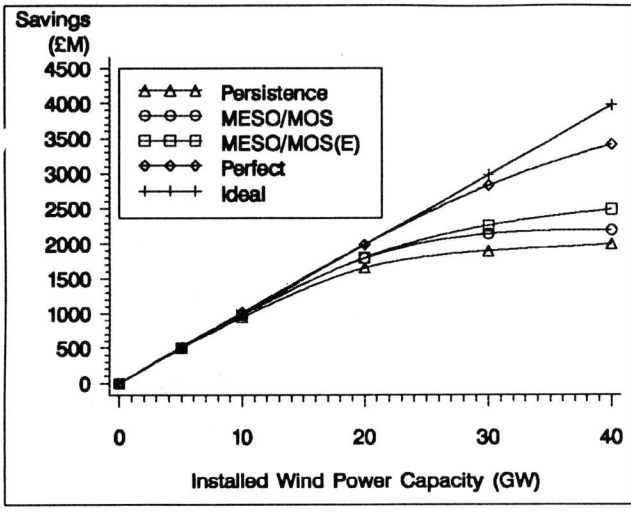

Figure 3: Fossil fuel savings for FY 89/90.

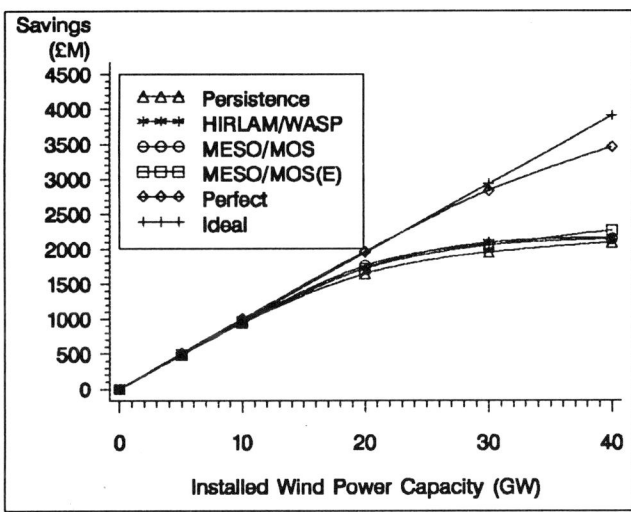

Figure 4: Fossil fuel savings for Dec 90 to Nov 91.

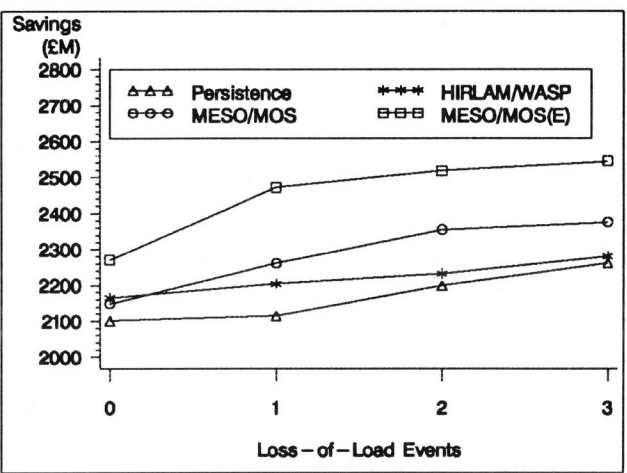

Figure 6: Fossil fuel savings vs. allowed loss-of-load events at 40 GW for Dec 90 to Nov 91.

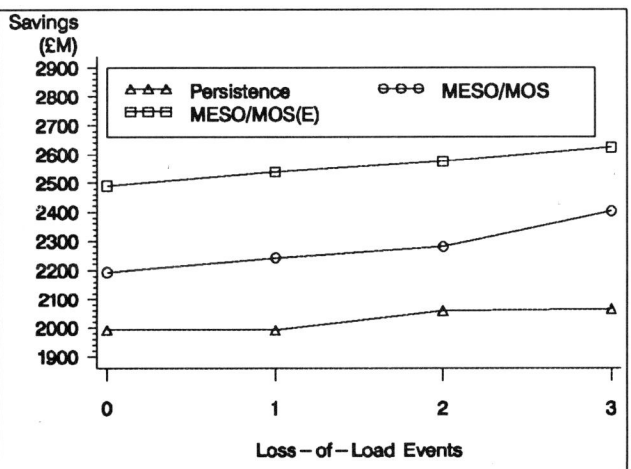

Figure 5: Fossil fuel savings vs. allowed loss-of-load events at 40 GW for FY 89/90.

On the effect of the utility grid characteristics on wind park power output fluctuations

A. I ESTANQUEIRO MSc, **R. F. AGUIAR, J. A. GIL SARAIVA** PhD,
Department de Energias Renovàveis, Portugal
RUI M. G. CASTRO MSc, **J. M. FERREIRA DE JESUS** PhD,
Instituto Superior Técnico, Portugal

SYNOPSIS In this paper a dynamic wind park model previously developed is used to simulate a two turbines wind park. The objective is to determine the influence of the utility grid characteristics in the performance of the park as well as on the power output and the on grid voltage fluctuations. The characteristics of the utility grid were modelled by the short circuit power values at the interconnection point to the park. These were taken as typical of a "weak" and a 'strong' grid. The results lead to the conclusion that this parameter does not seem to have a strong influence on the wind park power's output fluctuations, but on the other hand is determinant in what concerns on the fluctuations of the interconnection voltage busbar.

1 INTRODUCTION

Distributed System Generation (DSG) is one of the ways used to produce electric energy with a lower negative impact on the environment.

The potential investors in DSG are arriving to the conclusion that these power plants have to be of the order of the tens of megawatts so as to be economically interesting. So one may expect the rated power of these power plants to increase in the near future.

There is a lack of available 'tools' in what concerns the study of the DSG impact in the utility grid stability and performance.

The situation becomes more serious if a DSG component is a wind park or a wind turbine that delivers a continuously fluctuating power to the utility grid even in its steady state working mode.

The objective of the work that has been carried out is to develop a 'tool' that enables both the wind park investors and the utility grid technical staff to perform the necessary preliminary studies before connecting wind parks in windy but remote areas, which, most of the times, have a weak connection to the main power plants and are characterised by a low short circuit power value.

In a first step a dynamic model of a two wind turbine park was developed [1]. The model includes the utility grid near the wind power plant and enables to perform studies addressing the influence that wind power fluctuations have in the local consumers' voltage and frequency regulations.

Furthermore a Shinozuka method based wind model was developed and modified to enable the generation of cross-correlated wind synthetic series, so as to include the influence of the turbulence's smoothing effect in the power output of a wind park.

This paper concludes a preliminary study in order to assess the influence of the different parameters that affect the performance of a wind park in terms of the electrical output. This first step aims to achieve the parameters that may, or may not, be neglected in a future complete dynamic model of a wind park to be developed.

2 THE WIND PARK MODEL

The wind park time-dependent model used was described with detail in previous publications [1,2,3]. Nevertheless the main characteristics of the model are presented here. The model is based in a dynamic turbine wind series input and electric quantities' time series output model.

The evaluation of the behaviour of the wind turbine is accomplished through a characteristic equation of the wind rotor and through the differential equations that describe the generator and the grid.

The wind park model is an integrated model that includes the influence of the turbulence, the dynamics of the turbine, the electric interference between the generators, capacitors and transformers inside the park and the grid/consumers near the park. The integrated model includes sub-models that address the different phenomena:

i) a wind model with spatial and time correlation effects;

ii) a time-domain model of the WECS and electrical system;

iii) the characteristics of the utility grid in the inter-connection busbar to the wind park

The wind park model is able to simulate a wind park connected to the grid (with a topology as in Figure 1)

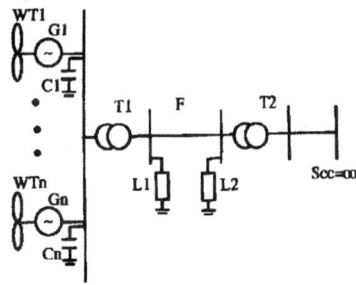

Fig. 1 - The wind park and grid's model scheme

A simple wind model to generate correlated wind time series was previously developed to assess the influence of the cross-correlation effects in the fluctuations of the power output [1,4].

The WECS model consists on the integration of a time domain wind turbine's model that includes the aeroelastic effects in the blades and accounts for the time variation of the torque and the shaft rotational speed [2,5] and an induction generator model that includes the saturation effects of the magnetic materials [6], thus addressing its non-linear working range.

These models were adapted to simulate the WECS typical behaviour with time-variable torque and slip and were first presented in [2,5]. They are now implemented in a software package named INDUSAT, which allows to perform several studies either in steady-state or transient conditions [3,7].

2.1 Wind Model

A dynamic model of a wind park must account for the effect of the different wind time series that affect the different turbines. Thus the knowledge of the simultaneous wind velocity time series at the various turbine locations is needed, including not just average wind velocities but also the superimposed turbulence. This is required by the low inertia of the electric system and by the existence of phenomena associated with turbulence effects on the turbine's power output, either in transient and in steady state.

A preliminary wind model was developed, based on the so called "Shinozuka method", i.e. the generation of time series from the amplitude and phase spectra. For the current case, the amplitude spectrum is obtained from a parametrised wind power's spectrum density (PSD). The phase spectrum is randomly generated for a (first) reference series; for the other

series, the phase values are dependent on the desired cross correlations among series [1,4].

The wind model is based on:
i) the Taylor hypothesis of "frozen" homogeneous turbulence;
ii) the assumption that each size class of eddies is represented in the power spectrum by the energy at the frequency f = L/v, where L is the eddy size (along wind) and v the mean wind speed; and
iii) the assumption that eddies with size less than the distance H between two certain sites will be too small to affect both of them at the same time - and therefore will bring a null average contribution to the cross-correlation.

Within these hypotheses, the Shinozuka method is used to generate cross-correlated wind speed series for any two places H apart taking the phase spectrum as equal for both places up to f'=H/v, and random thereon [4].

The well-known "Davenport spectrum" shape was used as the wind PSD for both places, with equal average speeds at 10 m high [1] and roughness coefficients of 0.008.

Samples from the model's stochastic wind series where taken as appropriate when yielding cross-correlations close to those estimated with exponential fits to field observations, with decay lengths of 200 m along-wind, and 50 m across-wind.

2.2 Wind turbine Model

To simulate the wind turbine performance the already mentioned time dependent model ,that uses as input the instantaneous wind's velocity, was used.

This model is based in the well know momentum "'strip theory". It takes the system as rigid and thus does not consider the aeroelastic effects in the blades

To describe the torque's characteristic of the turbine's rotor, a characteristic equation that accounts for the shaft's rotational speed and torque time-variation is used [2,5]

2.3 Induction generator and utility grid Model

The local loads were modelled as constant active and reactive powers for the computation of initial conditions and as a constant impedance for the wind park model. The transformers, the feeder and the a.c. system were modelled also as a constant impedance.

In order to obtain the initial conditions for the set of differential equations that accurately describe the system, a numerical model taking as input only the system parameters and the instantaneous wind speed was built.

This technique enables the computation- of the initial steady-state operating point through a modified power flow (to include the induction generator) which takes into account the electromechanical characteristics of the turbine/ generator groups to evaluate the rotor speed of each machine.

The interactions between the two turbine/generator groups inside the power plant and the conditions at the interconnection busbar were established through the short circuit power value in order to correctly model and determine the influence of this parameter in the electric power and voltage output fluctuations.

3 APPLICATION OF THE WIND PARK MODEL

The model described above enable us to assess the influence of parameters as the mean wind speed the turbulence and the grid characteristics in the electric output of the wind park in order to draw some conclusions on their effect on the stability of the park and also what happens in the local grid.

Most of the wind turbine/park models work in the frequency domain, but a time domain analysis is preferred by utility electric engineers: its results are quite important, when the study of electric performance through the electric output quantities, (voltage and currents) is regarded - even if only the steady state working mode is addressed. Another important advantage if one builds a complete differential equation model it is possible to perform studies of the transient working mode.

Time domain models put the problem of the representativity of just one simulation of wind time series with one corresponding input/electric power or voltage and current time series output. This was overcome by performing a large amount of simulations, thus regarding each simulation as a sample of a stochastic process.

The wind park model described above and illustrated in Fig. 1 was applied to the simplest configuration of both the wind park and the grid as shown in Fig. 2, where the local consumed power was set to zero, and only two turbines were considered. This was mainly due to the large amount of time needed to accomplish each simulation (1minute real time to each 1 sec simulation approximately).

The wind turbines are two small 12 m diameter prototypes with stall regulated rotor, described in previous publications [3,6]. The electric generator is an induction machine with 24 kVA rated power. Each wind turbine is locally compensated by a 4 kVAr capacitor bank.

The transformer T1 represents the step-up transformer in the wind park and is a 400V/30 KV, 100 kVA, 5% transformer. The 30 KV transmission line has 20 km of length and is represented by an impedance of 0.411+j0.275 Ω/km. T2 represents the substation transformer which steps-up the voltage from 30 KV to 60 KV being the rated power equal to 10 MVA and the short circuit voltage equal 8.35 %.

The objective of these studies is to determine, among the input parameters, which has the dominant effect in the wind park performance.

The effect that wind time series cross-correlation factors have in the wind park power's output was already addressed in a previous publication [1].

In this paper the model was used to assess the influence of the grid characteristics (via the short circuit power value in the interconnection busbar). Having in mind the necessity of representativity of the results, a large amount of simulations with a short circuit value characteristic of a strong and weak grid was performed.

The short circuit value typical of a weak grid was set to 5 times the rated power of the wind park (values lower than this were tested and the wind park has an unstable working mode and doesn't remain in parallel with the grid for medium to strong winds).

The short circuit value used to characterise a strong grid was taken as 20 times the rated power of the park. This value was chosen mainly because it is the lower limit stated in the Portuguese law, to enable the connection of a DSG to the utility grid. Above this value the grid is considered a 'strong' grid in Portuguese law terms.

The mean wind speed is, of course, determinant for the amount of energy produced but the importance of its influence on the voltage busbar fluctuations is still a bit vague. Therefore some simulations with different wind speed mean values (and uncorrelated series) were performed for the weak and the strong grid cases.

4 ANALYSIS OF THE RESULTS

The output of the simulations was analysed statistically both in time and frequency domains. The results are presented in Figures 3 to 10.

In the cases presented in figures 3 to 8, which correspond to the study of the performance under constant mean speed and variable cross-correlation factor and short circuit power values, the wind time

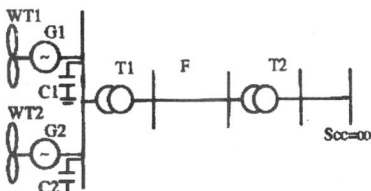

Fig. 2 - Topology of the grid to which the model was applied. The local consumers were not considered.

series used as input of the wind park were obtained after the same 'Davenport spectrum' with 7.0 m/s mean wind speed at 10 m high

In Fig 3 and 4 the difference in the behaviour of the busbar voltage and the power output respectively for both the strong and weak grids versus the cross-correlation factor of the input wind time series is illustrated.

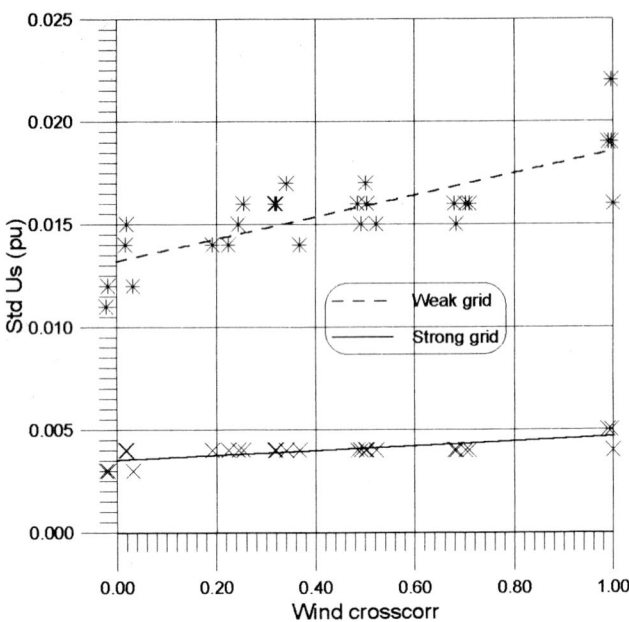

Fig.3 - Busbar voltage standard deviation versus wind cross correlation.
Comparison between strong and weak grid.
(Mean wind speed =7.0 m/s).

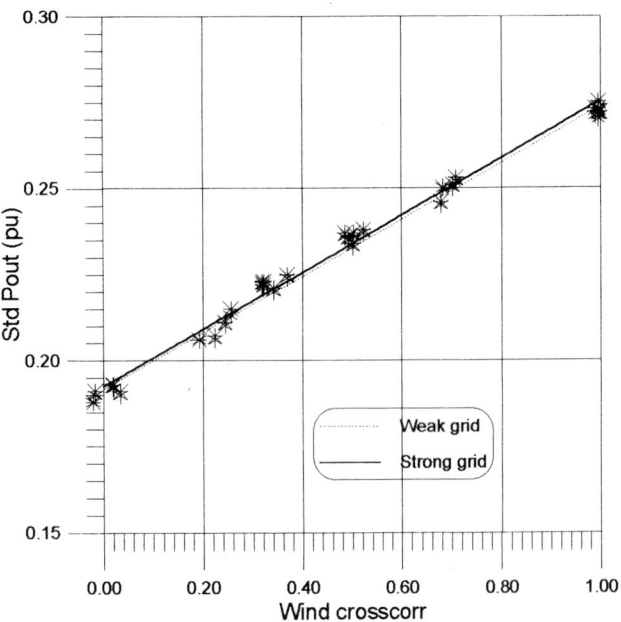

Fig. 4 - Power output standard deviation versus wind cross correlation.
Comparison between weak and strong grid.
(Mean wind speed =7.0 m/s).

In figure 5 are represented the PSD's of the power output of the wind park and the individual turbines, for the weak grid's case.

Figure 6 shows the comparison between weak and strong grid connection in terms of wind park power output fluctuations' PSD In both cases the PSD's correspond to the averaged spectra of the five samples simulated with a cross-correlation factor of 0.33.

Figure 7 contains simulation samples of the voltage time series for the two grid types, to illustrate the time behaviour of this quantity. The input wind time series (the same for the weak and strong grid simulations) are uncorrelated in this case.

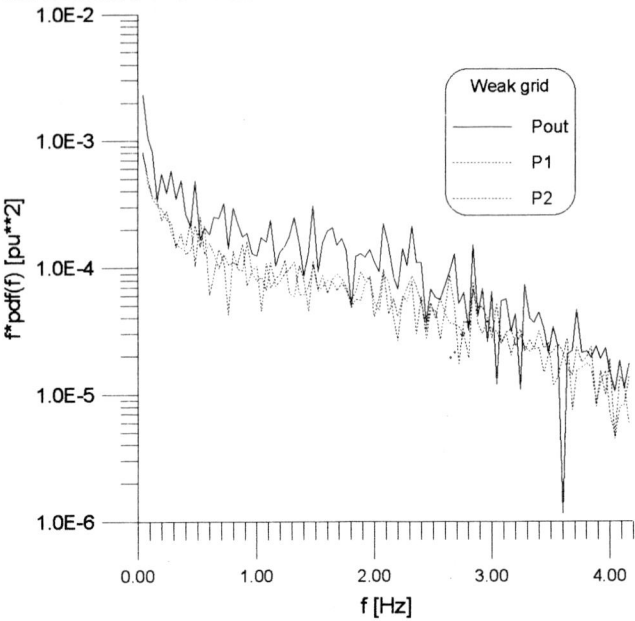

Fig. 5 - PSD of the wind park and the individual turbines' power output for the weak grid. (0.33 cross correlation, mean wind velocity =7.0 m/s)

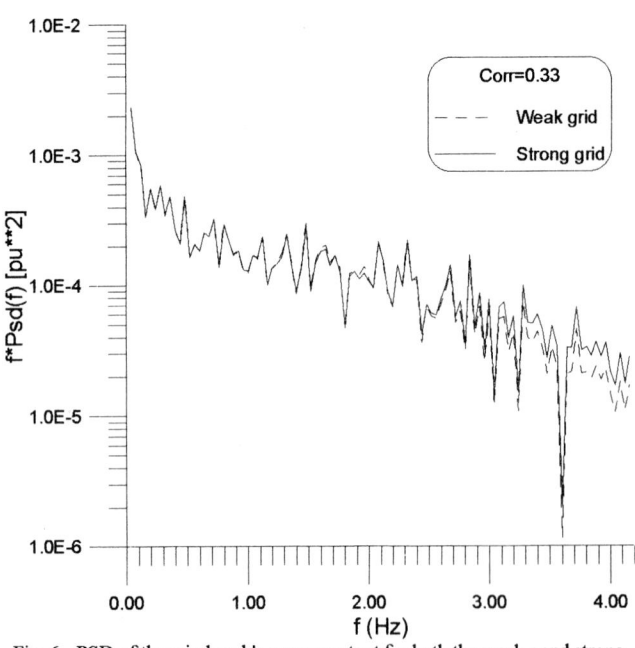

Fig. 6 - PSD of the wind park's power output for both the weak gand strong grids. (0.33 cross correlation, mean wind velocity =7.0 m/s)

Figure 8 shows the power output for both weak and strong grids, corresponding to the same simulations. The results for both the weak and the strong grids are presented but this is not obvious since the power output is pratically the same and the curves are overploted.

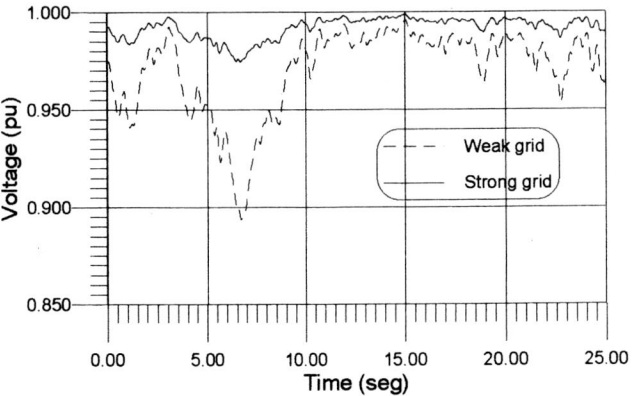

Fig.7 - Voltage fluctuation in the wind park busbar for strong and weak grids. The input wind time series are uncorrelated (Mean wind speed =7.0 m/s)

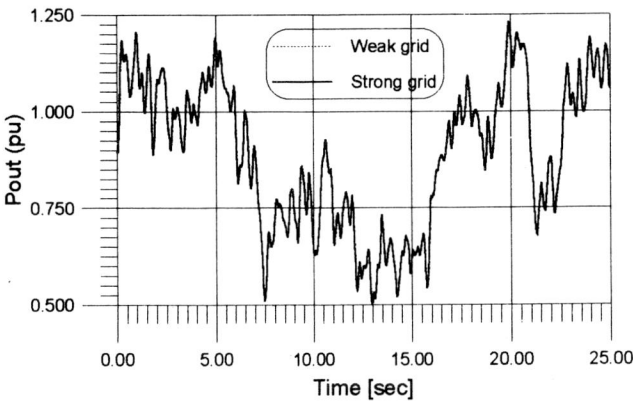

Fig.8 - Power output fluctuations for strong and weak grids. The input wind time series are uncorrelated (Mean wind speed =7.0 m/s)

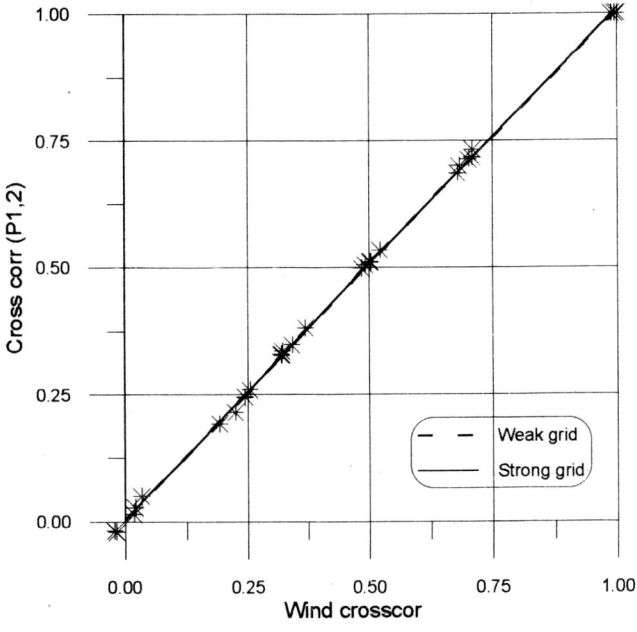

Fig. 9 - Cross correlation of the individual turbine power output versus the wind series input for both strong and weak grids (Mean wind speed =7.0 m/s)

In figure 9 the cross-correlation of the individual turbines' power output against the cross correlation of the input wind time series is presented. Again the two curves are not distinguishable, since this parameter has

no effect on the power output fluctuations and the curves are.

The mean wind speed is, of course, the most influent parameter in a wind park output and its effect is illustrated in figures 10, 11 and 12.

Figure 10 and 11 shows the standard deviation of the interconnection voltage busbar and of the power output, respectively, plotted against the mean wind speed. Figure 12 presents the mean voltage dependence on this parameter.

In figures 10, 11 and 12, for the weak grid case, results above 8 m/s of mean wind speed are not available, because the wind park presents, most of the time, an unstable working mode.

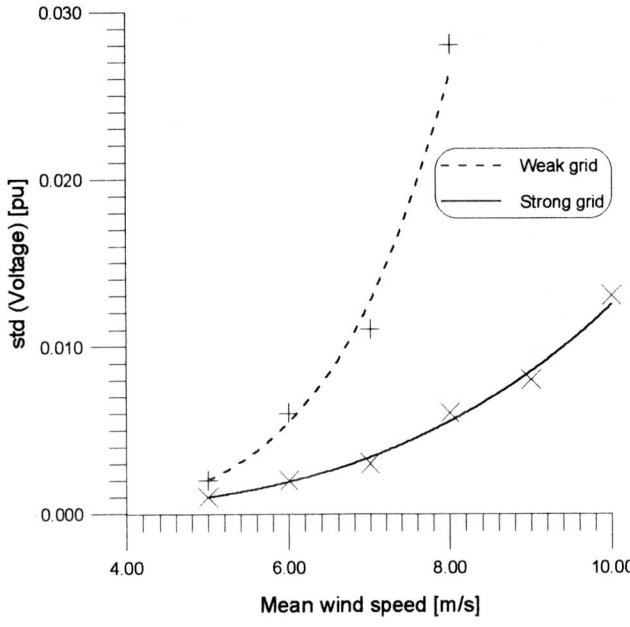

Fig. 10 - Standard deviation of the wind park's voltage busbar versus the mean wind speed for both weak and strong grids (uncorrelated wind time series).

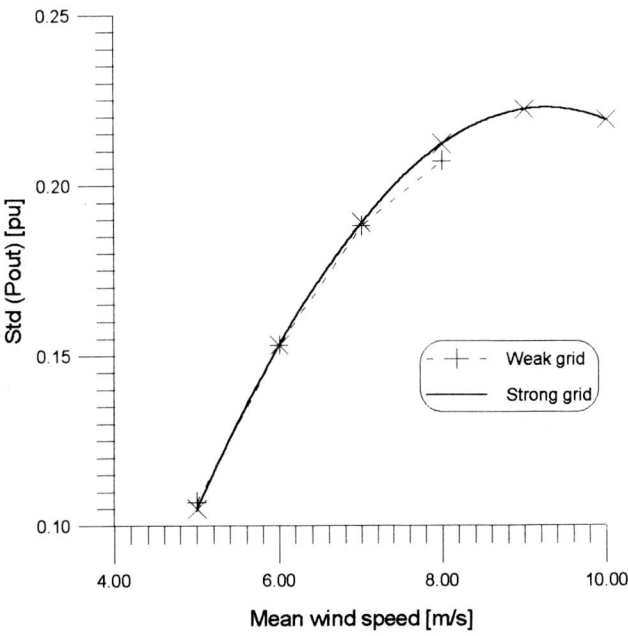

Fig. 11 - Standard deviation of the wind park power's output fluctuations versus the mean wind speed for both weak and strong grids (uncorrelated wind time series).

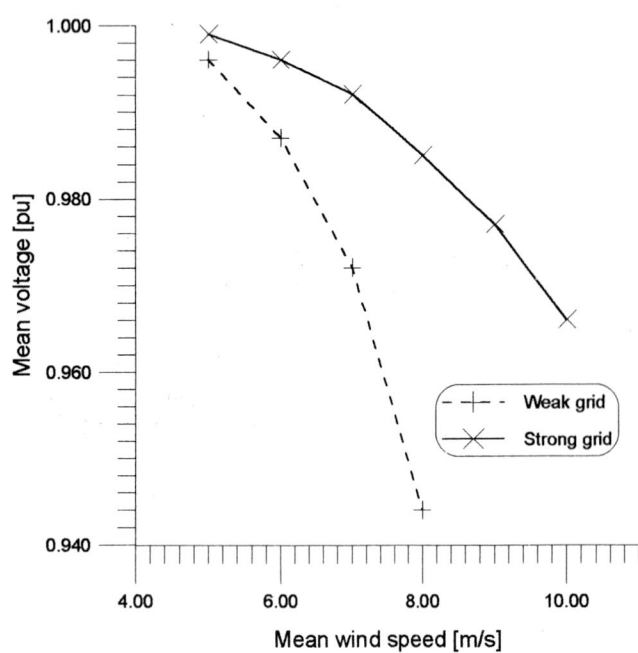

Fig. 12 - Mean voltage in the wind park busbar versus the mean wind speed for both weak and strong grids (uncorrelated wind time series).

5 CONCLUSIONS

The effect of the utility grid characteristics may be neglected when the electric power fluctuation is the only parameter to be studied. However, in terms of electrical analysis, the voltage fluctuation in the wind park's interconnection busbar is much more important than the power fluctuations since it affects the local consumers connected to the busbar. It may be easily concluded from figures 7, 10 and 12 that the grid characteristics strongly affect the voltage at the wind park's interconnection busbar, so this is a major parameter to include in future wind park models.

The smoothing effect of low wind cross correlation's factor in the wind park power fluctuations was already studied by many authors [1,8,9,10,11,12] and it is confirmed again here (Fig.5, 6). This is a quite important conclusion since most wind park models use mean wind speed values but neglect the turbulence effects. Figure 6 also points out a low amount of energy in the high frequency range (>2Hz) for the weak grid case. This may be explained by a higher smoothing effect in the weak grid due to higher fluctuations in the time series.

Finally, the mean wind speed has a strong influence in both the voltage amplitude and power fluctuation, as expected.

REFERENCES

[1] Estanqueiro, A. I et al; "The Development and Application of a Model for Power Output Fluctuation in a Wind Park", *ECWEC'93 Proc.* pp 798-901, Travemünde, 1993.

[2] Estanqueiro, A. I.; "Horizontal Axis Wind Turbines - A Behaviour's Model"., (in Port.) MSc Thesis, IST, Lisboa, April, 1991.

[3] Estanqueiro, A.I, Ferreira de Jesus, J.M:, Lalos, D.; "Transient Behaviour of a WEC under Different Load Conditions", in *"The potential for Small & Medium sized Wind Energy Applications in Mediterranean Countries"*, Rhodes, Greece, 1992.

[4] Aguiar, R. et al "Uma abordagem espectral à Geração de Séries Correlacionadas de Vento" in *"Congresso Ibérico de Energia Solar" Proc., (in Port.)*, Lisbon, April, 1993.

[5] Estanqueiro, A.I.; Ferreira de Jesus, J.M.; Gil Saraiva, J.A.; "WECS Unsteady Output Simulation", EWEC'91 Proc., pp 618-622. Amsterdam, 1991

[6] Rui M.G. Castro, "Induction Generators: It´s Use in Small Hydro Plants" (in Port.), MSc Thesis, IST., Lisbon, November 1988.

[7] Rui M.G. Castro, J.M. Ferreira de Jesus, "INDUSAT Basic Reference Guide V1.0 - Software Package for WECS Simulation", Instituto Superior Técnico, Lisbon, January 1993.

[8] Beyer, H. G. et al; "Fluctuations in the combined Power Output from Geographically Distributed Grid Coupled Wind Energy Conversion Systems - An analysis in the Frequency Domain", *Wind Engineering* Vol.14, nº3, 1990.

[9] Peltola, E. et al; "Simulations and Measurements on the Effect of a Wind Farm on the Local Grid", Paper D6 , *EWEA Special Topic Conference '92 on the Potential of Wind Farms*, 1992

[10] McNemey, G. and Richardson. R., "The Statistical Smoothing of Power Delivered to Utilities by Multiple Wind Turbines"; *IEEE/PES Winter Meeting paper 92 WM 024-0 EC*,1992.

[11] Taylor, G. J.; "Power Quality Measurements on a 24 m Windmill", *IEE Proceedings*, Vol. 134, Pt. A, nº 5, May, 1987.

[12] Smedman, A.; "Effect of Wind Speed Fluctuations on the Power Output of the Wind turbine in Kalkugnen, Sweden", *4th International Symposium on Wind Energy Systems*, Sept., 1982

[13] Ferreira de Jesus, J.M.; "A Model for Saturation in Induction Machines", *IEEE Transactions on Energy Conversion*, Vol EC-3, 1988

[14] Rui M.G. Castro, A. Eugénio Gomes, J.M. Ferreira de Jesus, "The Influence of Reactive Power Compensation in the Transient Behaviour of an Induction Generator", *Proceedings IFAC*, Seoul, August 1989.

[16] W. E. Feero, W. B. Gish, "Over voltages Caused by DSG Operation: Synchronous and Induction Generators", *IEEE T. Power Delivery*, January 1986.

[17] R. Dugan, D. Rizy, "Electric Distribution Protection Problems Associated with the Interconnection of Small Dispersed Generation Devices", *IEEE Transactions PAS*, Vol. PAS-103, Nº6, June 1984.

Coastal wind speed prediction

R. J. BARTHELMIE and **J. P. PALUTIKOF**
Climatic Research Unit, University of East Anglia, UK

SYNOPSIS The coastal region is of particular interest to siting of wind turbines. Offshore turbines are likely to be placed close to land to minimise connection costs while being at sufficient distances to limit visual intrusion. In the near-coastal zone, the influence of land on the wind field will generally still be important, depending on the site chosen. Here a number of methods of predicting the modification in the wind field in the coastal zone are compared and are applied to a small area of the Wash and north Norfolk where wind speed measurements from two near-coastal (land) sites are available for comparison.

1. INTRODUCTION

As wind passes over the coastal discontinuity changes in surface roughness, temperature and humidity occur. The difference in temperature can result in stability changes and/or development of land/sea breeze systems, while the change in roughness causes an internal boundary layer to develop.

In order to successfully predict the modification of the wind field, meteorological data from land and sea fetches should be available. However, the costs of offshore monitoring are much higher than those on land and these data are not generally available for offshore regions. Onshore, internal boundary layer theory with corrections for stability effects has been shown to work well (see, for example, Van Wijk et al., 1991; Bergström et al., 1988).

This paper concentrates on the coastal offshore zone. Here wind speeds are generally higher than those found over land, except in areas of significant topography. It is expected that the first batch of offshore wind turbines will be within 5-20km of the shoreline at sites with relatively low water depth. This proximity to land will reduce the costs of construction, laying power lines and transport for maintenance while sites should be far enough from the shore to avoid interference with leisure activities and to minimise the visual impact.

2. LAND AND SEA WIND SPEED OBSERVATIONS

A number of studies have been carried out comparing offshore with onshore wind speeds. These are summarised in Table 1. It is apparent that the data have been collected

Table 1. Increase of wind speeds from land to sea determined from observations

Wind speed over sea	Reference	Comments
$1.62 + 1.17 \, U_{land}$	Hsu (1988)	Linear regression of a large number of data sets.
$1.45 \, U_{land}$	Francis (1970)	Observations from Gorleston and Light vessel Smiths Knoll with temperature difference (air-sea) in the range -2.8 to 2.8°C and land wind speeds between 5.6 and 7.7m/s.
$1.05 \, U_{land}$ $1.12 \, U_{land}$	Lindley et al. (1980)	At 100m height 7.5km from shore " 20km from shore
$\approx 1.15 \, U_{land}$ up to $2.0 \, U_{land}$	Sethuraman and Raynor (1980)	$U_{land} > 10$m/s $U_{land} < 10$m/s

[1]Present affiliation: Meteorology and Wind Energy Department, Risø National Laboratory, Roskilde, Denmark.

under a number of different conditions and there is considerable scatter due to, for example, different roughnesses, stability, wind speed and topography. The predicted increase in wind speed is shown to vary between 5% (Lindley et al., 1980) and 45% (Francis, 1970).

3. INTERNAL BOUNDARY LAYERS

Although a number of models exist to predict the development of internal boundary layers in different conditions, the most accurate have a high data requirement (e.g. Melas and Kambezidis, 1992; Gryning and Batchvarova, 1990). It is the aim of the present work to develop a simple coastal discontinuity model which can be incorporated into a regional model of wind speed prediction.

A common way to describe the modification of wind flow as air passes over a change in surface roughness is by the development of an internal boundary layer (IBL). In the simplest possible case, after passing over the roughness change, the wind speed close to the surface gradually adjusts to the new roughness, while wind speeds aloft are unmodified. The transitional layer close to the surface is referred to as the IBL. As the fetch increases, the height of the IBL increases. Garratt (1990) gives a good review of internal boundary layer theory, observations and modelling and this includes determination of the IBL height.

In reality, a step change in surface roughness is often associated with additional changes in temperature (and possibly humidity) and therefore in stability. The change in turbulence generated mechanically at the surface and thermal effects should also be taken into account (Pal Ayra, 1988). Since there are relatively few data sets for classification of stability in the near-offshore area these effects have been disregarded here. However, it is known that stability is one of the most important factors influencing the development of the IBL. Generally, the presence of a stable boundary layer inhibits the growth of the IBL while in unstable conditions the IBL grows quickly.

3.1 Prediction of IBL heights

A number of equations for IBL development have been compared by plotting the increase in IBL height against fetch (Figure 1) using an upwind roughness of 0.1m (typical of land) and a downwind roughness of 0.0002m (used for sea). No stability corrections have been employed. Note that Bergström et al. (1988) specified use of their equation for only small roughness lengths. Some of the equations used here to predict IBL heights at fetches of up to 15km have not been tested at these distances.

The equations all predict IBL heights greater than 100m after fetches of 5km. Turbine hub-heights are generally less than 100m and so it follows that, for wind energy purposes, the wind field can be assumed to be fully readjusted within 5km of the coastal discontinuity.

However, this disregards any thermal effects. Under stable conditions the IBL grows much more slowly and the readjustment of the wind speed at heights of around 100m will occur over longer distances. Van Wijk et al. (1990) give examples of internal boundary layer development under different stability conditions. After a fetch of 5km under stable conditions the modelled IBL height has not reached 100m. Conversely, in unstable conditions the IBL grows quickly to a height exceeding 100m after 1km. Results so far available for stability of offshore areas are contradictory. Cleijne et al. (1991) compared the frequency of stability conditions at two offshore platforms in the North Sea. At the K13A platform unstable conditions occurred for more than half of the observed period. At West Sole, stable conditions were predominant. Korevaar (1990) also suggests that the mean air temperature in the North Sea is lower than the sea surface temperature giving unstable conditions. In coastal areas, with winds blowing from onshore, the land temperature is also important.

Bergström et al. (1988) make the point that turbulence conditions above and below the IBL height may be very different. This has implications for wind turbines sited in the coastal zone. If the IBL height is approximately the same as the turbine hub-height, the wind turbine will experience an increase in mean wind shear and the blades will pass through two regions which may have very different turbulence conditions. These extra stresses on the blades and nacelle may result in increased fatigue and therefore reduced lifetimes. Despite the fact that the IBL is a theoretical concept, these results suggest that there is a relatively narrow vertical band in which the wind and turbulence regimes must reach equilibrium, particularly in stable conditions. Additionally, the land/sea breeze may affect the wind regime in coastal areas.

4. MODELLING OF THE COASTAL WIND FIELD

In this section modification of the coastal wind field is investigated, using two different approaches in a relatively simple model and Risø's WAsP model (Mortensen et al., 1993). The results have been compared with observations from two near-coastal (land) sites.

4.1 900mb model

The Interpolated 900mb model is fully described by Barthelmie et al. (1991). It is based on the Moore 900mb model (Moore, 1982) and uses upper air data to predict surface wind speeds.

a) Original coastal modification method

The empirical approach used by Moore (1982) gives a mean surface wind speed at 10m over land of 4.1m/s and at 20m over sea 8.9m/s. With these values the wind speed at 50m height moving from land to sea recovers to 88% of its equilibrium (i.e. sea) value after 5km, 92% after 10km and 99% after 25km. At the land/sea boundary it is assumed that the coastline is 5km from the edge of the grid square. The modified coastal wind speeds are calculated for each grid square according to the distance from the coastline for each direction (in 45° bins) separately. The wind speeds are then combined into a mean wind field using speed and direction weighting factors. These factors are determined considering the mean wind speed and frequency of occurrence in each direction of the upper air data.

b) Internal boundary layer method

For comparison with the original method the equation for the development of the IBL in neutral conditions from Van Wijk et al. (1991) was selected:

$$h_i\left(\ln\left(\frac{h_i}{z_{o2}}\right)\right) = 2k^2x \qquad (1)$$

where:

h_i is the IBL height (m)
z_{o2} is the downwind roughness (m)
k is the Von Karman constant (0.4)
x is the fetch (m)

The method of determining the wind speed in each grid square is as follows. For each direction, the distance from the grid square to the coastal discontinuity gives an IBL height. If the IBL height is greater than the height at which the wind speed is to be predicted, the wind speed is assumed to be that of the equilibrium speed over the new surface. Over the sea the IBL should theoretically grow more slowly in neutral conditions (due to the lower roughness length). However, equation (1) uses only the largest of the upstream and downstream roughness values, which is the land value. The modified coastal wind speeds are calculated for each direction and averaged as above.

4.2 WASP

The Wind Atlas Analysis and Application Program (WASP) is fully described in Mortensen et al. (1993) and Troen and Petersen (1989). Essentially the model works by extrapolating from measured surface wind speeds taking into account topography, roughness changes and obstacles at the site to produce a regional wind climatology. The procedure can then be applied in reverse to predict the wind climate at a different site. It is therefore based on principles similar to those used in the 900mb model but has more thorough approach to the physics of the atmosphere. It can also be applied more generally because topographical enhancement is taken into account. The wind field at the coast is modified using internal boundary layer theory but is brought to equilibrium over a distance of 10km on either side of the coastline and has corrections for the different influences of stability over land and sea. Troen and Petersen (1989) point out that the width of the coastal zone depends on both climate and topography.

In the subsequent analysis WASP has been initialised with data from 1971-1980 from Coltishall which is available in the Wind Atlas (Troen and Petersen, 1989). In the prediction, no account has been taken of topography or obstacles at the two sites or roughness changes other than at the coast. The predicted wind speeds should therefore be comparable with results from the 900mb model in that they should represent wind speeds at the sites assuming flat topography in uniform roughness (set to 0.1m) depending only on the initialising wind climate and the distance of the sites from the coast.

5. VALIDATION AND TESTING

In order to test the models, a small area of the Wash and the adjacent coast of North Norfolk was selected. For the 900mb model this area was gridded into 1km squares. The mean upper air wind speed at Hemsby (over the period 1971 to 1980) was 10.3m/s and this value was used as input to the 900mb model with both the original coastal modification and the IBL methods. The roughness lengths over land was assumed to be 0.1m which is a reasonable approximation given that the primary land use in this area is agriculture.

This area was chosen for three reasons:

1. There is no significant topography in the southern and western parts of the Wash, either inland or at the coast. On the North Norfolk coast the topography is relatively simple.
2. Data from two near-coastal stations are available for validation.
3. The area was identified as a possible site for wind turbines by Lindley et al. (1980) and the first UK offshore turbine was to have been sited off the North Norfolk coast (Department of Energy, 1988).

5.1 Comparison with data from coastal stations

Short-term data from two near-coastal stations are available for comparison with model predictions. Here, the predictions are compared with observations at the 30m height. Wind data were collected over a three-year period (October 1988 to September 1991) at a site 4.5km from the North Norfolk coast (West Beckham). The second site, Sharpe's Bank, is located nearly three kilometres from the coast on the southern shore of the Wash estuary. The data period is shorter than for the first site (from December 1990 to November 1991). The locations of both sites are shown in Figure 2. The mean roughness lengths (determined from profile data) at West Beckham and Sharpe's Bank are 0.156m and 0.142m respectively. The value for land used in the model (0.1m) is therefore a reasonable approximation.

The mean wind speeds predicted by the models are shown in Table 2. At West Beckham, the IBL has no effect since at this distance from the coast we can expect the wind speeds in the IBL to be fully adjusted to the new roughness. However, using the original formulation the effect of the coastal discontinuity is felt over much larger distances and this gives slightly higher wind speeds close to the coast. The predicted wind speed using the IBL equation is higher at Sharpe's Bank than at West Beckham

Table 2. Observed and predicted 30m wind speeds at West Beckham and Sharpe's Bank (m/s)

Station	West Beckham	Sharpe's Bank
Observed	7.0	7.2
Predicted:		
900mb original	6.8	6.6
900mb IBL	6.2	6.4
WASP	5.7	5.7

because of it's proximity to the coast. The original formulation gives good results at West Beckham but under predicts at Sharpe's Bank (by 0.6m/s). The IBL version under predicts the wind speed by 0.8m/s at both stations. WASP gives low wind speeds at both stations, possibly because the initialising wind speed from Coltishall is relatively low. One possible reason for this is that the anemometer at Coltishall has a slower start up speed than those used at Sharpe's Bank and West Beckham resulting in the recording of a greater number of calms.

It is also important to remember that topography has not been used in the models here. No topographic enhancement of the wind speeds is expected at Sharpe's Bank. However, at West Beckham, the mean speed-up factor for 30m height is 0.03 (calculated using the method of Taylor and Lee, 1984). With a wind speed of 7.0m/s, the speed up is therefore approximately 0.2m/s.

5.2 The modelled wind field

We can now compare the percentage change in modelled wind speeds moving up- and down-stream of the coastal discontinuity. Here we have used three model runs which are shown in Figure 2. The first run moves in a north-south (N-S) direction through Sharpe's Bank, the second in a northeast-southwest direction (NE-SW) from a point on the west coast of the Wash, and the third in a north-south direction through West Beckham. The change in the wind speed with fetch has been normalised with the predicted wind speed at the coast (Figure 3).

In the Wash area the influence of the land surrounding the site in three directions is shown to reduce the increase of wind speed moving away from the coast using WASP. Where the site has a clear offshore fetch, the increase in wind speed is greater moving offshore and the decrease is larger in the onshore direction. The predicted decrease in wind speeds over land from the original method in the 900mb model is unaffected by the presence of land in three directions around the Wash giving about the same percentage change as a clear offshore-onshore roughness change. Over sea, the initial increase in wind speeds is bigger with a straightforward offshore fetch. There is no effect on the wind speeds predicted by the IBL method since the change in wind speeds is very rapid.

The underprediction of the wind speeds at Sharpe's Bank by the original formulation suggests that the modification of the wind field is taking place over distances which are too large. Conversely, the IBL version of the model predicts a wind field with a zone of adjustment only a few kilometres wide. Beyond this distance from the coast the wind speeds are at their equilibrium land or sea values. To some degree the wind field is brought to equilibrium over shorter distances by compounding the modified wind speeds from different directions. It is clear that the distance over which the wind speeds are modified would still be relatively small, however, since Figure 1 shows that the IBL heights exceed 30m after less than 2km. WASP predicts modification of the offshore coastal wind speed which lies between the two approaches in the 900mb model. Over land, the decrease in wind speed occurs over a smaller distance and is larger than that predicted by either of the methods used in the 900mb model.

The large differences between the predicted wind speed changes suggests that there is still a significant gap of knowledge of the wind in the offshore coastal zone. Validation of the different approaches can only be carried out when good quality data sets become available for this region.

6. CONCLUSIONS

The most common way to describe the wind speed adjustment after a change in surface roughness is to use internal boundary layer theory. Formulae given by a number of authors produce a wide range of predictions for the growth of the IBL after the same change in surface roughness. Perhaps this is not surprising given that the development and testing of these equations must have been carried out in a wide range of stability climates and topographic conditions.

According to the formulae examined the neutral IBL at heights of interest in wind energy studies (i.e. less than 100m) is fully developed after only a few kilometres. With the IBL method the use of 10km grid squares would give an abrupt change in wind speeds at the coast. Testing of the 900mb model was therefore carried out on a small area of the Wash which was gridded into 1km squares. Both versions of the 900mb model underpredict the mean wind speed at the two near-coastal sites with the original version giving slightly higher wind speeds. The wind speeds predicted by WASP for both stations were comparatively low. The location of the sites is not ideal, although there are advantages with the relatively simple topography. West Beckham is located at too great a distance from the coast for the effect of the IBL to be noticeable.

It is far from clear which of the three methods gives the best results with reference to the modification of offshore flow. The original formulation seems to be modifying wind speeds over too large a distance since the wind speeds at the coastal sites are underpredicted while the IBL method predicts an adjustment zone which is too narrow. WASP predicts modification of the offshore coastal wind speed which lies between the two approaches in the 900mb model but also underpredicts the wind speeds at the two sites.

The models need to be tested in other coastal areas which have relatively simple topography, preferably with data from stations which are located at shorter distances from the coast and in offshore areas. Unfortunately gridding the land data into 1km squares and applying the 900mb model is a fairly intensive task.

Since the potential for wind energy production in the near-offshore region (particularly in the UK and northern parts of Europe) is significant, detailed monitoring of both wind speeds and stability in the offshore coastal zone is essential if mean wind speeds, vertical profiles and turbulence are to be accurately assessed.

7. ACKNOWLEDGEMENTS

Eastern Electricity PLC and National Power funded the North Norfolk Wind Monitoring Project and Eastern Electricity PLC funded data collection at Sharpe's Bank. This work started under a Science and Engineering Research Council Studentship and has been continued with an EC Fellowship at Risø National Laboratory. We are also grateful to Paul Hannah for collecting and analysing data from Sharpe's Bank.

8. REFERENCES

Barthelmie, R.J., Palutikof, J.P. & Davies, T.D. (1991) Predicting UK offshore wind speeds. *Annales Geophysicae*, 9, 708-715.

Bergström, H., Johansson, P-E. & Smedman, A-S. (1988) A study of wind speed modification and internal boundary-layer heights in a coastal region. *Boundary-Layer Meteorology*, 42, 313-335.

Cleijne, J.W., Coelingh, J.P. & Van Wijk, A.J.M. (1991) *Description of the North Sea wind climate for offshore wind energy applications.* TNO Report 112324-22013. 91pp. TNO, Apeldoorn, The Netherlands.

Department of Energy (1988) *Harnessing the wind.* Available from the Energy Technology Support Unit, Harwell, Oxfordshire.

Elliott, W.P. (1958) The growth of the atmospheric internal boundary layer. *Transactions of the American Geophysical Union*, 39, 1048-1054.

Francis, P.E. (1970) The effect of changes of atmospheric stability and surface roughness on off-shore winds over the east coast of Britain. *Meteorological Magazine*, 99, 130-138.

Garratt, J.R. (1990) The internal boundary layer - a review. *Boundary-Layer Meteorology*, 50, 171-203.

Gryning, S.E. & Batchvarova, E. (1990) Analytical model for the growth of the coastal internal boundary layer during onshore flow. *Quarterly Journal of the Royal Meteorological Society*, 116, 187-203.

Hanna, S.R. (1987) An empirical formula for the height of the coastal internal boundary layer. *Boundary-Layer Meteorology*, 40, 205-207.

Hsu, S.A. (1988) *Coastal Meteorology.* 260pp. ISBN 0-12-357955-4. Academic Press Inc., London.

Korevaar, C.G. (1990) *North Sea Climate.* 137pp. ISBN 0-7923-064-3. Kluwer Academic Publishers, Dordrecht.

Lindley, D., Simpson, P.B., Hassan, U. & Milborrow, D. (1980) Assessment of offshore siting of wind turbine generators. *Proceedings of the Third International Symposium on Wind Energy Systems*, Aug 26-29, 1980. pp. 17-42. BHRA Fluid Engineering, Cranfield, UK.

Melas, D. & Kambezidis, H.D. (1992) The depth of the internal boundary layer over an urban area under sea-breeze conditions. *Boundary-Layer Meteorology*, 61, 247-264.

Moore, D.J. (1982) 10 to 100m winds calculated from 900mb wind data. *Proceedings Fourth B.W.E.A. Workshop*, pp.197-205. BHRA Fluid Engineering, Cranfield, UK.

Mortensen, N.G., Landberg, L., Troen, I. & Petersen, E.L. (1993) *Wind Atlas Analysis and Application Program* (WASP). Risø National Laboratory, Roskilde, Denmark.

Pal Arya, S.P.S. (1988) *Introduction to Micrometeorology.* 307pp. Academic Press, San Diego, California.

Panofsky H.A. & Dutton, J.A. (1984) *Atmospheric Turbulence: Models and Methods for Engineering Applications.* 397pp. John Wiley and Sons, New York.

Sethuraman, S. & Raynor, G.S. (1980) Comparison of mean wind speeds and turbulence at a coastal site and an offshore location. *Journal of Applied Meteorology*, 19, 15-21.

Taylor, P.A. & Lee, R.J. (1984) Simple guidelines for estimating wind speed variations due to small scale topographic features. *Climatological Bulletin*, 18, 3-32.

Troen, I. & Petersen, E.L. (1989) *European Wind Atlas.* ISBN 87-550-1428-8. pp. 656. Risø National Laboratory, Roskilde, Denmark.

Van Wijk, A.J.M., Beljaars, A.C.M., Holtslag, A.A.M. & Turkenburg, W.C. (1990) Diabatic wind speed profiles in coastal regions: comparison of an internal boundary layer (IBL) model with observations. *Boundary-Layer Meteorology*, 51, 49-75.

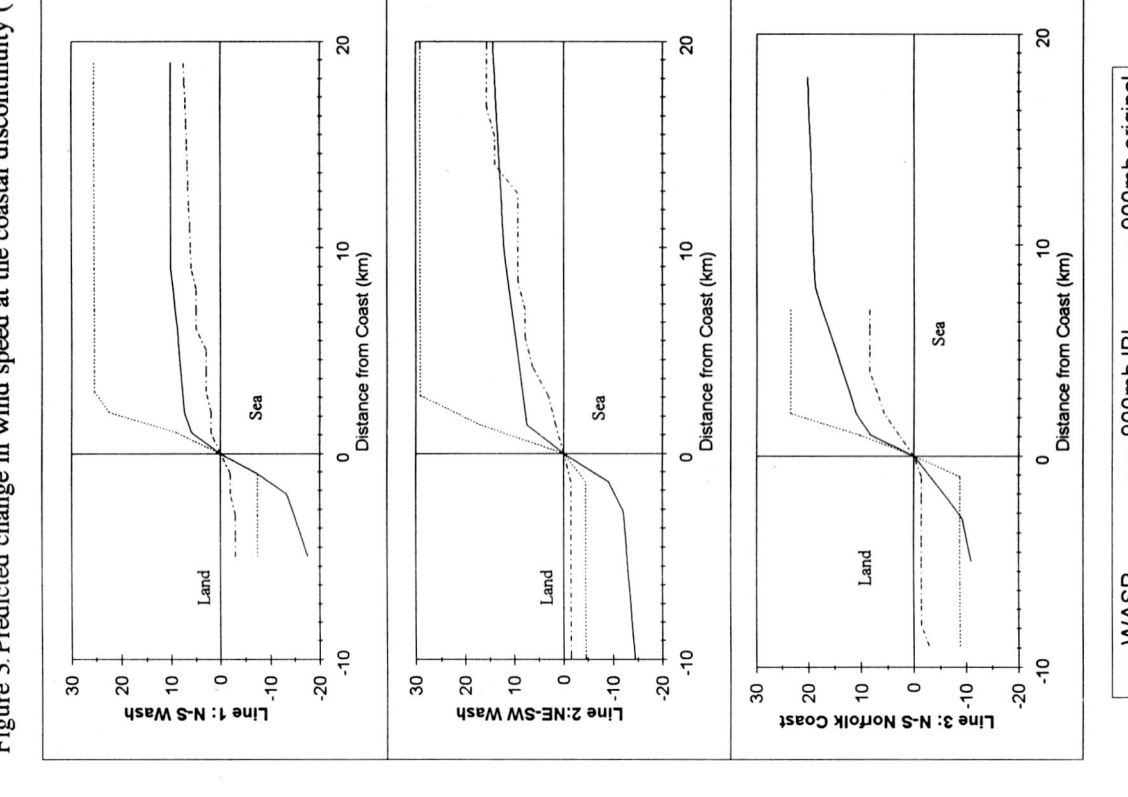

Figure 3. Predicted change in wind speed at the coastal discontinuity (%)

Line 1: N-S Wash

Line 2: NE-SW Wash

Line 3: N-S Norfolk Coast

—— WASP   — · — 900mb IBL   ····· 900mb original

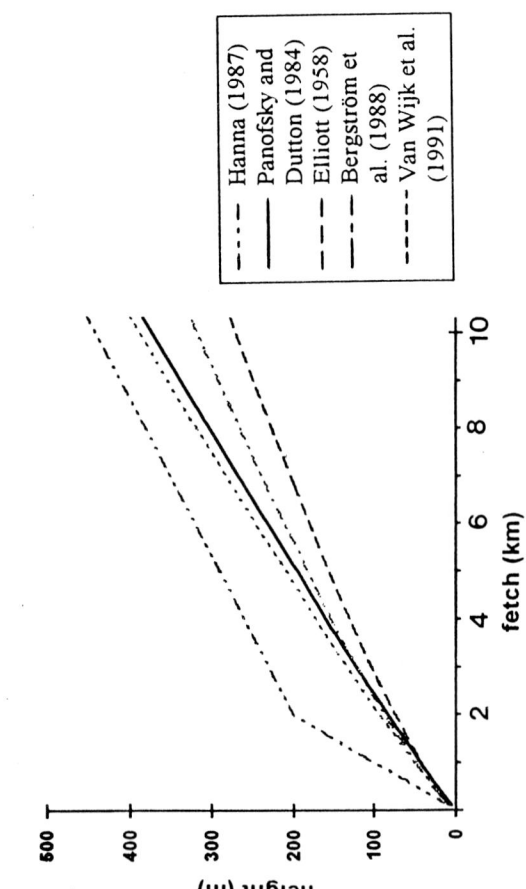

Figure 1. Development of the internal boundary layer

— · · — Hanna (1987)
—— Panofsky and Dutton (1984)
— — Elliott (1958)
— · — Bergström et al. (1988)
— — Van Wijk et al. (1991)

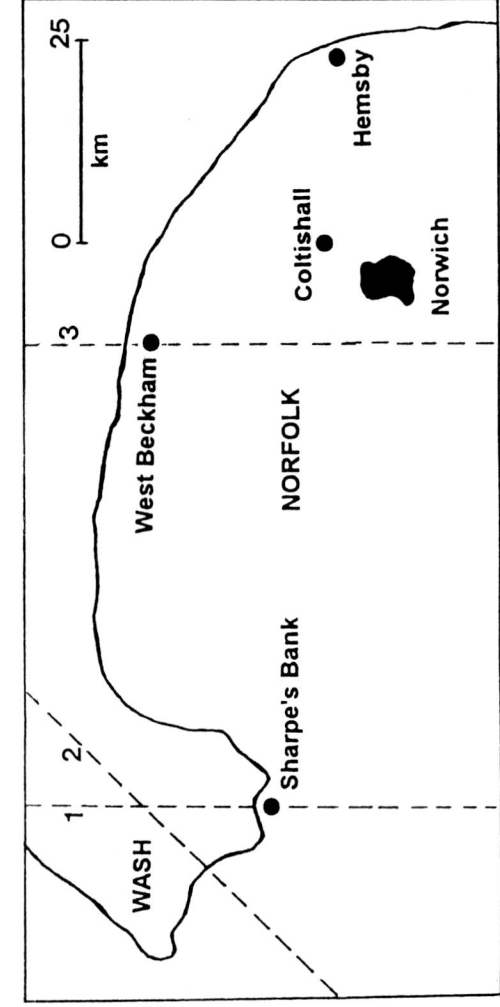

Figure 2. Location of the sites and directions used in the model testing

Monitoring at Vindeby: preliminary results

R. J. BARTHELMIE, M. S. COURTNEY, J. HØJSTRUP and **P. SANDERHOFF**
Risø National Laboratory, Denmark

SYNOPSIS Monitoring of turbines and meteorological conditions has been undertaken at the world's first offshore wind farm at Vindeby, Denmark to examine wind energy production offshore. Details of the site and description of the measurements to date are given. The full project will consist of meteorological measurements from one onshore and two offshore masts and structural and power measurements from two turbines in the array.

1 INTRODUCTION

Offshore siting of wind turbines will become more attractive as the availability of good land sites decreases. It is generally assumed that higher wind speeds offshore will at least partially offset increased siting costs. However, there is little information available about offshore wind speeds and many technical questions remain unanswered. There is therefore considerable interest in monitoring of the world's first offshore wind farm at Vindeby in Denmark. Additionally meteorological data will be obtained which will be useful in developing and testing models of the offshore and coastal wind climate.

2 SITE DESCRIPTION

Vindeby wind farm, located off the northwestern coast of the island of Lolland, consists of 11 Bonus 450kW turbines arranged in two rows oriented along an axis of 325-145°. Figure 1 shows the location of the wind farm. The most southerly turbine in the array is approximately 1.5km from land and the turbine spacing is 300m both along and between the rows. The water depth is between 2.1 and 5.1m. Electricity generation to the grid started in July 1991. For more details of the construction and operation of the wind farm see Olsen and Dyre (1993). One interesting problem detailed in this paper was the setting of the pitch of the blades. Initial pitching was set as on similar turbines on land and this had to be corrected due to over-production, the cause was thought to be lower turbulence offshore.

In order to study the meteorological aspects of the wind farm, and to provide information on wind flow in the coastal zone, three 45m meteorological masts have been erected, one on land and two offshore. The land mast is located nearly 2km south of the most southerly turbine in the array at approximately 16m from the coastline at 2m above sea level. The data collection hut is 10m east of the mast. The two offshore masts are placed at distances equal to the turbine spacing (300m), one to the west and one to the south of the first row. The locations of the masts with respect to the wind turbines are also shown in Figure 2.

2.1 Roughness and fetch

Figure 2 shows the terrain surrounding the land mast and roughness classified according to Wieringa (1986). To the south of the mast the terrain is mainly open farmland with a few scattered houses and trees, with open sea to the north. The topography is very flat with no enhancement of the wind speed expected. The coastline runs approximately along the line of 285-105°. Moving to the mesoscale, the definition of the terrain becomes more complicated due to the nature of the coastline. After examination of the fetch in the nearest 10km to the mast, directions between 290 and 65° can be classified as having sea fetch, while those between 135 and 195° have land fetch. The remaining directions have mixed fetch.

3 INSTRUMENTATION

Instruments have been installed at the same heights (with reference to sea level) on booms oriented in a north-south direction all three masts. On the land mast, wind speed is measured at four heights with turbulence, wind direction, absolute temperature, temperature differences, barometric pressure and solar radiation also being recorded. The sea masts have additional instrumentation for more detailed study of the wind climate offshore and humidity, sea surface temperature and wave heights will also become available.

Samples are taken at the rate of 20Hz and stored as half-hourly means with selected higher-resolution time series. Measurements made at the land mast are stored on a PC in the hut at the foot of the mast and are accessible remotely. Data from the two offshore masts are collected via turbines 4W and 5W and are connected via an undersea fibre-optic cable to a communications box close to the onshore mast. From this point it is connected to the hut and data will therefore also be accessible remotely.

Two wind turbines (4W and 5E in Figure 2) have been instrumented for basic structural measurements. On each turbine, one blade-root has been instrumented with strain gauges for bending moments in two directions. Additional strain gauges are mounted on the tower. Measurements of rotor position, turbine electrical output and yaw angles are also available. The objective of the turbine instrumentation is to make simultaneous measurements of the mechanical loads on the two turbines and the upstream turbulence for a variety of situations which basically fall into two classes:

- in the unobstructed low turbulence flow

- in the wake of one or more turbines

For a number of cases it will be possible to directly compare loads on one undisturbed turbine with measurements on the other turbine in a wake situation. Roughly 10% of representative time series will be stored to supplement the information from the on-line data reduction.

Data collection at the land mast began in January 1993 and at sea mast south in September 1993. Instrumentation of sea mast west and the two turbines is nearing completion.

4 PRELIMINARY RESULTS

4.1 Overview

The following analysis is based on measurements taken at the land mast over the summer period (May to September 1993). Unfortunately when the data are binned by hour or by sector this reduces the number of measurements from which averages can be taken..

Preliminary investigation of the effect of internal boundary layer development at roughness changes suggests:

1) the lowest anemometer is at sufficient height to be above the internal boundary layer which forms when the wind meets the coastline.

2) all the measurements on the sea masts should be within the internal boundary layer which develops as wind moves away from the coast.

The development of these internal boundary layers in different stability conditions and the problems of internal boundary layer development at multiple roughness length changes will be investigated as further data are collected.

4.2 Boom interference

Due to the alignment of the boom along the axis 350° - 170° major interference presumably introduced by the flow round the mast is experienced between 160 and 180° and between 340 and 0°. With the exception of the above mentioned directions the average difference as measured on the land mast is less than 0.1m/s in each sector bin. This does not present a problem at 48m or 38m since the top anemometer is free from obstruction and at 38m measurements from either instrument can be selected. However, it is clear that at 20m and 7m heights considerable interference to the measured wind speeds will be experienced when the wind blows from certain directions.

4.3 Diurnal Cycles.

a) Wind speed

A typical diurnal cycle in wind speed over land peaks during the afternoon and is minimum overnight. Over sea, little diurnal variation is expected.

Mean wind speeds from the land fetch (Figure 3a) have range of about 3m/s (at 48m) with a peak at 1000h and a minimum at 000h. This differs from the expected diurnal cycle. From the sea fetch (Figure 3b), the range measured at 48m height is smaller than that found over land (about 2m/s), peaking at 1900h with minimum speeds occurring between 700h and 1700h. It is clear then that the diurnal cycles in wind speed from land and sea fetches are very different and this will be the object of further analysis when additional data are available.

b) Wind direction

It was expected that the data set might be affected by sea/land breezes at this time of year but examination of the daily data shows relatively few directional shifts which could be associated with such systems.

c) Temperature

There should be little diurnal variation in temperatures over sea, while 'land' temperatures peak during the early afternoon and reach a minimum overnight. The temperature difference over land (46-10m) dT, is expected to be negative during the day and positive overnight. Neutral conditions are indicated by approximately -0.3°C (assuming dry or unsaturated conditions). Over sea, dT should show little diurnal variability.

Temperatures from the south (land fetch) (Figure 3b) show a diurnal cycle similar to that expected but with a peak during the late afternoon. The range of temperatures is just over 6°C, dropping to a minimum between 100 and 400h. The temperature difference ranges between 1.2 and -0.6°C. It becomes positive overnight peaking at 400h and dipping below zero from 700 to 1900h. The diurnal cycles of temperature and dT are inverted.

Temperatures from the sea fetch also show diurnal variability (Figure 3c) but the range is smaller than that found over land (less than 4°C). The maximum occurs between 1300 and 1700h with the minimum between 1000 and 1200h. A diurnal cycle is found in dT which is inverted compared to that shown by the land fetch and has a smaller range (0 to -0.4°C). Maximums are recorded between 1400 and 1600h with a minimum between 0500h and 0600h. Note that the average dT never becomes positive. The diurnal cycle of dT and temperature are very similar for the sea fetch with both peaking in the afternoon.

4.4 Sector analysis

The data have been divided into 10° bins where Sector 0 is 355 to 5°, Sector 1 is 5 to 15° and so on to Sector 35 which contains data from directions 345-355°. Sectors with a sea fetch are numbered 29 to 6 while Sectors 13 to 20 have a long land fetch. The remainder have mixed land and sea fetch.

Long-term measurements of the Danish wind climate (Troen and Petersen, 1989) suggest that wind speeds are expected to be most frequent and strongest from the southwesterly direction. This is found in the data set (Figure 4a) but there are also a surprisingly large number of relatively strong easterly winds. It is clear is that there is a much smaller difference in the measured wind speeds at 48m and 7m when the wind blows from a sea fetch reflecting lower roughness.

Measurements of absolute temperature and temperature differences also show variations by sector (Figure 4b). Highest temperatures occur in Sectors 14 to 25 which have land or mixed fetches. Positive temperature differences are found in Sectors 10 to 18 which presumably reflects the relatively high overnight values (indicating stable conditions) in the land fetch.

A plot of mean wind speeds and temperatures in each bin (Figure 4c) shows that there is an inverse relationship with higher temperatures corresponding to lower wind speeds and vice versa. This is likely to be a combination of synoptic and fetch effects.

4.5 Wind speed profiles

The data have been divided into four groups representing the different fetches and the mean wind speed profiles plotted. These are shown in Figure 5 together with 'best-fit' logarithmic profiles. Roughness lengths estimated using the profile data are also shown in Figure 5. The first three profiles are slightly stable with the last profile indicating slightly unstable conditions but none differ substantially from the logarithmic prediction.

The roughness lengths for the last three groups representing mixed and land fetches are close to those estimated from maps but the predicted roughness for the sea fetch is very low. The stability conditions at both land

and sea sites, and the change of the sea roughness with wind speed are undergoing further investigation.

4.6 Turbulence Intensity

Average turbulence intensities for each sector have been calculated using the standard deviation of wind speed at 48m and 7m where turbulence intensity, I, is given by:

$$I = \frac{\sigma_u}{U}$$

Figure 6 shows averaged turbulence intensities plotted by sector. Turbulence intensities at 7m are higher than those at 48m. There does not appear to be a big difference between turbulence intensities from the sea fetch compared with those at the land fetch at 48m. However, at 7m an increase is found in the land sectors. Two features at 7m are of particular interest. First, the large peaks in sectors 12 and 16 are most likely due to the presence of a number of obstacles in these directions: the data collection hut and trees. Second, the smaller peak in the northerly direction is probably due to mast interference effects. These are seen to a lesser extent in the turbulence intensity measured at 48m.

4.7 Ratios of wind speeds measured at land and sea masts

Measurements of wind speed and turbulence intensity at 48m height made simultaneously at the land and sea masts are compared in Figure 7. Here they are given as ratios of sea to land measurements. Perhaps the most surprising feature of this analysis is the relatively small difference in wind speeds from the south. It might be expected that the wind speeds measured on the land mast would be lower reflecting higher roughness while wind speeds at the sea mast would have recovered to 'over-sea' values but the values are close to one. Presumably the higher turbulence recorded at the sea mast in the north sectors is generated by the wind farm. The wind speeds measured at the sea mast are seen to be higher from 60° to 110° and from 210° to 240°. These directions also have lower turbulence recorded at the sea mast.

5 CONCLUSIONS

The monitoring project promises interesting and useful data for the study of the design and operation of offshore wind farms. The most interesting features of this preliminary analysis can be summarised:

1) The diurnal cycle of wind speed over sea peaks overnight as expected but a distinctive diurnal cycle in wind speeds over land was not found. Wind speeds from the sea fetch have a smaller range than those from the land fetch.

2) The difference between wind speeds measured at 48m and 7m is smaller for the sea fetch than for the land fetch

presumably reflecting lower roughness.

3) Diurnal cycles in temperature over land and sea are similar in shape but the peaks from the land fetch occur later than anticipated. The range in temperature from the sea fetch is smaller than that found over land.

4) While dT over land has an inverted cycle compared to absolute temperature, the diurnal cycle in dT over sea follows the pattern of absolute temperature very closely. The diurnal cycle is as expected from the land fetch indicating a small temperature inversion overnight. Conditions during the day are indicated as being close to neutral, the absence of more negative temperature differences indicating unstable conditions during the day is surprising considering that the data are from the summer months. As was postulated by Moore (1979) conditions from the sea fetch are shown to be most stable during the day but tending towards neutral overnight rather than becoming unstable. The range of dT is very small as would be expected.

5) There is an inverse relationship between temperature and wind speed when the data are subdivided by sector with higher wind speeds being associated with lower temperatures and vice versa. This is thought to be a combination of fetch and synoptic effects and will be the subject of further analysis.

6) Wind speed profiles are found to be reasonably well-predicted using a logarithmic profile. The roughness of the sea surface determined from profiles was lower than found in other studies of over-water roughness (e.g. Wieringa, 1986). Detailed assessments of stability conditions and roughness change with wind speed are presently being undertaken.

7) There is no significant difference between turbulence intensities from the sea fetch compared with those of the land fetch. Turbulence intensities at 7m are affected by local obstacles and flow distortion caused by the mast which can be seen to a lesser extent at 48m.

6 ACKNOWLEDGEMENTS

The monitoring project is funded by the Danish Governments Energy Research Programme (EFP 92) and by SK Power who have funded and installed the two offshore masts. We are grateful to the technical support team at Risø for their hard work in establishing the monitoring hardware. R.J. Barthelmie would also like to thank the EC for funding her fellowship and Risø National Laboratory for accepting her under the scheme.

7 REFERENCES

(1) Olsen, F. and Dyre, K. (1993) Vindeby off-shore wind farm - construction and operation. *Presented at the BWEA/DTI Joint Seminar: Prospects for Offshore Wind Energy, 29th June 1993, Harwell, UK.*

(2) Wieringa, J. (1986) Roughness-dependent geographical interpolation of surface wind speed averages. *Quarterly Journal of the Royal Meteorological Society,* 112(473), 867-889.

(3) Troen, I. and Petersen, E.L. (1989) *European Wind Atlas.* pp. 656. ISBN 87-550-1482-8. Risø National Laboratory, Roskilde, Denmark.

(4) Moore, D.J. (1979) Offshore Wind data. *In: Proceedings of the first BWEA Workshop, Cranfield,* pp. 199-207. BHRA Fluid Engineering, Cranfield, UK.

Figure 1. Location of Vindeby site

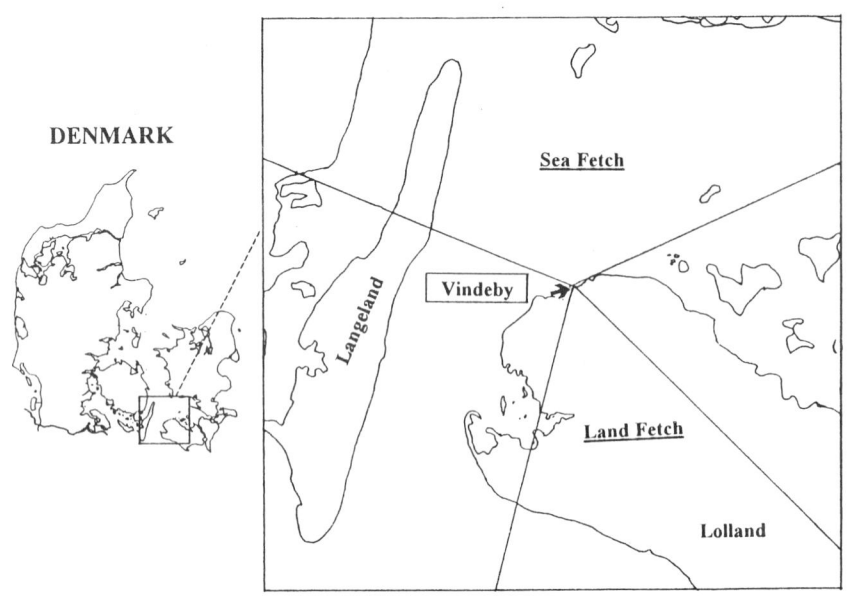

DENMARK

Sea Fetch

Langeland

Vindeby

Land Fetch

Lolland

Figure 2. Map of the site showing the turbine arrays, the location of the masts, and a roughness description (as classified by Wieringa, 1986) of the area around the land mast.

Roughness Class	Roughness	Description
	0.0002	Open sea
	0.03	Flat, open, low grass
	0.1	Crops, occasional large obstacles
	0.25	High crops, scattered obstacles
	0.5	Parkland, open housing
	1.0	Forest, dense housing

Figure 3. Diurnal cycles of wind speed at 48m height, temperature from 10m height and temperature differences (46-10m). The data are from two subsets with land (from directions 135-195°) and sea (from directions 290-65°) fetches.

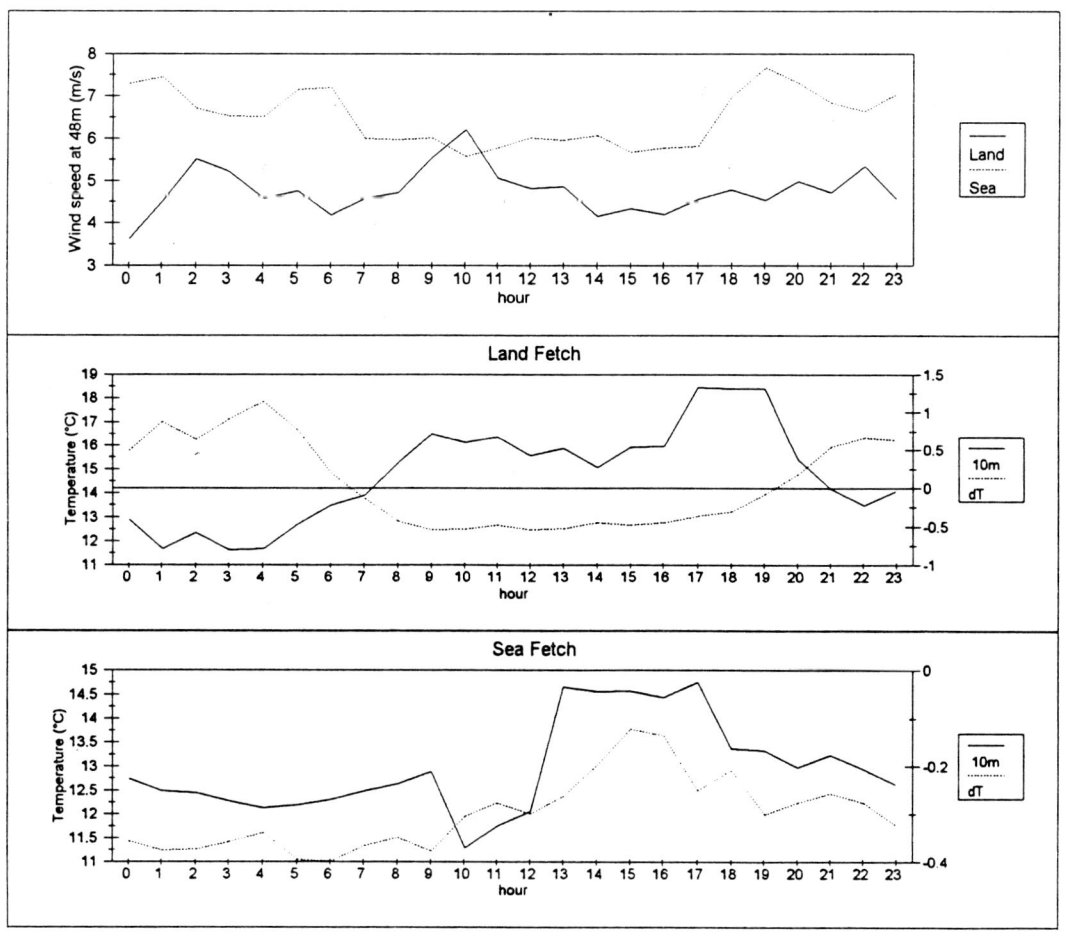

Figure 4. Average wind speeds, temperatures and temperature differences in 10° directional bins.

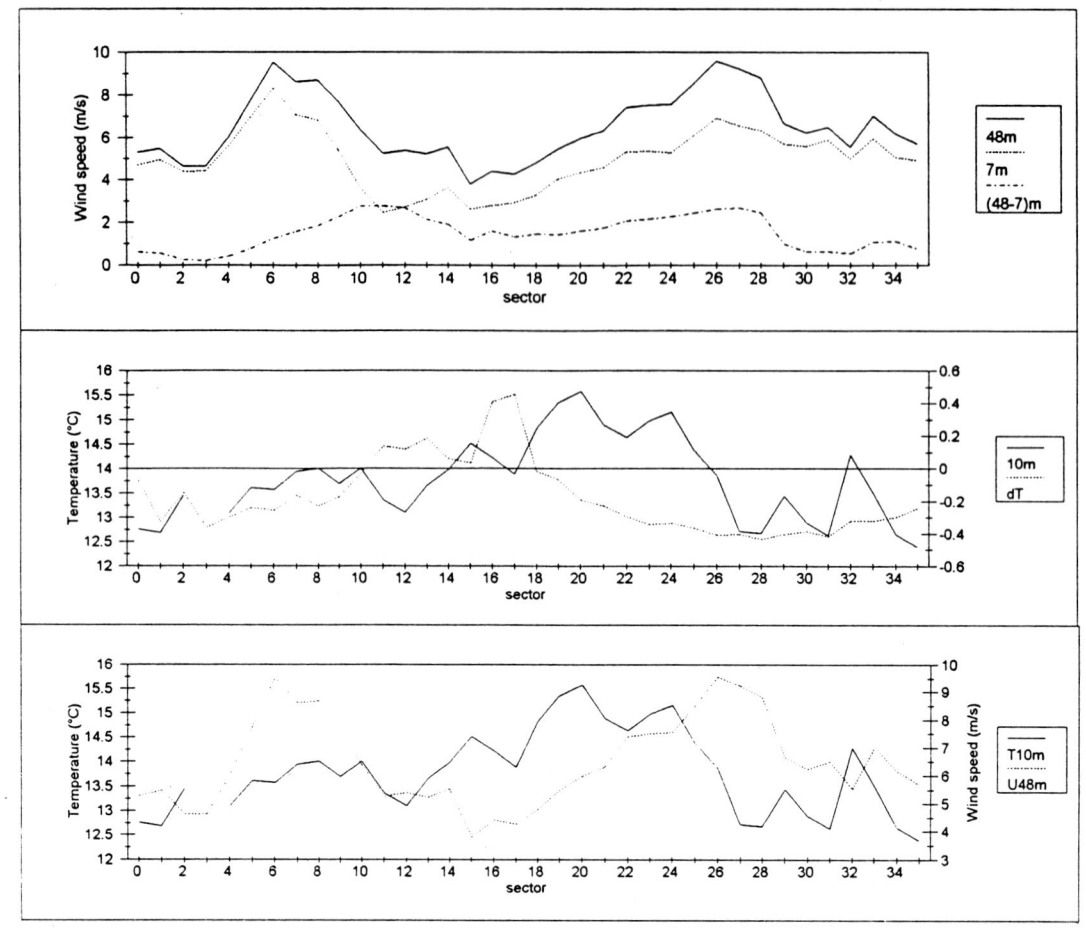

Figure 5. Wind speed profiles from different fetches. Solid line shows observations, dashed line is a logarithmic best fit. Roughness lengths (z_0) determined from profiles are also shown.

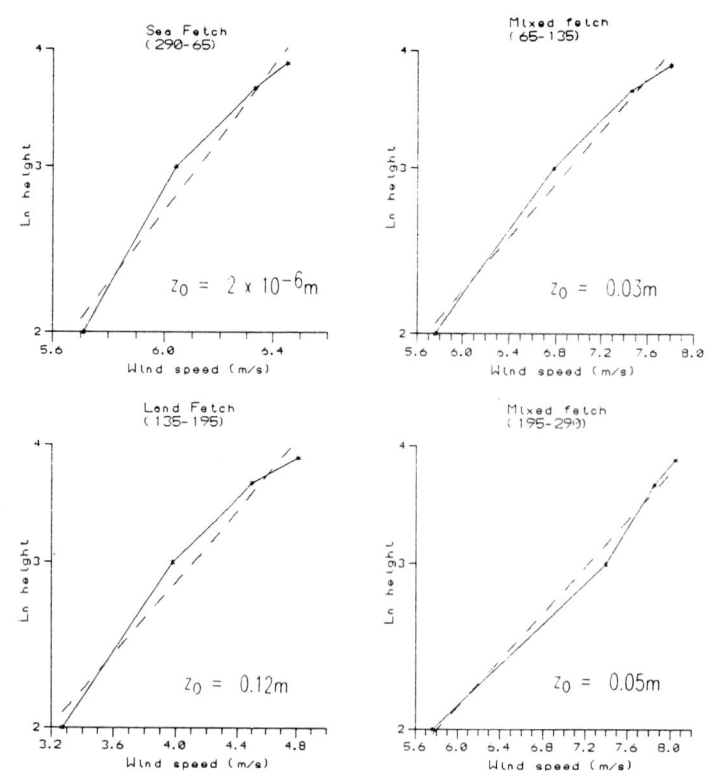

Figure 6. Average turbulence intensity by sector (10° bins) at 48m and 7m height.

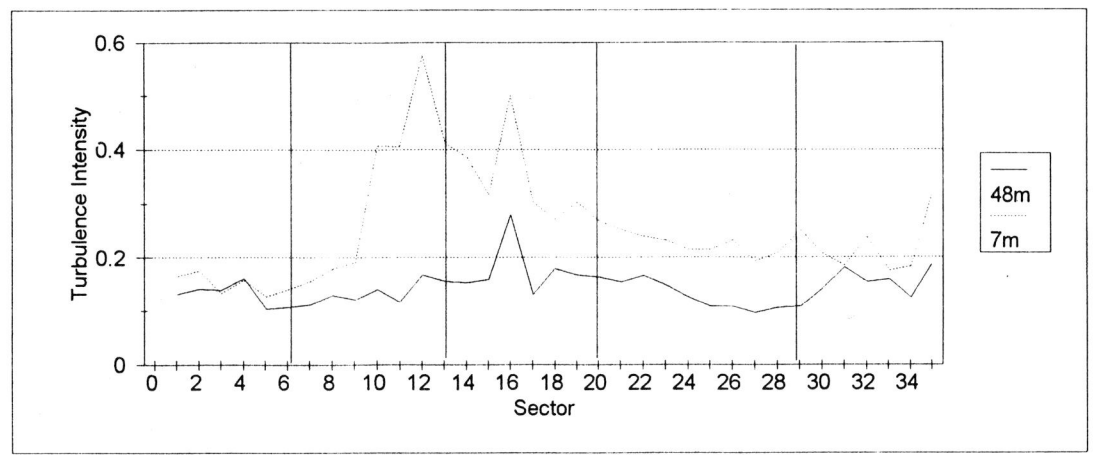

Figure 7. Ratio of wind speed and turbulence measured at the sea and land masts by direction.

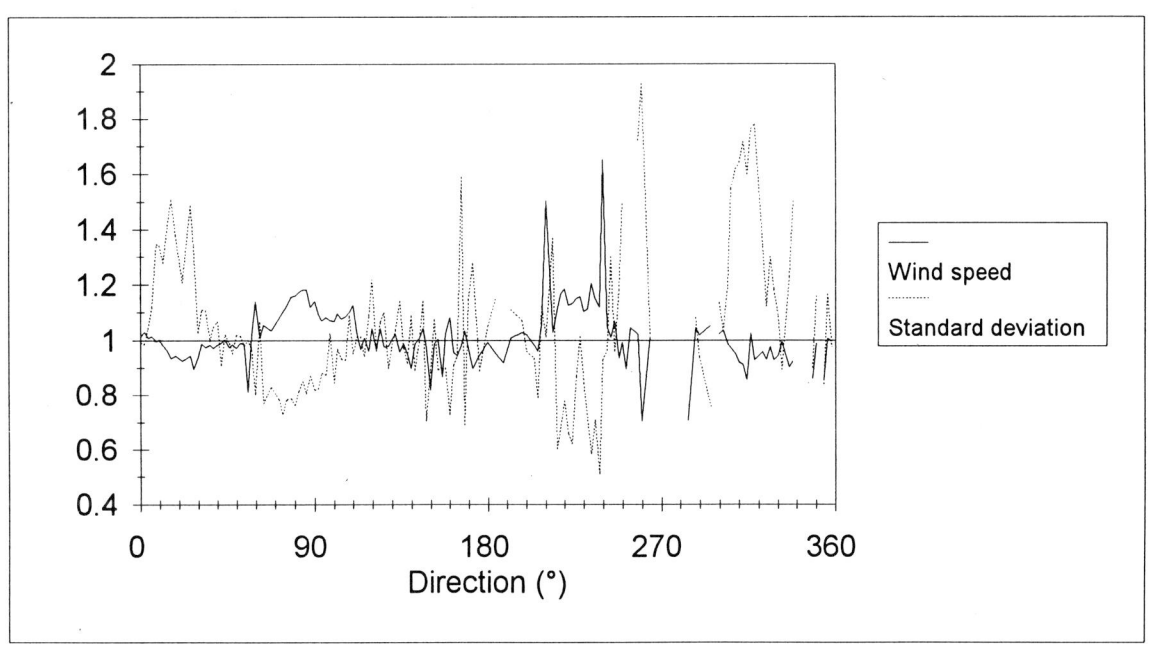

Estimating the wind energy potential

D. M. HOLLIS, MSc
The Met. Office, UK

SYNOPSIS In order to estimate the wind energy potential of a given site it is first necessary to know the local wind climatology. The UK Met Office currently uses two techniques for deriving the climatology at a location where there is no long observational record. These are a boundary layer flow model, WA^{S}P, and an in-house version of the widely used measure-correlate-predict (MCP) methodology. A description is given of how the Met Office has implemented these two approaches. Four case studies are then presented in which the relative accuracies of the two techniques are assessed.

1. INTRODUCTION

When assessing the suitability of a location as a site for wind energy generation there are a number of, sometimes conflicting, factors which need to be taken into consideration. These include the accessibility of the site, the predicted wind energy potential and environmental factors such as noise and the impact on the skyline. Perhaps the most fundamental of these is the wind energy potential since it is pointless to install a turbine on an otherwise ideal site if the expected output is insufficient to make it financially viable.

The first step in estimating the energy yield of a site is to determine, as accurately as possible, the local wind climatology. The basic characteristics that are required are the overall mean wind speed and the frequency with which any particular speed is likely to occur. The former will give a basic indication of whether or not a site is likely to be suitable while the latter can be combined with the power curve for a selected turbine to give a more accurate estimate of the potential of the site. The frequency with which the wind comes from any given direction is of less interest since turbines are designed to be able to rotate into the wind. However it can be important if a wind farm is planned since shadowing effects can be significant.

The most accurate way to derive the wind climatology of a proposed turbine site is to observe it directly i.e. to install an anemometer on the site and carry out measurements of the wind speed and direction for , say,

10 years. Clearly this is impractical. However such wind observations are already being made at 190 locations around the UK. The data for these and other anemograph stations that have now closed have been archived by the UK Met Office such that the database now contains over 160 records of 10 years or more. The existence of this archive makes it possible to estimate the climatology at an intermediate site where no long record exists. This may be achieved in three distinct ways.

The first of these is simply to identify an anemograph station that is in a geographically similar location to the test site. It is then assumed that the unadjusted climatology of the reference site is representative of the test site. However there are many cases where no suitable reference station exists. The second method therefore involves adjusting the reference climatology for differences in the terrain between the reference and test sites. This may be achieved using a computer model, based on physical principles, of how the airflow near the ground is affected by the underlying terrain. The third alternative is known as the measure-correlate-predict (MCP) approach. This involves taking measurements of wind speed and direction for a short period (6 months, say) at the test site and then correlating these with observations for the same period at a reference site. The derived relationships are then used in conjunction with a much longer period of data from the reference site to estimate the long term climatology at the test site.

Clearly before choosing one of the above methods it is desirable to have some idea of their relative accura-

91

cies and costs. In addition it would be useful to have guidelines on how to obtain the optimum results using a given technique and in which circumstances it is likely to perform well or badly. The remainder of this paper goes a small way to answering these questions. A brief description is given of how these estimation techniques have been implemented in the Met Office and this is followed by a number of case studies.

2. THE MET OFFICE APPROACH

The Met Office uses two techniques for estimating the local wind climatology. The first of these is the Wind Atlas Analysis and Application Program (WA^{S}P), a PC-based model available commercially from the Risø National Laboratory in Denmark. This numerical model is based on the linear theory of Jackson and Hunt (1). Details of the BZ flow model which forms the heart of WA^{S}P can be found, along with some verification statistics, in the European Wind Atlas (2). The principles of operation of WA^{S}P are straightforward. The primary input is a description of the wind climatology at the chosen reference site either in the form of a series of observations or, more conveniently, as a frequency table of wind speed against wind direction. This climatology is adjusted, using a description of the terrain surrounding the reference site, to give the expected conditions at a number of standard heights above flat homogeneous terrain. The terrain information can include orography, surface roughness and details of individual obstacles, as appropriate. The resultant adjusted climatology is termed a 'wind atlas'. Finally, a description of the terrain around the test site can be used to modify the wind atlas and give the required estimate of the test site climatology at a specified height above ground level.

There are thus three sets of information required to run WA^{S}P. These are the wind climatology at the reference site and the terrain descriptions at the reference and test sites. The first of these is obtained from the Met Office archive of wind data referred to above. Standard programs exist to produce a wind frequency analysis of speed against direction. For the purposes of WA^{S}P the wind direction is divided into twelve 30° sectors (centred on 0°, 30°, 60° etc) while the wind speed is divided into 50 one knot classes (0, 1, 2, ..., 49 knots). The class corresponding to calms is subsequently deleted because WA^{S}P requires a wind direction to be associated with every speed. Where possible the analysis is based on 10 years of hourly mean values. In areas where suitable data are sparse then shorter records or 10 minute mean values will be used instead although it is recognised that this will reduce the accuracy of the final result. The cut off at 49 knots is generally not a problem since reference stations are frequently located in lowland areas where an hourly mean speed of 50 knots is rarely experienced. If necessary the data can easily be grouped into two knot classes.

The terrain information used in WA^{S}P is obtained from two high resolution datasets each covering the whole of Great Britain. The first of these contains grid point values of the height of the ground above sea level. It has a horizontal resolution of 6 seconds of longitude by 3 seconds of latitude which is equivalent to approximately 100 metres in each direction. The heights are recorded to the nearest metre. The second dataset contains information about the local land use including point, line and area features. Programs have been written to extract values from these datasets and convert them into a form suitable for input to WA^{S}P. For the topographic data this involves interpolating from latitude-longitude coordinates to National Grid coordinates and then contouring the resultant grid using a proprietary software package. Of the land use data only areal features (forests, towns etc) are extracted because WA^{S}P is unable to deal with point or line elements. Appropriate roughness lengths are then assigned to these areas. The final dataset, for input into WA^{S}P, contains series of coordinate pairs that describe either height contours or the boundaries between areas of differing roughness lengths. The programs are designed to allow the user to specify separately the areas of topographic and land use data to be extracted and also, for the former, the horizontal grid point spacing and vertical contour interval. However the total amount of data that may be input is limited by WA^{S}P. Examples of typical areas and contour intervals are given in the next section. In all cases the surface roughness data have been restricted to a 10 km square centred on the point of interest as it is felt that the effects of roughness changes beyond this are relatively minor.

The MCP methodology used by the Met Office follows the general approach described in §1. Data for a short period from the test site are correlated with corresponding values from a reference site. Two sets of regressions are carried out. The first is of the test site wind speed against the reference site wind speed. The second is of the difference in wind direction between the two sites against the reference site wind speed. Separate calculations are done for each reference site wind direction (10°, 20°, ..., 360°) and thus a total of 72 regression equations are produced. If there are fewer than 10 pairs of values for any given direction then no equation is derived and a standard assumption is used instead i.e. that the speeds and directions at the two sites are equal. These equations can then be applied to a much longer period of data from the reference site. Thus for each observation it is possible to estimate the corresponding value at the test site. However it is known that if the wind speed is X knots at the reference site then it is not always Y knots at the test site. This natural variation is modelled by calculating the variance of the test site data about the regression line for 5 knot ranges in the reference site wind speed. The predicted value is then drawn from a normal distribution with a mean equal to the value given by the regression equation and a standard deviation based on the observed variability in the test site data. The program was originally written as a general purpose technique for estimating local wind climatologies and consequently all observations are included in the analysis i.e. there is no attempt to only model values that lie between the cut-in and cut-out speeds of a typical turbine. Although, for the purpose of wind energy, the program is used with hourly mean values it may legitimately be applied to 10 minute mean or gust data.

3. CASE STUDIES

The following is a preliminary assessment of the relative strengths and weaknesses of the two techniques used by the Met Office to estimate the local wind climatology. Four case studies are considered. In each the approach is the same. A test site that already has a long observational record is selected from the Met Office archive. A second station, the reference site, is then chosen and the data from there are used to estimate the climatology at the test site using both the WA^{S}P and MCP approaches. The predicted climatology is then compared with the observed climatology at the test site.

3.1 The Test and Reference Sites

a) Leeming and Finningley

The first test site is Leeming in North Yorkshire. It is located in the Vale of York, a broad valley some 25 km across. To the west are the Yorkshire Dales which rise to around 700m while to the east lie the North York Moors rising to 430m. Leeming itself is just 32 m above sea level. It is notable that there is a pronounced peak in the frequency of winds coming from the south-southeast which corresponds to the orientation of the Vale of York. For the WA^{S}P analysis two datasets were extracted. The first covers a 20 km square centred on Leeming and has a vertical contour interval of 10m while the second covers an 80km square and has a 50m contour interval.

The climatology at Leeming was estimated using data from Finningley, South Yorkshire, for the period 1983-92. Finningley lies approximately 100km to the south-southeast of Leeming. It is surrounded by relatively flat low-lying countryside such that it is necessary

to go some 40km west before there is any appreciable gain in height. Consequently the climatology is essentially unaffected by major orographic features. A 30km square of topographic data with a 10m contour interval was used for input into WA^{S}P.

b) Lowther Hill and West Freugh

The second test site was Lowther Hill, Strathclyde, situated at 727m above sea level in the Scottish Lowlands. Data covering an 11km square and with a 20m contour interval were used to describe the topography around Lowther Hill. The wind climatology at this site was estimated using data from West Freugh, Dumfries and Galloway, for the period from March 1982 to February 1989. West Freugh lies near the coast and is some 95km west-southwest of Lowther Hill. The topography around West Freugh was described using a 20km square of data and a 10m contour interval.

c) Great Dun Fell and Carlisle

Great Dun Fell, Cumbria, is the highest of the four test sites at 847m above sea level and is situated close to the highest point of the Pennines. It lies on the top of an escarpment that runs northwest-southeast. To the southwest the ground falls away to around 100m above sea level in the Eden Valley over a distance of 10km while to the northeast the slopes are more gentle. Beyond the Eden Valley lies the Lake District, an extensive area of high ground rising to over 950m in places. The Great Dun Fell topography was described using data for a 14km square and with a 20m contour interval.

The reference station used was Carlisle, Cumbria, located 45km northwest of Great Dun Fell and at just 26m above sea level. Wind data was used for the period 1985-92. Although the area within 15km of Carlisle is relatively flat, the wind climatology is noticeably affected by the Solway Firth and the funnelling effect of the Scottish Lowlands to the north and the Lake District to the south. Since it is impractical to input such large scale topography into WA^{S}P it was decided to use a 22km square of data with a 10m contour interval thus describing just the immediate area around Carlisle.

d) Bala and Valley

The last test site is Bala, Gwynedd, located at the northeastern end of Lake Bala in North Wales. The lake is around 1km wide and 5km long and is orientated southwest-northeast. Thus the anemometer at Bala lies at the bottom of a narrow, well defined valley in the middle of a large area of high ground. For the WA^{S}P analysis an 11km square of data and a 20m contour interval were used. The climatology at Bala was estimated using data from Valley, Anglesey, for the period 1978-87. Valley, which lies 95km northwest of Bala, is a well exposed site being both close to the sea and away from any areas of high ground. Data were extracted for a 20km square and using a 10m contour interval.

3.2 The Mean Wind Speed

Table 1 shows the observed and predicted mean wind speeds in knots for each pair of sites.

Table 1 Mean wind speeds

	WA^{s}P	MCP	Observed
Finningley	-	-	8.1
Leeming	7.8	8.8	8.2
West Freugh	-	-	10.3
Lowther Hill	16.7	18.1	19.9
Carlisle	-	-	8.5
Great Dun Fell	13.4	21.8	21.9
Valley	-	-	13.0
Bala	7.6	8.3	7.9

From this it can be seen that for a lowland site in relatively open terrain such as Leeming, it can be better to simply take the unadjusted mean from a reference site in similar terrain. Despite this the errors introduced by using either of the estimation techniques are relatively small.

For hill top sites such as Lowther Hill and Great Dun Fell it is clear that some adjustment is essential and that MCP performs significantly better than WA^{s}P. Of the two WA^{s}P estimates the one for Lowther Hill appears to be by far the better being 16% too low compared with 39% for Great Dun Fell. However it is notable that both estimates are approximately 60% greater than the observed speeds at their respective reference sites. This represents about the maximum speed-up that may be predicted using a linear model such as WA^{s}P i.e. the greater accuracy at Lowther Hill is due at least partly to the observed speed at West Freugh being proportionately higher than for Great Dun Fell and Carlisle. Whether better results could be achieved using a more sophisticated model is unclear. The theoretical basis of such models is that the degree of speed-up at the top of a hill or ridge is related to the gradient of the upwind terrain. Figure 1 shows a cross-section from the Eden Valley to Great Dun Fell. Taylor and Lee (3) state

that the speed-up at the top of a ridge may be estimated by 2h/L where h is the height of the hill and L is the distance from the summit to the point where the terrain has dropped by half the hill height. If this is applied to Great Dun Fell then the wind speed at the top is predicted to be 68% greater than that in the valley, comparable with the results from WA^{s}P. Thus, although Great Dun Fell is at nearly 850m above sea level, simple boundary-layer flow theory does not predict such high speeds as are actually observed at the summit. A possible explanation is that air coming over the Lake District (which lies upwind with respect to the prevailing wind direction at Great Dun Fell) does not in fact descend the Eden Valley before reaching the Pennines i.e. the wind at Great Dun Fell is more closely related to the geostrophic wind than to the wind in the valley and consequently the slope of the terrain is not a significant factor in determining the observed speed. A model such as WA^{s}P is clearly incapable of coping with this type of situation and thus the poor results are not so surprising.

The situation at Bala is almost the reverse with the test site being much more sheltered than the reference site at Valley. However both models perform well and predict the observed speed at Bala to within around 5%.

3.3 The Wind Speed Distribution

For wind energy applications probably the most important aspect of the climatology is the frequency distribution of wind speed. In the case of WA^{s}P a Weibull distribution is fitted to the input data and consequently the output statistics reflect this. No such assumption of distribution is necessary in the MCP approach. Both techniques can be used to estimate the frequency with which the wind speed may lie in any given range. For the purposes of this study it was decided to keep things simple and group the data into Beaufort classes i.e. Force 1 (1-3 knots), Force 2 (4-6 knots) etc. Speeds of Force 8 or more (34 knots or more) were grouped together.

In the case of Leeming and Finningley the observed distributions are already very similar. Neither technique makes a significant alteration to the Finningley climatology and so not surprisingly both estimates are very good.

A more rigorous test is that of Lowther Hill and West Freugh. Although Force 4 is the most common class at both sites the complete distributions are very different as would be expected from an inspection of the mean speeds alone. Nevertheless both estimation techniques perform well. Of the two the MCP approach is more successful and in Figure 2 the estimate obtained is compared with the two observed climatologies.

The MCP estimate for Great Dun Fell is also very good despite the even greater disimilarity between the site and Carlisle. In contrast the distribution generated using WA^{S}P is a long way from what is observed at Great Dun Fell and is probably closer to the original distribution at Carlisle. This is consistent with the very poor estimate of mean wind speed that was obtained using WA^{S}P.

Finally both techniques are reasonably successful at predicting the speed distribution at Bala, particularly for the higher wind speeds. At lower speeds the MCP approach is slightly the better of the two. This level of agreement was reflected in the good estimates of the mean speed.

In general the MCP technique is more successful than WA^{S}P at estimating the wind speed distribution. This is probably due mainly to the greater flexibility of the former as compared with the rigid assumption of a Weibull distribution necessary in a model such as WA^{S}P. Despite this the only really poor result using WA^{S}P was for Great Dun Fell.

3.4 The Wind Direction Distribution

Although of less significance to wind energy assessments, it is worth looking briefly at the ability of these two techniques to model the frequency distribution of wind direction.

It was noted earlier that at Leeming there is a significant peak in the frequency of winds coming from the south-southeast in addition to the more ususal peak for westerly winds. Figure 3 shows the estimate obtained using WA^{S}P. It can be seen that WA^{S}P has left the Finningley climatology virtually unaltered and has failed to reproduce the second peak. This result was the same using both the 20km and 80km terrain datasets for Leeming. It is likely that this peak is due not only to the channelling of the wind between the high ground either side of Leeming but also to the formation of katabatic winds in the Vale of York. This may partly account for the poor result as models such as WA^{S}P are generally unable to deal with thermal effects of this sort. In fact it seems to be generally true that, except in sharply defined terrain such as at Bala, WA^{S}P is unable to make more than relatively minor changes to the direction distribution. Even at Bala, although the peak direction is correctly predicted, the associated frequency is still much lower than is observed (19% instead of 33%).

The MCP approach is again on the whole more successful than WA^{S}P. It correctly predicts the existance of a second peak at Leeming and the high percentage of along valley winds at Bala. Figure 4 shows a plot from the MCP program of the mean difference in wind direction (Valley minus Bala) against the wind direction at Valley. From this it can be seen that the magnitude and sense of the differences are consistent with the wind being deflected by the topography around Bala.

4. CONCLUSIONS

The sites used in the case studies were deliberately chosen to test the limits of the two estimation techniques. Despite this it has been shown that both models are capable of producing satisfactory results in difficult terrain. Overall MCP performs better than WA^{S}P. Although this is not surprising, the greater accuracy must be weighed against the additional cost of monitoring a site for perhaps 6 months.

In order to draw more general conclusions about the strengths and weaknesses of such estimation techniques it is necessary to look at a much larger sample of sites. The Met Office will shortly be undertaking a detailed study of this sort which will be funded by ETSU. The aim will be to arrive at some general guidelines on how to optimise the accuracy of each technique and to identify those situations where a given approach should be used with caution.

REFERENCES

(1) JACKSON, PS and HUNT, JCR, Turbulent wind flow over a low hill. Quart. J. R. Met. Soc., 1975, 101, 929-955.

(2) TROEN, I and PETERSEN, EL, European Wind Atlas, 1989, pp656 (Risø National Laboratory).

(3) TAYLOR, PA and LEE, RJ, Simple guidelines for estimating wind speed variations due to small scale topographic features. Clim. Bull., 1984, 18, 3-32.

Figure 1 Cross-section from River Eden to Great Dun Fell

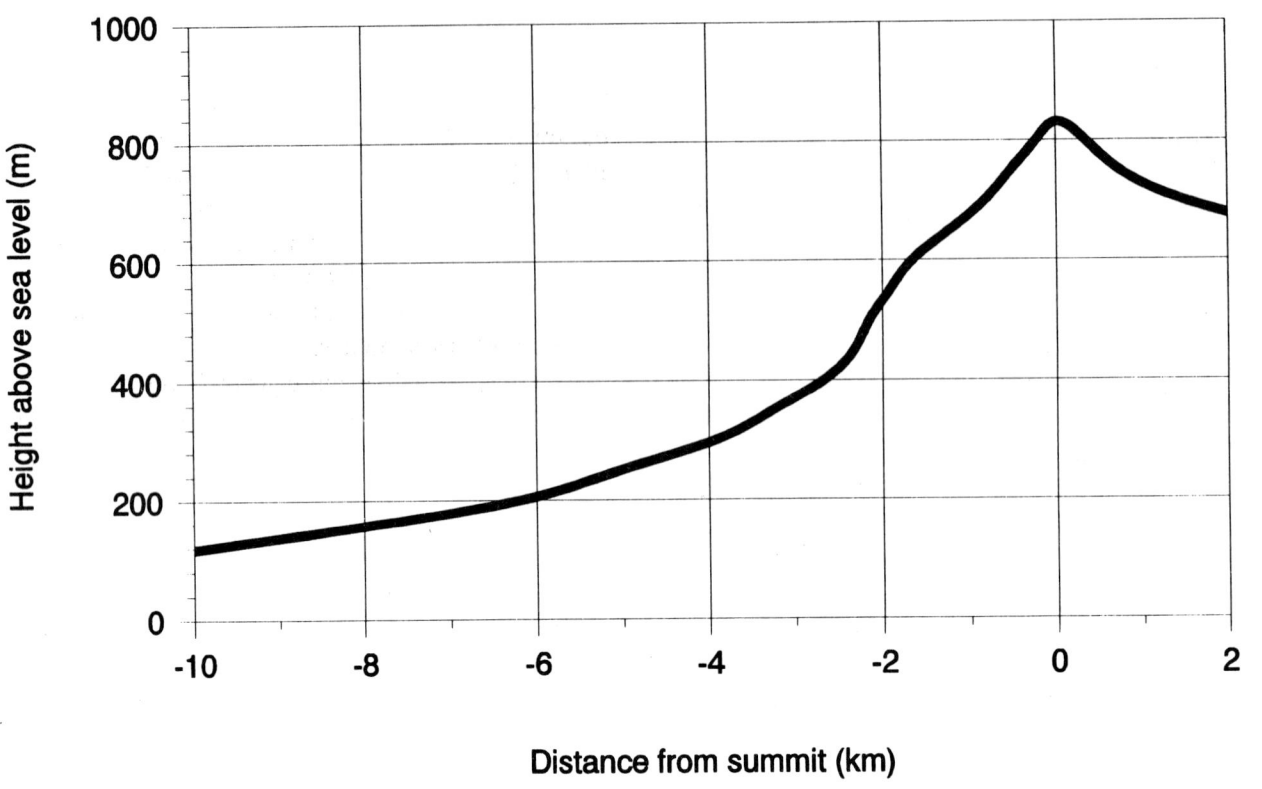

Distance from summit (km)

Figure 2 Wind Speed Distribution at Lowther Hill

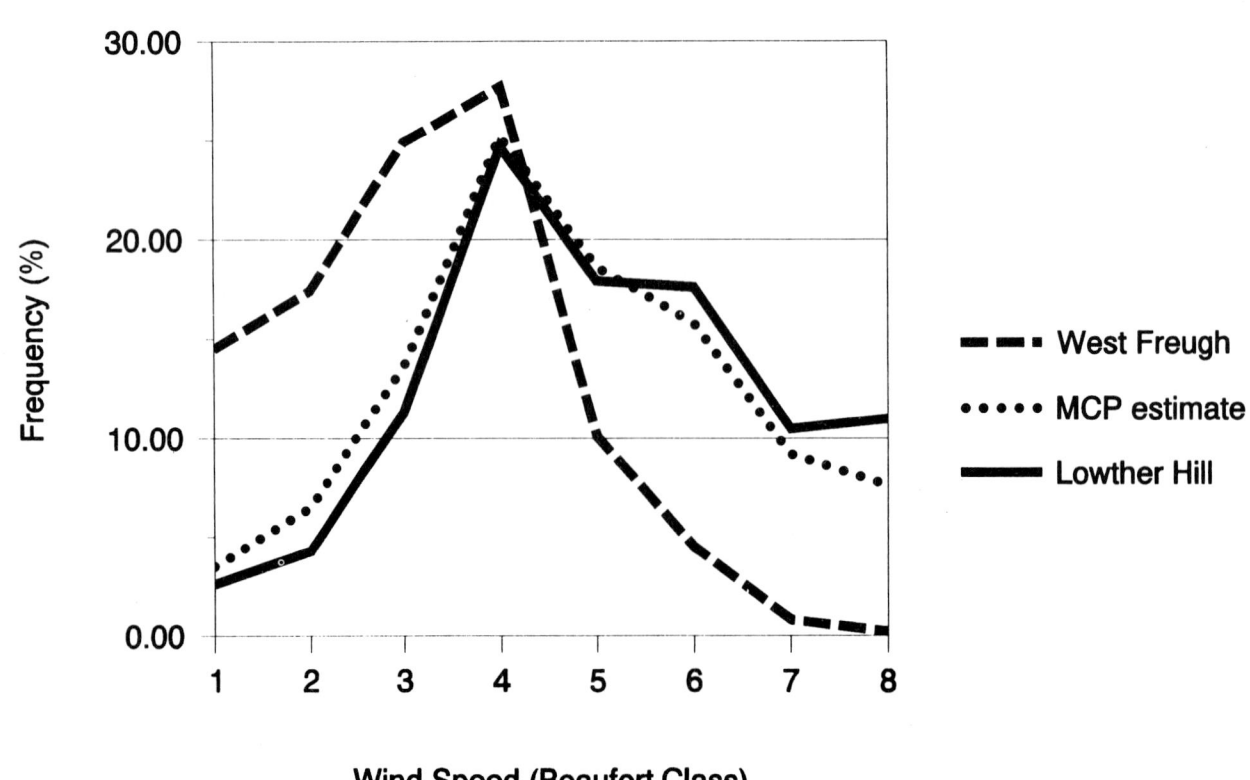

Wind Speed (Beaufort Class)

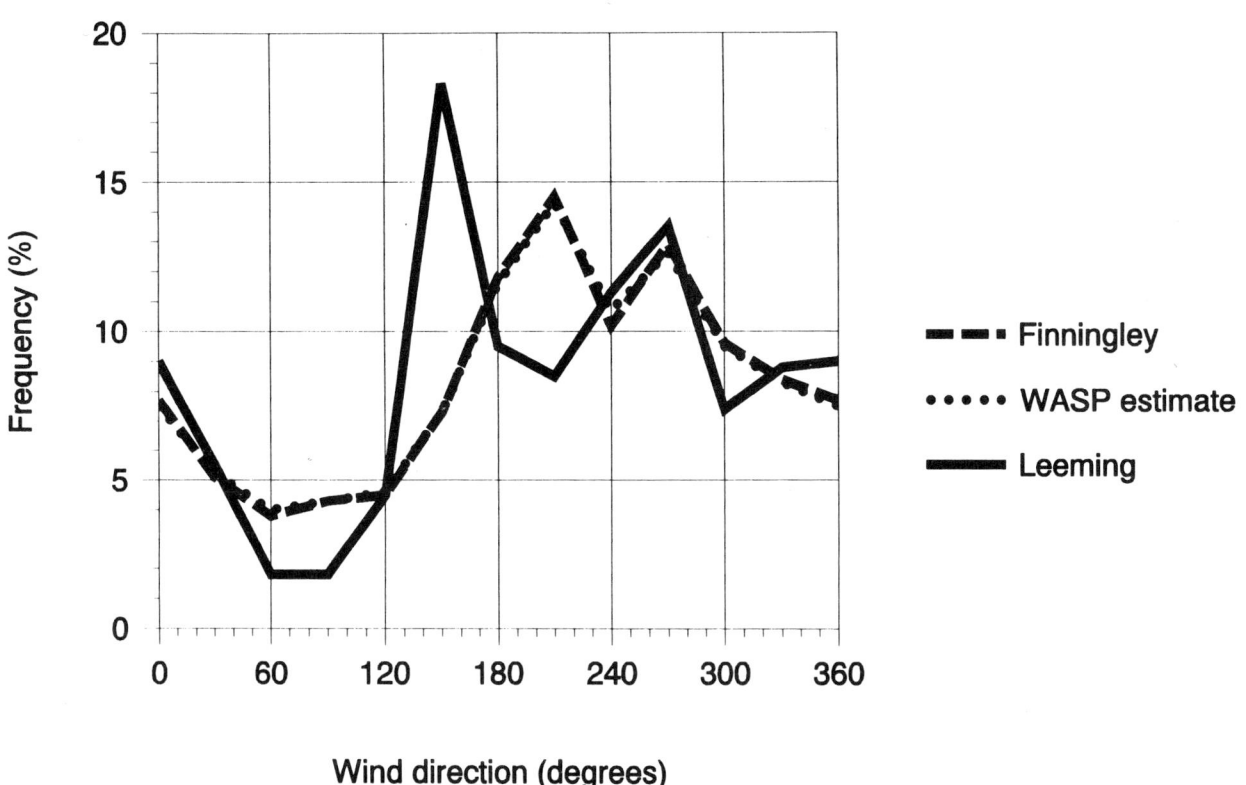

Figure 3 Wind Direction Distribution at Leeming

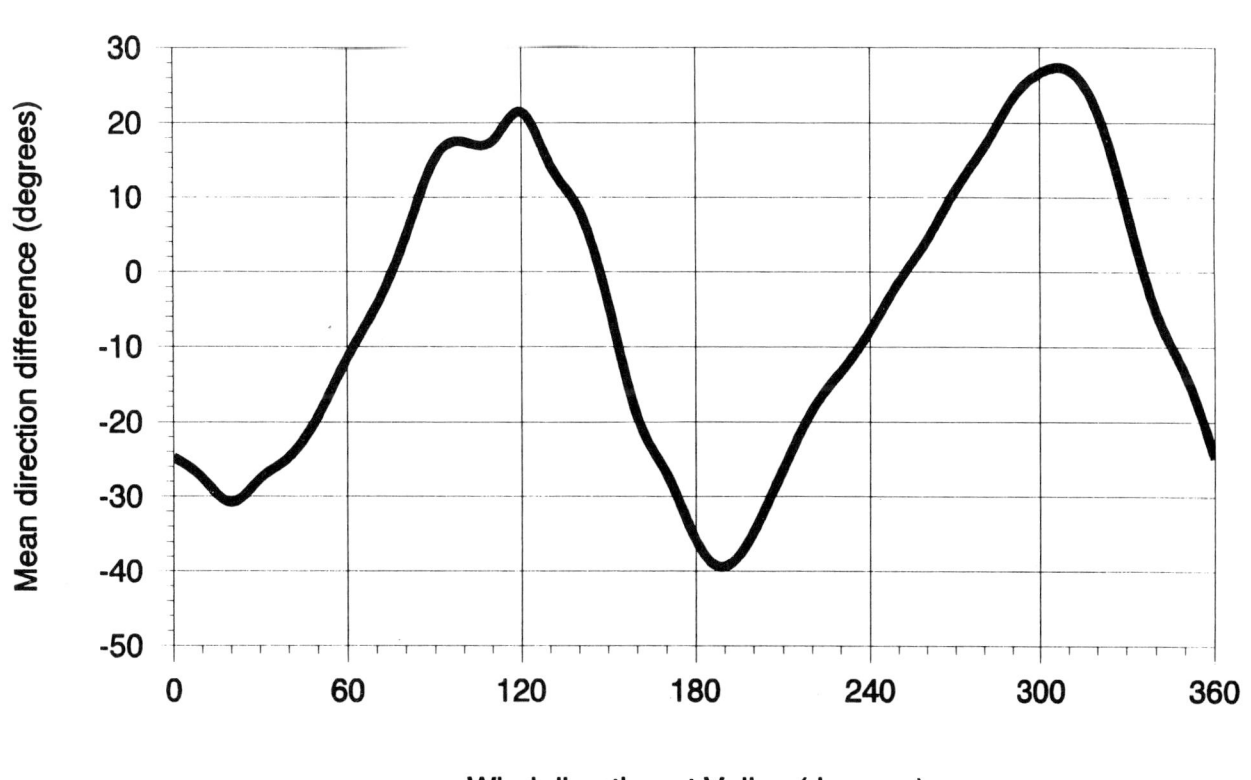

Figure 4 Mean Direction Difference (Valley - Bala)

On the turbulent structure of wind turbine wakes over rough terrain

C . G. HELMIS, K. H. PAPADOPOULOS, A. T. SOILEMES, P. G. PAPAGEORGAS, and
D. N. ASIMAKOPOULOS
Athens University, Greece

SYNOPSIS The near wake turbulent structure downwind of a medium-sized,horizontal axis wind turbine at a distance of one rotor diameter is discussed. Observations show that the turbulent field varies from the edge to the center of the wake and strongly depends on the turbine's power coefficient. Consideration of the perturbation produced by the tower construction is crucial in the interpretation of results.

NOTATION

D	turbine rotor diameter (=17m)
$R=D/2$	
H	hub height (here,25m)
f	rotation rate of the blades
λ	tip speed ratio $(=2\pi fR)$
U	wind speed
c_p	turbine power coefficient
u,v,w	longitudinal, lateral and vertical wind component
σ_x	standard deviation of parameter x
$\overline{u'w'},\overline{v'w'},\overline{u'v'}$	shear stresses
I_U	turbulent intensity ($\sigma U/U$)

1 INTRODUCTION

An experimental campaigne was conducted on the Samos Island Wind Park in August,1991, with the purpose to reveal the mean and turbulent structure of the near wake region of a single wind turbine operating in complex terrain.

Published field measurements usually concentrate on the decay rate of the velocity deficits in the far-wake region in relation to power production optimization of wind farms (1)-(6). Wind tunnel studies of the near-wake region are not an accurate representation of the real flow situation (7). The length of this region is usually overestimated (6),(8) suggesting improper parameterization of the turbulence produced by the turbine.

Measurements at distances as close as 1D downstream are very limited (6),(9). The turbine operation leads to turbulence increase through tip and hub-vortices degeneration and shear generation at the edges of the wake. This extra turbulence is important mainly in the near wake region. The present study reports on the turbulence generated by the turbine (both integral quantities and spectral analysis) and on the effect of the tower shadow.

2 SITE DESCRIPTION-INSTRUMENTATION

Samos Island lies in the eastern Aegean Sea. During the warm, dry season it is under the influence of northerly strong winds (the 'etesians'). The Wind Park has been established in the western part of the island 390m above mean sea level. It is situated on the top of a ridge confined by the island's two major mountain ranges (Fig.1).

To the north, the ridge drops along a 17° slope for the first 200m in all directions encountered during the experiment. The Wind Park comprises nine three-bladed, horizontal axis Vestas WM19S wind turbines. The turbines reach their rated 100KW power at 13m/s. The cut in and cut out wind speed is 3m/s and 27m/s, respectively. The 24m-high tower is of a concrete conical type with a base diameter of 1.9m narrowing to 1.1m. The blades' rotational rate is maintained at f=48r/min through the operational range of incident wind speeds. Consequently, c_p will be related to U via its dependence on λ. The maximum c_p value is attained at hub-height wind speeds of 8-10m/s.

Measurements comprise two instrumented masts: a 12m-high upwind one and a 30m-high downwind one (Fig.2). The instrumentation concerning the wind measurement is given in Table 1. Additionally, platinum resistance thermometers were installed at 5 levels ($A_3,A_6,A_{12},B_{15},B_{25}$).

Measurements cover the period 16/8/91-24/8/91, including a sensors' intercomparison test. Due to unrecoverable malfunctioning of the B_{25} propeller anemometer, data from only the propeller pointing to the prevailing wind direction (350°) are presented.

The masts' instruments were sampled and data stored once per second. Measurements were taken with the wind turbine both in and out of operation (the wake and non-wake dataset, respectively). Atmospheric stability ranged from slightly unstable to slightly stable.

Table 1 Instrumentation of Masts A,B

Mast	Wind Speed	Wind Direction
A	3m,6m,12m	6m
B	10m,15m19m	19m,29m
	25m,29m	

* $A_3,A_6,A_{12},B_{10},B_{19},B_{29}$: cup anemometers.
* B_{15},B_{25} : 3D propeller anemometers.

A detailed analysis of the field measurements (10),(11) has shown that i) the wind turbine operates in a region of increased wind speed and low turbulent intensity (5%-10%), ii) the A_{12} anemometer may be considered a reliable reference for the wind speed and I_U upstream of the wind turbine and iii) the wind directions (measured at mast A) for which the levels of mast B lie inside the wake are 330°-360° for B_{15} and greater than 327° for B_{19},B_{29}, while the B_{10} level lies outside of the wake.

3. WIND TURBULENCE IN THE WAKE

Figure 3 presents the mean normalized, with respect to the A_{12} anemometer, turbulent intensities at mast B, based on 1-min samples. Background (when the turbine rotor is parked) I_U values increase with height through mast B (Fig.3) and exceed those of mast A. This is due to the speed-up of the wind flow over the hill leading to a wind speed maximum at around 15m and to a decrease of σ_U at heights below 15m, attributed to rapid distortion processes (10). Besides, the presence of the tower\nacelle structure leads to the maximum in I_U along the central directions. For the wake dataset, I_U attains a broad (at 19m ,29m) or a rather narrow (at 15m) maximum. Substracting the two curves would be simplistic, because the extra turbulence induced by the turbine assists in the faster dissipation of the tower-related perturbation of the wind field, indicating nonlinear interaction.

The differences of the wake and no-wake curves for B_{19} and B_{29} probably demonstrate a marginal reduction of I_U in the centerline, with evidence of a small increase toward the blade tips for B_{19}. A clear increase of turbulence occurs at B_{15}, which is always at the wake boundaries and is affected by tip vortices. Although the frequency response of the propeller anemometer (at B_{15}) is better than that of the cups this does not seem to account for the observed features, as it is shown in the next section. Taylor (6) noticed that σ_U in the center of the wake was hardly increased above ambient level and attributed it to the flat velocity profile observed in the same region. These observations are in agreement with our results (Fig.4b), where the existence of a central flat region is obvious. This region is extended at 29m, where (Fig.3d) I_U is never increased above its background level. On the other hand, at 19m an increase in I_U for directions greater than 355° is apparent in accordance with the velocity ratio profile (Fig.4a). The B_{19},B_{29} levels are close to symmetrical with respect to H, therefore the solid curves in Fig.4a,b should be similar. However, the 'peaked' pattern at 19m is caused by a combination of a pattern similar to that of

29m and of the tower shadowing (B_{29} lies above H). The preceding results seem to indicate that the shear-related turbulence dominates over the mechanical turbulence produced by the rotor. Out of the wake, but close to its boundary, I_U (Fig.3a) is reduced relative to the background value. The observed reduction of turbulent intensity at B_{10} for the wake dataset may be related to diffusion of the tower-perturbation by the extra turbulence at the wake periphery.

The mean crosswind profile of $\overline{u'w'}$ for 15-min samples of the B_{15} propeller anemometer, which usually lies close to the wake boundaries, is presented in Fig.5. In accordance with (2), there is a significant increase of $\overline{u'w'}$ starting at 332°, where B_{15} enters the wake, and becoming maximum at 354° implying downward transfer of momentum, that is along the vertical gradient of horizontal momentum.

Table 2 presents results for a case (20/8/91) when data were available both before (OFF: 10:57-11:31) and after (ON: 11:50-12:24) starting the parked turbine. The wind direction was always close to the centerline (OFF:351°,ON: 352°).

Table 2 Parameters variation at mast B for two consecutive periods when the wind machine is in and out of operation.

	A_{12}		B_{10}		B_{19}		B_{29}	
	OFF	ON	OFF	ON	OFF	ON	OFF	ON
U(m/s)	15.7	14.7	14.7	14.3	13.9	11.2	13.8	11.1
σ_U(m/s)	1.45	1.51	1.61	1.71	1.65	1.37	1.54	1.15

				B15			
	U	σ_u	σ_v	σ_w	u'w	'v'w'	u'v'
	(m/s)				(m/s)²		
OFF	14.2	1.46	1.35	0.89	0.076	-0.189	0.685
ON	12.9	1.78	1.42	1.04	-0.688	-0.290	0.976

Confirming the conclusions drawn from Fig.3, it is seen that 30 per cent, 20 per cent reduction in σ_U at 29, 19m levels, respectively, are observed after the wind turbine is set in operation. Substantial increases of 25 per cent for σ_u and σ_w are observed at 15m, accompanied by virtual constancy of σ_v. All three covariances at the same level are remarkably increased. Downward momentum flux is implied at the lower boundary of the wake, as was also found in (2).

4 SPECTRAL ANALYSIS

Wake spectra of the three wind components demonstrate a "transfer" of energy from low to high frequencies (2). Spectra in the near-wake region will be affected as follows (6):
a) at the wake boundaries, intense tip vortices increase spectral amplitudes around the blades rotational frequency (and its harmonics).
b) shear generated in the process of wind energy extraction should produce turbulent motions on the order of the wake extent.
c) low frequencies will lose energy in accordance with the velocity deficits due to the extraction of wind power.

The present data are unable to resolve effect (a).

Composite wind speed spectra for three directional groups for the no-wake and wake datasets are formed. Because the wake deficit is mainly determined by the incident wind speed (11), each directional group is further subdivided according to the A_{12} wind speed (Table 3). It is noted that the B_{15} spectra refer to the longitudinal component u.

Table 3 Data categorization for spectral analysis.

	Wind Direction	Wind speed (A_{12})
Set 2A	328°-338°	<15m/s
Set 3A	338°-348°	<15m/s
Set 3B	338°-348°	15-20m/s
Set 3C	338°-348°	>20m/s
Set 4A	348°-358°	<15m/s
Set 4B	348°-358°	15-20m/s

For the 2A set (Fig.6) B_{15},B_{19},B_{29} more or less lie at the wake boundaries and possess increased energy (about 50 per cent) at mid frequencies in agreement with the results of (2). At smaller length scales the effect fades possibly due to the poor resolution of the anemometers.

Moving to set 3A (Fig.7) B_{19}, B_{29} are removed from the boundaries to the interior of the wake and the observed differences between the wake and nowake curves merely reflect those for the A_{12} level. However, B_{15} is now located at the lowest wake boundary and shows a net (ie accounting for the non equivalence of the wake and no-wake datasets as shown by the A_{12} spectra in Fig.7) increase of energy in all frequencies.

For the 4A set (the central one) mast B vertically scans the wake centerline and the spectral contents have accordingly been modified: the B_{19}, B_{29} levels (Fig.8) suffer a clear decrease of their mid frequencies and the u spectra at B_{15} at the wake edge show an even more (compared to Fig.7) intense increase of the high frequency portion. The spectrum for the wind component along the unique performing propeller (pointing into 350°) of the B_{25} anemometer at hub height is also included in this figure, because for the 4A set the wind direction was roughly along its axis. It is also noted that the observed values of the standard deviation of wind direction rarely exceed 10° over the selected averaging period. Increased energy in the smaller length scales at H is observed, resembling the behaviour at the wake boundaries.

At strong winds the wake deficit is weak (11). Comparing spectra of the 3A, 3B sets (Fig.7,9) major changes are seen. B_{15} does not reflect any modifications, since the wake is now weak. The B_{19},B_{29} levels, while unaffected for the 3A set, now demonstrate a spectacular fall in high frequencies. The same holds true at 19m,29m for the 4B set (Fig.10). Because these directions refer to the central part of the wake, the 15m level again enters its edges and reestablishes the increase in small scales, as seen in previous figures.

Spectral analysis under very strong winds (set 3C,Fig.11) confirms the pattern emanating from the preceding discussion: as the wake becomes weaker again, the B_{15} spectrum maintains its background character. The B_{19}, B_{29} spectra show a marginal decrease of the high frequencies.

5 DISCUSSION

The previous exposition of spectra seems adequate to suggest a consistent pattern of turbulence in the near wake region. The value of background measurements in proper distinction of the exact effects of the wind turbine operation is first emphasized; otherwise conclusions are probably disputable. It is here suggested that the turbulence field is a strong function of the position in the wake and of the operational c_p-λ range of the wind turbine.

Published results demonstrate a high frequency energy enhancement linked to degeneration of tip vortices and/or to shear generated (2) turbulence at the wake boundaries. The present results suggest that the high frequency portion of wind speed variance is always enhanced at the wake periphery, although as the wake weakens under strong winds the effect is accordingly reduced. The low frequency portion is unaffected at the wake boundaries. A slight increase of high frequency energy is apparent at hub height, yet for stronger winds it is implied that the spectrum resembles the background one.

Differences from past experimental experience arise when the wake interior is considered. In the moderate wind range spectra are mostly modified close to the centerline (set 4A), where a wide band reduction of wind speed variance is observed. The possible source (6) is reduced shear in the relatively flat central region of the wake (Fig.4). When winds are sufficiently high that an important portion of the incident wind energy remains unused, the wake deficit is weak limiting the importance of the characteristics discussed above.

The low frequency energy in the wake interior is reduced from background values at the moderate wind speeds due to the energy extraction process. This common feature is reversed at the highest wind speeds (Fig. 9,10,11). The wind turbine then operates in the region of decline of the c_p coefficient ($dc_p/dU<0$). In this wind speed range it is obvious that the greater the wind speed the smaller the deficit. Therefore, for any kind of low frequency wind speed fluctuation (so that the response of the rotor is sufficient to follow it) the downstream observed fluctuation will be enhanced (the minimum in the fluctuation cycle will be lessened more than the corresponding maximum). This could result in a low frequency energy increase.

6 CONCLUSIONS

The previous analysis revealed interesting features of the turbulent wind field in the near wake region. Interpretation of results is performed by reference to both upstream and background (rotor parked) downstream measurements. Turbulence varies from the boundaries to the center of the wake. Increased turbulent levels are observed near the blade tips and possibly around the hub height for all wind speeds. In the wake interior, decreases of turbulence are observed in mid frequencies due to shear reduction caused by a flat crosswind velocity profile. The low frequency portion reverses behavior in high wind speeds, where an increase in energy is observed. This is probably due to the shape of the turbine c_p-λ characteristic curve.

ACKNOWLEDGEMENTS

The authors wish to thank the Greek Ministry of Industry, Energy and Technology (General Secretariat of Research and Technology) for financial support.

REFERENCES

(1) BAKER, R.W. and S.N. WALKER Wake measurements behind a large horizontal axis wind turbine. *Solar Energy*, 1984, __33__, 5-12.

(2) HOGSTROM, U., D.N. ASIMAKOPOULOS, H.D. KAMBEZIDIS, C.G.HELMIS and A.SMEDMAN. A field study of the wake behind a 2MW wind turbine.*Atmospheric Environment*, 1988, __22__, 803-820.

(3) TAYLOR, G.J., J.HOJSTRUP and E. LUKEN. Full-scale wake data from the NIBE wind-turbines. Proceedings Euroforum- New Energies Congress 1988, Germany,Vol.3,755-758.

(4) LARSEN,A. and P.VELK. Wind turbine wake interference. A validation study. European Wind Energy Conference, 1989, Glascow, Scotland, 482-487.

(5)ELLIOTT, D.L. and J.C. BARNARD. Observations of wind turbine wakes and surface roughness effects on wind flow variability. *Solar Energy*, 1990, __45__, 265-283.

(6) TAYLOR, G.J. Wake measurements on the NIBE wind-turbines in Denmark. Part 2:data collection and analysis. CEC Contract EN3W.0039.UK(H1),1990.

(7) AINSLIE,J.F. Calculating the flow field in the wake of wind-turbines. Proc. Int.Conference on Wind Farms, 1987, Leeuwarden, Netherlands.

(8) ALFREDSSON, P.H., J.A. DAHLBERG and P.E.J. VERMEULEN. A comparison between predicted and measured data from wind turbine wakes. *Wind Engineering*, 1982, __6__,149-155.

(9) PAPACONSTANTINOU, A. and G. BERGELES. Hot wire measurements of the flowfield in the vicinity of a HAWG rotor.*J.Wind Eng. Ind. Aerodynamics*, 1988, __31__, 133-146.

(10)PAPADOPOULOS,K.H.,A.T. SOILEMES, P.G. PAPAGEORGAS, C.G. HELMIS, D.N. ASIMAKOPOULOS and H.D. KAMBEZIDIS. Surface wind flow characteristics over complex terrain in relation to wind energy applications. ISES Solar World Congress, Budapest, Hungary, August 23-27,1993.

(11) HELMIS C.G., D.N. ASIMAKOPOULOS, K.H. PAPADOPOULOS, P.G. PAPAGEORGAS , A.T. SOILEMES and H.D. KAMBEZIDIS. An experimental study of wind turbine wakes over complex terrain. ISES Solar World Congress,Budapest, Hungary, August 23-27,1993.

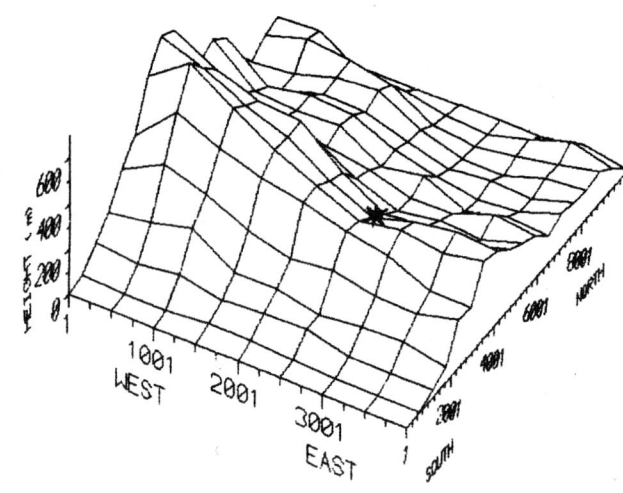

Fig.1 A three dimensional topographic sketch of the area around the Wind Park (denoted by a star).

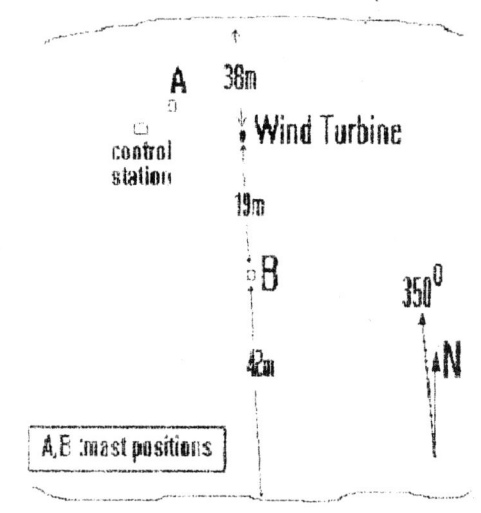

Fig.2 Instrumentation deployment over the experimental site.

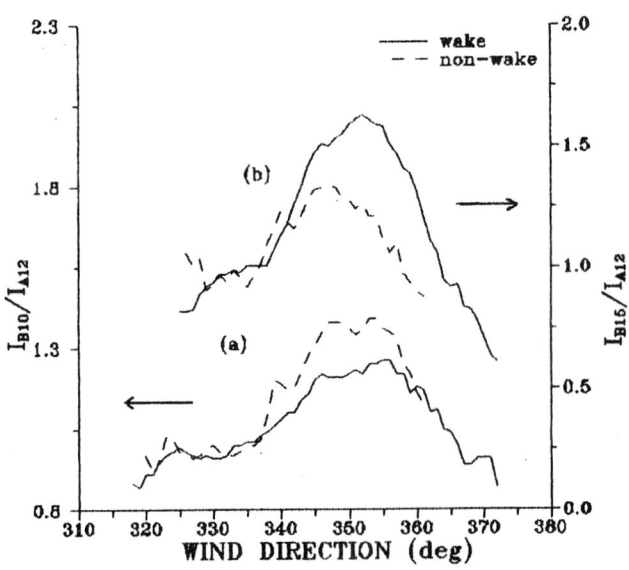

Fig.3 (continued overleaf).

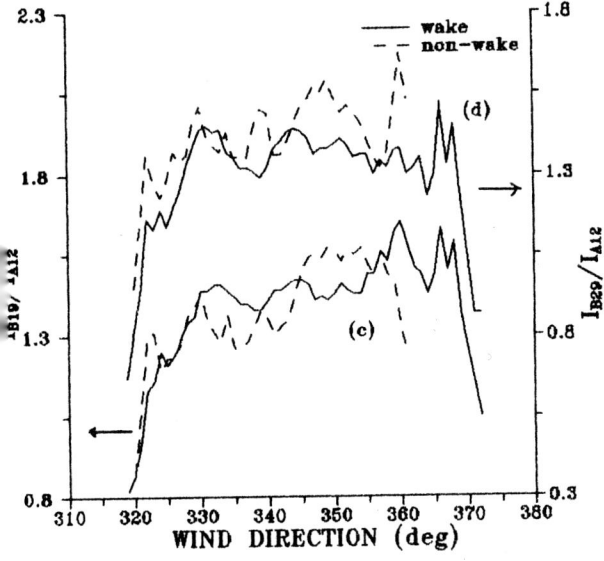

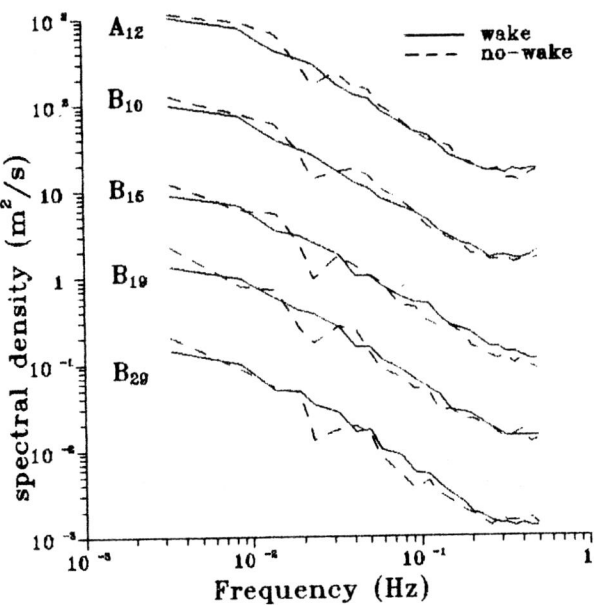

Fig.3 Normalized, with respect to A12, turbulent intensities at mast B for the wake (solid lines) and no-wake (dashed lines) dataset: (a) for B10, (b) for B15, (c) for B19 and (d) for B29.

Fig.6 Composite wind speed spectra at mast B for set 2A of Table 3. The reference A12 spectra are also included. For clarity purposes A12, B10, B15, B19, B29 spectra are multiplied by 10,1,0.1,0.01 and 0.001, respectively.

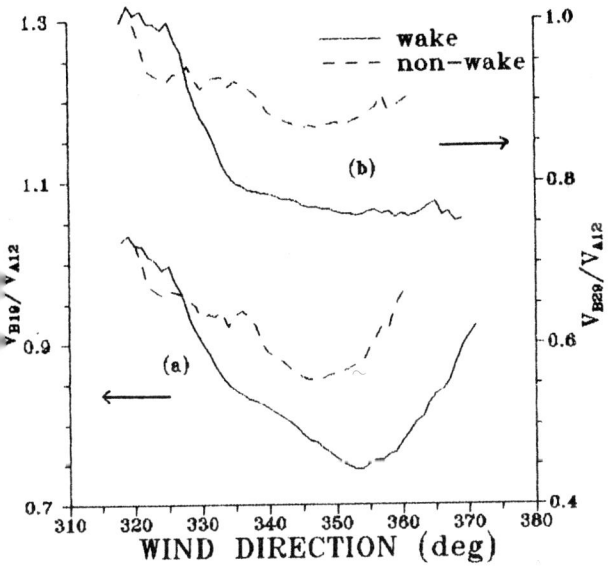

Fig.4 Wind speed ratios at B19 (a), B29 (b).

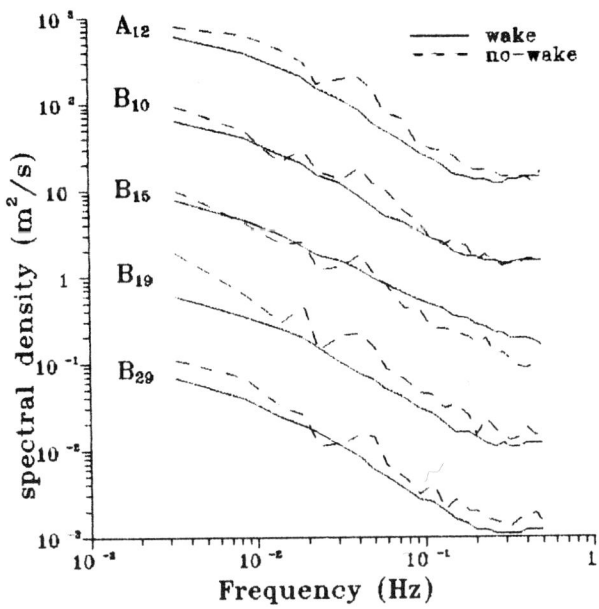

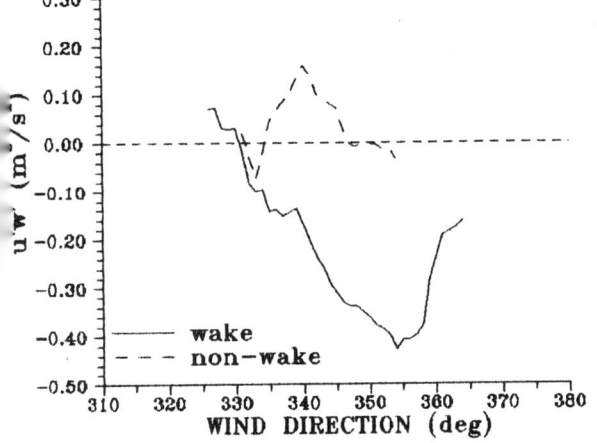

Fig.5 Vertical momentum flux measured at B15.

Fig.7 As in Fig 6 for set 3A of Table 3.

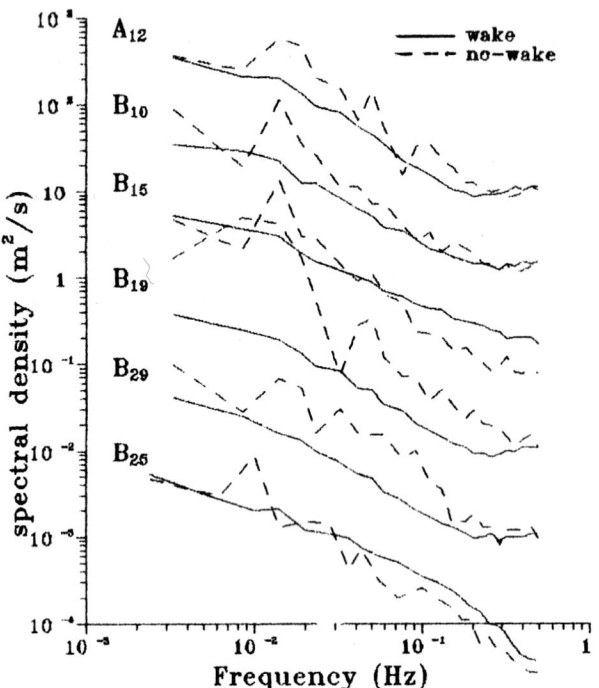

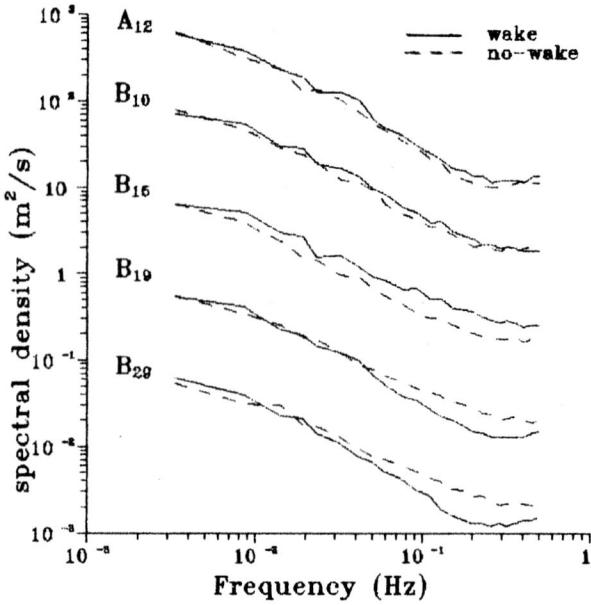

Fig.10 As in Fig 6 for set 4B of Table 3.

Fig.8 As in Fig 6 for set 4A of Table 3. Additionally, the B25 spectra multiplied with 0.0001 are included (see text).

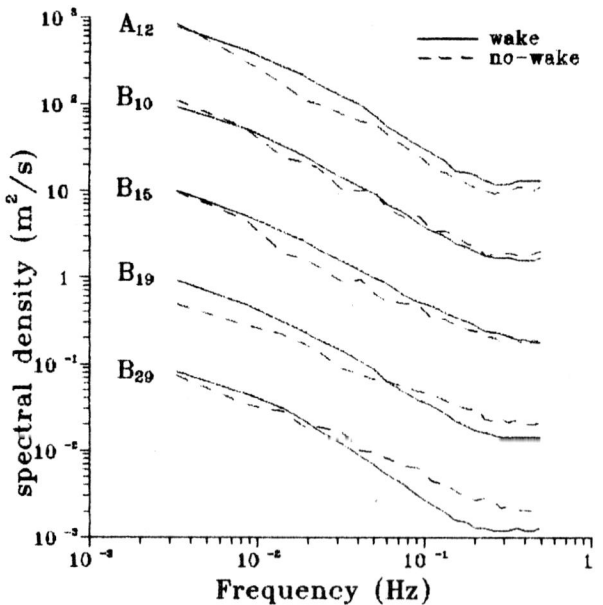

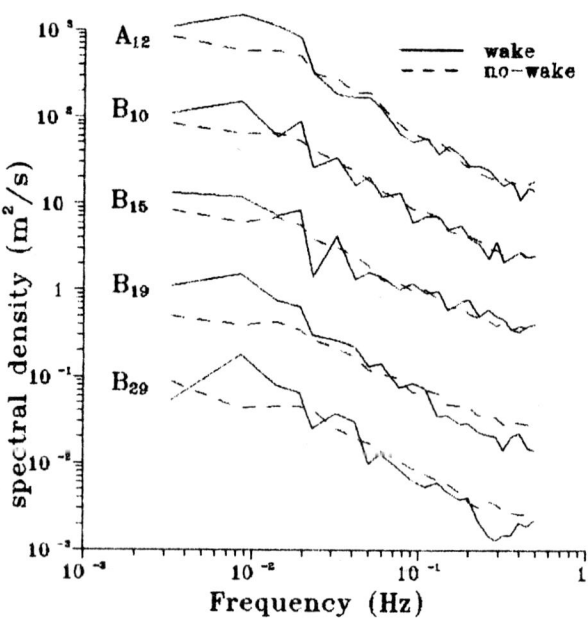

Fig.9 As in Fig 6 for set 3B of Table 3.

Fig.11 As in Fig 6 for set 3C of Table 3.

A database to aid the commercial development of renewable energy in the North East region of England

G. JENKINS, R.HARRISON, and **R.FOSTER**
University of Sunderland, UK
N. JOHNSON
Northern Electric Generation Ltd, UK

SYNOPSIS A database has been developed to identify renewable energy projects in the North East of England which may have good commercial prospects for near-term development. To this end it is used to perform a pre-feasibility assessment on identified prospects to assess their likelihood of success. Some of the prospects identified are now being taken forward to the feasibility study stage.

1 INTRODUCTION

The database has been developed as part of the "Study of Renewable Energy Resources in the North East Region of England", (Ref 1). This study has been jointly funded by Northern Electric Generation Ltd, Northumbrian Water Ltd, and the Department of Trade and Industry and has been carried out by the University of Sunderland and the Energy Technology Support Unit (ETSU). The database has been constructed using DataEase database software. It now consists of 124 entries, 41 of which are for wind energy.

2 THE AREA OF STUDY

The study area is defined by the boundaries of Northern Electric's Authorised Area. As shown in Figure 1:

3 RENEWABLE ENERGY TECHNOLOGIES STUDIED

The study has identified those technologies which may be important in the North-East for electricity generation. These fall into four broad groups:

- wind energy

- hydro-power (run of river schemes, reservoir schemes and schemes identified within the Northumbrian Water water supply and sewage treatment system)

- bio gases - landfill gas, sewage gas and farm gas

- waste combustion - municipal and general industrial waste, special industrial wastes (eg tyres) forestry wastes and farm wastes (particularly poultry litter).

4 DATABASE DEVELOPMENT

Studies of renewable energy resources lead on naturally to prospects for commercial development. However, for real developments to occur, a number of stringent criteria have to be satisfied:

- the resource conditions need to be suitable, small schemes are usually less attractive than larger ones for a number of reasons.

- the schemes must be compatible with environmental constraints

- it must be possible to connect to the electricity grid without undue cost penalties

- the economics of the developments must be favourable.

A large number of possibilities will be rejected using these criteria when viable schemes are being selected. This presents a problem since individual renewable energy developments are usually small in commercial terms. To attempt a detailed analysis of a large number of schemes is an expensive exercise which would divert effort away from the successful schemes. Applying simple criteria to screen the initial prospects could help to reduce this risk. The database has been developed to carry out this screening.

The database operates by scoring schemes under a number of headings reflecting the viability criteria outlined above. These are the major critical success factors:

- **physical resource factors:** relate to the amount of energy which a scheme would produce

- **environmental factors:** relate to the potential acceptability of the developments in environmental and planning terms

- **grid factors:** relate to the integration of the scheme into the existing electrical grid

- **cost factors:** relate to the costs of electricity production from the scheme.

The attributes which are used for scoring must be significant so that they can differentiate good prospects from poor prospects but also they must be consistent with the level of information which is available during the early assessment stage. These are stringent requirements which are difficult to satisfy and so in many cases broad assumptions have been used to substitute for expensive accurate information.

A number of subsidiary critical success factors were identified for use within the database:

- **resource quality:** a measure which indicates how good the resource is at a site, such as calorific value for combustion fuels, windspeed for wind turbines and hydraulic head for hydro schemes

- **resource quantity:** a measure of the likely size of the scheme, using quantities such as annual waste tonnage for combustion schemes, average water flow rate in hydro schemes, and number of wind turbines for wind farm developments

- **environment/planning designation:** this relates to the landscape designation of the site under investigation, for example National Parks.

- **environment/ecology factors:** these are designations affecting the site such as wildlife, archaeological and geological designations

- **technology specific environmental factors:** such as visual impact, air and water quality, noise and other factors

- **grid - primary substation factors:** a measure of the distance of the scheme from the primary and the fault level of that substation

- **grid - secondary substation factors:** this relates to distance of the scheme to the nearest point of the distribution system which has sufficient current carrying capability.

- **cost factors:** an overall assessment of the likely cost of the scheme including elements such as capital and maintenance costs, expressed as a unit cost of electricity.

Every scheme is given a score from 1 to 5 for each of these subsidiary factors and these are then used to calculate four main critical success factors in the areas defined above. These are then, finally, combined to give an overall score for the scheme. It is possible to give different weights to the success factors when the final scores are being calculated. For simplicity the scoring system uses only scores between 1 and 5.

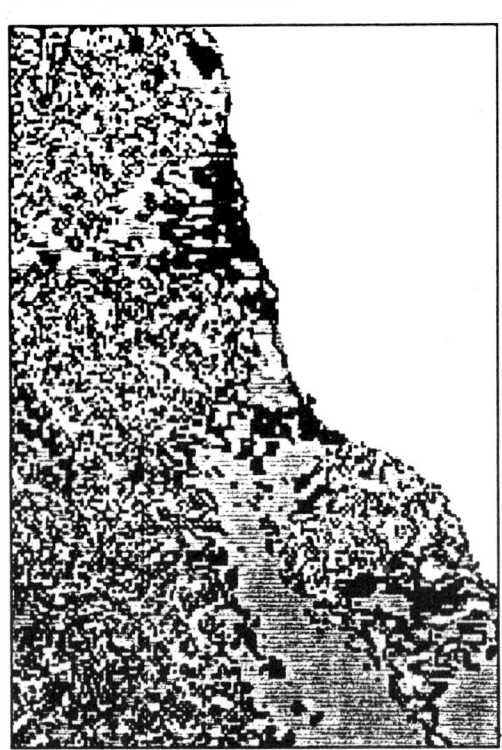

Map showing Windspeed in the North East

5 PHYSICAL RESOURCE CRITICAL SUCCESS FACTORS

Different physical resource critical success factors are appropriate for each technology. However, in each case (apart from landfill gas) two separate resource factors can be identified. The Resource 'Quality' factor relates to specific energy capture and the Resource 'Quantity' factor relates to the size of the scheme. The result is that both taken together give an indication of the installed power and the energy produced. (Table 1).

It is possible to identify further factors which define the physical resources more precisely but these usually require information which is not available at the pre-feasibility level.

5.1 Wind Energy Resource Factors

This study uses the ETSU wind model (Ref 2) which gives annual mean wind speeds in each one by one kilometre square. A map of these windspeeds at 25m above local ground level in the North East is shown in Figure 2.

The electricity output from a wind farm is determined by:-

- the number of wind turbines,
- the wind turbine machine power rating,
- the capacity factor at which the machine operates.

Wind turbine power rating is determined by rotor diameter and site design wind speeds. Typical commercial wind turbines available at the present time, are designed for annual mean wind speeds of about 6.5 m/s and are rated at about 400 kW.

The capacity factor at which the machine operates is determined by the relationship between site annual mean wind speed and the wind speed for which the machines are designed. Capacity factors of 28% are obtained at site AMWS of 6.5m/s (at 25m height), rising to 40% at 8 m/s. Sites with wind speeds below 6.5 m/s are generally considered to be non-viable. Only the best hilltop sites in the region have wind speeds above 8.0 m/s.

The number of wind turbines is determined by the site area and the rotor diameter. Wind turbines are typically spaced 10 diameters apart. This gives about 8 machines of 35 m diameter per square kilometre, when some allowance is made for unusable areas of most sites.

Thus, this wind resource at a wind farm site can be represented by two factors.

- resource quality factor; the wind speed, which determines the capacity factor (Table 2).

- resource quantity factor; the number of machines, which is limited by the site area (Table 3).

The overall physical resource factor is obtained by combining the scores for these factors with equal weightings. Thus, a 2 square km site with a wind speed of 7.0 - 7.5 m/s obtains a quality score of 3 and quantity score of 3 and overall resource score of 3.

TABLE 1. Physical Resource Critical Success Factors

Technology	Resource Quality	Resource Quantity
Wind	wind speed	number of wind turbines
Hydro	net head	mean flow
Landfill gas*	calorific value	gas production
Sewage/gas farm	calorific value	gas production
Waste combustion	calorific value	waste throughput

* n.b. a further factor is required to allow for the finite lifetime of gas production at landfill sites

TABLE 2   Wind energy resource quality factor scoring system (windspeed)

Windspeed m/s (at 25 m)	Score
< = 6.5	1
> 6.5 - 7.0	2
> 7.0 - 7.5	3
> 7.5 - 8.0	4
> 8.0	5

TABLE 3   Wind energy resource quantity factor (number of machines based upon standard 400 kW turbines)

Number of machines	Score
<=3	1
4 - 12	2
13 - 20	3
21 -28	4
>28	5

6 ENVIRONMENT CRITICAL SUCCESS FACTORS

Most new developments or extensions to existing developments must be submitted to the relevant planning authority for approval. There are two aspects to this. The first aspect of the process is analysis and comment on proposals by planning professionals including any requirements for environmental assessment. This is a rational procedure which takes account of legislation and guidance, established planning practice and also the policy of the planning authority.

The second stage which may overlap and interact with the first stage is a political process whereby individual groups and organisations may attempt to influence the planning decision to obtain a sympathetic outcome. It is not possible to codify the planning process so that the likelihood of different outcomes can be predicted. Planning professionals give rational and consistent advice but local political forces may influence the outcome.

At the pre-feasibility level all that is possible is to attempt to anticipate the reaction which planning professionals may have to developments, this again is difficult in the renewable energy field, where so often proposals involve unusual technologies being located in areas which may have had little industrial development. Also renewable energy is only just being included in development plans, so there is little guidance available from this source.

The advice given by planning professionals is affected by two aspects of the proposal.

- the location - eg whether the proposed development lies within a designated area, or whether the proposal is already allowed for in the Structure Plan.

- the specific environmental impacts of the technology

While it should be recognised that these two aspects will tend to interact, for purposes of formulating a scoring system in the database, these aspects have been explicitly separated. Two success factors have been defined, in discussion with planning professionals;

- a site designation factor which is independent of technology and depends only upon location

- a technology specific factor which is independent of location and depends only upon the technology.

6.1 Site designation subsidiary critical success factors

These are divided into two groups;

- **landscape planning area designations.** National Parks and Green Belts are designated by statute and are areas within which there is a presumption against any development, various areas of landscape value can be designated locally. Industrial sites are designated for development in local structure plans and the presumption is in favour of development. A landscape designation map of County Durham is shown in Figure 3 as an example.

- **special ecological and other areas.** Within this category are internationally designated nature reserves and national nature reserves at one extreme and undesignated areas at the other. Archaeological and special geological sites are included within this.

The scoring system gives a low score to a site which is heavily restricted in terms of development potential and a high score to sites which are in non designated areas.

6.2 Technology specific environmental impact factors

Renewable energy developments all have different environmental impacts. Thus waste combustion plants may be visually intrusive, may give rise to emissions and there may be significant amenity disturbance due to traffic. In discussion with planning professionals, four main environmental factors have been identified and each technology has been scored for each of these.

- **visual impact** relates to the effect of the development on the aesthetic quality of the landscape within which it is located.

- **air and water quality** indicates the water and/or air pollution produced by the scheme when it operates

- **amenity/noise/traffic** measures the amenity damage caused by the noise of the plant in operation or by the road traffic to and from the site

- **ecology/other** indicates the disturbance caused to natural ecosystems.

For windfarms, visual impact is anticipated to be the greatest problem, more so than for other renewables. Noise may also affect siting considerations, such as proximity to housing. Air and water quality is unaffected, and the ecology of the site is only slightly affected.

Figure 3 Landscape designation map of County Durham (by courtesy of Durham County Council)

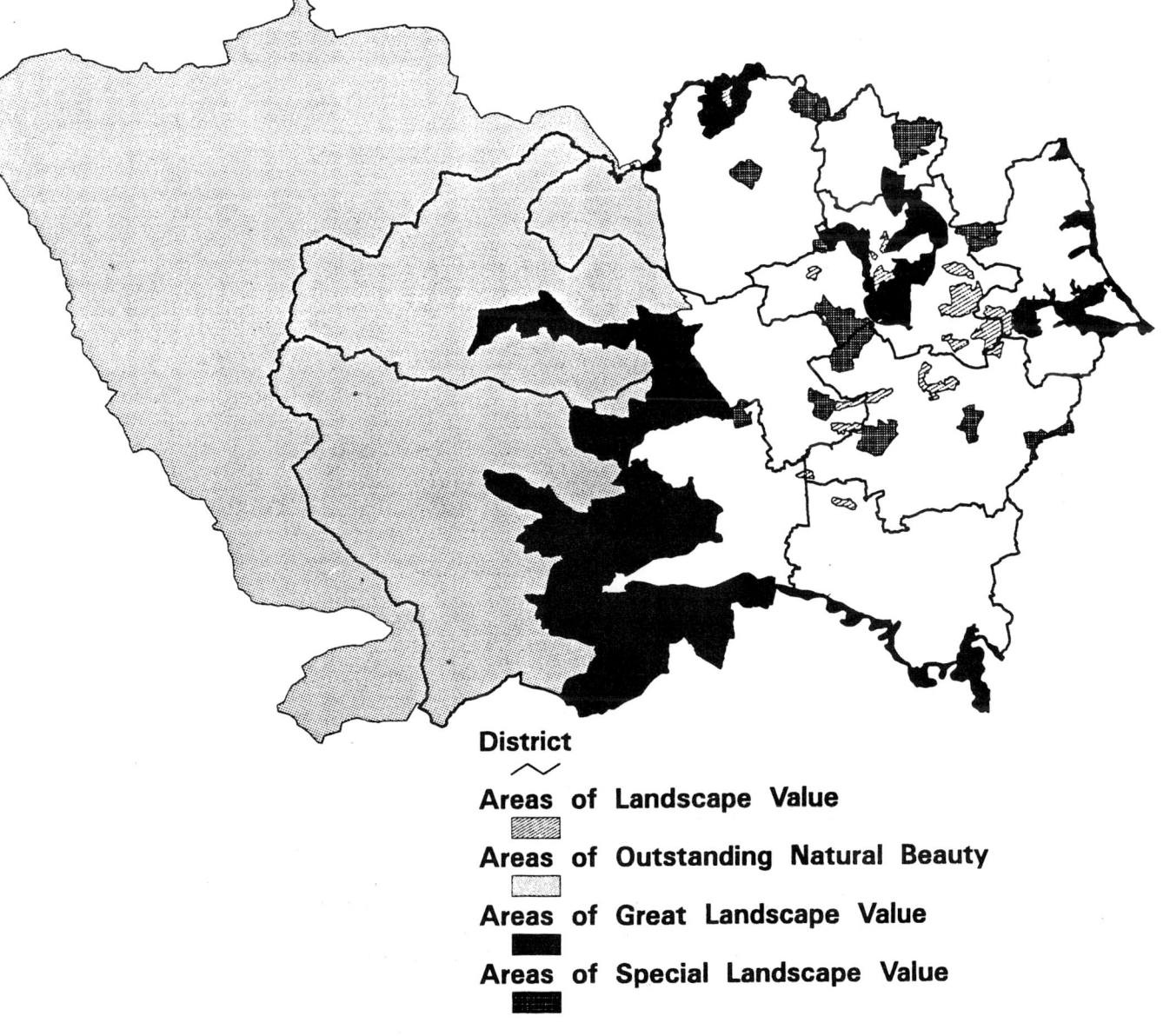

District

Areas of Landscape Value

Areas of Outstanding Natural Beauty

Areas of Great Landscape Value

Areas of Special Landscape Value

7 ELECTRICITY DISTRIBUTION SYSTEM CRITICAL SUCCESS FACTORS

7.1 The electricity distribution system

This system consists of two subsystems; a primary system which distributes large quantities of power at high voltages from the National Grid to primary substations and a secondary system which distributes smaller volumes of power at lower voltages to premises. The secondary power lines in the Northern Electric Area are mainly at 11 kV and 20 kV.

As a general rule of thumb schemes with a capacity greater than 2MW are connected to the primary distribution system at a primary substation and those 2 MW or less may be connected to the secondary distribution systems at a secondary substation.

Thus for schemes greater than 2 MW a site close to a primary substation is preferable since an electrical line is required to be built between the site and the substation. The cost of such a line is related, amongst other things, to the length of the line.

For schemes of less than 2 MW a suitable point of connection into the secondary system is required such as an existing electrical line of sufficient current carrying capability. Such areas are identified, as a general rule, by a local load of at least 100 kW per square kilometre.

Another factor to consider is the effect of the generator under fault conditions on the distribution system. The fault level capability of the switchgear situated at primary substations is dependent upon the margin between its rated capacity and the electrical load attached to that substation. A primary substation which has a small margin between its rated capacity and local load has a low fault level capability and vice versa. Such fault ratings affect all schemes and not simply those greater than 2 MW. The fault infeed of a generator connected into the secondary system is still 'seen' by switchgear at its associated primary substation.

7.2 Scoring

Three Critical Success Factors are used to quantitively assess the connection of a scheme to the distribution network for any one scheme, two of these factors will be calculated as shown in Table 4. These are:

- **Primary Distance** The distance from the site to the nearest primary substation. A computer program has been developed, whereby given site coordinates the distance to the nearest primary substation is calculated.

- **Secondary Distance** The distance from the site to a point on the secondary distribution system which has a line of suitable current carrying capability to allow connection. A computer program has been developed, whereby given site co-ordinates a grid of 1 km squares surrounding the specified site is created with each square indicating the maximum loading of the distribution system in that area. The program utilises a large data set, the Authorised Area of Northern Electric. This program can be utilised to give an indication of the proximity of an electrical line of suitable current carrying capability.

- **Fault Level** capability of the associated primary substation. In calculating the Primary Distance the program identifies by name, the nearest Primary Substation. It is a reasonable assumption that the nearest primary is also that to which the scheme will be electrically connected. From the Primary Name a score, based upon the primary's fault level capability, is given and used in the CSF calculation.

7.3 Weighting

Since the type of substation required for connection is dependent upon scheme size and will subsequently affect the priority given to the above CSFs, two alternative weightings as illustrated in Table 4 are used:

8. COST CRITICAL SUCCESS FACTORS

8.1 Economic Assessment Rules

A variety of approaches can be taken to the assessment of the economic performance of renewable energy developments using discounted cash flow techniques.

- Unit cost
- Net present value
- Internal rate of return
- Payback time.

The calculation of unit cost is the most suitable as it does not require information on selling price of electricity.

Depending upon the purpose of the calculation there are still a number of approaches which can be used to calculate the unit cost. These can be divided into two basic groups.
- **Financial assessment.** This requires a knowledge of all costs together with details of

project financing and taxation regime which are specific to the organisation.

TABLE 4   Weightings for Grid CSF

	Scheme Size	
CSF	≤ 2MW	> 2MW
Primary Distance	0	50%
Secondary Distance	30%	0
Primary Fault	70%	50%
Total	100%	100%

- **Economic assessment.** This requires a knowledge of all costs but does not take account of details of financing and taxation.

The purpose of the cost success factors is to rank the project, not to determine accurate unit costs. This would be carried out in a project-specific financial analysis for the investors. Hence the simplest approach is the most suitable, consequently, the approach of **economic assessment** is adopted here. Important factors are lifetime and discount rate.

- **Definition of lifetime.** This should be taken to be the operating lifetime of the system. This is taken to be 20 years for all schemes and all technologies except landfill gas.

- **Choice of discount rate** A discount rate which represents a reasonable average net interest rate for large organizations could be used. This is adjusted to reflect the level of technological risk which may be associated with some of the technologies.

8.2   Scoring system for costs

A universal scoring system across all technologies making no allowances for different technologies can be devised. However, it is also possible to define a scoring system which reflect the way in which previous NFFO orders, supported particular technologies by paying different prices for their output.

The rules for the NFFO calculations in any future round are not known at the time of writing and the scoring system does not necessarily reflect the unit costs of the future round but it does ensure that the costs are ranked in generally the same way as they would have been in the last round of the NFFO.

8.3   Outline costing rules for wind energy
An essential part of the cost calculations are the costing rules used to estimate capital (investment) and operating costs. The following simple rules for a wind farm cost model have been derived from the available sources.

Capital cost

- Supply, install and commision wind farm = £1,000/kW installed.

Operating costs

- Operation and maintenance cost = 1.5% of capital cost per annum.

- Land Rental based on 2.5% of gross revenue = £4/kW installed per annum.

- Rates = £14.9/kW installed per annum.

- Insurance = 1.0% of capital costs per annum.

Overall operating costs approximate to 4.5% of capital costs per annum.

8.4   Energy outputs for cost band calculation
The energy output is calculated from the capacity factor determined from the site annual mean winspeed, assuming a standard 400kW turbine with planned maintenance.

9 OVERALL WEIGHTING
The database has a degree of built in flexibility. the internal weightings can be changed between sub-CSF's which have been shown previously. The overall external weighting between the four categories of resource, environment, grid

and cost, is open to change by the user. Currently they are equally weighted. These could be altered as experience on the use of the database results is obtained. For example, the environment category may be given greater weight, the grid may be de-emphasised by this method.

10   DATA ENTRY

Each scheme entered onto the database consists of 6 screens of data.

The upper main screen contains site name and technology type, unique site data reference number and overall score results. The lower main screen: Ordnance Survey Grid reference, planning authority name and details, landowners name if known and site operators name if known.
The resource screen for wind required annual mean windspeed and number of standard turbines.

The environment screen requires information on whether the site is new or existing, the landscape planning designation and ecological designation.
The grid screen requires data from the sub-station primary and secondary distance computer program built into the database, namely the nearest primary sub station reference number, the primary distance, the secondary distance, the secondary load size and the OS grid reference of the nearest 100 kW load square.

The cost score screen requires the scheme lifetime, discount rate, annual energy output, scheme rated capacity, the capital cost, operation and maintenance cost, transport of fuel, fuel charge, other charges (including rates).

11   SOME RESULTS FROM THE USE OF
          THE DATABASE

The database currently contains data on the following prospective schemes;

wind	39
landfill and waste gas	30
hydro electricity	37
incinerators	16

The extensive local knowledge of the project partners has been used to select these initial sites. The database has then been used to rank sites in each category. A number of the best sites are now undergoing feasibility studies by the project partners with a view to further development of the most promising schemes.

For wind schemes a total rated power of approximately 342 MW installed has been identified in these 39 schemes in the region. The list of sites

identified is by no means exhaustive, although most of the 8 m/s plus sites outside restricted areas have been included. This list includes some sites known to be under consideration by other developers.

In conclusion, it can be seen that the North East of England contains a substantial wind resource. It is expected that the results of this study and development of the database will lead to a substantial increase in the use of wind energy and other renewables in the region in the near future.

In addition, the technique developed enables a rapid assessment of potential renewable energy schemes in any area, which may be of interest to utilities throughout the UK. The technique could also be used as an assessment tool for funding agencies for a "first cut" through a large number of projects.

References

1) "Study of Renewable Energy Resources in the North East of England". ETSU/Northern Electric/Northumbrian Water, 1993 (in preparation).

2) Burch, S.F., & Ravenscroft, F., Computer Modelling of the UK Wind Energy Resource, ETSU WN 7055, 1992.

Wind energy resource in the South West of England

HAYDN SCHOLES,
CSM Associates Ltd, UK
TARIQ MASOOD
South Western Electricity Plc, UK

SYNOPSIS: A renewable energy resource study of South West England has been undertaken by South Western Electricity Plc (SWEB) and the Department of Trade and Industry (DTI). A range of renewable energy technologies were investigated with a view to estimating resource size and determining which schemes might be appropriate for development. This paper describes the methodology and software developed to undertake the regional onshore wind resource analysis. It includes resource/ constraint analysis, identification of potential wind farms, assessment of their impact on the electrical distribution system and estimation of their probable generation costs. Finally a summary of the results of the resource analysis is presented.

1. BACKGROUND

The joint SWEB/DTI South West regional renewable energy study was undertaken by staff from SWEB and the DTI's Energy Technology Support Unit (ETSU) with certain elements contracted to CSM Associates Ltd based in Cornwall. Of the various renewable energy technologies investigated in the study, onshore wind has probably the most immediate potential and is possibly the most significant. The study defined four levels of resource.

Technical Resource

This is the renewable energy resource which could be exploited if technical ability is the only limitation. All environmental, planning and other practical considerations are ignored.

Accessible Resource

The accessible is the quantity of the technical resource limited by basic practical conditions (physical constraints) and environmental and institutional constraints. For most technologies this measure of resource is very large and therefore unlikely to be fully exploited. This definition should not be confused with the recent Renewable Energy Advisory Group's (REAG) definition of accessible resource. REAG's accessible resource definition includes an additional economic constraint limiting it to that which can be produced at less than 10p/kWh for an 8% discount rate.

Economic Resource

The economic resource is the extent of the resource if economic constraints are placed on the resource in addition to all the previously defined constraints. Unlike the other resource levels the economic resource is multi-valued, i.e. the resource size varies with generation cost. The economic

resource must also be qualified by the criteria, eg discount rates, and methods used to evaluate it.

Regional Resource

This is the quantity of the resource that SWEB considers likely to be developed by the year 2000 after taking account of technical, environmental, planning, economic and institutional criteria applied separately to each technology. The regional resource will normally be a small fraction of the accessible resource. It should be stressed that this resource assessment is based on a large element of judgement and uncertainty.

A suite of analysis software was developed by CSM Associates Ltd to assess resource levels for the onshore wind resource. There are four stages in the assessment process; a resource/ constraint analysis, a cluster analysis to identify potential wind farms, an electricity distribution grid impact analysis and finally an economic analysis of the remaining potential wind farms.

2.0 RESOURCE/CONSTRAINT ANALYSIS

2.1 Inputs

For an area the size of the SWEB region it was deemed most appropriate to compute resource levels using a Geographical Information System (GIS) on a square kilometre grid basis. A GIS code called SITE was used for this purpose. SITE was originally developed by the Camborne School of Mines to estimate the size of Hot Dry Rock geothermal energy resources and was adapted to perform the onshore wind analysis. Ancillary codes for digitising, rasterisation and data format conversion were developed as required.

The starting point of the resource analysis is an estimate of the annual mean wind speed (AMWS) for each square kilometre in the region. This data is normally derived from computer models. In the case of the SWEB region it was provided by ETSU who had already undertaken wind speed modelling exercises using the NOABL computer model. The second major input is the type of wind turbine to be employed. In reality many types of turbine will be utilised but to simplify the problem a "typical" wind turbine was defined with characteristics averaged from a range of commercially available machines. Its specifications are shown in the following table.

TABLE 1 - Wind Turbine Specifications

Installed capacity (C)	400kW
Rotor diameter (d)	34 m
Minimum spacing	333.3m ($10d$)
Availability (a)	95%
Average array efficiency (e)	90%
Ex-factory cost	£275/m² of swept area
Installation/commissioning cost	15% ex-factory cost
Site infrastructure cost	30% ex-factory cost
Grid connection cost	Site specific
Operating & Maintenance costs	1.5% capital cost
Reactive power/meterage charges	0.15 p/kWh
Land rental	2.0% gross revenue
Rates	£2.50/installed kW
Insurance	1.0% capital cost

The third major input to the resource analysis is the number of turbines that can be sited in each square kilometre. This is a function of some initial population and any constraints to siting. Each constraint reduces the number of turbines by the fraction of the square kilometre covered by that constraint. The constraints are organised to be mutually exclusive. This implies a constraint hierarchy. In general nationally designated constraints were assumed to take precedence over local ones. Each level of resource is the result of the application of an appropriate set of constraints to the previous level.

Not all the square kilometres in which wind turbines can be sited will be sufficiently productive, there will be some with wind speeds that are too low to generate power economically. In order to put the resource estimates into an appropriate context, two wind speed thresholds were adopted. These can be interpreted as "probably" economic (7.0 m/s and above) and possibly economic (6.5 m/s and above).

2.2 Technical Resource Estimation Method

The technical resource calculation assumes that the only constraint to siting is a minimum spacing criterion. The number of turbines that can be sited in any square kilometre (N) is limited only by wake interaction effects. The separation between machines is expressed in terms of the rotor diameter (d). For sites with no dominant wind direction a spacing of $10d$ is usual. For sites with highly directional wind fields spacings as low as $5d$ have been used. As the analysis can not be site specific at this stage, the $10d$ spacing was adopted, i.e. for a 34m diameter machine, $N = 9$.

The power from a single wind turbine is computed as follows:-

$$P = C \ L(V) \ e \qquad \text{kW}$$

Where C is the installed capacity, e is the array efficiency (including other losses), and $L(V)$ is the load factor for a specific annual mean wind speed V.

The power available from any single square kilometre is

$$P = C \ L(V) \ e \ N/1000 \qquad \text{MW}$$

and the annual energy output E is

$$E = C \ L(V) \ e \ N \ a \ 8760/1000 \qquad \text{MWh}$$

where a is the machine availability factor. For the technical resource $N = 9$ for all squares. The total technical resource is the sum of the resource in the individual square kilometres in the region.

2.3 Accessible Resource Estimation Method

The accessible resource is the technical resource limited by physical, environmental and institutional siting constraints. The effect of these constraints is to reduce the population of wind turbines in each kilometre square within the region and consequently the overall resource size.

2.3.1 Physical Constraints

Two methods for defining more realistic initial populations of turbines due to physical constraints were are available at this point. The first method was based on the technical maximum population, 9 per sq km reduced by the following constraints:-

    A roads
    B roads
    Towns
    Villages
    Woodland
    Rivers
    Railways
    Quarries
    Airfields
    Electricity distribution grids
    Radio/TV transmitters and repeaters

Isolated dwellings are not taken in to account with this method hence it overestimates the resource.

The second method was derived from a turbine siting study undertaken by the Institute of Terrestrial Ecology (ITE) for ETSU. The study examined 32 different land classes taking random samples of each class throughout the UK land mass (12 per class) and assessed the number of turbines that could be sited on each sample. It took into account isolated dwellings and some linear features such as roads and railways. ITE land class data was available for the SWEB region and hence it was possible to define an initial turbine population in each square kilometre. The following additional constraints were then applied to estimate the resource:-

Towns
Villages
Woodland
Quarries
Airfields
Electricity distribution grids
Radio/TV transmitters and repeaters.

It should be noted that neither approach accounts for microwave communication links which are thought to be a significant problem to wind farm placement in the South West. It proved difficult to acquire information about this constraint within the time scale of the study. The resource could therefore be over-estimated, however the effect of this constraint may be ameliorated by judicial positioning of wind turbines within any one wind farm.

2.3.2 Environmental and Institutional Constraints

Social, environmental and aesthetic considerations will limit wind farm development in certain areas. Unlike the physical constraints, institutional constraints can be considered to be "soft" and possibly amenable to change. Again the effect of each constraint is to reduce the number of turbines that are present in each square kilometre. The following environmental and institutional constraints were considered:-

Areas of Outstanding Natural Beauty (AONBs)
Areas of Great Landscape Value (AGLVs)
National and Regional Parks
Heritage Coast
National Trust land
National and Local Nature Reserves (NNRs, LNRs)
Sites of (Special) Scientific Interest (SSIs, SSSIs)
Green belts
Environmental Sensitive Areas (ESAs)
Metalliferous mining, China clay and Quarrying consultation zones
National, County and District planning considerations
CEC Special Protected Areas (eg Ramsars)

Historic monuments were ignored in the analysis, as the data consisted primarily of cairns, standing stones and small scale features which were unlikely to directly impact turbine siting. Certain conservation areas belonging to voluntary bodies were ignored as the sites were very small and usually covered by SSSIs, NNRs, & ESAs. Military zones were also ignored as in many parts of the South West they coincide with other constraints. The remaining areas with the exception of target ranges could in any case be used for wind turbine siting. One remaining constraint to wind turbine siting is planning consent. A point of debate in this area is whether wind farms would be allowed in AGLVs. The first commercial wind farm operating in the South West is sited in an AGLV, however applications in other AGLVs have been rejected. Resource estimates were therefore computed both excluding and including AGLVs.

2.4 Economic Resource Estimation

The total economic resource is the accessible resource, however it may not be cost effective to exploit all of it. The exploitation of wind energy is very much a site specific issue. To accurately estimate the economic resource and subsequently the regional resource, it is necessary to consider the characteristics of individual wind farms, in particular their power output, capital cost and running costs. Although most of this information was available in parametric form, the site specific costs associated with connecting a particular wind farm to the electrical distribution system needed to be assessed. In addition there was a question whether the distribution system could accept all the potential resource when it was manifested as a finite number of wind farms at specific locations. To progress the analysis further a means of identifying potential wind farms in the mass of available resource was required.

3.0 IDENTIFYING POTENTIAL WIND FARM SITES

3.1 K-Mean Clustering

The method selected to identify individual wind farms, K-mean clustering, attempts to assign a large number of data points (the square kilometres) to a fixed, and smaller, number of clusters (the wind farms). The method is implemented in a computer program in three distinct phases. In the first phase (thinning) outlying and isolated data points are eliminated. The second phase (clustering) then groups the remaining data points into the most favourable clusters. This second phase is iterative. The third phase allows the user to modify the solution found in the second phase using some simple wind farm design rules.

3.2 Thinning

The thinning phase is conducted as follows. All the squares are examined and those containing less than a specified minimum are discarded. Next the number of turbines within a specified distance of the square currently under consideration is determined. If the number of turbines found is greater than a user supplied threshold then the square is retained, otherwise it is dropped from the analysis. This procedure is performed for each square in turn.

3.3 Clustering

The clustering approach is based on a non-hierarchical K-mean clustering algorithm due to Anderberg [1]. The actual method used is an iterative version of the method of MacQueen [2]. The process consists of two steps. The first is to assign the starting positions of the cluster centres. Then each square is added to the cluster which is least adversely affected by having that square added to it. This is done until all the squares have been assigned. This defines MacQueen's method. Measures are taken to ensure that the initial cluster centres are well dispersed throughout the data set, even if the data is supplied in a well ordered form. The second step consists of going through the data considering each square. If it proves advantageous, in terms of the defined cost function, to swap a square from one cluster to another then this is done. This process continues until a complete pass is made through the data without a single swap taking place. This is the iterative modification to MacQueen's method.

3.4 Modification

The wind farm identification problem requires additional constraints to be placed on the clustering. The first constraint is that no wind farm should exceed a maximum number of turbines. The second limit is that very diffuse wind farms will not be economic and so should be split into smaller denser farms. Both of these constraints are catered for by the program. After the initial clustering has been performed the user is informed which is the largest farm produced and also the sparsest. The sparseness is defined in terms of the length of electrical cable connections between squares and the total number of turbines in the farm. The user can then set limits for these two constraints. The data is then re-examined and the largest and sparsest farms split up if necessary.

In each case a new cluster centre is introduced as the most outlying point in each farm. The data is then re-clustered until no further swaps take place. A compromise is needed between only splitting the biggest farm and splitting all the farms which are too big. The former would take too long and the later would introduce too many new farms. The compromise is that a farm can only be split if it is more than a defined distance away from any previously split farms in this iteration. Once this splitting and re-clustering has taken place the user is informed of the current state of the largest and sparsest farms and how many exceed the requested limits. This continues until the user terminates the process or all the farms are within the originally requested limits.

4.0 WIND FARM INTERACTION WITH THE DISTRIBUTION SYSTEM

4.1 Objectives

Once the potential wind farms had been identified their interaction with the distribution system was examined. The objectives of this analysis were:-

(a) To identify the points on the existing 33kV/132kV grids and substations where wind farms could be connected.

(b) To identify those connections that could be accepted in capacity and fault level terms without reinforcement of the distribution system.

(c) To identify the amount of reinforcement required to accept the remaining wind farms (or some proportion of them)

(d) To assess the cost of connection in each case.

4.2 The Distribution System

The distribution system in the SWEB region is very complex both in topological terms and in its ability to accept new generators. To make the problem tenable some simplifying assumptions were applied. The first of these was that no connections at less than 33kV were allowed. Ignoring the 11kV network significantly reduces the complexity of the problem. The second was that there were unlikely to be any wind farms large enough to require direct connection to the 132kV grid. The problem was therefore effectively reduced to the 33kV distribution system.

The following basic data about the distribution system was supplied by SWEB, the topology of the 33kV, 132kV and 400kV networks, the location of sub-stations, bulk supply points, junctions and breakpoints, transformer capacities and fault levels of each component of the network. This information was input into a purpose written distribution system analysis program, TOPOL. The program builds, links and checks the distribution system topology and capacity, then allows the most favourable connection for a wind farm be identified and costed.

4.3 Connecting Wind Farms to the Distribution System

TOPOL can be interrogated about connection options and costs for specific wind farm sites simply by giving the site location and installed capacity. The reduction in available capacity at substation and bulk supply point (BSP) transformers and on the 33kV lines is accumulated so that further wind farms use the newly adjusted capacities in determining their connectivity. The following parameters were used in the grid connection algorithm.

TABLE 2 - Grid Connection Parameters

Search radius for a 11kV substation	10km
Search radius for a 33kV line segment	30km
Search radius for a BSP	30km
Power output of single wind turbine	0.41MVA

Summer line ratings (worst case condition) were used to limit cable capacities. Six different ways to connect a wind farm were then considered:-

A single dedicated 11kV circuit
Two dedicated 11kV circuits in parallel
Connection to existing 33kV circuit including 0.1" dia line
Connection to existing 33kV circuit excluding 0.1" dia line
Dedicated 0.15" dia 33kV line to a BSP
Two dedicated 0.15" dia 33kV lines to a BSP in parallel

4.4 Connection Costs

The cost is computed for each different method and the cheapest available option is used. Costs were based on the following formula which varies for each connection type. An allowance for a proportion of underground cabling and an indirect connection path is made in the cost/km values.

$$\text{COST (£k)} = \text{fixed charges} + \text{cost/km} \times \text{distance to wind farm}$$

The order the farms are presented in can have a profound effect upon connectivity and the capacity accepted by the distribution system. Initial tests with the data sorted by both increasing and decreasing installed capacity gave results that were not in line with expectations. Finally it was decided to initially order the wind farms by lowest generation cost based on a parametric connection cost of 15% of the capital cost of the farm.

The results of the above analysis is a set of connection costs for each of the constraint scenarios. From this the total connectable resource is computed along with that which is unconnectable, i.e. the amount of grid reinforcement required. Typically less than 5% of the wind farms were rejected. The remaining wind farms, with their new site specific connection costs, were then subjected to the following economic analysis.

5.0 ECONOMIC ANALYSIS

5.1 Inputs

The primary inputs to the economic analysis are the capital cost and annual running costs of each wind farm. These consisted of the ex-factory capital cost of the "typical" wind turbine in terms of cost/m^2 of swept rotor area, plus other costs expressed as percentages of the capital cost, installed capacity, power produced or the gross revenue. To these the site specific site connection costs derived in the grid connection analysis were added. The annual output of each wind farm can be computed from the output of the individual square kilometres that make them up.

5.2 ETSU's Discounted Cash Flow Methodology

The one remaining and possibly most important factor is the method of assessing generating costs. Normal practice within ETSU is to use a simple discounted cash flow (DCF) approach which assumes that the project is fully equity funded. The DCF is carried out over the full operational lifetime of the plant, typically 25 years for a wind farm. Although many more approaches to investment appraisal are available [3] and are more sophisticated than DCF analysis this simplistic approach allows a reasonable comparison to be made between different renewable energy schemes with very different economic characteristics.

5.3 A Commercial Discounted Cash Flow Methodology

An alternative market related approach was considered taking into account that commercial projects would typically make some use of loan finance. Although the exact method of investment appraisal will vary from organisation to organisation, for the purposes of the study a standard "commercial" DCF methodology was adopted consisting of the following elements. Projects would be undertaken with a 20% equity, 80% loan funding split. These loans are paid off during the first 10 years

of the project. The interest rate for the loan is a function of its size, i.e. base rate +5% up to £5M, base rate +4% up to £10M and base rate +2% over £10M. This DCF is carried out over the "commercial" lifetime of the project which in the case of a wind farm would be 15 years.

5.4 Implementation

The above differences apart the two DCFs are substantially the same. A one year build period is assumed. Annual costs and revenues begin in the year following. The revenue stream is modified by varying the unit sale price of electricity (p/kWh) until the net present value (NPV) of the project's DCF is zero or slightly positive. The unit price of electricity at this point is the break-even generation cost at a particular discount rate. It should be noted that this is an iterative process. The analysis was undertaken using Microsoft's Excel spreadsheet. Both types of DCF were implemented in a single Excel macro function, (Estimate). This function is called repeatedly by another macro (Breakeven) which implements the iterative solution. DCFs were undertaken for lifetimes of 15 and 25 years and discount rates of 8%, 15% and 25%.

6.0 WIND FARM PROXIMITY CONSTRAINTS

Finally an exclusion zone, in addition to those constraints already described, was applied to investigate the sensitivity of the resource to a proximity constraint. An arbitrary minimum distance of 5km between individual wind farms was chosen. This spacing was used purely as an example. In practice each site would be reviewed on an individual basis and at certain sites it is unlikely that an exclusion zone would necessarily apply. However, it can have a significant effect on the resource. If this separation criterion is applied the connected regional resource is reduced by approximately two thirds. The algorithm used scans a list of wind farms and compares their co-ordinates. Farms lower down the list within the current farm's exclusion zone are discarded and the size of the resource reduced accordingly. Again the order in which the farms are presented to the algorithm is significant. For consistency, the same ordering criteria as applied to the grid connection analysis (lowest generation cost first) was used.

7.0 RESOURCE ESTIMATES

Several different scenarios were considered in the course of the overall resource analysis, both to understand the behaviour of the computer models and to investigate the sensitivity of the resource estimates to different constraints. The following table summarises the results and is taken from the report on the study issued by ETSU and SWEB, Renewable Sources of Electricity in the SWEB Area - Future Prospects [4]. It is based on a 6.5 m/s wind speed cut off threshold and a nominal 5km exclusion zone. As discussed earlier there are a great many uncertainties in the resource estimation process. The results in table 3 should therefore be considered indicative rather than definitive.

TABLE 3 - Estimated Wind Resource within the SWEB Region above a mean wind speed of 6.5 m/s

Resource Description	Installed Capacity (MW)	Annual Output (TWh/y)	%age of Technical Resource
Technical	12 270	31.7	100.0
Technical, after physical constraints included	7 399	15.6	49.2
Technical, after physical constraints and exclusion from designated areas	2 891	5.3	16.7
Technical after physical constraints and exclusion from designated areas including AGLVs	1 679	3.38	10.7
The above resource after connection to the distribution system	1 614	3.26	10.3
Remaining resource after distribution system connection and application of nominal 5km exclusion zone	552	1.14	3.6
Regional resource likely to be developed by the year 2000	150-300	0.3-0.6	0.95-1.90

8.0 SUMMARY

A coherent suite of software for regional onshore wind resource analysis has been developed. It has been successfully applied in a regional renewable energy study of South West England allowing the technical, accessible and economic resource levels to be comprehensively estimated. The software has subsequently been used in similar regional studies of North Wales and Scotland.

ACKNOWLEDGEMENTS

The work described in this paper was carried out as part of a joint ETSU/SWEB study. The views and judgements expressed in the paper are those of the authors and do not necessarily reflect those of ETSU or the DTI.

REFERENCES

[1]   Anderberg M.R., 1973, Clustering analysis for applications. pp359, Academic Press Inc (London) ltd.

[2]   MacQueen J.B., 1967, Some methods for Classification and Analysis of Multivariate Observations. Proc. Symp. math. Stastist. and Probability, 5th, Berkely, vol 1, 281-297, AD 669871, Univ. of California Press Berkley.

[3]   Lumby S. 1988, Investment appraisal & financial decisions, Chapman and Hall

[4]   South Western Electricity Plc and the Energy Technology Support Unit, Renewable Sources of Electricity in the SWEB Area - Future Prospects, 1993.

A comparison of physical and statistical methods for estimating the wind resource at a site

L. LANDBERG PhD(a), MSc, BSc, MDaMS, IKAK and **N. G. MORTENSEN** PhD(a), MSc, BSc, MRYC
Risø National Laboratory, Denmark

SYNOPSIS This paper will attempt to cast some light on the ever ongoing dispute between the followers of physical methods for wind resource estimation such as WA^{S}P (Wind Atlas Analysis and Application Program), and the followers of statistical methods such as Measure-Correlate-Predict (MCP).

It is demonstrated that, for sites in complex terrain with only a few months worth of data, the outcome of the estimate of the wind resource when using MCP is very sensitive to which months and how many the correlation has been calculated for.

It is also shown that WA^{S}P is performing quite well, despite the fact that the assumptions underlying the flow model are violated.

1 INTRODUCTION

When one wants to estimate the wind energy resource at a given potential windfarm site, with no or few measurements, one has to link these measurements to measurements of a long duration from another (near-by) site. The idea behind this being that within a certain distance - given by the local meso-scale conditions - the overall wind climate is the same. To obtain this link, two methods can be used:

1. A physical method, i.e. a method based on a physical model of effects affecting the two sets of measurements.

2. A statistical method, i.e. a method based on statistical correlations between the two time-series.

Here we will make some general statements with respect to the two types of methods. We have chosen to concentrate on one representative of each of the two types of methods: for the physical method WA^{S}P (Mortensen et al, 1993a) is used and Measure-Correlate-Predict (see eg Derrick, 1993) is used for the statistical method. Both these methods are widely used in the wind resource estimation community.

2 GENERAL REMARKS

Since this study is based on a specific set of data from an area with complex terrain we would like to start out with a few general remarks about the two models under discussion, a summary is given in Table 1.

2.1 WA^{S}P (Wind Atlas Analysis and Application Program)

WA^{S}P is a PC-program used all over the world to estimate wind resources. Its major advantage is that it can generalise a long-term meteorological data series (collected at e.g. an airport) to be valid not only at the site where it has been measured, but in an area around the measuring site. The size of this area depends on the gradients in the geostrophic wind (such as eg in northern Europe) and on the local flow regimes (such as eg in southern Europe). The way the generalisation is done is by correcting the data series for effects which only affect the measuring site, but are not of more general nature. These local effects are: shelter from near-by obstacles (as houses, wind breaks etc), the effect of roughness and

119

changes in roughness (e.g. from water to land), and the effect of orography (e.g. speed-up on hill tops).

The major disadvantage is that the program does not (at present) include thermally driven local effects, as see-breezes (caused by the different heating of land and water), ana- and katabatic winds (winds caused by heating and cooling of the surface, respectively). Another disadvantage is that it is not possible, beforehand, to give a solid estimate of the size of the region where the calculated wind climate is valid. Hence one has to base the estimates on prior knowledge obtained from the terrain type one is operating in.

2.2 MCP (Measure Correlate Predict)

MCP can like WA^{S}P be run on almost any PC, and it is used in many places, especially in the UK. The advantage of the method is that one can put up a mast (preferably 30 m high) at the proposed wind turbine site for only a few months, correlate the measured time-series with a near-by long-term time series (measured at e.g. a synoptic station), and then, when the correlation has been established, use the long-term time-series to estimate the wind resource at the proposed site.

The disadvantage is that if the long-term and the on-site time-series do not have a high correlation coefficient then the resulting estimate is not very reliable. Another disadvantage is that the resource is only valid at the location of the on-site mast and at the height of the measurements. So to use this sort of technique at eg a wind farm a program that can calculate the wind resource at different places (as WA^{S}P) must be used anyhow.

3 AN EXAMPLE

As an example of the use of the two methods, data from 10 meteorological masts located in the northernmost part of Portugal will be used. These data have been measured as part of the University of Porto's contribution to the CEC-funded JOULE project "Wind measurement and modelling in complex terrain". A detailed description of the stations and their surroundings, the instrumentation used and the data obtained so far is given by Restivo and Petersen (1993), a map is shown in Figure 1.

For the purpose of this investigation six stations were selected, representing three different degrees of topographical complexity along a 50-km long profile: from the Atlantic coast in the W to the more than 1000-m high mountains of Arada-S. Macário in the

Table 1: An overview of the two methods' pluses and minuses.

WA^{S}P	+	works with no on-site measurements
	+	works at any height
	+	works at any place
	−	may give inexact results in very complex terrain
	−	does not include local thermally driven effects
MCP	+	can generalize a short on-site measured time-series, if a long (climatological) is available
	−	if on-site data are wrong, generalization is wrong
	−	can only predict at the height of the measurements of the on-site mast
Both	−	if long time-series is wrong (or non-representative) on-site resource is wrong

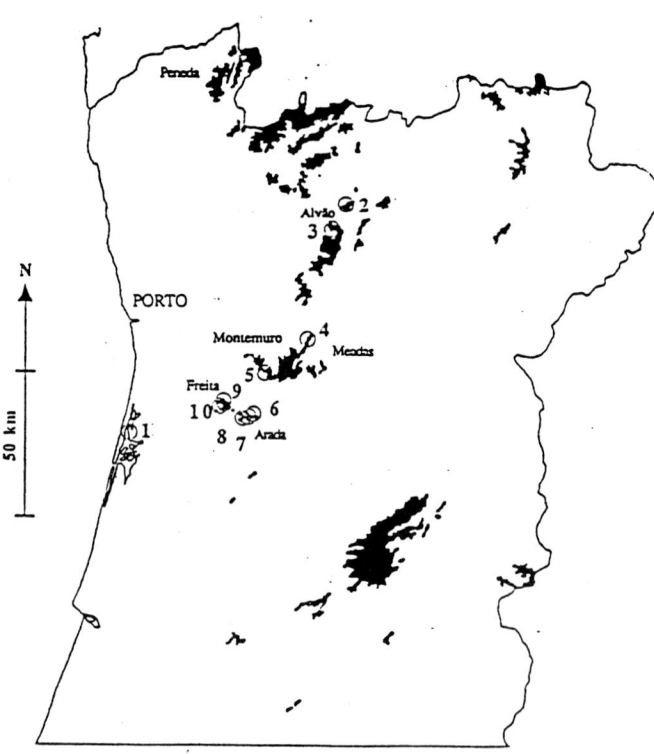

Figure 1: A map of the stations used in this study from Restivo and Petersen, 1993.

E. In the following we will refer to the stations by the numbers 01, 10, 09, 08, 07, and 06 (going from W to E). It must be stressed that in this type of terrain (like the one chosen here) the assumptions behind the flow model in WA^{s}P with regard to the slopes of the terrain involved are severely violated at some of the sites. For the MCP, simple guidelines as to where to use the method do not exist, but it is expected that this case also is very close to the scope of validity of MCP. The purpose of this exercise is thus to see how well the two methods perform in a real case, *in spite* of these violations.

3.1 The data

At each station the wind is measured at 10 m agl. The instrumentation is as follows

- A cup anemometer (Risø 70) measuring 10-minute averages and 3 sec gusts, consecutively. Each anemometer has been individually calibrated.

- A direction sensor (Aanderaa Instruments 2750) giving the 'instantaneous' wind direction.

The data are collected using an Aanderaa Instruments 2990 data storage unit. The measuring period is from June 26th 1991 to July 23rd 1993, ie more than two year's worth of data are available.

3.2 WA^{s}P

The WA^{s}P analysis is much along the lines of the analysis presented in Mortensen et al (1993b); except for the fact that more data are now available. As can be seen from Table 2 - where data from each of the 6 stations have been used to predict the wind climate over a flat grass field at 10 m height - there is more or less the expected agreement between the different predictions (in the case of identical overall wind climates they should all be identical). The reasons why there are the minor differences, especially between stations 6 and 7 and the rest, can be any of the following: 1) The masts are situated in different wind climates, 2) the WA^{s}P-model is not able to simulate properly the flow in the very complex terrain around stations 8, 7 and 6, and/or 3) the maps used in WA^{s}P are not covering an area big enough to include all orography having influence on the flow at the sites. In the following sections we will discuss some of these aspects.

Table 2: Estimated mean wind speeds (ms^{-1}) and mean wind power densities (Wm^{-2}) 10 m a.g.l. over a uniform, flat surface of $z_0 = 0.03$ m (roughness class 1).

Station	01	10	09	08	07	06
Wind speed	3.7	3.9	3.9	3.8	2.9	3.1
Power density	72	82	83	85	33	39

3.2.1 Inter-predictions

To see whether the masts are situated in different wind climates the data from the stations are used to predict each other, see Table 3 and cf. Mortensen et al (1993b). As can be seen from this table stations 1, 10, 9 and 8 predict each other quite well, but turning to stations 6 and 7 different results are obtained. It is very hard to find an explanation for this, one could be that the two stations, which are located furthest away from the coast, are actually in a different wind climate (eg one dominated by other thermally driven local flows than the rest of the stations). This has the consequence that station 8 which is only 5.6 and 3.2 km away, respectively, from stations 6 and 7 is in the wind climate ranging all the way from the coast and 50 km inland. This means thus, that the wind climates generated by WA^{s}P in this example are valid as far as 50 km away. In some cases, however, the climate is valid out to only 3 km. This stresses the fact that when in complex terrain the WA^{s}P model should be used with utmost care.

Table 3: Measured and estimated mean wind speeds (ms^{-1}) and mean wind power densities (Wm^{-2}) at 10 m a.g.l. for the six stations. The power densities have not been corrected for the influence of different air densities (which is of the order of 10 %).

	01	10	09	08	07	06	Meas.
01	**4.1**	4.2	4.3	4.2	3.2	3.4	4.3
	108	105	120	111	48	53	115
10	5.6	**5.4**	5.5	5.5	4.2	4.5	5.5
	278	**218**	243	262	99	130	215
09	5.6	5.9	**5.8**	5.5	4.2	4.6	5.8
	275	311	**271**	236	125	136	269
08	6.5	6.7	6.4	**5.9**	4.7	5.2	6.0
	511	518	410	**295**	213	225	290
07	6.3	6.9	7.0	6.9	**5.2**	5.4	5.3
	351	499	514	555	**194**	223	195
06	5.3	5.8	5.8	5.6	4.3	**4.5**	4.6
	217	293	281	281	115	**128**	128

3.2.2 Sensitivity to the size of the map

In complex terrain it is very important that all terrain significant to the flow model is present in the maps used as input to WA^{S}P. To see the sensitivity to this, it has been tried to vary the size of the maps from 1×1 km to 8×8 km in steps of 1 km. The result is shown in Figure 2. As can be seen from this figure it is necessary for the maps to cover areas at least 6×6 km for the calculations of WA^{S}P to converge. The explanation of why stations 6 and 7 have a much higher decrease (and as a consequence, a higher sensitivity to the area of the map) can also be found in the paper by Mortensen et al (1993b) Table I, where it is seen that the change in 'orographic complexity' (σ_z) is much more rapid for stations 6 and 7.

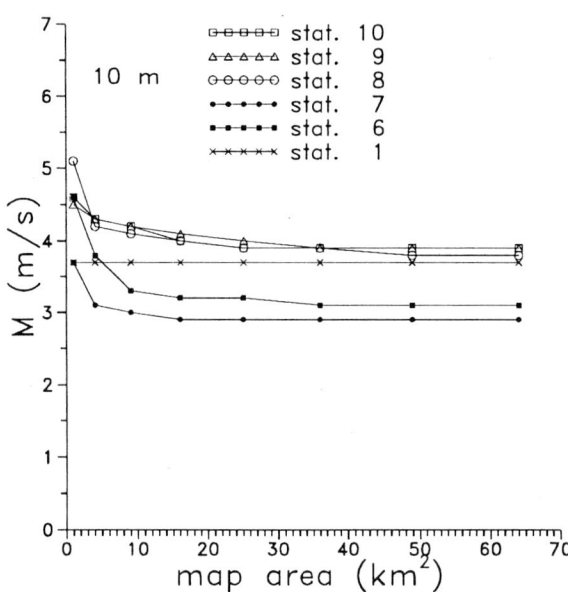

Figure 2: The calculated mean wind, M, at 10 m agl over a flat grass field, cf Table 2, plotted versus the area of the orographic map used in the WA^{S}P program for the 6 selected stations.

3.3 Measure-Correlate-Predict

Several have applied the Measure-Correlate-Predict (MCP) method to the resource estimation problem (see eg Derrick, 1993). Basically what the method does is that it establishes a relation between a short on-site time-series and a long time series eg from a near-by synoptic meteorological station. This relation can then be used to estimate the long-term wind climate at the new site. Normally the relation is taken to be linear, ie

$$u_{\text{on-site}} = a + b u_{\text{long-term}} \qquad (1)$$

Sometimes the relation is a simple scaling, ie without an offset ($a = 0$).

To use the MCP method a good correlation between the on-site and the long-term data is cardinal. In the following we will therefore investigate this matter.

3.3.1 Cross-correlations

Studying the 15 possible cross-correlations (assuming westerly flow) between the 6 stations it is found that the general shape of the correlation function is as shown in Figure 3. The correlation function is defined as the cross-correlation between the two time-series with different lags plotted against the lag. The cross-correlation is given by

$$r_{xy}(k) = \frac{c_{xy}(k)}{\sigma_x \sigma_y} \qquad (2)$$

where

$$c_{xy}(k) = \frac{1}{N} \sum_{t=1}^{N-k} (x_t - \bar{x})(y_{(t+k)} - \bar{y}) \qquad (3)$$

and x_t the value of the time-series at time t, σ_x and σ_y are the standard deviations, $\bar{x}$ and $\bar{y}$ the mean values, and $k(= 0, 1, 2, \ldots, K)$ the lag.

In this study we only consider wind speeds (or rather 10 minute averages) higher than 3 m/s, to exclude the very weak winds with ill-defined directions, since they are often very site-dependent (and therefore uncorrelated) and insignificant for wind energy purposes. The cross-correlations are also calculated with all wind directions in the same bin. Since all the correlation functions generally decrease it is concluded that when calculating the linear regression it is not necessary to take any time-of-flight lag into account. The correlations are shown in Table 4, and as can be seen the correlation between the coastal station (station 01) and the rest of the stations is low, but indicating, none the less, that some correlation exists between the stations. This can most likely be explained by the fact that the distances vary from approx. 40 to 50 km.

3.3.2 The regression

To study the sensitivity of the regression two experiments have been carried out:

1. Each of the stations have in turn been considered as the station with the long term time-series. The data from the other stations have then been used as the on-site data. At each

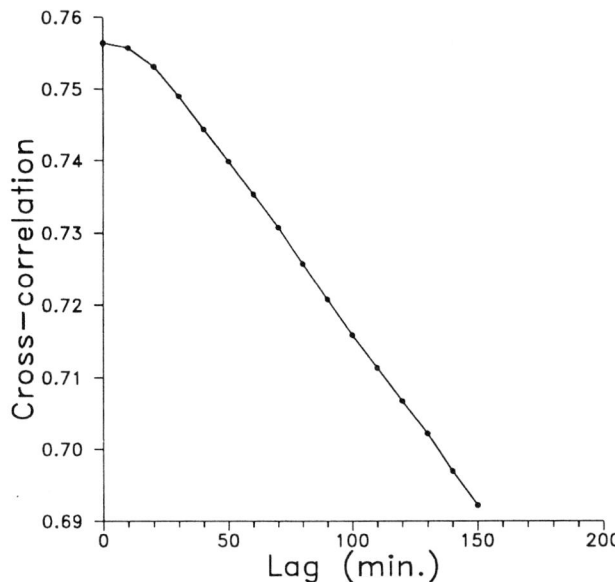

Figure 3: The cross-correlation function for station 10 and 07.

Table 4: The cross-correlations for the selected stations (lag 0 min). The stations are listed from west to east going from left to right.

	01	10	09	08	07	06
01	1.00	0.67	0.63	0.71	0.60	0.56
10		1.00	0.70	0.82	0.76	0.80
09			1.00	0.86	0.87	0.89
08				1.00	0.88	0.91
07					1.00	0.91
06						1.00

of the 'on-site' stations 7 'measurement campaigns' have been carried out, the campaigns lasted for 1, 3, 5, 7, 9, 11 and 12 months, respectively. The idea of this experiment is to see the sensitivity to the length of the measuring period and also the dependence on geographical location.

2. At each of the stations the measuring campaign described under item 1 is shifted in time with step lengths of one month. As an example consider the 3 months measuring campaign. The first 3 months period spans months 1 to 3, the second spans months 2 to 4, the third months 3 to 5 etc. The idea here is to see the annual variation of the result of the regression.

In all the calculations above, the linear regression is calculated sector by sector in 12 sectors.

The results for station 10 (a mountain station) using data from station 01 (the coast station) are shown in Figure 4. A number of conclusions can be drawn from this figure:

1. As the length of the measurement campaign increases the scatter of the MCP estimate is reduced.

2. For a measuring campaign lasting one year the scatter is still ± 0.5 m/s. Note, that in practice, campaigns lasting from 3 to 8 months are used.

3. The actual mean value at station 10 is never reproduced by the method, the method seems to converge to some other value (approx. 5 m/s). This is quite puzzling (and it is found for all the stations), the explanation being that the correlation coefficients are calculated using only winds higher than 3 m/s, which means that if the wind below 3 m/s do not fall on the calculated regression line, differences can be expected. When using all wind speeds in the regression the actual mean (5.5 m/s) is found as the convergence value. This case - contrary to normal procedure - suggests that all wind speeds should be used when calculating the correlation.

For reference, the mean of the predictions, using station 1 as the long-term reference, where the correlation has been calculated over a period of 12 months, are shown in Table 5.

4   DISCUSSION and CONCLUSIONS

This study has shown that both WA^{S}P and MCP are able to make useful predictions of the long-term

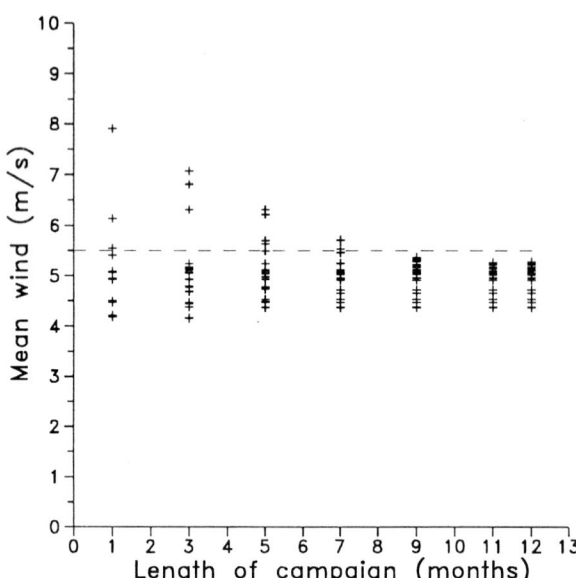

Figure 4: The predicted mean wind (m/s) at 10 m agl for station 10 using station 1 as the long-term reference station plotted against the length of the on-site measurement campaign at station 10. For each of the campaigns several parts of the time-series have been used.

Table 5: The mean wind speed (m/s) and the standard deviation (m/s) of the MCP-predictions using station 1 as the long-term reference for each of the stations, where the correlation has been calculated over a period from 3 to 12 months compared to the measured mean speed.

	10	09	08	07	06
mean, 3 months	5.1	5.6	5.7	4.9	4.2
std. dev. 3 months	0.8	0.7	0.8	0.7	0.7
mean, 5 months	5.0	5.5	5.6	4.9	4.2
std. dev. 5 months	0.6	0.5	0.6	0.5	0.5
mean, 9 months	5.0	5.4	5.5	4.7	4.0
std. dev. 9 months	0.3	0.3	0.3	0.3	0.3
mean, 12 months	5.0	5.4	5.5	4.7	4.0
std. dev. 12 months	0.3	0.3	0.3	0.3	0.3
Measured	5.5	5.8	6.0	5.3	4.6

wind climate at the selected sites in the studied area. The uncertainties for each model can, however, be quite large, especially for MCP. It must be borne in mind that the area under study is quite complex, and most likely dominated by other effects than just the complexity of the terrain. These effects could be thermally driven circulations on the meso- scale, and none of the two methods are able to include this. It is therefore necessary to take other models and concepts into account. The most obvious is to run a meso-scale model such as the KAMM (KArlsruhe university Meso-scale Model). This is now being done at University of Karlsruhe, Germany, in co-operation with Risø National Laboratory, Denmark, in the CEC-sponsored project mentioned in the beginning of this paper. The results of the very preliminary studies are very promising. In these studies KAMM is run with a horizontal resolution of 5×5 km^2 for northern Portugal (the area where the stations discussed in this study are located). More firm results from this model are expected in the middle of 1994.

ACKNOWLEDGEMENTS

We would like to acknowledge the late Prof. Antonio Restivo, who died suddenly and very unexpectedly a few months ago, for all his help and generous co-operation over the many years we worked with him.

We would also like to thank Dr Erik L Petersen, Risø National Laboratory, for many helpful discussions. The work was supported by the CEC under contract JOUR-0067-C.

REFERENCES

Derrick, A, 1993: *Development of the Measure-Correlate-Predict strategy for site assessment.* In proceedings of the European Community Wind Energy Conference, Travemünde, ed AD Garrad, W Palz and S Sheller. p 681-685.

Mortensen, NG, L Landberg, I Troen and EL Petersen, 1993a: *Wind Atlas Analysis and Application Program (WAsP). Vol. 2: User's Guide.* Risø-I-666(EN)(v.2). Risø National Laboratory, Roskilde, Denmark. 133 pp.

Mortensen, NG, EL Petersen and L Landberg, 1993b: *Wind resources, part II: Calculational methods.* In proceedings of the European Community Wind Energy Conference, Travemünde, ed AD Garrad, W Palz and S Sheller. p 611-614.

Restivo, A and EL Petersen, 1993: *Wind measurement and modelling in mountainous regions of Portugal, Preliminary results*. In proceedings of the European Community Wind Energy Conference, Travemünde, ed AD Garrad, W Palz and S Sheller. p 603-606.

Troen, I. and E.L. Petersen (1989). *European Wind Atlas*. ISBN 87–550–1482–8. Risø National Laboratory, Roskilde. 656 p.

Determination of areas of similar mean wind velocities in complex terrain using the example of the Swabian Alb in Germany

J. ALBIGER and **V. REICH**
University of Stuttgart, Germany
A. BÖHRINGER and **H. MÜH**
Energie-Versogung Schwaben AG, Stuttgart, Germany

In foothills like the Swabian Alb located in the South of Germany it is difficult to find suitable areas for an installation of wind energy converters. The topography and the land utilisation influence the wind conditions and the areas available for establishing wind farms. This paper describes a procedure which allows to determine the areas of similar mean wind velocities as well as the wind potential (areas on which an installation of wind energy converters is possible) based on wind measurements at several locations, a digital height model and data about the land utilisation. With a developed computer model the windspeed for every point in the investigation area is determined under consideration of the topography and the land roughness. Finally, areas of similar wind velocities and suitable sites can be identified. Also first results are presented obtained by applying this model to the Swabian Alb.

1 INTRODUCTION

In Germany the wind energy use has gained more significance during the last few years. This is manifest among others in the government support progamme of this renewable energy source (250 MW Windprogramm). Although at the moment wind energy converters are being installed almost exclusively at the North- and East sea coast and in the North German plane, the inland areas will become more important, since the best situated locations at the coast will be in use soon. Against this background it is the objective of this paper to describe a method for the determination of the mean wind velocity distribution on a low mountain range. Since the topography and the land roughness influence the wind conditions significantly, both have to be considered in a spacially high resolution. The methical approach is applied to the Swabian Alb, a low mountain range located in the South of Germany.

2 WIND MONITORING PROGRAMME

In the course of a monitoring programme of the Energie-Versorgung Schwaben AG (EVS) [1] wind measurements was carried out at different locations for 6 to 12 months. Figure 1 shows the position of the monitoring stations indicated in the Gauss-Krueger coordinate system. The High Value is equal to the distance to the aquator and the Right Value indicates the distance to the 9 degree meridian, added with an offset of 3 500 000 m.

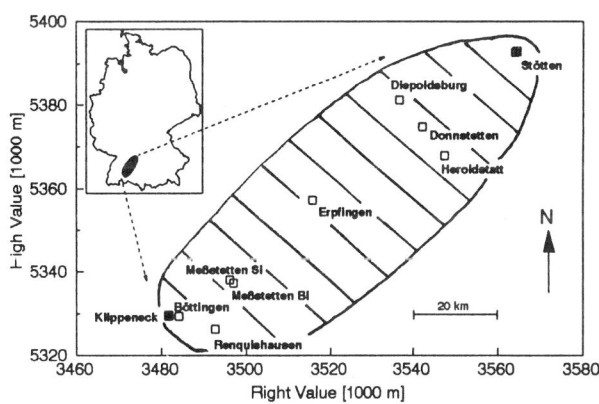

Fig. 1: Geographical Position of the Monotoring Stations.

2.1 Determination of Long Term Mean Windspeeds

The wind conditions at the different stations are monitored only for limited and partly different times. This is the reason why they are neither comparable to each other nor representative for long term wind conditions. Our objective is to transfer these measured values into expected long term means. For that purpose the means of the monitored windspeeds of one period are determined for every wind direction and are set in relation to the corresponding means of the German

Weather Service (DWD) station Stötten of the same period. From this monitoring station of the DWD long term measured windspeeds are avilable [2]. This ratio shows how much higher or lower the windspeed in comparison to the according value of the DWD-station is. Provided that this ratio does not depend on the chosen time period and is approximately constant, the long term conditions at the short term monitoring stations can be estimated. After having carried out this correlation separately for every wind direction, the long term average wind distribution with the corresponding windspeeds can be given. This serves as basis for the calculation of the long term windspeed compressed for all directions. Table 1 indicates these long term windspeeds of the different EVS-stations and of the DWD-stations Klippeneck and Stötten. The velocities range from 3.7 m/s for Renquishausen to 4.6 m/s for Donnstetten.

Table 1: Long Term Mean Windspeed and Height above Sea Level of the Different Stations.

	Long Term Mean Windspeed	Height above Sea Level
Monitoring Stations (EVS)		
Böttingen	4,2 m/s	993 m
Diepoldsburg	3,9 m/s	805 m
Donnstetten	4,6 m/s	838 m
Erpfingen	3,8 m/s	800 m
Heroldstatt	4,2 m/s	816 m
Meßstetten-Bl	4,2 m/s	962 m
Meßstetten-Si	4,0 m/s	978 m
Renquishausen	3,7 m/s	885 m
Monitoring Stations (German Weather Service)		
Stötten	4,5 m/s	734 m
Klippeneck	3,7 m/s	973 m

2.2 Analysis and Intepretation of the Results

The Black Forest located at the West of the Swabian Alb is a large scale wind flow obstacle which influences the wind conditions over the investigated terrain. At wind blows from West, the Swabian Alb is situated in the wind shadow of the Black Forest. This effect becomes evident if the mean windspeeds from West are plotted over the Right Value of the respective monitoring station (see figure 2). With Right Value increasing the distance to the flow obstacle becomes bigger. This leads to an increase of the mean windspeed. The hatched area in figure 2 indicates the range of the windspeeds. Although the stations Diepoldsburg, Donnstetten and Heroldstatt are situated at lower height above sea level than e.g. Klippeneck or the stations in Meßstetten, the higher West windspeeds are

not measured over the higher located West-Alb, but over the East part of the Swabian Alb.

Since the west wind blows with a higher velocity and more often than wind from other directions, the shadow effect described above also influences the windspeed averaged for all wind directions. In spite of the lower heights of the East Alb the mean windspeeds are higher than of the higher situated West-Alb.

Figure 2 also shows that windspeeds at monitoring stations of which some are only a few kilometres away from each other differ considerably. This, among other factors, is due to the influence of the local topography and surface roughness. The mechanical and thermal roughness change the turbulence and the wind profile. The topography can channel and accelerate but also block and slow down the current.

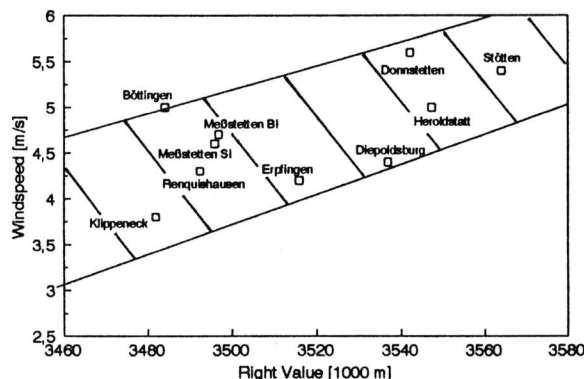

Fig. 2: Mean West Windspeeds for Different Monitoring Stations on the Swabian Alb.

3 METHODICAL PROCEDURE

In order to determine the regional wind conditions based on these wind data, it is necessary to quantify local specific effects which influence the wind conditions significantly. Therefore, the topography and the land roughness have to be modelled in a spacially high resolution. After that, the equations have to be formulated which describe the relation between the terrain form and the wind current. Now the wind conditions can be transferred to the vicinity around the station taking also the geographical height into consideration. This is carried out separately for different wind directions in order to consider large scale effects.

3.1 Topography

Digital height modell of Baden-Württemberg [3] is taken as the basis for the modelling of the topography of the investigation area. In this height model the area of the Federal State Baden-Württemberg is divided in square elements with an edge length of 50 m. The height above sea level for each element is given in meters. The Right and High Value of the Gauss-Krüger coordinate system determine the geographical postion of the square. Based on these data, it is possible to describe the terrain in a spacially high resolution. As a result we get the data base necessary to calculate the local effects of the topography on the wind conditions.

3.2 Terrain Surface

Among the topography the structure and the attributes of the surface have a strong influence on the wind conditions. To model the ground surface LANDSAT-TM Satellite photos are evaluated. Based on so called training areas of which precise land utilisation data are known spectral values delivered by the satellite in a 30 x 30 m resolution are ordered into the classes forest, residential area and agricultural area. Subsequently, the satellite data can be transformed into information about the land surface. Hereby it has to be considered that the borders between the single land utilisation data are fluent. In reality there are more than three classes. This can cause mistakes during the data classification. In order to correct these inconsistancies, an algorithm searching for mistakes investigates the transformed data before next steps are made.

Because of the banking of the satellite the classified and corrected land utilisation data have to be rectified. Then the data are transformed into the Gauss-Krüger coordinate system. In order to obtain the same resolution as the height modell, the resolution is enlarged to 50 x 50 m. Finally the data are stored in the same format like the data of the height modell to simplify the access.

3.3 Small Scale Wind Conditions

The influence of the topography and the surface structure on the wind conditions can be described by equations of the fluid mechanics. First of all, it is started with a two-dimensional description of the terrain. For a certain angle in the horizontal plane a vertical sectional plane is assumed. This cross section is a two-dimensional contour line whose heights and attributes concerning the land utilisation data serve as data basis for the equations. With the results of these equations an amplification or a diminution factor for the windspeed can be calculated in dependence of the

topography and surface roughness. The sectional plane is revolved gradually on the vertical axis in the investigation point. In doing so new contour lines are generated which are evaluated in the same way. For one wind direction the factors of the contour lines lying in the sector are averaged. The average characterizes the investigated direction sector.

In order to determine the influence of the topography, a stationary incompressible two dimensional flow is assumed which is composed of a basic flow and a part due to the topographical terrain. Based on the theory of the thin wing [4] and the simplifying presuppositions (friction-free flow, which always follows the surface, small ratio between the height of a topographical feature to its horizontal extents) a connection between the terrain shape (here the contour line) and the change in the windspeed is given.

Out of that the surface roughnesses are considered. The classified land utilisation data readable for every contour line are subjected to the roughness model. For the different land utilisation data we take the roughness lengths indicated in the European Wind Atlas [5]. The roughness length of the elements on the contour line are averaged. Hereby elements with a smaller distance to the investigation point are more weighted by a factored function. If the measured windspeed are related to a velocity which would result from a relatively smooth surface (roughness length equal to 0.03 m) a degree of the change in speed is obtained due to the given roughness.

3.4 Restrictions to Install Wind Energy Converters

In order to determine the wind potential, both the long term wind conditions and the areas on which an installation of wind energy converters is possible have to be known. The linking of both these data leads to areas available for an installation. For the evaluation of such areas, the exclusion of areas is concentrated on residential areas and forest, protected nature and military zones and traffic roads. Moreover, the following safety distances are considered [6]:

Forest	200 m,
Residential area	500 m,
Roads	50 m,
Protected Nature Zone	200 m.

The data about the forest and the residential zones belong to the land utilisation data already used in the roughness model (see chapter 3.2). The other data concerning the roads, military and protected nature zones are collected from different sources. They have then to be adopted to the format of the data base already existing [7, 8]. An algorithm checks whether an 50 * 50 m element lies within one of the listed safety distances. Should this be the case the algorithm marks it as not available for wind energy use.

3.5   Verification of the Model

The developed programme system SimWind combines the procedure described above and the data with a graphical output. The entire wind model is based on some simplifying suppositions, because always a compromise has to be found between the efforts and the obtainable accuracy. Therefore, the results obtained by SimWind have to be verified first.

Supported by the graphical output, the land utilisation data can be easily verified by comparison with usual land maps. In the same way the traffic roads, military and protected nature zones are verified. To check the wind model, the single monitoring stations are successively taken out of the station network. Therefore the wind model does not find the information about the long term wind means at the position of the station taken out. In this case the wind conditions have to be calculated with the aid of the remaining stations. Table 2 shows the comparison between the long term means based on data measured at the station (left column) and based on the interpolated results of the other stations (right column). In other words, the right column contains the prediction of the remaining stations for the corresponding station.

Table 2: Comparison between Measured (compare table 1) and Predicted windspeeds.

Long Term Means	Measured	Predicted
Monitoring Stations (EVS)		
Böttingen	4,2 m/s	3,9 m/s
Diepoldsburg	3,9 m/s	4,2 m/s
Donnstetten	4,6 m/s	4,2 m/s
Erpfingen	3,8 m/s	3,8 m/s
Heroldstatt	4,2 m/s	4,2 m/s
Meßstetten-Bl	4,2 m/s	4,3 m/s
Meßstetten-Si	4,0 m/s	3,9 m/s
Renquishausen	3,7 m/s	3,7 m/s
Monitoring Stations (DWD)		
Stötten	4,5 m/s	4,2 m/s
Klippeneck	3,7 m/s	4,0 m/s

Table 2 indicates that the measured and predicted means correspond quite well. The relative deviation amounts to maximally 10 %. At several stations the windspeeds are forecasted with in accuracy of 0,1 m/s. On the one hand, the good accordance is due to the compact network of stations. On the other hand, the simplifying equations concerning the small scale effects (local topography and roughness) seem to represent the real wind conditions sufficently well.

4   RESULTS

The procedure described in chapter 3 is applied to the area of the Swabian Alb. The programme system determines the long term windspeeds for any point in the spacial resolution of 50 * 50 m. Based on the calculated wind conditions at every point, areas of similar windspeeds can be deduced. Under consideration of the given restrictions (see chapter 3.4) and a minimum windspeed of 3,5 m/s only 2,5 % of the investigated area remain available for an installation of wind energy converters; this means 130 km^2 of an entire investigation area of 5300 km^2. These 130 km^2 can be subdivided into different windspeed classes. As shown in Figure 3, about 119 km^2 of the wind potential are indicated by windspeeds lower than 4,0 m/s. Only 11 km^2 are characterized by windspeeds larger than 4,0 m/s.

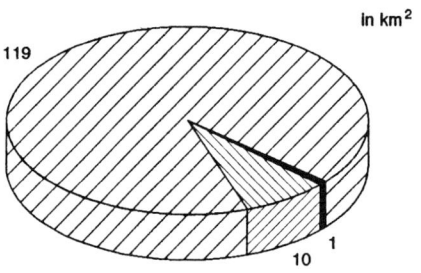

In km^2

☐ 3,5 - 4,0 m/s   ◨ 4,0 - 4,5 m/s   ■ > 4,5 m/s

Fig. 3:   Wind Potential Areas of the Swabian Alb.

Although most of the investigated area is not usable for the wind energy concerning today's economical boundary conditions, there are a lot of suitable sites. With the help of maps showing the areas of similar windspeeds, these good locations can be easily identified. Out of that, the information about suitable sites are collected in a file. After sorting the sites concerning the predicted mean windspeed, a ordered list of locations allows to exploit the given wind potential in a most efficient way in terms of the wind conditions.

5 FINAL VIEW

The objective of the paper was to describe the procedure to identify spacially high resoluted areas of similar mean windspeeds in complex terrain. Based on this method and the resulting computer model, suitable sites can be determined. The procedure was developed at the example of the Swabian Alb and shall be transformable to other low mountain ranges.

The applications of the methodical to the Swabian Alb shows:

- that the developed computer model is a very useful tool for an analysis of the general wind potential (areas available for installing wind energy converters) and for a detailed identification of suitable sites,
- that about 130 km^2 of the investigated area (5300 km^2) are characterized by windspeeds larger than 3,5 m/s and are available for an installation of wind energy converters,
- that there are suitable sites where the mean windspeed exceeds 5,0 m/s.

In a further step it is planned to simulate a generation of electricity on the available areas. In dependence of the wind energy converter spacing, the number of wind converters, which can be installed on the Swabian Alb, will be determined. Since the wind conditions are already known, the energy output can be calculated by using power curves of different wind energy converters. By this means it shall be possible to predict the energy output for both the entire investigation area and single sites.

Literature

[1]   Böhringer, A.
       Auswahl und Beurteilung von geeigneten Standorten für Windkraftanlagen auf der Schwäbischen Alb
       7. Internationales Sonnenforum 1990, 9. - 12. 10. 1990, Frankfurt, Tagungsbericht Band 3, S. 1789 - 1795
[2]   Christoffer, J.; Ulbricht-Eissing, M.
       Die bodennahen Windverhältnisse in der Bundesrepublik Deutschland
       Berichte des Deutschen Wetterdienstes Nr. 147
       Selbstverlag des Deutschen Wetterdienstes, Offenbach a. M., 1989, 2. vollständig neu bearbeitete Auflage
[3]   Landesvermessungsamt Baden-Württemberg (Hrsg.)
       Digitales Höhenmodell Baden-Württemberg
       Landesvermessungsamt Baden-Württemberg, Stuttgart, 1990
[4]   Karamcheti, K.
       Principles of ideal-fluid aerodynamics
       John Wiley and Sons, New York, USA, 1966
[5]   Troen, I.; Petersen, E. L.
       Europäischer Windatlas
       Riso National Laboratory, Roskilde, Dänemark, 1990
[6]   Kaltschmitt, M.; Wiese, A. (Hrsg.)
       Erneuerbare Energieträger in Deutschland
       Springer-Verlag, Berlin, Heidelberg, New York, 1993
[7]   Landesamt für Straßenwesen Baden Württemberg (Hrsg.)
       Straßendatenbank Baden-Württemberg
       Landesamt für Straßenwesen Baden-Württemberg, Stuttgart, 1990
[8]   Landesanstalt für Umweltschutz Baden-Württemberg
       Räumlichkeitsinformations- und Planungssystem (RIPS) des Umweltinformationssystems Baden-Württembergs, Karlsruhe, 1993

The influence of transmission system disturbances on the dynamic behaviour of a wind park

RUI M. G. CASTRO and **J. M. FERREIRA DE JESUS** PhD
Instituto Superior Técnico, Portugal
ANA I. L. ESTANQUEIRO MSc, **R. AGUIAR** and **JORGE A. G. SARAIVA** PhD
Departmento de Energias Renováveis, Instituto Nacional de Engenharia e Tecnologia Industrial, Portugal

SYNOPSIS: In this paper an integrated non-linear model of both the wind park and the interconnection network is presented. The model is an integration of models previously developed for each component of the system, namely wind turbines, induction generators, reactive power compensation system, transformers, feeder and possible local loads connected to the feeder. In order to draw some conclusions regarding the interaction between wind parks and the existing a.c. system, some case-studies have been performed using the developed models.

1 INTRODUCTION

In 1988 the Portuguese government issued legislation which allows independent producers to generate electrical energy and obliges the public electrical utility to purchase the energy produced by these producers.

Clearly, at that time wind power was not competitive regarding other renewable energy sources. The limited knowledge of the wind energy potentials, the little experience with the technology and, as a consequence, an incorrect evaluation of the risks by potential plant owners were the main reasons found to explain the very few projects presented in the area of wind energy.

Recently, this situation has changed. Some studies were performed in this area which allowed a better knowledge of both wind energy resources and technology and contributed to a more confident evaluation of this form of energy by private investors (1). Also, the experience with the operation of wind parks in Portugal (namely in the islands of Madeira and Açores), together with the expected reduction in WECS costs, will certainly contribute to attract more investors to the wind energy area.

In these conditions, some wind parks are currently about to be installed in Portugal and a boost is expected in the near future. Issues such as the interactions between the machines inside the wind park and the coupling to the existing a.c. system of an intermittent and non-dispatchable energy source were brought up to date, thus the necessity of possessing the appropriate tools to correctly assess both wind park steady-state and transient behavior became a must.

In order to assess the interaction between the wind farms and the utility grid to where they are connected, some computational tools have been developed with the aim of assisting both wind park and distribution system planners and designers.

These computational tools are based on models that are able to accurately simulate the behavior of wind parks under transient situations. Two types of models have been produced:

A. Models that are able to simulate the transient behavior of each WECS belonging to a park as well as the transient behavior of the park with respect to the utility system, henceforth denoted by detailed wind park models.

B. Models that are able to simulate the relevant dynamics of the park with respect to the utility system, henceforth denoted by wind park reduced order models.

These models address different problems and will be able to assist designers and planners in different but complementary issues. Whereas the detailed wind park models will assist in the design of the configuration of the wind park, the reduced order models will assist in the interconnection between the wind park and the utility grid.

This paper is concerned with the development and application of wind park detailed models above denoted as type A. models, type B. models being the subject of a companion paper "A wind park linearized model" (2).

In this paper an integrated non-linear model of both the wind park and the interconnection network is presented.

The model is an integration of models previously developed for each component of the system, namely wind turbines, squirrel cage induction generators, reactive power compensation system, transformers, feeder and possible local loads connected to the feeder.

In order to draw some conclusions regarding the interaction between wind parks and the existing a.c. system, some case-studies have been performed using the developed models. For instance, the influence that an internal power plant fault or a disturbance in the feeder has in the a.c. system, as well as the impact of utilities normal operation practices (as it is the case of a tripping in the utility grid breakers followed by automatic reclosure) in the behavior of the wind park, have been assessed.

2 DEVELOPMENT

The system studied is presented in Fig. 1

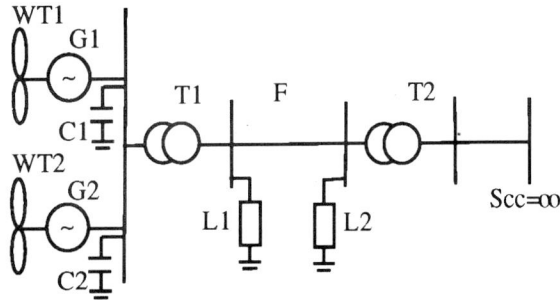

Fig. 1 - System studied.

The wind park studied was composed by two *24kVA*, *400V* induction generators, squirrel cage type, each one driven by a wind turbine.

Each reactive power compensation system was sized in order that the admittance of the capacitor bank equals the slope of the linear part of the no-load magnetization characteristic of the induction generator. This size was chosen as a result of previous studies in this domain (3). Moreover, this size obeys to the Portuguese legislation.

The wind park is connected to a *30kV* feeder through a step-up transformer. This transformer feeds a local load of *5kVA, cosφ=0.9ind*. A substation is represented by another step-up transformer, which has a *15kVA, cosφ=0.9ind* local load connected to the *30kV* side.

The short-circuit power at the interconnection point was assumed to be *20* times the wind park rated power, thus the a.c. system is represented by a reactance with the appropriate value.

Models have been developed to describe the behavior of the different subsystems that constitute the system studied. Also, a wind model able to generate correlated wind time series was developed, currently addressing only the wind turbine side-by-side configuration (4).

The overall model was implemented in a computer program called *INDUSAT* (5) which allows to perform several studies either in steady-state and transient conditions.

A wind turbine model was built after the blade aerodynamic characteristics and taking into account the effects of variable rotational speed, thus enabling a torque/speed characteristics to be obtained (6).

As the prediction of induction generator steady-state and transient performance requires proper account of saturation effects, a model based on *Von der Embse* circuit theory was developed (7).

The local loads were modeled as constant active and reactive power for the computation of initial conditions and as a constant impedance in the transient studies. Both the transformers, the feeder and the a.c. system were modeled also as a constant impedance.

Due to its crucial importance in the simulation of the behavior of the system, a dedicated algorithm was developed in order to obtain the steady-state operating point, i.e. the initial conditions for the set of differential equations. This technique appeals for classical power flow concepts modified in a suitable way, to calculate the relevant quantities, based only on the system parameters and on the instantaneous wind speed.

It is a current practice to describe the overall system in a reference frame rotating at the synchronous frequency *50Hz* imposed by the utility grid, due to the useful simplifications in the system's representation this approach implies. In steady-state conditions the synchronous frequency actually equals the air-gap field frequency of each one of the induction generators composing the wind park. However, in fault or isolation situations leading to both the wind park and the existing a.c. system become isolated systems, the frequency is no longer imposed by the utility grid but is actually dictated by the behavior of the turbine/generator groups.

This clearly indicates that a technique to evaluate each induction generator air-gap field frequency during transients is required, so that the relevant quantities involved can be accurately calculated. Moreover, the reference frame speed must be updated accordingly, in order to a get a systems's representation in a coherent reference frame.

In order to overcome this situation, a method for evaluating the induction generator air-gap field frequency during disturbance situations was developed. This technique is based on the fact that a change in the phase of a stator quantity means that a change in its frequency has occurred, thus enabling to evaluate the frequency through a dedicated algorithm.

As far as the reference frame speed is concerned it should reflect the changes occurred in the induction generators air-gap field frequencies. As a wind park is generally composed of a quantity of induction generators possessing equal characteristics but running in different operating points, it was thought that the reference frame speed should be set to the mean of the generators frequencies.

In order to assess the validity of the models developed some test results were obtained both in purpose-built simulators and in experimental sites. The comparison between computer and test results was found satisfactory (7),(8).

3 APPLICATIONS

The computer program *INDUSAT* was used to obtain a couple of results showing the impact of some disturbances in the feeder on the transient behavior of the wind park. For this purpose, some case-studies were selected in order to portray situations which are typical when the interaction between the park and the network is to be assessed.

Case-study I shows a situation where a fault in the feeder occurs at $t=0.5sec.$ followed by an opening of the feeder breakers at $t=0.7sec.$ and an automatic reclosure $300msec.$ later. These characteristics have been chosen because they correspond to normal practices in the operation of a grid with dispersed generation. To simulate the wind input to the turbines two wind time series with a cross-correlation factor of 0.5 and an equal mean wind speed of $7m/sec.$ were selected. These characteristics apply to a $25sec.$ time scale, but actually we have concentrated our study in a window of $5sec.$ time scale in which the two wind time series have mean wind speed of $6.74m/sec.$ and $6.51m/sec.$, respectively.

Case-study II portrays a similar situation than case-study I but with the difference that the cross-correlation factor between the two input wind time series has been set to 0.7 maintaining the mean wind speed equal to $7m/sec.$ However, in the selected $5sec.$ time scale the two wind time series have mean wind speed of $6.96m/sec.$ and $7.18m/sec.$, respectively.

Figs. 2 display the induction generators *emfs* obtained for case-study I (Fig. 2a) and case-study II (Fig. 2b) simulations. It is apparent from these Figs. that in case-study I the system will reach a stable situation, whereas in case-study II an unstable situation is obtained (demagnetization state).

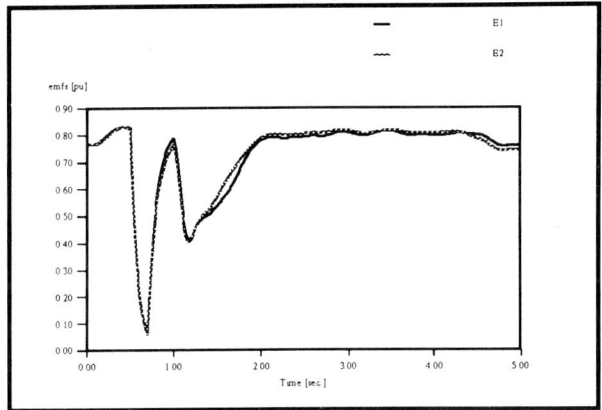

Fig. 2a - Induction generators *emfs* (case-study I).

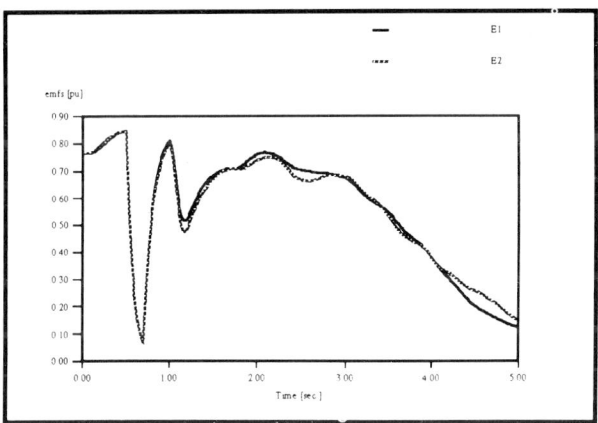

Fig. 2b - Induction generators *emfs* (case-study II).

These results indicate that, as far as the transient behavior of the wind park is concerned, both the wind's mean speed and turbulence play a very important role. Taking the case of the mean wind speed, which is easier to account for, one can remark that the mean wind speed "seen" by the park in case-study I is less than in case-study II what can partly explain the different results achieved from a stability point of view.

The role of the wind characteristics may be even more important than the fault attributes. In order to show this conclusion different fault characteristics have been simulated. Case-study III portrays a situation in which a more serious fault has been simulated with the same wind conditions of case-study I. In the current case-study a fault in the feeder occurs at *t=0.5sec.* followed by an opening of the feeder breakers at *t=0.8sec.* and an automatic reclosure *500msec.* later. This procedure has been accomplished in order to try to superimpose the effect of the fault characteristics on the effect of both the wind's mean speed and turbulence.

The objective of case-study IV is identical, this time by simulating a less serious fault. With the wind conditions of case-study II a fault in the feeder has been simulated at *t=0.5sec.* followed by an opening of the feeder breakers at *t=0.6sec.* and an automatic reclosure *200msec.* later.

The results obtained are presented in Figs. 3, by displaying the *emfs* of the induction generators for case-study III (Fig. 3a) and case-study IV (Fig. 3b).

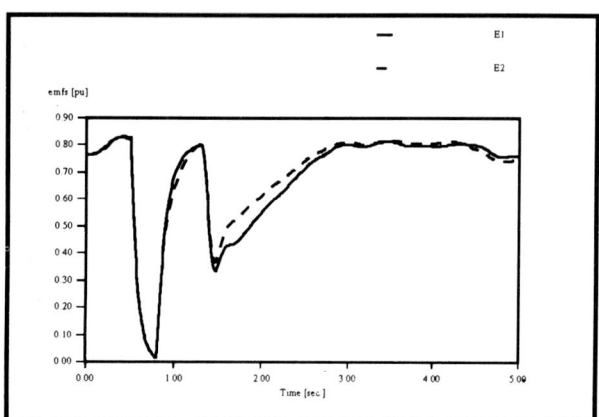

Fig. 3a - Induction generators *emfs* (case-study III).

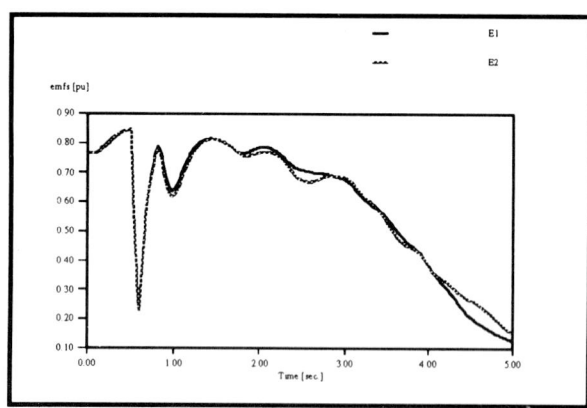

Fig. 3b - Induction generators *emfs* (case-study IV).

This results clearly show that the effect of both the wind's mean speed and turbulence may actually dictate the dynamic behavior of the wind park. In fact, in spite of these situations have been performed with the aim of reinforcing the effect of the fault characteristics, the same results concerning the stability of the system have been achieved.

4 CONCLUSIONS

A detailed non-linear wind park model has been presented, together with a model describing both the transmission system and the utility grid. The model is based on the individual models previously developed for each component of the system, namely wind turbines, induction generators, reactive power compensation system, transformers, feeder and possible local loads connected to the feeder, thus resulting in an integrated full model implemented in a computer program.

In order to draw some conclusions regarding the interaction between wind parks and the existing a.c. system, some case-studies have been performed using the developed models. The case-studies address mainly fault in the feeder simulations, in cases where the wind input time series of the turbines have different mean wind speeds but are related through cross-correlation factors.

As statistical analysis was clearly out of the scope of this paper, only some specific simulations were performed. However, the results obtained allow the conclusion that both the wind's mean speed and turbulence effects may, in some situations, be more important than the fault characteristics in what concerns the dynamic behavior of the wind park.

5 REFERENCES

(1) J.M. Ferreira de Jesus, Jorge A.G. Saraiva, Rui M.G. Castro, Ana I.L. Estanqueiro, Ricardo Aguiar, "Strategic Study of Wind Energy in Portugal", Report prepared for NATO SfS Programme III, Lisbon, Nov. 1992.

(2) Rui M.G. Castro, J.M. Ferreira de Jesus, "A Wind Park Linearized Model", BWEA15, British Wind Energy Association Annual Conference, York, Oct. 1993.

(3) Rui M.G. Castro, A. Eugénio Gomes, J.M. Ferreira de Jesus, "The Influence of Reactive Power Compensation in the Transient Behaviour of an Induction Generator", Proceedings IFAC, Seoul, Aug. 1989.

(4) R. Aguiar, Ana I.L. Estanqueiro, Jorge A.G. Saraiva, "Uma Abordagen Espectral à Geração de Séries Correlacionadas de Vento", (in Port.), VI Congresso Ibérico de Energia Solar, Lisboa, Abr. 1993.

(5) Rui M.G. Castro, J.M. Ferreira de Jesus, "INDUSAT Basic Reference Guide V1.0 - Software Package for WECS Simulation", Instituto Superior Técnico, Lisbon, Jan. 1993.

(6) Ana I.L. Estanqueiro, "Horizontal Axis Wind Turbines - A Behavior's Model", (in Port.), MSc Thesis, Instituto Superior Técnico, Lisboa, Abr. 1991.

(7) J.M. Ferreira de Jesus, "A Model for Saturation in Induction Machines", IEEE Transactions on Energy Conversion, Vol.EC-3, Sep. 1988.

(8) Ana I.L. Estanqueiro, J.M. Ferreira de Jesus, D. Lalos, "Transient Behavior of a WECS Under Different Load Conditions", European Seminar on The Potential for Small & Medium Sized Wind Energy Applications, Rhodes, Jun. 1992.

(9) Ana I.L. Estanqueiro, R. Aguiar, Jorge A.G. Saraiva, Rui M.G. Castro, J.M. Ferreira de Jesus, "The Development and Application of a Model for Output Fluctuation in a Wind Park", 1993 European Community Wind Energy Conference and Exhibition, Travemunde, Germany, Mar. 1993.

(10) Ana I.L. Estanqueiro, J.M. Ferreira de Jesus, Jorge A.G. Saraiva, "WECS Unsteady Power Output Simulation", 1991 European Wind Energy Congress and Exhibition, Amsterdam, Oct. 1993.

Assessing the impact of grid integrated renewable energy sources on the economics of an expanding power system

S. J. WATSON, BSc, ARCS, PhD and **J. A. HALLIDAY**, BSc, PhD, CEng, MBCS, FRMetS, MIEnvSc
Rutherford Appleton Laboratory, UK.
A. G. TER-GAZARIAN, BSc, MSc, PhD, SRA(VAK)
Moscow Power Engineering Institute, Russia
S. C. DAVIS
Trinity College, Cambridge, UK.

Synopsis: This paper assesses the economics of an expanding large scale power system which could accommodate renewable energy sources such as wind power, both from the point of view of long term planning using a deterministic model and of hour to hour system control using a time domain simulation model. The results will illustrate a planning tool which may be used as a guide to the economics of installing wind power into a system such as the England and Wales National Grid.

1 Introduction

Two basic models are used in this work to assess the economics of wind integration. Firstly, the Reading University/Rutherford Appleton Laboratory grid simulation model [1, 2, 3] is used to simulate the hourly running of a grid where conventional plant is scheduled ahead to take account of the system demand and intermittency of renewable energy sources, and on-line spinning reserve capacity is carried to meet unforeseeen shortfalls in wind power generation and peaks in consumer demand. Wind power forecasts are made by site enhanced Met. Office Numerical Weather Prediction Forecasts [4, 5]. Output from the grid simulation model is then used to generate a 'penalty' cost function resulting from the intermittency of wind power. This penalty includes added costs in the form of extra fuel burnt by conventional power stations to provide spinning reserve. Secondly, a deterministic model is used to evaluate the total capital and running costs of a dispersed and expanding power system.

Preliminary results are presented which demonstrate how the optimum growth of wind power in the England and Wales National Grid may be assessed using the two models. Two scenarios are examined which show both the importance of regional wind speed and of limits on power flows between different regions of the grid.

2 The National Grid Model

Details of this model have been widely reported in previous papers [1, 2, 3, 4, 5] and so only a brief description of the important points will be given here.

The Reading University/Rutherford Appleton Laboratory National Grid Simulation Model (NGM) has been updated to reflect the plant mix of the 1992/93 England and Wales National Grid. Recent research using this model [4, 5] made use of 1989/90 system demand and observed and forecast wind speed statistics. It is important to use concurrent load and wind speed statistics to take account of correlations in the two data sets. These data have been used in the present case by scaling the 1989/90 system load data by the ratio of the total electrical energy consumed in 1992/93 to that consumed in 1989/90. The model was used to find the total fossil fuel costs for the year 1992/93 for a number of penetrations of total installed wind power capacity between 0 GW and 40 GW. Wind power was assumed to be forecast using site enhanced Numerical Weather Prediction (NWP/MOS) forecasts [4, 5]. The fuel savings as a result of installed wind power in the grid were then calculated by subtracting the fossil fuel costs for a given amount of installed wind power from the fossil fuel costs incurred when there is no wind power installed in the system. In addition, the 'ideal' fossil fuel savings I^W were calculated by multiplying the unit cost for power generated by conventional power sources when there is no wind power installed in the grid by the number of units of wind power that could have been generated for each penetration were conventional power displaced. The 'ideal' savings are those which would result if wind power were as a conventional power source, *i.e.* it could be switched on and off at will (no 'operating' penalty) and all available wind power could be absorbed by the grid (no 'discarding' penalty). The fossil fuel savings were then fitted by a polynomial, $G^W(X^W)$ as a function of total installed wind power capacity, X^W in order to provide an analytical function to be used by the deterministic model as described in the next section.

Previous work [3] has suggested that the variability of wind power can be reduced by spreading out wind turbines over a large area. The NGM was run assuming that wind power was spread out over ten sites throughout the U.K. Earlier research [3] has suggested that fossil fuel savings could be reduced by up to 50% if all wind power were concentrated at one site instead of spread out over ten sites. To simulate this reduction in fossil fuel savings, an arbitrary analytical function of the variance of installed wind power capacity at the ten sites within the U.K. was devised (see § 3.5).

3 The Deterministic Model

Although the NGM simulates the installation of wind power at a number of sites throughout the U.K., it is essentially a *one-node* model, *e.g.* all conventional power generation and demand are assumed to be at one point. In reality, the England and Wales (or indeed any national) grid consists of a network of interconnected load demand and power supply centres. In theory, one could construct a network which would model the smallest detail of this network down to individual households. In practice, this would require an enormous amount of data and and a very unwieldy model. The best approach is to construct a schematic of the grid consisting of the major interconnected load and generation centres throughout England and Wales. This can then be used to deduce macroscopic results for the purpose of economic forward planning.

To this end, the deterministic model described in this paper makes use of information generally available about the England and Wales grid [6]. The model simulates power flows between 14 regions of the National Grid connected by the 'Super-Grid' *i.e.* the 400 kV and 275 kV network. At present, power flows from Scotland and France have been ignored but will be included in the near future.

The purpose of the deterministic model is to minimise a cost function which is a function of capital and running costs subject to a number of power flow, load demand and capacity constraints. The model is an extension of work detailed in [7].

Table 1: Demand and capacity (1992/93) in the 14 node regions of the England and Wales National Grid. *Excluding 2100 MW of pump storage.

Node	Region	Dem. (MW)	Cap. (MW)
1	North	2716	4295
2	Yorks	5923	10866
3	N Wales & W Lancs	4236	*8888
4	E Lancs	2805	0
5	Notts	312	5349
6	W Midlands	6660	6245
7	Anglia	4671	1497
8	West & Wales	4485	6698
9	Thames	2561	8471
10	Outer London	5597	2230
11	Inner London	3164	132
12	S Coast	3613	1176
13	Wessex	1298	1718
14	Peninsular	1398	0

3.1 The 14-node Schematic

The National Grid was divided into 14 regions (Fig. 1). Mean inter-node connection lengths were determined from maps of the 'Super-Grid' [6]. Power transmission limits between nodes were assumed. Table 1 lists some details of the 14 node regions.

3.2 Load and Generation Curves

A power system must be able to meet the maximum possible demand. The demand curve used in this model consists of three steps: (1) the night-time summer trough, (2) the winter daytime plateau and (3) the winter 'tea-time' peak as illustrated in Fig. 2a. The durations of these three steps are ten, ten and four hours respectively. If desired, a load margin above the winter peak could be added to ensure system reliability.

Coal, oil and combined cycle gas turbine power plant are able to generate at all times in order to meet the required demand (see Fig. 2c). Open cycle gas turbines (OCGT's) are assumed to only operate during the winter peak. Operating costs for each node are calculated taking an average appropriate to the plant mix in that node. Nuclear plant is assumed to generate at constant output at all times and operating costs are neglected. The load in each node is adjusted by the installed nuclear capacity in that node accordingly. New installed wind power plant is assumed to generate only during the plateau period (see Fig. 2b). This was considered reasonable within a three step approximation as wind speed is greatest during the day and the bulk of wind generated power will be during this period. Existing storage plant is neglected at present as the installed capacity is small compared to the total capacity of the grid. However, new installed storage plant is assumed to be charging during steps one and two and discharging during the peak at step three (see Fig. 2d).

3.3 The Cost Function

The cost function, f is a function of running and capital cost of new power plant and new transmission lines. The capital cost of existing plant and transmission lines is assumed to have been written off and thus does not enter the equation. The function is of the form:

$$
\begin{aligned}
f = {} & C_{jk}^{P}(X_{jk}^{P}, a_{j}^{P}, d_{j}^{P}) \\
& + C_{k}^{W}(X_{k}^{W}, a^{W}, d^{W}, v^{\max}, v^{\min}, \lambda) \\
& + C_{l}^{T}(X_{l}^{T}, a^{T}, d^{T}) \\
& + R_{ijk}^{P}(P_{ijk}^{P}, P_{ijk}^{PE}) \\
& - S^{W}(X_{k}^{W}, X^{W}) \cdot G^{W}(X^{W}) \\
& + I^{W}(X^{W}) \qquad\qquad (1)
\end{aligned}
$$

where all summations over subscripts i, j, k, l etc. in this and all subsequent equations are implicit. Note that:

$$
X^{W} = X_{k}^{W} \qquad (2)
$$

The terms on the rhs of Eq. 1 are in order:

1. C_{jk}^{P} — the yearly capital cost of newly installed conventional power plant type j (including storage plant), in node k for installed capacity X_{jk}^{P}, amortisation period a_{j}^{P} and discount rate d_{j}^{P}. This term is the product of the newly installed capacity multiplied by the amortised yearly capital payback per unit of installed capacity.

2. C_{k}^{W} — the yearly capital cost of newly installed wind power plant in node k for installed capacity X_{k}^{W}, amortisation period a^{W} and discount rate d^{W}. This term is also a function of the amount of land in a node with a particular mean wind speed parameterised by the maximum and minimum long term mean wind speeds ($v^{\max}$ and $v^{\min}$, respectively) and the rate of decrease in mean wind speed per turbine λ (the 'land use parameter') as more land is used to site more turbines in a node. This is described in more detail in §3.4.

3. C_{l}^{T} — the yearly capital cost of newly installed transmission lines in a node connection l for installed power carrying capacity X_{l}^{T}, amortisation period a^{T} and discount rate d^{T}. Similarly, this term is the product of the newly installed carrying capacity multiplied by the amortised yearly capital payback per unit of installed carrying capacity.

4. R_{ijk}^{P} — the yearly running cost of new and existing conventional power plant type j (including storage plant) in a node k for generated power from new power plant P_{ijk}^{P} and existing power plant P_{ijk}^{PE} during load duration step i. This term is the product of the generated power multiplied by the average calculated running cost per unit of generated power (taking account of plant efficiencies).

5. $S^{W} \cdot G^{W}$ — the fossil fuel savings due to the use of NWP/MOS forecasts to predict wind power. The term $G^{W}(X^{W})$ is the polynomial fit to the fossil fuel savings as a function of total installed wind power capacity X^{W} as described in §2 above. The function $S^{S}(X_{k}^{W}, X^{W})$ is the reduction factor in fossil fuel savings due to the 'non-diversity' of wind power over all nodes. This will be explained below in §3.5.

6. I^{W} — the 'ideal' fossil fuel savings as a function of the total installed wind power capacity X^{W}.

The difference between terms 6 and 5 give the 'penalty' cost of wind power over and above a conventional power source.

3.4 The Capital Cost of Wind Power

The yearly capital payback of wind power was assumed to be a function of the wind speed in a given region (node). Firstly, a base capital cost per installed unit of wind power was assumed. This was amortised appropriately to give a yearly capital payback per unit of wind power B^{W}. This factor was then modified by an exponential function, as shown schematically in Fig. 3, which takes into account the mean wind speed per installed unit of wind power as more wind turbines are sited in a region (node). Clearly, one would site the first turbines in the windiest available areas. Subsequent turbines would be sited in the next windiest available area and so on. This function depends on the maximum wind speed, the minimum wind speed and the rate of decrease of wind speed per installed turbine in the node. Wind power is proportional to the third power of the average wind speed. This gives the function:

$$\frac{\partial C_k^{\mathrm{W}}}{\partial X_k^{\mathrm{W}}} = \frac{AB^{\mathrm{W}}}{v_k^3} \qquad (3)$$

where

$$v_k = v_k^{\min} + (v_k^{\max} - v_k^{\min})e^{-\lambda X_k^{\mathrm{W}}} \qquad (4)$$

and A is a normalising constant which assumes that the wind turbine capital cost is for a machine rated for an average wind speed of 7.7 m/s. C_k^{W} is obtained by integrating Eq. 3.

It should be stressed that this function is arbitrary and was simply chosen as an analytical method of simulating land use by wind turbines.

3.5 The 'Non-Diversity' Function

To simulate the reduction in fossil fuel savings from wind power when wind power is spread out over fewer sites, the 'non-diversity' function was formulated:

$$S^{\mathrm{W}} = m\left(1 + \left(\frac{1}{m} - 1\right)\sqrt{1 - \frac{\sigma^2}{n(n-1)}}\right) \qquad (5)$$

where

$$\sigma^2 = \left(\frac{nX_k^{\mathrm{W}}}{X^{\mathrm{W}}} - 1\right)^2 \qquad (6)$$

The constant m gives the maximum fractional reduction in fossil fuel savings and is assumed to be 0.5. The quantity σ^2 is the 'non-diversity parameter' and is akin to a normalised variance of the wind power over the total number of nodes n (in this case 14). Fig. 4 shows a schematic representation of the function in Eq. refeqn:spread.

This function, too, is somewhat arbitrary but was chosen to represent analytically the more rapidly decreasing reduction in fossil fuel savings with decreasing diversity of wind power.

3.6 Constraints

The minimum value of the cost function is subject to a number of constraints. The first set of constraints relate to the power balance in each node during each step of the load duration curve. These constraints are effectively equalities which can be summarised:

$$\begin{aligned} D_{ik} = \ & P_{ik}^{\mathrm{W}} + P_{ijk}^{\mathrm{P}} + P_{ijk}^{\mathrm{PE}} \\ & - P_{ikl}^{\mathrm{T}} - P_{ikl}^{\mathrm{TE}} + \epsilon_l P_{ikl}^{\mathrm{T}'} + \epsilon_l P_{ikl}^{\mathrm{TE}'} \end{aligned} \quad (7)$$

where D_{ik} is the load demand in node k during load duration step i, P_{ik}^{W} is the generated wind power, $P_{ijk}^{\mathrm{P(E)}}$ is the conventionally generated power by new (P) or existing (PE) power plant type j (charging of storage plant represents negative power generation), $P_{il}^{\mathrm{T(E)}}$ is the power transmitted *from* node k down the new (T) or existing (TE) transmission line l during load duration step i, $P_{il}^{\mathrm{T(E)}'}$ is the power transmitted down the new (T) or existing (TE) transmission line l to node k during load duration i and ϵ_l is the efficiency of the transmission line l.

The next set of constraints limits the power generated from the new power plants to the newly installed capacities:

$$P_{ijk}^{\mathrm{P}} \leq X_{jk}^{\mathrm{P}} \qquad (8)$$

and

$$P_{ik}^{\mathrm{W}} \leq X_k^{\mathrm{W}} \qquad (9)$$

Similarly, the power flows in the new transmission lines must be limited to the newly installed transmission capabilities:

$$P_{il}^{\mathrm{T}} \leq X_{il}^{\mathrm{T}} \qquad (10)$$

and

$$P_{il}^{\mathrm{T}'} \leq X_{il}^{\mathrm{T}} \qquad (11)$$

Constraints on storage ($j = \mathrm{s}$) during charge ($i = 1, 2$) and discharge ($i = 3$) give:

$$\epsilon_{j=\mathrm{s}}^2 P_{i=1,j=\mathrm{s,k}}^{\mathrm{P}} + \epsilon_{j=\mathrm{s}}^2 P_{i=2,j=\mathrm{s,k}}^{\mathrm{P}} - P_{i=3,j=\mathrm{s,k}}^{\mathrm{P}} = 0 \quad (12)$$

where $\epsilon_{j=\mathrm{s}}$ is the one-way efficiency of the storage medium.

3.7 Bounds on variables

All existing power flows are bounded by the existing installed capacities as detailed in Table 1. At present, the installation of all new plant and new transmission lines is excluded. Total new installed wind power power plant is limited to 40000 MW with no more than 10% of this figure in any one node.

3.8 The Optimisation

The cost function in §3.3 is minimised subject to the constraints in §3.6 with the bounds in §3.7 using the NAG library subroutine E04UCF [8]. The number of variables used is quite large. Following the description in §3.2, in order to reduce the number of variables slightly, the wind generated power was assumed to equal the installed capacity for step 2 (the only step in which wind power is assumed to generate). In addition, the running costs per node were averaged over all existing power plants, assuming that OCGT's were only operational during the third step peak period.

4 Running the NGM and Deterministic Models

Firstly, the National Grid Model was run for the 1992/93 grid to determine the fossil fuel savings for penetrations of wind power up to 40000 MW. These savings and the ideal savings were then used to determine the wind power penalty to be used by the deterministic model. The deterministic model was run to minimise the cost function for the present day plant mix to find: (a) the target capital cost for wind power in each region and (b) the optimum penetration of wind power in each region for a given wind power capital cost.

The maximum, minimum and land use parameter were determined very approximately from NOABL determined maps of wind speed over the U.K. [9]. These estimates are quite crude and are hoped to be improved upon when more detailed maps of the resource are obtained.

5 Results

It should be stressed that all the following results assume a capital payback period for wind power (a^{W}) of 25 years and a discount rate (d^{W}) of 8%. *No* subsidies have been included.

5.1 Target Capital Costs

The deterministic model was run several times gradually increasing the capital cost of wind power and observing when the optimised installed wind power in each node fell to zero. This determined the target capital cost for when it is economical to install wind power in a particular region (node).

Table 2 lists the target cost in each node. It should be noted that the results are quite sensitive to the maximum and minimum wind speed and land use parameter estimated for each region. The maximum and minimum wind speeds for inner and outer London were set artificially low to reflect the scarcity of land for building wind farms in this area as well as the effect of urbanisation on the

Table 2: Target cost for wind power in the 14 regions (nodes) of the England and Wales Grid.

Node	Cost (£/kW)	Node	Cost (£/kW)
1	1800–1900	8	1800–1900
2	1100–1200	9	200–300
3	1800–1900	10	<100
4	1800–1900	11	<100
5	200–300	12	300–400
6	300–400	13	1800–1900
7	300–400	14	1800–1900

average wind speed. In reality, one would not consider building a large amount of wind power plant in these regions and the deterministic model can reflect this by bounding the appropriate installed capacity variables to not exceed zero.

5.2 Optimum Installed Capacities

Present day estimates of the capital cost of installed nameplate wind power plant are in the region of £1000/kW. Assuming this capital cost the minimum value of the cost function gives an optimum installed wind power capacity of 8400 MW (a penetration level of 13%). The spread of this capacity is fairly even within nodes 1,3,4,8,13 and 14 at a level of between 1000–1600 MW in each of these nodes as is illustrated in Fig. 5. It is clear that the areas in which wind power are recommended to be installed are closely related to the average wind speeds in those areas. If the capital cost of wind power plant were to come down to similar levels as conventional plant (around £500/kW) and fossil fuel costs were to double, the optimum installed wind power capacity is predicted to be 19300 MW (a penetration level of 25%). This is shown in Fig. 6. There is now a larger variation in the amount of capacity installed in a number of the nodes. Note that it now becomes economical to install wind power in other regions where the average wind speed is lower. Also, there is a large increase in the capacity installed in the East Lancs area (limited only by the 4000 MW bound). This is due to the relatively high wind speeds in the Pennines, the lack of conventional power plant in this region and the ability to export a relatively large amount of power from this region.

6 Conclusions

A methodology for assessing the optimum growth scenario for a power system has been described.

This methodology is a combination of a simulation model in the time domain to assess the reduced savings from wind power as a result of 'discarding' and 'operating' penalties and a deterministic model to assess optimum system planning by minimising a cost function subject to power system and wind speed constraints.

The two scenarios presented are preliminary tests of this model and should not be interpreted as definitive predictions. They do, however, illustrate some of the features of this methodology. More work needs to be done on assessing the spread of wind speed in each region as this is found to be an important factor in the optimisation. Eventually, the results will be extended to allow the planning of both wind power and conventional plant allowing for load growth, plant retirement, increased storage and fuel cost changes. This will involve iterating between the NGM and the deterministic model. This methodology could be extended to other renewables such as tidal or solar if sufficient data were available. It is also possible to impose constraints or costs on fossil fuel emissions to include future environmental constraints.

Whilst appreciating that there may be other factors which will influence the optimum planning of wind power, it is hoped that this methodology might at least help the planning of wind power at the macroscopic level.

7 Acknowledgement

This project is funded by the Commission of the European Community's Directorate General XII for Science Research and Development under the JOULE II programme of work contract number JOU2-CT92-0191.

References

[1] Bossanyi, E.A., 1983 *Wind Engineering* **7**, 233–246.

[2] Bossanyi, E.A. and J.A. Halliday, 1983 Proc. 5th British Wind Energy Association Conference, Cambridge University Press, 62–74.

[3] Halliday, J.A., 1988 PhD. thesis, University of Strathclyde. Published as Rutherford Appleton Laboratory internal report RAL T 075.

[4] Watson, S.J., J.A. Halliday, and L. Landberg, 1992 Proc. 14th British Wind Energy Association Conference, Mechanical Engineering Publications Ltd, London, 291–297.

[5] Landberg, L., S.J. Watson and J.A. Halliday, 1993 Proc. European Community Wind Energy Conference, Lübeck-Travemünde, Germany, H.S. Stephens and Associates, Bedford, 677–680.

[6] National Grid Co. plc., 1993 *NGC Seven Year Statement for the years 1993/4 to 1999/2000.* Published by National Grid Co. plc.

[7] Ter-Gazarian, A.G. and N. Kagan, 1992 *IEE Proc. C* **139**, 499–504.

[8] *NAG Fortran Library — Mark 15 User Guide Vol. 4.*

[9] Burch S.F. and F. Ravenscroft, 1992 Department of Trade and Industry report ETSU-WN-7055. Crown Copyright.

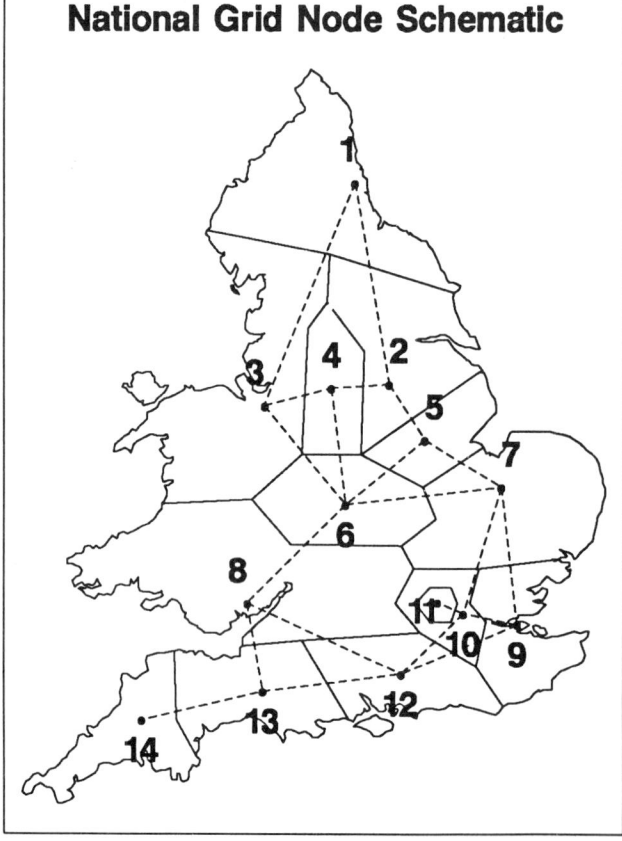

Figure 1: 14 node schematic of the National Grid.

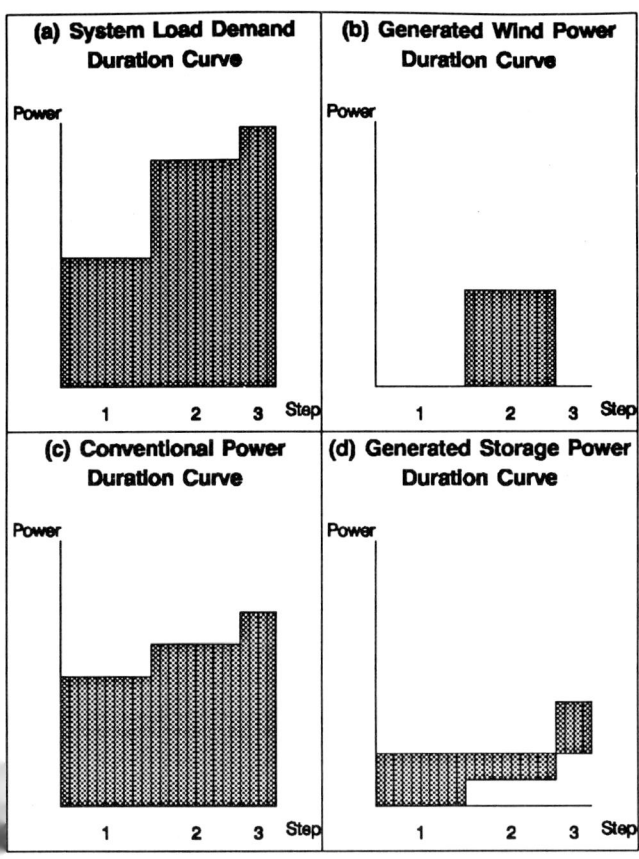

Figure 2: Power plant and load duration curves.

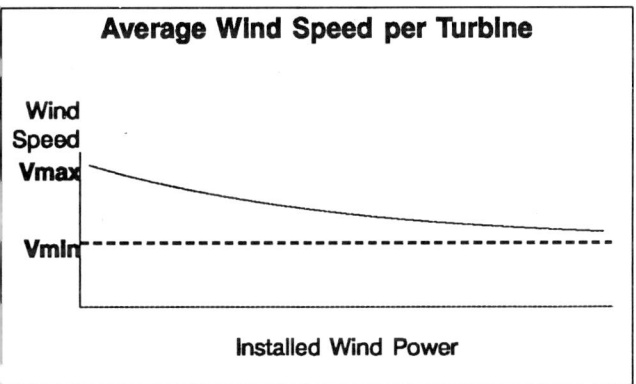

Figure 3: Exponential function of mean wind speed per node.

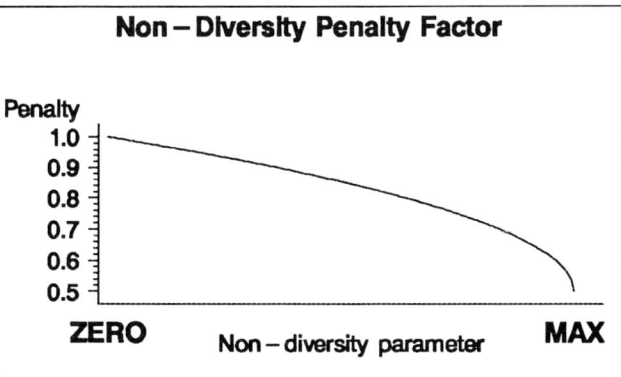

Figure 4: The 'Non-diversity' function.

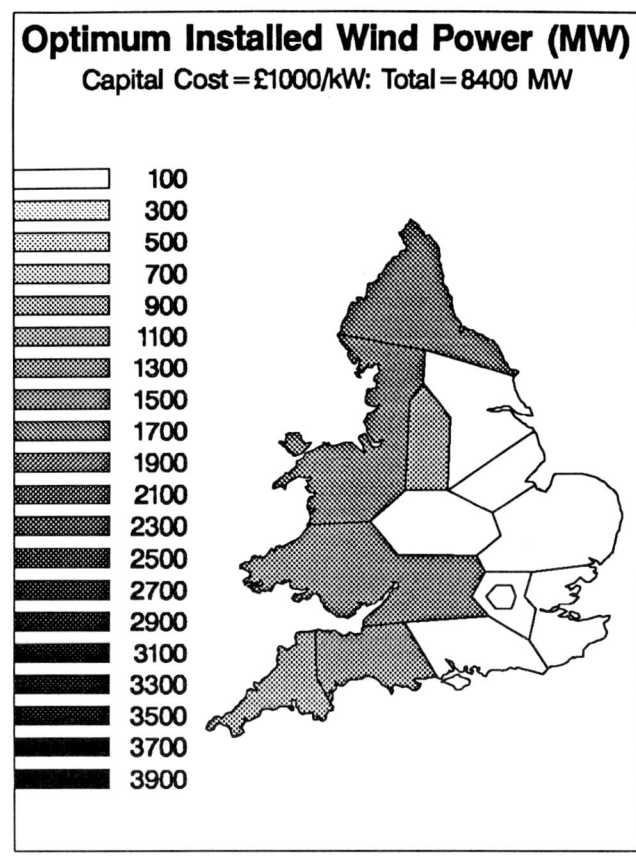

Figure 5: Optimum installed wind power capacity for capital cost of wind power plant=£1000/kW.

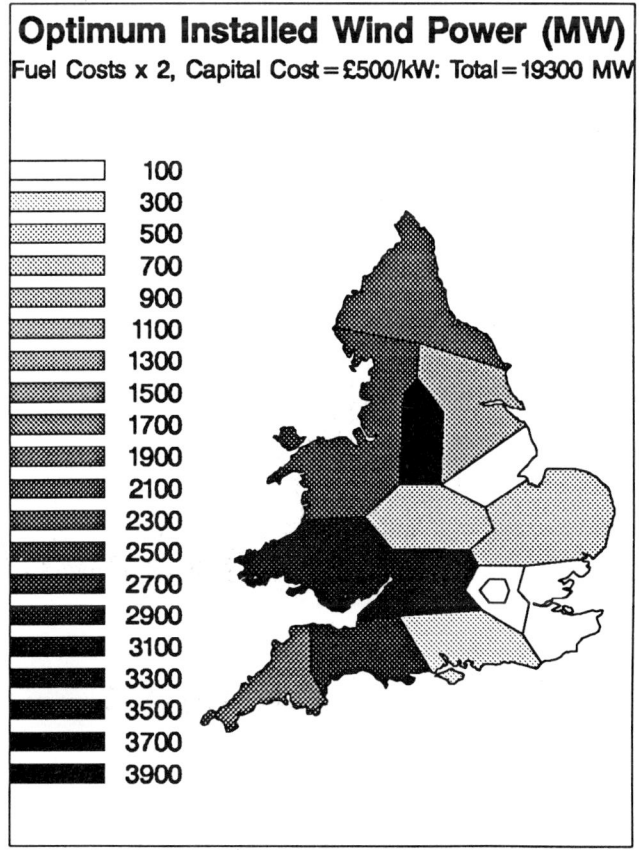

Figure 6: As Fig. 5 but for capital cost of wind power plant= £500/kW, fossil fuel costs doubled.

Power quality measures for isolated power systems

PER NØRGAARD, POUL SØRENSEN, JOHN O. TANDE
Risø National Laboratory, Denmark

ABSTRACT: Existing power quality standards and defined measures - defined mainly by the power utilities and for applications in large, integrated power systems - are in general related to specific requirements and problems relevant for the large systems, and therefore not always sufficient for describing the power quality parameters characteristic for isolated power system applications. The problem is clearly illustrated by the most different nature of the frequency variations for large, stiff systems and those for isolated systems respectively.

The paper focus on special power quality problems related to integration of wind energy into isolated power systems, illustrated by examples from measurements performed e.g. on three isolated diesel based 5-10 MW power systems in Cape Verde.

Statistical methods and measures for quantification of power quality parameters are discussed and defined, and their application and relevance for the evaluation and design of the integration of wind energy in Cape Verde is demonstrated.

1 INTRODUCTION

The term 'power quality' is not well-defined, but covers a number of different aspects, standards and measures. However, the power quality (whatever it is) is important both to the consumers and to the utilities, supplying the power.

For the consumers point of view a minimum power quality is required, in order to ensure proper operation of the equipment connected to the grid. The utilities, on the other hand, have to design the power supply system, in order to fulfil the power quality requirements.

In general, the power quality is defined and specified by the utilities in terms, that applies for specific problems in large systems with well-defined power generating units and consumers (consuming power).

The existing power quality terms does not always cover the specific needs, e.q. in isolated power systems and/or in systems with distributed power generating units.

In general, modern power generating wind turbines are designed for connection to grids with high power quality. In case of abnormal line voltage or frequency the wind turbines internal safety system (the controller) automatically disconnects the wind turbine in order to protect the wind turbine and to protect the grid.

This paper will focus on the power quality requirements for a standard wind turbines connected to a weak grid in an isolated power system.

2 POWER QUALITY

From a connection (consumers) point of view the power quality is described by:
- the voltage,
- the frequency,
- the waveform (harmonics).

The description of the voltage includes:
- The power values of the voltage (rms-values/steady-state),

- The amplitudes of the line- of phase voltages (short term variations),
- Voltage transients (peak-values),
- 3-phase symmetry/asymmetry.

2.1 International standards

A number of international standards defines parts of the power quality - e.g.

CCITT	Voltage levels
IEC 555	Disturbances
IEC 868	Flicker meter

In addition, national, EEC and international organizations are at present working on further recommendations and standards. The EEC work in CENELEC is based on the philosophy that the power supply is to be understood as a free market product.

3 GRID VARIATION EFFECTS

Below the impact of voltage and frequency variations in the grid on a standard wind turbine with induction generator will be analyzed. To make it clear: for this type of power generating unit, the voltage and the frequency are controlled by the grid, while the power (or current) is determined by the wind turbine.

3.1 Voltage variation effects

The impact on the wind turbine caused by voltage variations.

The wind turbine has to be protected against extreme over and under voltage. Over voltage (in specific transients) can cause damage of the wind turbines electrical equipment, while under voltage can cause malfunction of the control and safety system. Also smaller changes of the voltage will influence the operation, but the power output will be affected only slightly.

The active power output from the generator, P_g, is given by the mechanical power input, P_m, reduced by the generator loss, P_{lg}:

$$P_g = P_m - P_{lg} \qquad (1)$$

The electrical power output from the generator is given by

$$P_g = U \times I \qquad (2)$$

With increasing grid voltage, U, the current, I, and

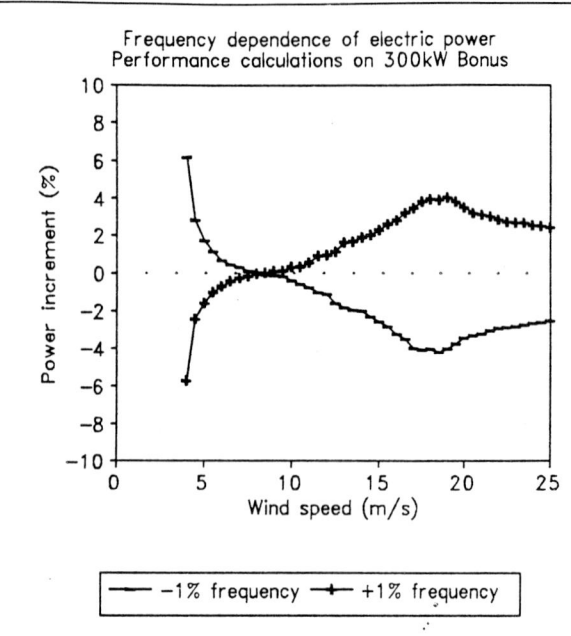

Figure 1: Frequency variations - steady state. The power output (relative to nominal power curve) as a function of the wind speed for ± 1 % frequency variation (relative to nominal) for a 300 kW wind turbine (Bonus 300 kW).

thereby the generator losses, P_{lg}, will decrease. In addition, the slip in the generator

$$s = \frac{\omega - \omega_0}{\omega_0} \qquad (3$$

and thereby the rotational speed, ω, will decrease, and the mechanical power input, P_m, will be reduced. The result is a reduced output power. With decreasing voltage the opposite effect is observed.

3.2 Frequency variation effects

The wind turbine is very sensitive to the frequency. The grid frequency controls the rotational speed of the rotor, and thereby the aerodynamic properties. In addition, a constant change of the frequency will result in power fluctuations due to the inertia of the rotor. These two effects are described below.

Steady state operation

The operation of the wind turbine is designed for nominal frequency. At other frequencies in specific the aerodynamic characteristics will be changes.

For stall regulated wind turbines this results in changed power output. The power coefficient curve as a function of the tip speed ratio

$$u_t = \left(u/v_{tip}\right)^{-1}$$

where u is the wind speed and v_{tip} is the speed of the blade tips (at nominal rotation), will be almost unaffected, corresponding to a scaling of the power coefficient curve along the wind speed axis.

Figure 1 presents an example of the effect on the power output for a stall regulated wind turbine. A 1 % change of the frequency leads to up till 4 % change in the power output.

Dynamic properties

The additional mechanical power due to the inertia of the rotor and caused by the variation of the rotational speed is

$$\Delta P = \Delta T \omega = I \dot{\omega} \omega \qquad (5)$$

where T is the torque, I the inertia, ω the angle speed and ω' is the angle acceleration.

3.3 A sample wind turbine

The effects are illustrated on a sample wind turbine with fixed rotational speed, stall controlled power regulation and induction generator. Table I lists the specifications.

Table I: The sample wind turbine - main characteristics and specifications.

Sample wind turbine Main characteristics and specifications		
Operation characteristics	:	Fixed speed Stall regulated Induction generator
Rotor		
diameter	:	30 m
nominal speed	:	33 rpm
weight	:	6 tons
inertia	:	200,000 kgm^2
Generator		
nominal voltage	:	3 × 415 Vac
nominal frequency	:	50 Hz
nominal power	:	300 kW
nominal speed	:	1500 rpm
Gear ratio	:	1 : 45
Inertia (@ generator shaft)	:	100 kgm^2

At the nominal grid frequency f_0 = 50 Hz, the synchronous speed of the generator is ω_0 = 1500 rpm = 157 rad/s.

A ramp change of the grid frequency of 1 % per second

$$\frac{df}{dt} = \pm 0.5 \; Hz/s \qquad (6)$$

results (as a first order approximation) in an angle acceleration at the generator shaft of

$$\dot{\omega} = \pm 1.5 \; rad/s^2 \qquad (7)$$

A rotor inertia of I_r = 200,000 kgm^2 will at the generator shaft with a gear ratio of 1:45 be transformed to I_g = 100 kgm^2. The power change will then be

$$\Delta P \approx \mp 25 \; kW \qquad (8)$$

The result are summarized in Table II.

4 MEASUREMENTS IN CAPE VERDE

In Cape Verde[1] a DANIDA[2] supported wind energy project is prepared. On three islands with diesel based power supply systems wind energy will be installed with capacities to the limit of simple grid-connected wind turbines (i.e. the installed wind power capacities will not affect the normal operation strategy of the diesel power plant). Risø, ElsamProjekt[3] and CarlBro International[4] is Danida's technical advisers in the project. The wind turbine supplier is in process of being selected, and the first batch of wind turbines is planned to be installed in 1994.

Table II: Summary of the effects on the sample wind turbine.

Parameter	Changes	Effects (Power output)
Voltage Steady state	± 10 %	slip: ± 0.6 % power: ± 2 %
Frequency Steady state	± 1 %	± 4 %
Dynamic (ramp response)	± 1 % / sec	∓ 9 %

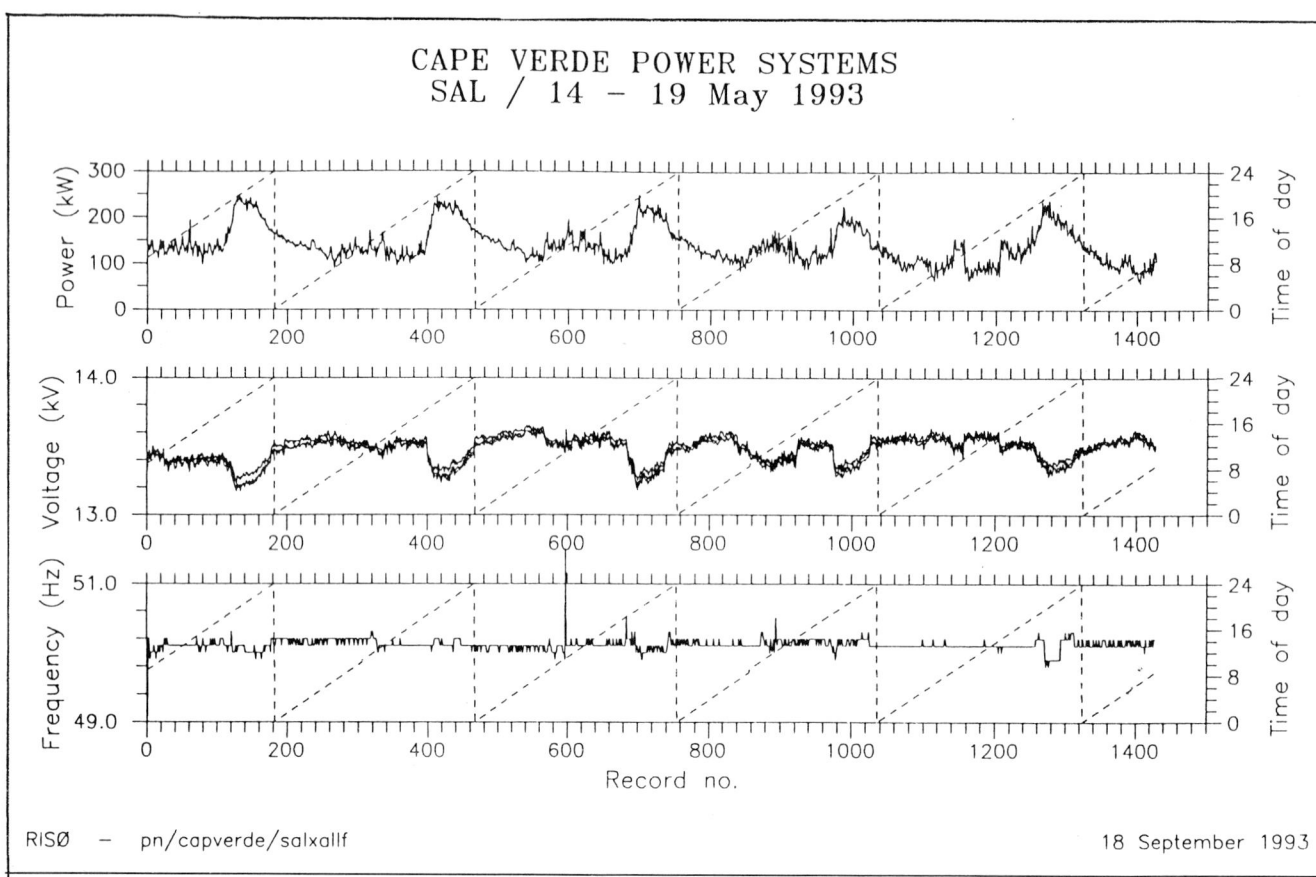

Figure 2: Data from Sal island, Cape Verde, May 1993. The electrical parameters are measured with a VIF Energy Analyzer. Duration: 5 days, sample rate: 5 minutes.

In parallel, a feasibility study for high penetration wind power into the systems is prepared. One assumption for the project was that the introduction of the wind power into the power supply systems should not cause a lowering of the power quality for the consumers. As part of the feasibility study the electrical parameters should be measured prior to and after the installation of the wind power, and the impact of the wind power should be assessed. The power quality should therefore be defined in terms of quantified measures based on measurements of the electrical parameters.

Together with ElsamProjekt and Electra[5] preliminary measurements have been carried out in May 1993. Standard electrical parameters was measured simultaneously at several nodes in the supply system. Supplementary measurements will be performed later this year.

Figure 2 presents selected parts of the data collected from the power supply system on Sal island relative far from the power plant.

5 MEASURES FOR VOLTAGE AND FREQUENCY FLUCTUATIONS

As a supplement to the existing methods two methods (also presented by Bindner and Lundsager at ECWEC'93, Ref. [1]) are presented below to analyze the magnitude and the characteristics of the voltage and frequency variations:

1) the standard statistical distribution function and

2) the standard rainflow counting method[6]

The two methods are complementary, as the standard statistical distribution will express the mean value and the 'width' of the variation (the deviation), while the rainflow counting express the number versus magnitude distribution.

The two methods has been used to analyze the characteristics in rapid variations of the voltage and frequency in small power systems using measured data from Risø's Wind/Diesel Test Facility (Ref. [4]) and data from Fröya Wind/Diesel System (Ref. [3]). Both systems operating in stand-alone mode (only the wind turbine generates power).

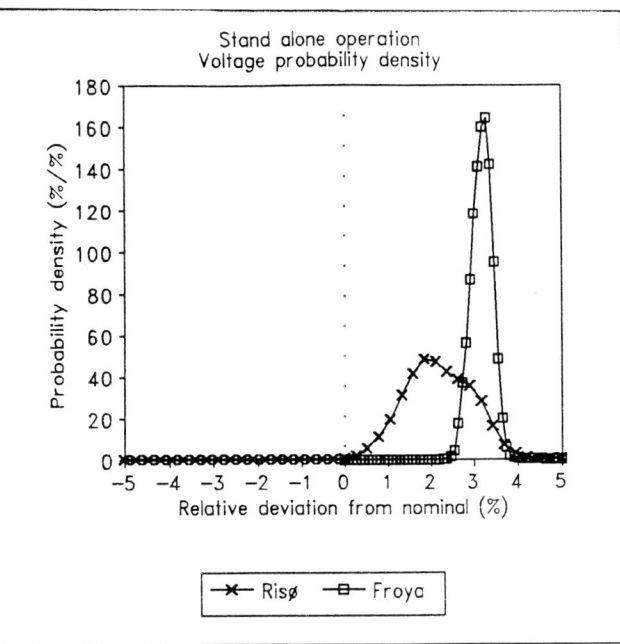

Figure 3: Voltage variations - statistical distribution. Data:
1) Risø: Risø Wind/Diesel Test Facility, date 22/2-93, 120 minutes, sample rate 8 Hz;
2) Fröya: date: 28/11-92, 30 minutes, 10 Hz.

5.1 Voltage fluctuations

Figure 3 and Figure 5 presents the results of the analysis of the voltage fluctuations. The graphs clearly demonstrate the difference of the two systems. The Fröya system has a more narrow distribution and a corresponding lower counting of large voltage fluctuations - a higher power quality regarding the voltage fluctuations. The voltage mean values are slightly above nominal.

5.2 Frequency variations

The statistical distribution and the rainflow counting methods can be utilized on the frequency variations as well.

Figure 4 and Figure 6 presents the results of the frequency analysis of the data from the Risø and the Fröya systems.

Again, the Fröya system seems to have a better power quality performance regarding the frequency fluctuations, however, not as clearly as for the voltage.

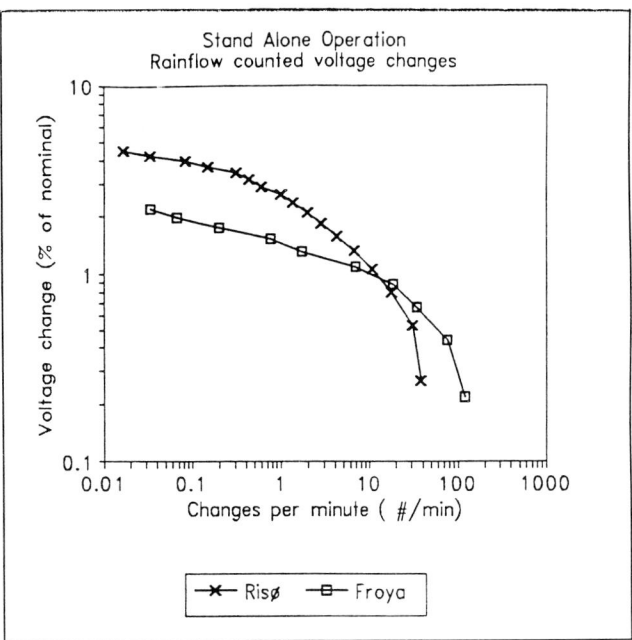

Figure 5: Voltage fluctuations - rainflow counting. Data: see Figure 3.

6 CONCLUSION

Two standard statistical analysis methods have been presented and used for evaluation of the power quality regarding voltage and frequency fluctuations. The two methods describes different characteristics of the fluctuation, that are relevant for the operation of a wind turbine.

REFERENCES

[1] K.Uhlen, H.Bindner, P.Lundsager et. al: "Engineering design tools for wind-diesel systems. Presentation and validation of the Modular Dynamic Model." ECWEC '93, Travemünde, 1993.

[2] H.Bindner, K.Uhlen, P.Lundsager: "Power quality and grid stability of simple wind-diesel systems: results from Risø's experimental system". BWEA'92, 1992.

[3] K.Uhlen et. al: "Design and operation of the full scale Norwegian wind/diesel laboratory model". EWEC'89, 1989.

[4] P.Lundsager, P.Nørgaard: "The 55/30 kW experimental wind/diesel system at Risø National Laboratory". Risø M-2717, 1988.

[5] IEC 868: Flicker Meter. CEI/IEC, 1991 + 1990 + 1986.

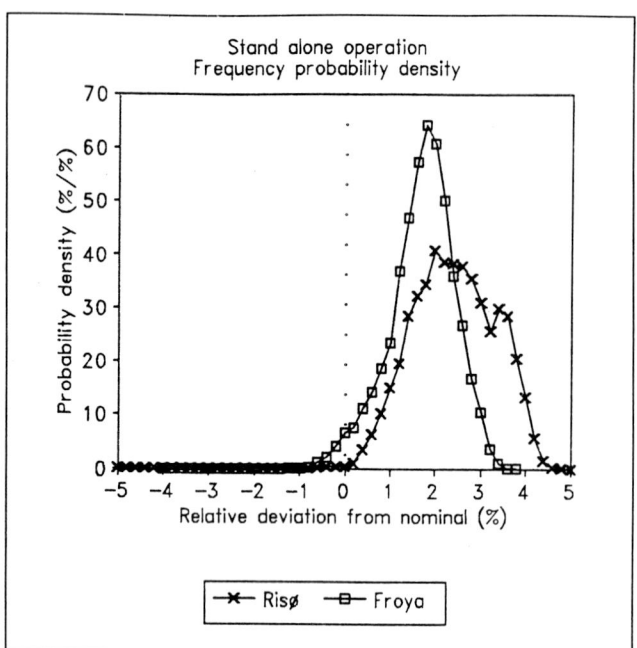

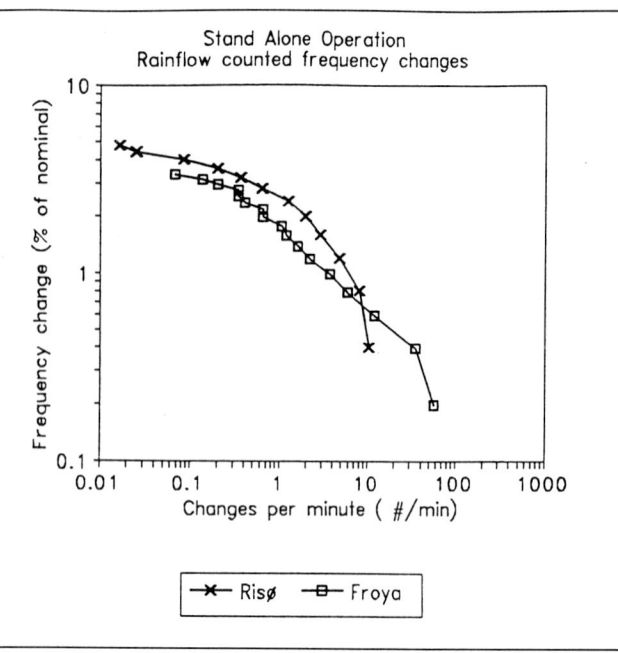

Figure 4: Frequency fluctuations - statistical distribution. Data: see Figure 3.

Figure 6: Frequency fluctuations - rainflow counting. Data: see Figure 3.

[6] IEC 816: Guide on methods of measurement of short duration transients on low voltage power and signal lines. CEI/IEC, 1984.

[7] IEC 555: Disturbances in supply systems caused by household appliances and similar electrical equipment. CEI/IEC, 1990 + 1982.

[8] DEFU KR 77: Nettilslutning af vindmøller (Grid-connection of Wind Turbines). DEFU, DK, 1988.

[9] DEFU RK 16: Spændingskvalitet i lavspændingsforsyningsnet (Power quality in low voltage power supply systems). DEFU, DK, 1987.

[10] IEA Recommended Practice for Wind Turbine Testing and Evaluation. Vol. 7: Quality of Power, Single Grid-connected WECs. International Energy Agency, 1984.

Notes:

1) The Republic of Cape Verde (islands west of Africa).

2) DANIDA: Danish International Development Agency.

3) ElsamProjekt, Power Station Engineering, Denmark.

4) CarlBro International, Consulting Engineers, Denmark.

5) Electra: the public Cape Verdian Utility Company.

6) 'Rainflow counting' is a general statistical method to analyze numbers and magnitudes of 'cycles', e.g. used in the structural engineering to express the amount of vibrations and to evaluate expected lifetime of vibrating constructions.

The financing of wind energy in the UK compared to other countries and its implications

CATHERINE MITCHELL
University of Sussex, UK

SYNOPSIS: The ability to obtain finance for wind energy will have major impacts on the type of developments which occur. Wind development in the UK is of large, high wind speed sites developed by large, well established companies. Small scale independent or community development is negligible in the UK. This is a direct impact of the NFFO support mechanism.

Introduction

This paper compares the means of financing wind energy in the UK with Denmark, the Netherlands and Germany. 'Financing' is understood as the process of obtaining capital to bring a project to fruition. In general, the means of financing a renewable energy project in a country is the sum of the 3 major inputs:

1. Government policy:
    R&D policy
    Demonstration policy
    Market support mechanisms
    Planning policy
2. The national financial system
3. Utility policy and its regulation

These factors will differ in every country and together fuse to create a different financial mechanism with different implications for the development of wind energy. The way that wind energy develops in a country is therefore not divorced or seperate from these factors but is the result of them.

Because of the constraints on space, this paper will only briefly describe the tariffs, subsidies and financing systems in place in each country (details can be found elsewhere[1]) and concentrate on discussing their implications.

1. DENMARK

Denmark's success in wind energy has been based on a number of ingredients, painstakingly explored by Peter Karnoe[2] of the Copenhagen Business School. These ingredients are based on two important factors:

1. the clarity and long term nature of government support for wind energy (not its generosity)

2. the ease of obtaining finance which itself depends on a number of factors.

The Danish Government has been supportive to wind energy, since it first provided 30% capital grants for wind turbines in 1979 (although these have now been phased out)[3]. Wind generators receive a guarantied 85% of the average small consumer price of electricity[4]. Non-utility owners[5] are refunded the carbon dioxide tax and energy tax (which is levied on all electricity bills) including VAT payable on the energy tax while utility generators only have the carbon dioxide tax refunded. In general, all costs associated with running a wind turbine are tax free and in addition, single owners, unlike co-operative owners, are allowed to depreciate their turbine against tax over 20 years[6]. Finally, tax free investment has been the greatest incentive for co-operative investors[7].

In 1985, it was agreed by the Danish Energy Agency and the Danish Electricity Association (DEF) that utilities should be the main developers of wind energy in the Denmark. Due to last until 1995, the agreement was updated at the end of 1992 to include a clear statement of utility responsibilities towards renewables concerning grid connections. Although the utilities are currently appealing against these responsibilities it appears that the rulings are in favour of wind energy. Utilities pay for wind energyv projects out of their own revenues; at a nominal rate of interest. Independent developers are restricted by residence (they can only invest in turbines situated in their own locality or a neighbouring locality) and in consumption (they can only invest in a turbine share equivalent to 150% of their own consumption). Despite this, between 1985-1993 individuals and co-operatives installed 270 MW compared to the utilities 100 MW[8].

Building societies make loans at the equivalent of the bank base rate; while banks lend at 1-2% over their bank rate. Borrowers therefore borrow as much as possible from the building societies and top it up with a bank loan. Providing the wind turbine is in a good site most single or co-operative owners are able to make up a loan for 100% of the total capital costs (TCC). If the site is not so good, (eg on the Eastern islands) it may be that 5-10% of the total capital costs will be in cash form with the bank and/or building society loaning the remaining 85-90%.

Thus on a good site, most single owners do not have to contribute equity. Banks, in general, consider the turbine as collateral. In addition, there is an acceptance of co-operative financing in Denmark for co-operative projects which are in the majority. While, the majority of capital is provided by 'commercial' banks or building societies a number of other means of financing are available. For example, Denmark has many 'ethical banks', which offer bank loans at lower rates of interest than commercial banks. Banks and building societies are aided by an efficient and centralised tax system which enables them to check individuals, their tax returns and to

confirm their earnings. Long-term, freely available data about turbine reliability and wind speeds also aid the financiers; but it is experience which reduced their perception of risk of wind energy.

2. THE NETHERLANDS

The Dutch Government established the National Environmental Policy Plan[9] (NEPP) in 1990 which incorporated 3 main planks of support for renewable energy: the 'MAP' payment, premium prices (which can be paid for out of the MAP money) and the TWIN programme.

Within the NEPP, each economic sector, including the power sector, agreed informally to attempt to reduce their sectors carbon dioxide emissions. Each utility is able to increase their small consumer charge by a maximum of 2% to pay for any extra cost incurred by the emission reduction strategies; this is known as the MAP payment. The utilities are required to buy all renewable electricity offered to them. Each utility sets the amount that they pay for renewable energy; and this differs quite markedly from region to region (between 10 and 20 cents/kWh compared to the average cost of electricity from fossil and nuclear sources of about 8-8.5 cents/kWh[10]). The difference is paid for out of the MAP money as are discretionary grants to individual or co-operative developers and utility investment. Utilities, therefore have considerable power within the wind energy industry and it is therefore no surprise that about 70% of wind energy installations in the Netherlands are developed by utilities or joint utility/private companies. Additional utility development can be paid for out of their revenues with a return on investment equivalent to their cost of capital. However, some utilities have obtained finance in other ways, eg project finance or utility bonds.

In parallel to this, the Government has established the TWIN programme which is run by NOVEM, the Dutch

Energy and Environment organisation which is intended to support the policies established within the NEPP. This is an oversubscribed, rolling programme which distributes a set annual grant derived from general taxes on a first come first served basis. A developer is eligible for a maximum grant of 35% of total capital costs (known as the national Investment Grant NIG), providing that the project technology is on an approved NOVEM list.

Independent schemes are eligible for NIG grants; they may receive an enhanced regional electricity payment and/or a grant via their utility MAP payment. Over 100 single turbines have been installed by farmers in the North of the Netherlands. Rabobank of the Netherlands was established as, and continues to be, an agricultural co-operative bank; 90-95% of farmers still bank with them and are often on the co-operative board which decides bank lending policy. The majority of the loans for farm turbines are through the Rabobank with the land taken as collateral. In addition, there are a number of co-operatively financed turbines which have a less defined path of finance although they are eligible for the NIG and like the single owned turbines may receive a premium payment and/or a utility grant. However, as with Denmark, 'commercial' banks are prepared to have co-operative accounts and in addition there are a number of 'ethical' banks which lend at slightly lower rates of interest than commercial banks. There is no limit on independent investment or where investors live, as in Denmark.

3. GERMANY

Germany established the most comprehensive support system for renewable energy in Europe[11] with the Grid Feeding Act (Stromeinspeisegesetz) of 1991 (whereby utilities are required to buy wind electricity at a fixed percentage of the consumer price) and the 250 MW Wind Programme in 1991 (which paid 6pf/kWh to those given contracts through the rolling and now finished programme). The programmes link in with Government subsidised bank loans via the Deutsche Ausgleichbank[12,13], Lander (Regional Government) grants and federal support for information agencies. However, this determined effort may suffer from the increasing demands placed on Germany as a result of re-unification.

In addition, Germany appears to have the most flexible and wide range of financing mechanisms available to renewable energy of the 3 countries examined.
The Banks which organise the DAB loans for a project (for which they are paid a 1% margin) are usually prepared to loan the rest of the necessary capital at about 1% over the commercial bank rates. Banks may or may not require some external collateral (in addition to the collateral of the turbines) to the project depending on the project situation and size but this is unlikely to exceed 25% of the TCC of the project. Not all commercial banks in Germany are prepared to lend to wind energy projects; but there are enough which do. Germany also has a number of ethical banks which lend money at low rates of interest; fund managers which collect investment for specific projects[14] and a number of equity gathering institutions[15] which exploit the German tax incentives for investment in certain industrial sectors, which includes energy efficiency, renewable and alternative energy. A few windfarms have also been project financed.

Although some utilities have invested in renewable energy projects, less than 20% of wind energy projects have been developed by utilities[16]. They are, like Denmark and Holland, owned by their local authorities or municipalities. The costs of the windfarms are paid for by their consumers and they are expected to make a low but unspecified rate of return on their investment.

4. The UK

The Electricity Act of 1989 established the Non-Fossil Fuel Obligation which obliged the regional electricity companies (RECs) to pay premium prices until 1998 for nuclear and renewable electricity in England and Wales. There have so far been two NFFO Orders. One in 1990 totalling 102 MW declared net capacity (DNC) and the other in 1991 totalling 472 MW DNC of which nearly 85 MW DNC was wind capacity.

The 1990 NFFO contracts were awarded on a cost-justification basis while the 1991 NFFO contracts were awarded by competitive bidding within technology bands. This latter process requires tranches, rather than a rolling application programme, so that a number of projects can come together to allow competitive bidding. The recent announcement of a 1993-4 Order appears to be based on a competitive bidding system albeit for a much welcomed 15-20 years[17].

The majority of wind projects in the UK have been project financed. Banks are normally prepared to lend 70-80% of the total capital costs at 1-2% over the base rate in addition to a fee of 1-2% of the total capital costs. The other 20-30% requires equity investment. The rates of return on investment have rarely been below 20% IRR and often above 30% IRR.

Obtaining finance has proven to be increasingly difficult as the size of the project is reduced. There has been a financing gap between £0.25-0.5m (the maximum that local banks feel comfortable in lending) and between about £2m (the minimum level for a select few project finance teams. In addition, the UK lacks alternative methods of obtaining local bank loans for renewables; as yet co-operative type bank loans have not occurred. As a result there is almost no co-operative type of development in the UK; very few single turbines and almost no small scale independent development.

The RECs have no responsibilities towards renewable energy development beyond their obligation to process their regions NFFO applications. Some RECs have developed into major renewable energy investors; generally for high rates of return not least because of possible required justification to their shareholders but also because of their responsibilities towards their dividends.

4. COMPARISON

There are four main differences between the UK and the compared countries. Firstly, the UK NFFO system is intended to support commercial development of renewables while the other countries support mechanisms (eg TWIN and BMFT) is to stimulate installation and is not concerned with a hurdle project rate of return.

Secondly, the utility regulation and responsibilities towards renewables differ. In Denmark and the Netherlands, utilities have been given primary responsibility for installing wind energy. In addition, the Dutch, Danish and German utilities require much lower rates of return on their investments than the UK RECs, primarily because the latter are privately-owned, commercial companies which have to provide dividends for their shareholders and justify their investment against alternative options. Finally, the utility responsibility and charges with respect to grid connection have been established in Denmark and are being established in Germany.

Thirdly, the UK has a high average project cost of capital, compared to the other countries, despite very similar bank base rates. Subsidised loans in Germany and the availability of grants in the Netherlands is particularly helpful in decreasing the average cost of capital.

It should be noted that capital is available in Denmark, Germany and the Netherlands at these rates for small scale private developers for 85% plus of the total capital costs of the project and often 100%. Thus the 8.4% in Denmark for private developers is a worst case scenario; while the 8.3% + fee for a £5m+ project in the UK is a best case scenario.

Thus the fourth main difference is that the UK has few options for finance other than commercial banks and, in particular, small projects or developers in the UK find it difficult to obtain finance. Local 'commercial' banks in the UK have not yet agreed to co-operative accounts, as used widely in Denmark and the Netherlands, nor do they accept the turbine as collateral or part collateral. Together, this means, in effect, that in the UK collateral will be equal to the total capital costs of the project. This compares, for example, with Germany where collateral will equal 25% of the total capital costs and of course means that more total investment may occur for the available collateral.

Table 1   The Average Project Cost of Capital to Renewable Energy Developers

Country	Interest rate (% real)

Denmark

utilities	5-7
private	5.5

(based on 60% from Building Society 40% from banks, 1% over base rate)

private	8.4

(based on 100% from banks 2% over base rate)

Netherlands

utilities	2.6

(based on 25% capital grant, 25% MAP payment 1% over base rate)

utilities	3.1

(based on 25% capital grant, 25% MAP payment, 2% over base rate)

private	6.2

(based on 25% capital grant, 1% over base rate

Germany

private	3.3

(based on 75% DAB grant, 25% bank load at
1% over base rate)

UK

£5m +	8.3 +

fee
(based on 20% equity at 20% IRR, 80% bank
loan at 1% over base rate)

	10.1+

fee

(based on 20% equity at 25% IRR, 80% bank loan at 2% over base rate)

As of June:
Danish inflation rate of 1.1% and a bank base rate of 7.5% and a building society rate of 6%.
Dutch inflation rate of just under 3% and a bank base rate of 7.25%.
German inflation rate of 4.3% and a bank base rate of 8.5%.
UK inflation rate of 1.6% and a bank base rate of 6%.

5.   DISCUSSION

Table 1 above illuminates the important differences in the real project cost of capital. This, of course, implies that the cost/kWh delivered from the same turbine with the same output would be much higher in the UK compared to the other countries; provided the electricity payment was the same. However, high electricity prices should in part offset the effects of the high cost of capital.

Table 2 Overview of Tariffs for Wind Energy and the Small Domestic Consumer Electricity Prices (ecus)

Country	Wind Payment*	Small Domestic Consumer price**
Denmark	8.31	11.2
Netherlands	3.5-6.9	8.42
Germany	11.83	13.3
UK	13.64	10.9

Source: van Zuylen, van Wijk, Mitchell, 1993.
*payment made per kWh of wind electricity by utilities
** the average price paid per kWh of electricity by the domestic consumer.
1 Up to 150% of electricity consumption
2 Depending on fuel price
3 Excluding BMFT payment
4 Wind tariff of 1991 NFFO, only until 1998, then pool price.

The high average project cost of capital is not primarily because of the banks (although UK project finance bank fees appear higher than their European counterparts) but because of (1) there being an equity requirement and (2) the rates of return expected of that equity investment. The concept of equity in the compared countries is completely different. In some, but not all, cases equity is required but not at particularly high rates of interest. The UK support mechanism is therefore profitable for investors and to a lesser extent to project financiers. The opportunities for these investors or bankers, in the compared countries, are much poorer; if not non-existent.

Thus, not only is the average project cost of capital lower in these countries, but they have more flexible local bank lending, which is of particular help to the smaller projects. Much of this is to do with the banks experience of lending to renewable energy projects and it is likely that local UK banks will become more flexible as they make more loans to renewable energy projects (and are asked to make loans which are particularly useful; such as co-operative accounts). Loan guarantees for the first few demonstration local bank loans would be likely to raise bank confidence in small renewables projects, and would be most suitable in areas of good wind and water resources such as Wales and Scotland.

As mentioned above, wind energy in the UK has primarily been project financed. They have often required an ex-nationalised industry equity investor within the project (the RECs, the water companies; the generators). These investors require a high rate of return on investment. A high project cost of capital in conjunction with the NFFO, a competitive system with limited availability of contracts, has led to the development of large sites with high wind speeds to allow a low bid (against their competitors to ensure, or to try and ensure, a scarce contract) and to a lesser extent pay for the high costs of capital. This results in areas with wind speeds below a certain level remaining undeveloped by the large companies because they are less able to bid in at competitive level or make the necessary returns to pay for the high costs of capital.

However, these lower wind speed areas are also undeveloped by smaller developers/communities because of their difficulties in obtaining finance. Thus wind development in the UK is, in general, following the large company, large site, high wind speed route. One or two developers are trying to follow a lower wind speed, smaller scale approach but it is very much against the trend.

6. CONCLUSION

The UK has a high average cost of capital for renewable energy. This has been offset to some extent by the high premium price. However, high costs of capital in conjunction with competitive bidding for a constrained capacity leads to development of large, high wind speed sites. Lower wind speeds are unable to produce the revenue to allow low bids or to meet financial repayments. Smaller projects or developers have problems finding finance. However, the NFFO system provides good investment and project financing opportunities.

Nevertheless, the prospects for wind energy in the UK would be bleak if the high costs of capital were not offset by a high premium price, as was hinted at in Timothy Egger's 21 July 1993 announcement

7. REFERENCES

[1] The most detailed account is Foreign Financing Systems, prepared by C Mitchell and G MacKerron for the Commercialisation Section of ETSU, 1993; van Zuylen, van Wijk and Mitchell, 1993, A comparison of the Financing arrangements and tariff structures for wind energy in European Community Countries, ECWEC; 'European Windpower - Getting a grip on tariffs and subsidies' and 'Project finance examined' in Power in Europe Numbers 152 and 153 (together a very short version of the ETSU paper).

[2] Peter Karnoe, 1990, Technological innovation and industrial organisation in the Danish Wind Industry, Entrepreneurship and Regional Development, 2, 105-123; Ulrik Jorgensen and Peter Karnoe, 1991, The Danish wind-turbine story - technical solutions to political visions, paper presented at workshop on 'Constructive Technology Assessment, Twente, Denmark, Sep 20-22, 1991; Peter Karnoe, Danish world leadership in wind technology: the result of a Danish recipe?, Working Group 10, 10th EGOS Colloquium, Vienna, Austria, July 15-17 1991; Peter Karnoe, 1990, Entrepreneurial organisation and the accumulation of knowledge in modern wind technology. Rosneas Workshop, May 20-23, 1990.

[3] Danish Association of Windmill Manufacturers, 1991, Windpower in the 1990's.

[4] The small domestic consumer price is currently around 45 ore where 100 ore = 1 Danish Kroner (Dk) see Paul Eric Morthorst, The Economics of Private Windmills (Privatejede vindmollers okonomi, Danish Energy Agency, June 1991.

[5] non-utility owners are private investors and divide into 2 main categories (single owners of turbine(s) and multiple or co-operative owners) which benefit from differing tax and ownership regulations.

[6] Paul Eric Morthorst, The Economics of Private Windmills (Privatejede vindmollers okonomi, Danish Energy Agency, June 1991.

[7] ibid

[8] op cit, Ref 3

[9] NOVEM, 1992, National Support Programme for the application of wind energy in the Netherlands (TWIN) 1991-1995, Utrecht, Netherlands.

[10] E van Zuylen and A van Wijk, 1993, Tarieven en subsidiebeleid voor windenergie in Nederland en het buitenland, January 1993, Ecofys, Utrecht Netherland.

[11] H Wagner, 1992, Renewable Energy Development in Germany, presentation to the International Symposium on Solar Energy, Kanazawa, Japan, September 5-6 1992.

[12] The Deutsche Ausgleichsbank, 1991, The Jahresbericht, Bonn, Germany.

[13] Deutsche Ausgleichsbank, 1993, 016 GER: ERP Loans for Energy Saving, Bonn Germany.

[14] GLS,GKG, Windkraftfonds 1(2,3), Bochum, Germany.

[15] eg Beratung Verwaltung Treuhand, Munich, Germany.

[16] van Zuylen, van Wijk and Mitchell, 1993, A comparison of the Financing arrangements and tariff structures for wind energy in European Community Countries, ECWEC.

[17] NFPA, statement by the Minister for Energy, T Egger, 21 July, 1993.

Financial structures for wind farms

J. H. JOHNS
Ernst & Young, UK

SYPNOSIS

A summary of the structures used to finance wind farms. The impact of the 3rd tranche of the NFFO. An evaluation of cash returns using Ernst & Young's wind energy model. Suggestions for further developments in financing including the use of specialist financial instruments and tax planning techniques.

1 SUMMARY OF STRUCTURES USED

The financial structures put forward to finance wind farms under previous rounds of the NFFO were influenced by a number of factors:

(a) the high level of risk perceived by some financial institutions due to, inter alia:

  (i) the lack of track record of the industry in the United Kingdom (notwithstanding experience in Denmark and the USA in particular); and

  (ii) the view that wind energy schemes were unusual, at the leading edge of technology, and therefore required high margins of comfort (and reward);

(b) the relatively low number of proven operators able to finance developments from their own reserves;

(c) difficulties in obtaining planning permission, which when combined with the fixed end date of the 2nd round NFFO, meant that for many projects the full length of the NFFO contract was not available, adversely affecting cashflow, making it difficult to retire debt and achieve adequate returns for investors;

(d) the high capital cost of projects in the initial stages, leading to roll up of interest on debt, and tax inefficiency caused by early losses;

(e) uncertainty as to the likely price of electricity after 1998; and

(f) in the light of the factors identified above, the need for sophisticated tax planning and structuring to enhance investor returns and to accelerate cashflow.

Fortunately a relatively small number of committed individuals (landowners and developers), companies (either with wind energy experience overseas or involvement in electricity power generation in the UK), and manufacturers, saw through a number of practical as well as financial and legal problems, so that today a number of commercial wind farms are already in operation. Their achievement is to be congratulated. As financial adviser to Windelectric Limited and Cornwall Light and Power Limited. I am pleased to be associated with two such schemes.

The principle credit must go to landowners, developers, planners and manufacturers, however I would argue that careful financial engineering of projects has in many cases been vital, to provide acceptable rewards to investors and to debt providers, and to improve gearing efficiency, minimising cashflow risk.

As a consequence financial structures have often been complex (except perhaps where finance has been provided wholly at a corporate level). Perhaps because the industry is in its infancy the preferred structures have involved the participation of several parties in an individual project either by way of a formal corporate grouping or joint venture. This trend is likely to continue, however, it must be recognised that the requirements of many corporate investors are likely to increase rather than diminish. The learning process is complete for many and return requirements of 25% to 35% are quoted ever more frequently. The implication is that for many mid range projects new sources of institutional funding are likely to be required, with the need for some institutions' past reluctance to invest to be overcome, as the benefits of existing projects become clear. This is largely an issue of education and publicity by the BWEA and other bodies as indicated by the returns to our survey is already having its effect.

2 KEY FACTORS OF EXISTING SCHEMES

The key factors of successful schemes have been:

(a) the availability of an equity source (usually 10 to 20 per cent of total funds), provided by a combination of:

    (i) landowner/developer involvement (either by cash, and accumulated costs, usually planning, or injection of wind rights);

    (ii) specialist industry investment (often to gain experience of wind farm operation); and

    (iii) venture capitalist investment (if desired return criteria can be met)

(b) the availability of term debt finance, (up to 80 per cent) with cashflows based on researched wind speeds, demonstrating repayment usually within one year of the expiry of the NFFO contract. This has often led to an unspoken requirement for an average wind speed of 8 m/s, although we have financed projects at lower wind speeds;

(c) in appropriate circumstances the use of mezzanine finance, in some cases with warrants to provide enhanced returns;

(d) close attention to project management, the guarantee of output from reliable manufacturers and other measures taken towards the minimisation of risk; and

(e) a strong management/ownership team motivated by appropriate refinements to the financial structure.

3 CURRENT POSITION OF INSTITUTIONS: THE NEW NFFO ROUND

For the purposes of this paper we have surveyed a total of 25 leading financial institutions, the majority of whom have in the past expressed interest in the sector: 19 of the 25 institutions replied. The survey has not included corporate investors such as REC'S, who understandably prefer to deal with projects on a case by case basis. The results of the survey are set out in Fig 1.

The results indicated that the majority of equity to date has been provided by corporate investors such as REC'S and experienced operators using joint venture structures. The participation of financial institutions has been particularly important in respect of mid range projects.

We also asked respondents to comment on the financing of future projects in the light of the third NFFO tranche (see Fig 2). It is interesting to note that none of the financiers surveyed stated 15 to 20 years as their preferred period of investment, indicating that cashflow returns from projects under the 3rd round, need to be sufficient to allow repayment of debt in the 10 to 15 year period (ideally with 10 years).

It is also significant that no single venture capitalist, including those who already have investments in the industry, wishes to leave their funds in for periods in excess of 10 years (ie an exit is required).

As a consequence future projects are likely to:

(a) continue to seek equity support from corporate investors such as REC'S and specialist joint venture developers;

(b) require an increasing proportion of mezzanine and operating lease finance. In particular from those houses with wider experience of project finance able to provide longer finance periods eg 15 years - being at the lower end of typical turbine lives;

Fig 1	Venture Capitalists	Mezzanine finance	Banks
Past transactions			
Number surveyed	11	5	9
Replies	9	3	7
Percentage who have provided finance	11%	33%	43%
Number of investments made	1	2	7
Size of projects	£6.5 million	£3 to £9 million	£5 to £30 million
Secured debt provided as a percentage of total finance	66%	Up to 80%	70 to 100%
Mezzanine debt			
- amount of loan (typical)	£1 million	£1.6 million	None
- period of loan	8 years	8 years	None
- coupon	16%	14%	None
Equity stake takes	30%	None	None
Projected equity return from investment	25% minimum	-	-

Fig 2	Venture Capitalists	Mezzanine finance	Banks
Future projects			
Interested in future projects	22%	100%	100%
Size of projects	£2 to £10 million	£0.5 to £100 million	£3 to £20 million
Secured debt as a percentage of capital cost	Up to 80%	Up to 80%	70% to 80%
Mezzanine debt			
- amount of loan	Up to £5 million	Up to £20 million	None
- period of loan	8 to 10 years	Up to 15 years	None
- coupon	3 to 8% over LIBOR	Varies	None
Equity stake taken	Up to 20%	Varies	None
Expected equity return	Min 20%	20-30%	None
Preferred term of investment	to 10 years	10 to 15 years	5 to 10 years
Percentage who would consider lease finance	0%	66%	43%

Other comments

Venture Capitalist	Time is the biggest single factor which reduces our rate of return. Therefore projects over 10 years are very difficult to justify.
Bank -	The limiting factor to date has been the NFFO structure with the end of the subsidised payment period in 1988.
Bank -	Even though the turbines are expected to work commercially for up to perhaps 20 years, it should not be assumed that finance would be available for such a period.
Bank -	We are finding the skills we have developed in the UK are proving useful abroad.

(c) continue to require tax planning and other sophisticated techniques to enhance returns.

In this respect the various initiatives instigated by ETSU to investigate the feasibility of investment funds being formed specifically for the benefit of the renewable energy and wind energy sectors are to be welcomed. Such funds are feasible and will, I suggest, come to fruition in the shorter rather than longer term.

In particular I would suggest that the statutory and financial characteristics of split capital investment trusts are particularly suitable as a mechanism to attract funds from private investors - who are used to returns of between 10 and 15% (for example by way of the current crop of secured tenancy residential assured tenancy BES schemes) as compared to the 15% to 22% returns generated by a typical wind project (before the benefits of financial and tax structuring).

4   FINANCIAL CHARACTERISTICS OF A 3RD ROUND SCHEME

The financial returns of projects to date, have in many cases been at the margin of returns required by mezzanine and venture capital institutions, necessitating considerable innovation by financial advisors to achieve appropriate structures to obtain funding. This, notwithstanding the relatively high pence per kwh obtained in many 2nd round contracts.

As indicated by one reply to our survey; the 7 year capped period of the 2nd round, together with uncertainty over electricity prices after that period, contributed to the difficulties of many schemes which did not raise finance. The 3rd round NFFO with its extended period considerable assistance in remaining uncertainty. However due to convergence, prices will be lower, with the need to retire debt well before the expiry of the contract period posing its own challenges. I discuss other possible developments in financing later in this paper, but would first like to explore Ernst & Young's windenergy model of cashflows for a typical development with capital investment of £10m conducted under the terms of the 3rd NFFO.

Figure 3 shows financial return measured by way accumulated cash surplus or deficit. The model has been used to examine the effects of changes in output, and a number of price and the interaction of these variables with a number of financial structures. Any form of global analysis is by its very nature simplistic and I cannot over emphasise the need for a detailed and project specific business plan, based on reliable data obtained over the entire season, combined with thorough sensitivity and financial analysis and attendance to technical risks.

The substantial cash balances shown by the model at year 20 are of course not available to the lucky project owner. They represent the funds available to purchase replacement turbines (a not insignificant amount), a margin of comfort to deal with areas of risk, monies to reward mezzanine finance providers and third party equity financiers, both in terms of running returns and exit reward - and finally the reward for owners.

The amounts shown are absolute values, which have not been discounted to provide for the effects of inflation.

In broad terms transactions providing cash balances after 20 years in excess of £35 million (for an initial investment of £10 million) are likely to give sufficient scope for the successful financial structuring of a project.

As can be seen from the diagrams presented, the following critical factors need to be in place:

(a) average wind speeds in excess of 7.0 to 7.5 m/s at current industry standard output levels and if promised efficiency improvements come to fruition;

(b) a minimum average price of 8 pence per kw/h throughout a 20 year period;

(c) at least 5% equity finance with a further 10 - 20% in the form of mezzanine finance;

(d) providing a return of 18 - 20% for the whole project at 4% inflation; and

(e) retirement of mezzanine/preference finance before years 12 to 15;

BANK BALANCE VARIED BY UNIT PRICE

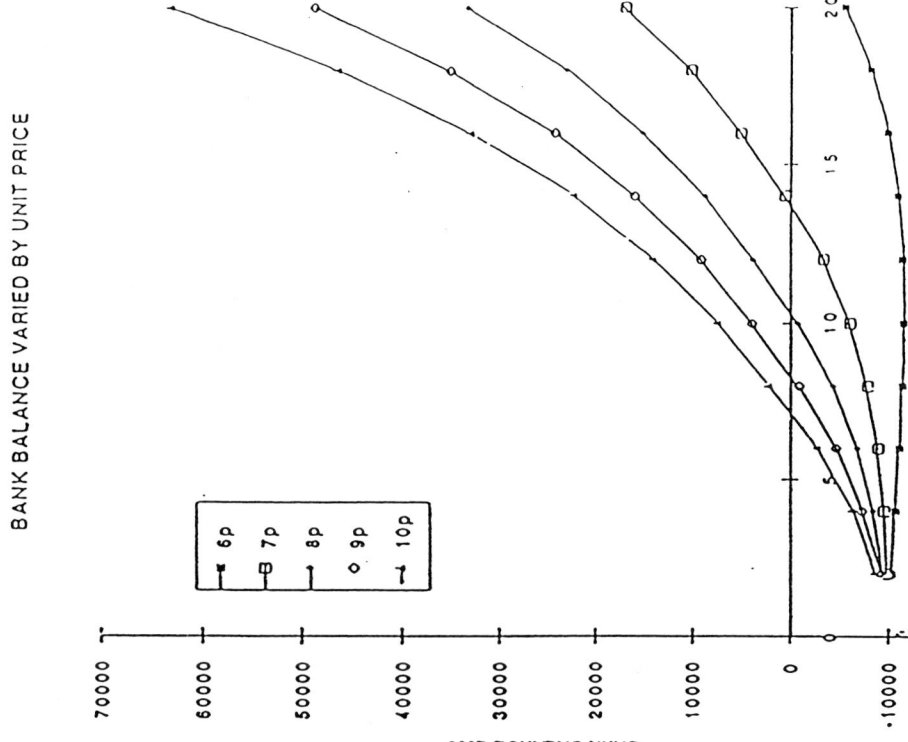

BANK BALANCE VARIED BY POWER OUTPUT/WIND SPEED

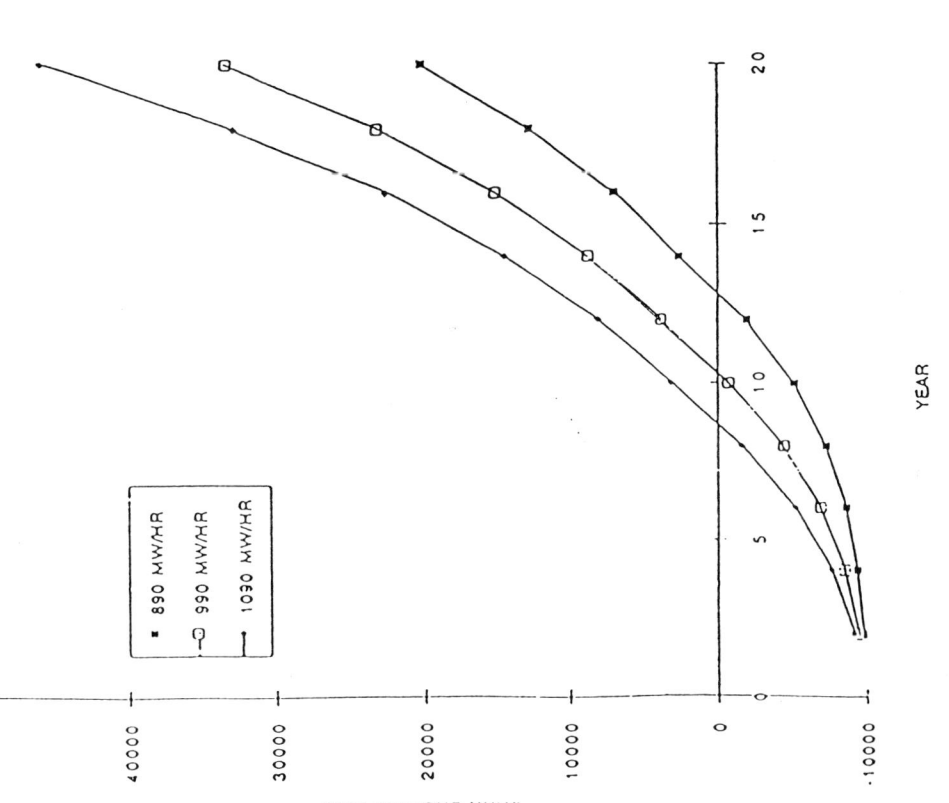

NOTE THE POWER OUTPUT IS DEPENDENT ON THE EFFICIENCY OF THE TURBINES USED 990 MW/HR EQUATES TO A WIND SPEED OF APPROXIMATELY 7 TO 8 M/S USING STANDARD TURBINES WITH APPROPRIATE MARGIN OF COMFORT

165

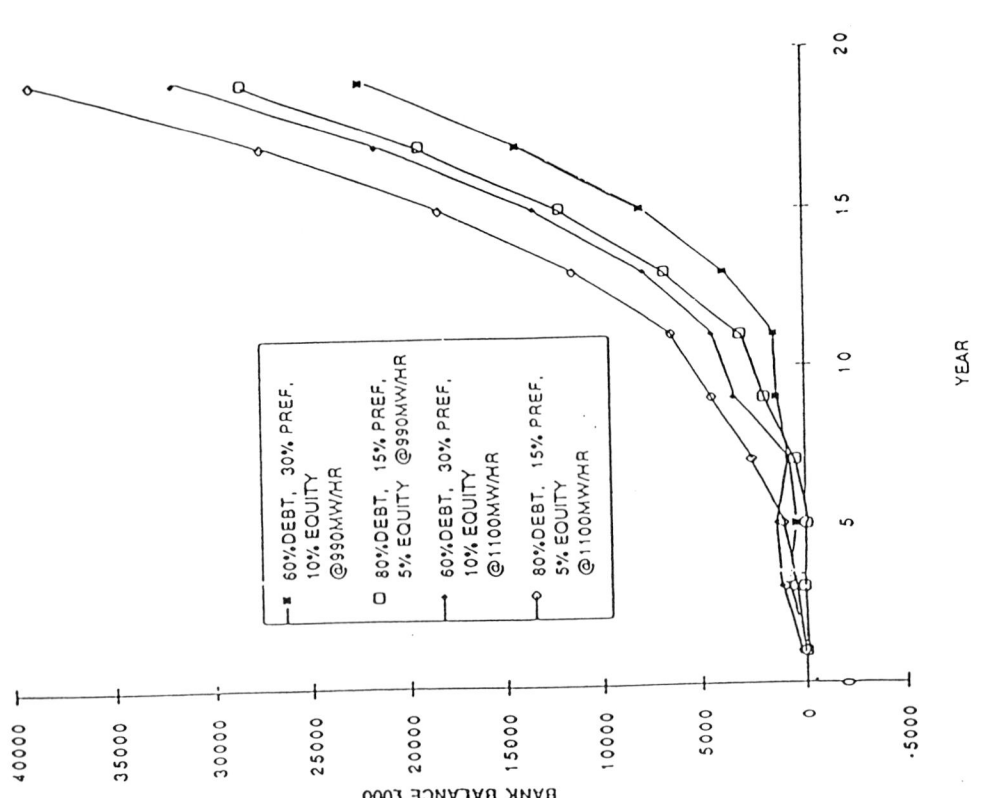

BANK BALANCE VARIED BY GEARING, POWER OUTPUT AND
UNIT PRICE

BANK BALANCE VARIED BY GEARING

The model assumes repayment of ordinary shareholders from internal funds rather than by way of sale or flotation at a deemed exit of 6 to 8 p/e. The model has been run without the benefits of tax structuring and other techniques. These opportunities are invariably project specific and in our experience can enhance returns by a further 3 to 5%.

Of course when reviewing the trends shown by the model difficulties occur when thoughts of practical considerations come into play. For example mezzanine financiers and preference share holders may require retirement of their debt in years 8 to 10 rather than 12 to 15. In many scenarios which satisfy the criteria set out above, detailed cashflows would show that higher returns would be required to satisfy the needs of all investors and debt providers. On the other hand special factors, such as the injection of significant equity and wind rights by landowners, may substantially improve the economics of a project.

Notwithstanding these understandable comments, a clear recommendation arises from our survey and the results of our conceptual model. Namely that the placing of split period bids should be encouraged for 3rd round contracts. These would have a higher fixed price for the first 10 years of contract and lower fixed price for the second 10 years of a contract. This would allow earlier debt retirement and meet the Minister of Energy's requirement that NFFO prices should move towards the pool price for electricity. As a consequence a greater number of projects would be financed with more likelihood of operational success.

5   FUTURE DEVELOPMENTS

The demands of the new NFFO require (quite rightly) further innovations in all aspects of the industry. Technological innovation is likely to produce increasing turbine efficiency and promote manufacturing development thereby reducing capital cost. Financial engineering also has its role.

We believe that fruitful areas of investigation include the use of:

(a)   deep discount bonds;
(b)   qualifying index linked securities;

(c)   extended period operating leases (combined with interest rate swaps); and
(d)   specific tax planning techniques.

If a number of issues are further researched there may be the prospect of the obtaining long term funds for the 15 to 20 year period at rates on or below the 8% to 9% commonly quoted (with margin) for 10 year fixed deals. We are currently achieving these rates for a number of long term construction projects and believe that the techniques used are transferable to the renewable energy arena, although the comparison of bricks and mortar security to that provided by wind turbines is a moot point.

6   CONCLUSIONS

To summarise, I am positive about the next round of the NFFO. However, I believe that further hard work is required by all of us who are committed to the industry. I believe that the NFFO contracts should allow split pricing weighted to the earlier years. I believe that it is possible to solve the technical problems associated with the formation of renewable energy investment trusts. I also believe that it is possible to develop a new generation of financing mechanisms for the industry to provide the support and security of development it needs.

The industry is no longer in its infancy with teething problems, it now faces the excitement and tribulations of an adolescent seeking the rewards of adulthood, as it realises that at some time it will need to move out of the shelter of its parents to the real world of market forces.

The wind industry has learned to walk, it now needs to run.

Realisation of wind park projects by cooperatives

HENNING HOLST, Dipl.-Ing.
WINKRA GmbH, Germany

Many of the recent wind farm projects in Northern Germany have been established by cooperatives. The following paper tries to give the reasons for this and gives some background information about the general conditions for the application of wind energy in Germany. It may be used as a guideline for similar projects outside Germany.

The final breakthrough for the application of wind energy in Germany was caused by the issue of the "Einspeisegesetz" at 1 st January 1991. This federal law obliges the utilities to purchase kWhs produced by renewable energies at a fixed price. Operators of wind turbines, whether single turbines or wind farms are paid 90 % of the average customers price of all German utilities (in 1993 this is a price of 16.7 Pf/kWh = 7 P/kWh).

In connection with a funding programme of the federal ministry for research and technology, the "250 MW wind programme", giving an additional 6 Pf/kWh (2.4 P/kWh) towards the first 10 years of operation of wind turbines, the government created a good incentive for private investors to install wind turbines.

Alternatively some of the provincial governments offered support of up to 30 % of the total investment in order to get the applications for wind energy on their way.

Because of improved economy of the 500 kW turbine size which is generally installed in wind farms today, the practice for funding wind farm projects changed. Today subsidies are preferably given to larger projects with a high output of electricity. The number of machines receiving federal and/or provincial support has been reduced significantly. While in 1991 almost all projects received subsidies towards most machines, the amount of machines subsidised has been reduced in accordance with the revenue of the projects. In practice today federal and/or provincial support may be given to a third or a quarter of the installed machines. This has to be negotiated with the funding administrations of the provincial governments. Furthermore the provincial investment subsidies today

are given according to noise level, performance and output of the machines and do not exceed 17 % for a machine.

Among the provincial governments Schleswig-Holstein is also considering to support projects by buying shares or offering lower interest rates for part of the investment.

Financing a wind farm in Germany requires 10 – 20 % equity from the owner. Up to 75 % of the capital costs of the project is available from banks at special renewable energy interest rates (about 7 %). The rest is financed by open-market bank loans. The banks regard the wind farm as guarantee for 60 – 70 % of the total investment, and the investor must find guarantees for the remainder.

In the course of these favourable conditions the federal and provincial funding bodies recieved an enourmous amount of applications for their wind energy-programmes.

In the windy areas along the German coastline the planning authorities were subsequently faced with building applications for thousands of wind turbines either as single machines or in small wind farms. In order to determine wind farm areas the provincial governments such as Schleswig-Holstein issued planning requirements for wind farm installations. These regulations were issued to concentrate wind turbines in suitable areas. Subsequently the former single applicants had to get together to establish jointly owned projects, the "coops".

The tools to determine wind farm areas are based in the provincial planning law and in the federal building regulations.

In general the planning sovereignty lies with what in British terms would be the parish councils. The county councils are responsible for applying the planning law and have to consider the objections of the public bodies such as National Park offices, landscape protection authorities etc. But the final decision for the planning permission lies again with the parish councils. This favours of course wind farm projects by inhabitants of these small local communities. This is the reason why most projects in Schleswig-Holstein have been established by local investors, either as single investors or by coops. In other provinces local utilities started to lease land from local landowners to establish projects on their own. So far we have had very few projects established by investment companies. The biggest project of an investment company for example was started before the issue of the "Einspeisegesetz" and has been the only project of this kind since.

One critical issue in the planning of wind farms in Germany is the noise immission. The noise level of a wind farm may not exceed 45 dB(A) at night at the nearest dwelling or 40 dB(A) at the nearest hamlet.

In the planning phase the noise immission has to be calculated according to the VDI regulations 2714 and 2058. In practice a security deduction of 3 dB(A) is made for wear of the machines over their expected livespan.

These regulations have to be taken into consideration in the planning phase. Most operators therefore seek the consulting aid either of a manufacturer or an independent consultant. Most of WINKRA's clients today are cooperatives of local investors.

We see the advantages for projects of coops as follows:

- better chances for price reductions in negoiations with manufactures if many machines of one type are being installed
- fairer distribution of subsidies possible
- cost saving concepts for grid connection
- noise regulations are easier to be met for uniform planning in wind farm areas
- public acceptance higher if locals are participating in the projects
- uniform land lease contracts with the involved landowners
- less work for the involved planning authorities
- most economic setup of machines possible

- lower maintainance and insurance costs
- public acceptance higher for plausible setup patterns

In practice the steps towards a cooperatively owned wind farm are:

- shaping a setup of local investors
- planning permission in cooperation with parish council, county council and provincial government bodies
- foundation of a company in cooperation with lawyers and accountants
- planning of setup and type of machines
- application for subsidies (federal and provincial)
- negotiations for grid connection
- building permission
- negotiations with banks
- negotiations with manufacturers
- construction of wind farm
- commissioning
- operation

WINKRA is working as a consultant for many cooperatively owned projects in Germany:

- On the North sea island of Pellworm a local cooperative is planning a wind farm of 4 MW
- A couple of smaller coops on the mainland are establishing wind farms up to 3 MW with our expertise
- On the Baltic island of Fehmarn a cooperative of 34 islanders is building Germany's largest wind farm this year a project of 34 machines (17 MW)
- In Ostfriesland we are consulting a project of 70 machines (41 MW) handled by 3 companies, partly locally owned.
- Negatiations for further large projects of local investors are on their way.

With rising local pressure group action against large wind farm projects in Germany, jointly owned projects of many local investors will gain even more importance in the future. We are convinced that locally owned projects will raise the acceptance of wind energy. We think these cooperatively owned projets are applicable for other countries especially in the EC. As a joint effort of environmentally concerned people they offer a good opportunity for each individual to actively pay a share against global warming.

Benefits to the local community: the limits of proper planning gain

G. M. TRINICK, BA, (HONS)
Bond Pearce Solicitors, UK

SYNOPSIS That one person's perceived benefit from a development is another person's adverse impact is a truism, but at the same time a real issue in planning systems. Wind energy development brings this issue into sharp focus because the clean energy benefits it brings, and which are clearly advocated by the developer, may not be appreciated by those for whom the turbines will represent a major physical feature into the foreseeable future. In attempting to bring local communities some tangible benefit from a wind energy installation developers can easily fall foul of the ill defined boundary rule of what is not proper planning gain, or to use the new term advised in guidance from the DOE, community benefit. When does a financial offer or the provision of some facility become the purchase of a planning permission. What are the proper mechanisms by which such benefits can be secured? This paper will examine these issues. It will attempt to define a conceptual and legal framework for community benefit in the English and Welsh planning system. It will by example indicate ways in which benefits can be secured. Finally, it will suggest that the existing proper boundaries of community benefit need some revision to accommodate the proper concerns of the responsible developers.

INTRODUCTION

A familiar story in the planning world of the 1980s, regrettably not apocryphal, is that of the large housing or retail developer with full pockets negotiating his way to a planning permission with expansive promises of benefits to the community. Examples include the provision of football stadia of a quality suitable for Manchester United for relatively small towns, or the construction of expensive all weather running tracks. The provision of new facilities for the community was sometimes extensive and by any definition, unrelated to the development for which planning permission was being sought. While research by the Department of the Environment between 1987 and 1990 revealed only a handful of cases where the planning system had been abused in this way, such cases did tend to attract publicity.

On the other hand the world of planning control is also full of examples of community benefits which seem reasonable, and reflect a genuine concern on the part of the developer to mitigate the effects of his proposal in the minds of the receiving community.

A common sense approach, based on an instinctive regard for proportionality and decency, easily distinguishes between the grotesque and the reasonable. However, the English and Welsh planning system, as operated conscientiously and correctly by Planning Officers, often fails to make any such distinctions. It must be wary of the disproportionate offer of gain to the community, and this can condemn the genuine and responsible approach which recognises that the gains a development will bring will cause people who live close to it anguish as expectations of ordered continuity without major change are disturbed.

Against this background I will begin by describing the legal ground rules and the approach to community benefit advised by the Secretaries of State for the Environment and for Wales.

THE GROUND RULES

The Town and Country Planning Act 1990 envisages that certain benefits may be secured as part of a development proposal by means of what is known as a Planning Obligation. Specifically Section 106 of the 1990 Act provides that:

> '*Any person interested in land in the area of a local planning authority may, by agreement or otherwise, enter into an Obligation (.....'a Planning Obligation)';*
> - *Restricting the development or use of the land in any specified way;*
> - *Requiring specified operations or activities to be carried out in, on, under or over the land;*
> - *Requiring the land to be used in any specific way; or*
> - *Requiring a sum or sums to be paid to the authority on a specified date or dates or periodically.'*

This Obligation may be created by an agreement between the developer and the planning authority, or by an unilateral undertaking given by the developer. Either in an agreement or in an undertaking there will be covenants to do or not to do certain things. Whether by agreement or undertaking the Planning Obligation is enforceable by the planning authority. An important rule is that the owner of the land or the operator of the development cannot secure a variation or discharge of the Obligation within 5 years of its creation except with the agreement of the local authority. After 5 years he can make an application to vary or discharge the Obligation as if it were a planning condition (that is to say a planning application) and he can take any refusal to a planning appeal.

The history of Planning Obligations lies in a gradual transfer of responsibility for the provision of infrastructure from the public to the private sector. Until the 1970s when housing land values in particular started to rise rapidly, it was usual that the public sector would provide the off-site needs of new developments. All the private developer had to do was to deal with matters within his site.

The fact of more money becoming available to private sector developers through rising land prices was accompanied by the beginnings of a new policy towards the capital funding of public authority projects. Less and less money has become available to the public sector and it has become Government policy to look more to the private sector. This combination of factors has led to the position that it is perfectly normal for a private developer to provide not only facilities on site, but also off site infrastructure such as road improvements. From there it is only a small step to ask the private sector to do that little bit more in providing benefits to the area, whether or not in return for a more favourable attitude to a planning application.

Since these trends first emerged the issue of community benefit, or planning gain as it was known until the 1990 Planning Act, has been keenly debated and contested in the courts.

The current provisions within the 1990 Act relating to community benefit include some amendments made by the Planning and Compensation Act 1991. When the 1991 Act was in Committee within the House of Commons the then Minister for Planning Sir George Young gave a useful indication of the Government's approach to community benefit/planning gain:

> *'A planning gain would do more than merely provide facilities that would normally have been provided at public expense. It would provide facilities that the public purse could never have afforded..... Conservative members believe that there is no reason why the public sector should provide all the schools, community centres and infrastructure. A mixed economy, with the energy of the private sector being added to the resources of the public sector, is a process that I would hope one would want to encourage'*

Given the commercial benefits and financial rewards which can be achieved through the grant of a planning permission the temptations represented by a logical extension of Sir George Young's comments, both to hard pressed local councils and to developers, are all too obvious.

The limits to proper community benefit are advised by the Secretaries of State, and developed and interpreted by judges in the courts. In DOE Circular 16/91 (Welsh Office Circular 53/91) the Secretaries of State advise in detail on the proper scope, content, and flavour of Planning Obligations. In particular 5 tests have been formulated by the Secretaries of State. The test of reasonableness of seeking a Planning Obligation depends, says the Circular, on whether what is required:–

'–   *Is needed to enable the development to go ahead (eg car parking); or*

–   *In the case of financial payment will contribute to meeting the cost of providing such facilities in the near future; or*

–   *Is otherwise so directly related to the proposed development and to the use of the land after its completion, that the development ought not to be permitted without it (eg educational, recreational, sporting or other community provisions, the need for which arises from the development); or*

–   *In the case of a mixed development to secure an acceptable balance of uses; or*
      to secure the implementation of Local Plan policies for a particular area or type of development (eg affordable housing in a large residential development); or

–   *Is intended to off set the loss of or impact on any amenity or resource present on the site prior to the development, for example in the interests of nature conservation. The Secretaries of State have specifically welcomed initiatives taken by some developers to create nature reserves and carry out other such positive activity.'*

Overlying the need to satisfy one of these five tests the benefit provided must, says the advice in the Circular, "fairly and reasonably relate in scale and kind" to the proposed development.
Prior to the 1990 Act the courts had developed a way of looking at the propriety of community benefit which was sometimes expansive and sometimes restrictive, but which in general was leading towards greater freedom for the developer. Lawyers will say that the reported cases were contradictory, and so they were, but I detected a flavour coming through the case reports which reflected the words of Sir George Young which I have already quoted.

Now it seems that DOE Circular 16/91 has retreated towards a purist stance. How will the courts interpret the latest advice, bearing in mind that it is the role of the court to reflect the will of Parliament and only to provide a gloss of interpretation where the wording of an Act leaves room for more than one meaning? If the language of Parliament is clear than there is no scope for interpretation by the court. Again, if the language of Parliament is clear then it would be quite wrong for the court to develop a creative and interpretative role based on the advice of the Secretary of State. Such advice cannot amount to more than a commentary and to an expression of policy.

Turning to wind energy development there is little doubt, on specific advice within DOE Circular 16/91, that the creation of compensatory habitats for plant life or an agreement regulating agricultural practices in an area of environmental sensitivity, would in principle be perfectly proper. Whether or not such benefits could be regarded as fairly and reasonably relating to the development would depend on their physical extent, and on whether or not the benefits covenanted could be related to a provable impact.
Again, covenants in an agreement to remedy any interference with television reception or to improve off–site access roads so as to take construction traffic would be perfectly in order.

However, under present planning law and practice the payment of money to a community for purposes not related directly to the development, presents a more difficult case. Referring back to what Section 106 of the 1990 Act actually says payment of money to a local authority is specifically encouraged, but that particular part of the section does not say to what end that money should be applied. Bearing in mind what I have said about the respective roles of the courts and the Secretaries of State it would seem open to the judges to sanction, in the context of planning control, the payment of money for more general community purposes, where the payment could be said to reflect not the physical impact of a development, but rather the wish to a developer to give a community some benefit from a development which would not automatically bring with it local economic or social benefits.

Indeed I and others have wondered if Parliament's intention was precisely that judges should be allowed some freedom to develop the law relating to payments of money under Planning Obligations so as to reflect the continually changing balance between the provision of infrastructure by the private and public sectors. Parliament may also have had in mind that Obligation should be allowed within limits, to reflect local concerns that developers offer very little local benefit in return for the impact of their developments. Therefore in theory at least the Courts may have some continuing freedom to reflect both fiscal and social concerns. I will return to this point towards the end of my talk.

What emerges very clearly from the advice of the Secretary of State and from the decisions of the Court is the continuing difficulty of applying a clear and objective set of criteria at a local level. It is asking a bit much of a community, especially in a close knit rural area, to ignore tangible benefits on offer which may not otherwise be affordable. To ask a planning committee of local people to separate out the merits of an application from benefits which may be irrelevant to such merits is utterly pointless.

THE WAY FORWARD

None of what I have said really matters until someone challenges a planning permission issued in obvious return for disproportionate or otherwise suspect community benefit. It is important to note that the challenge, possibly by a group of local people aggrieved at the decision, would be to the permission itself and not to the Planning Obligation. The Obligation is at one level simply a private contractual document between the developer and the local authority. Provided that a local authority acts within its general powers in concluding an Obligation the document itself cannot be challenged. Indeed if there is a unilateral undertaking the local authority is not a party and therefore cannot act outside its powers.
However, at another level the Planning Obligation is a public document because it is intended to serve a planning, and therefore a public purpose. If the court feels that a permission has been granted in improper circumstances, specifically in return for benefits which relate in no way to the development, then the permission itself can be struck down and made void. No developer can afford to ignore that possibility.

There are three ways forward. Firstly, wind energy promoters can continue to stretch a point in negotiating with local authorities and therefore to ignore judge made law to a greater or lesser degree. In doing so developers often follow their consciences in dealing with the local community. However, there is always the risk of a judicial challenge.

It is worth bearing in mind a more relaxed approach to community benefit exists within Sections 111 and 137 of the Local Government Act 1972. I am not going to provide the wording of these Sections, but Section 111 broadly enables a local authority to do anything "calculated to facilitate" the discharge of any of its functions. Section 139 permits a local authority to receive gifts of property (which includes money or land) for the benefit of inhabitants within its area. The use of these sections in the context of a planning application is the Council can properly receive money towards the provision of a community facility. It could also take the conveyance of land so as to provide such a facility.

In contrast, under Section 106 of the 1990 Act the Council would be in some difficulty in receiving land although, as I have said, the position may be slightly more fluid in the case of money: under both sections a local authority is empowered to receive money for the benefit of the community.

In the circumstances of cooperation which obviously exist if a resolution for the approval of the application has been made, the provisions of the Local Government Act 1972 which I have mentioned may assist. However, it still remains the case that a challenge made to the validity of a planning permission on the basis that it was improperly granted could result in the revocation of the permission.

A second approach to the return of benefit to the community to compensate for the impact of a development lies within the law relating to compensation. Again this is complex area and I can only scratch the surface today. Infrastructural developments such as new public roads and airports carry with them a package of measures designed to compensate neighbours for the undoubted impact of these developments. A well established system defines entitlement to compensation and provides methods of valuing the adverse impact of the development. The Land Compensation Act 1973 specifically provides for the payment of compensation for nuisance caused by new development, where that development is carried out under statutory powers and there is a statutory immunity from an action for nuisance in the courts. Should these principles of compensation be extended to wind energy development? It has been suggested.

I think that the answer must be no. The only impact of wind energy development which could be susceptible to an objective system of compensation is noise emission. However, even assuming that the measurement of noise from wind turbines can be thoroughly standardised, who will define the level of received noise at which compensation should be payable? The level might have to vary to reflect not only background noise levels, but also the quality of noise being received. In addition does the wind energy industry really wish to acknowledge that we have a problem with noise? The whole thrust of our approach to date has been, quite correctly, that if certain standards are achieved noise emissions are acceptable. Again, the introduction of statutory compensation may enable turbines to be sited closer to houses, but we will then get into problems of visual domination. For these reasons, and on any more detailed analysis there are probably others, I do not think that a scheme of compensation is going to solve our difficulty.

What is really required is a new approach to community benefit which recognises, I suggest, certain principles. Planning law and practice should recognise that where a development may be permitted in line with national policy and possibly development plan policy, but in circumstances which are wholly exceptional, then those circumstances ought also to be recognised in the offers which a developer makes to mitigate the effects of his development. Specifically major structures are not normally permitted on the tops of hills. The encouragement given to wind energy development has little to do with normal land use criteria: it relates directly to Government energy policy which is aimed at reducing harmful emissions and at achieving diversity and security of energy supply. These are exceptional circumstances in land use terms, but Government energy policy is a powerful and valid planning consideration. How many other forms of development would have achieved permissions close to a National Park, within a rural part of an Area of Outstanding Natural Beauty, or more generally in prominent rural locations.

In contrast no proposal for a food retail store or a large residential development could ever be regarded as exceptional. Normal land use principles will always apply. The only comparisons to be made with renewable energy development, and perhaps more generally energy development, in applying the principles of planning control are with the extraction of minerals, possibly the deposit of waste and with the provision of new infrastructures such as roads.

I think that planning law and practice ought also to recognise that the communities or people on which wind energy development impacts are almost always in rural areas, some more remote than others. People are unused to change and do not expect change. Many, especially in south west England, will have moved to remoter locations to avoid change. Indeed all normal and long standing planning advice would point to the maintenance of the status quo and to a very restrictive attitude to any development. Suddenly Government energy policy points, for the best reasons which I fully support, to the possibility of what by any definition is a major development. Those who live in areas attractive to wind energy developers are unexpectedly confronted with possibilities they could never have foreseen. Legally, and perhaps philosophically, it may be argued that no person is entitled to the preservation of his view, or that people living in rural locations enjoy such environmental advantages in comparison with urban dwellers that they ought to absorb the impact of wind energy development without seeking the total preservation of their amenities. However, I cannot help but have some sympathy with adverse local reactions based on these considerations. I feel that planning law ought to recognise such local concerns in the field of community benefit.

It would be irresponsible of me to put forward these thoughts unless I recognised that achieving a sensible way forward is going to be extremely difficult. The problem is that we are not dealing with absolutes, but with a constantly shifting balance of factors which themselves will change through time as fiscal influences and social perceptions change. On the one hand planning control needs to recognise such shifting sands by giving planning authorities freedom for manoeuvre. On the other hand to give such freedom of manoeuvre will open the way to abuse. The Secretaries of State could probably have done little more than they did in giving advice in DOE Circular 16/91, and of course this is advice and not law. Its restrictive advice was inevitable, given the impossibility of defining a more liberal position which did not open the door to the simple purchase of a planning permission.

In summary therefore my feeling is that judge made law will have to be allowed to develop and to reflect the concerns of the time. I certainly hope that the apparent freedom which Parliament has allowed in the payment of money for community benefit be will given life and a reasonably liberal structure by the courts.

In this short contribution I have only scratched the surface of a complex and delicate subject. I hope that a real debate develops and that a constructive solution emerges. What is certain is that to apply rigid planning precepts, as advised in Circular 16/91, to a wholly new and unexpected form of development is politically difficult at a local level, and in my view the road to a short lived industry. We must come to terms with our receiving communities in the benefits which we offer in return for local impacts. We must do this if we are to stand any chance of fulfilling the Government's renewable energy targets, and at the same time making a useful contribution to the Government's commitment to a cleaner environment.

Environmental assessment of wind energy projects

P. J. RADMALL, MA, BPhil
EPL, UK

SYNOPSIS The background to Environmental Assessment (EA) is described and its application examined in the context of the unique characteristics of wind energy projects. A number of key problem areas are identified as priorities for future consideration.

1. INTRODUSTION

Environmental assessment (EA) has been used widely since its formal introduction into the UK planning system in 1988, To date, only the larger wind energy projects have been subject to EA, but in the wake of the government's Planning Policy Guidance Note on Renewable Energy (PPG 22), it seems likely that the majority of commercial wind farms could soon fall within this requirement. The challenge will be for developers, consultants and local authority planners to ensure that EA responds to the unique demands of the wind energy sector, whilst maintaining its integrity as an analytical tool.

This paper provides an introduction to the process and legislative context of EA, reviews its specific application to wind energy projects, and identifies a number of aspects requiring particular attention in future EA practice..

2. BACKGROUND

The concept of EA originated in the USA during the mid-1960s, where it was implemented at a federal level by the Environmental Protection Act, 1968. It was adopted elsewhere - for example, in Australia and Canada - during the 1970s, and was the subject of a European Commission(EC) Directive in 1985. The provisions of the Directive were applied to the UK by the Town and Country Planning (Assessment of Environmental Effects) Regulations, 1988 (referred to hereafter as the Regulations).

In simple terms, the Regulations grafted EA onto the existing planning system administered mainly by local authorities. Development which lies outside this system - for example, forestry - was the subject of separate regulations.

The Regulations divide projects into two types: Schedule 1 projects, for which EA is mandatory; and Schedule 2 projects, for which the need for EA is determined by the local planning authority, Schedule 1 covers major industrial or infrastructural projects such as motorways and large power stations. Schedule 2 covers a wide range of other developments which may be capable of causing significant impact, depending on their location and scale.

Wind farms were listed in neither category, since at that time their viability in the UK was unproved. However, PPG 22 signalled the government's intention to regard them as Schedule 2 projects when they are sited within or close to areas of designated natural amenity, or when they consist of more than ten turbines or 5MW installed capacity. In view of public concern over the perceived environmental effects of wind energy, together with the potential sensitivity of many windfarm locations, EA could in practice become virtually mandatory in the majority of cases.

3. THE EA PROCESS

Guidance on EA has been published by the Department of the Environment, and a considerable literature is available. Regard for the environmental effects of proposed projects has long been implicit in development control. However, EA requires a particular discipline of approach which must be grasped by all participants if it is to facilitate the consent procedure. The key tasks within EA are summarised below.

- Defining the project, and thereby its potential sources of environmental impact. This should include all aspects of construction, operation and decommissioning, together with related or consequential developments such as visitor centres or connection to the grid.

- Reviewing alternatives to the project, where these have been considered by the proponent.

- Describing the baseline environment, i.e. that which is likely to exist at commencement of the project. This will in most cases correspond to the existing environment, although any changes during the planning phase, or during the operational life of the project, should be anticipated.

- Consulting with relevant bodies, in particular those with statutory environmental responsibilities. Issues raised through consultation should be addressed in the EA when they are properly an environmental concern.

- Scoping the breadth of the EA, i.e. identifying those issues likely to be of concern.

- Predicting and evaluating impacts. The requirement is to identify the likelihood of significant impact. Significance will vary according to the subject under assessment, and is invariably a cause of debate amongst EA practitioners. However, a useful shorthand for determining significance is to combine the magnitude of impact with the value of the resource or the sensitivity of the receptor affected.

- Mitigating those impacts, by incorporating appropriate measures into the project design. The effectiveness of these measures should be demonstrated, and any residual impacts assessed.

- Presenting the findings of the assessment in the form of an Environmental Statement(ES), together with a Non Technical Summary.

The above tasks must be undertaken for any EA, regardless of the scale, type or location of project to which it is applied. Therein, however, lies the potential difficulty in its successful adaptation to the wind energy sector.

4. FEATURES UNIQUE TO WIND ENERGY PROJECTS

Wind farms present a combination of features very different from other forms of development, from conventional energy projects, and even from other renewable energy technologies.

They require very limited permanent landtake, thereby permitting existing land use to continue after construction, This restricts the potential to affect terrestrial resources such as ecological sites, archaeological remains or high-grade farmland. The construction period is relatively brief, which helps to mitigate potential nuisances such as noise, dust or traffic. Moreover, construction is essentially a fabrication task; it requires no major earthworks apart from foundations for towers and trenching for cables.

Unlike combustion technologies, wind farms produce no airborne emissions. This, quite justifiably, is promoted as their major environmental benefit, particularly in comparison with the generation of greenhouse gases from conventional fossil fuel technologies. Furthermore, windfarm operation is virtually automatic, and produces very little off-site traffic or waste. However, wind farms are likely to generate a higher proportion of spectator traffic than other energy projects.

Windfarm design and siting are extremely site-specific; this characteristic is shared with other natural energy technologies (e.g. wave and hydro), but not with combustion projects, The tolerances on turbine layout can be small, which often limits the ability to consider alternatives. Moreover, unlike other technologies, wind farms will seek out sites of high exposure, which will also be extremely visible. Conventional on-site mitigation of visual impact (i.e. by screen planting or mounding) will not be feasible in many cases, due to the risk of impeding airflow.

Renewable energy in general, and wind energy in particular, tend to be emotive issues, capable of attracting divergent views from a range of interest groups during consultation. Some of these views may be predicted; for example, Friends of the Earth are likely to be supportive, the Countryside Commission and RSPB more equivocal. The opinions of local residents and planning authorities may be very difficult to anticipate.. The EA must respond to such views without being diverted from its primary purpose as an analytical tool rather than a promotional vehicle..

Finally, the wind energy sector is characterised by a wide range of participants, from corporate players to local landowners. The scale and capitalisation of projects reflects these differences, and will influence both the scope of their potential impact and the level of analysis that developers can afford.
The EA must demonstrate flexibility in response to these variables, without compromising its statutory content.

5. THE SPREAD OF EFFECTS

The potential environmental impacts of wind energy have been well documented in a number of studies, papers and research projects, and do not require detailed repetition. The interaction of these impacts within the EA process, however, has received little attention, and requires participants to be aware of the perspective from which impacts are to be assessed, particularly with regard to the evaluation of cumulative and net effects.

The first problem is that of the geographical extent over which adverse impacts are likely to be projected. On this basis, impacts may be divided into three types:

- On-site impacts, such as land use, archaeology and terrestrial ecology.
- Immediate off-site impacts, such as noise, bird strike and traffic.
- Extended off-site impacts, such as visual and electromagnetic interference.

These differences may be reflected both in the level of attention devoted to each impact in the EA, and in its ability to be contained by mitigation. On-site impacts can in most cases be avoided or minimised by small changes in layout or by careful construction practice. Unless the tolerances for relocation of turbines are extremely limited, detailed investigation or remediation of ecological or archaeological sites is unlikely to be necessary.

Immediate off-site impacts can be of sufficient concern to influence the granting of planning consent or the conditions attached to it. Both operational noise and bird strike may be addressed by changes in layout, to ensure that minimum distances are maintained from nearby dwellings and from breeding sites, feeding grounds and flight paths. Although the whole issue of bird strike remains highly contentious, developers are well advised to avoid unnecessary confrontation by building appropriate safeguards into project design. Construction impacts such as noise and dust can be dealt with by applying conventional mitigation practices (e.g. wheel-washing, restricted hours of working). Traffic impacts may also be managed so that they remain acceptable, particularly in view of their temporary nature during construction, although local authorities may in some cases seek planning gain in the form of improvements to the road network. Spectator traffic is more difficult both to predict and to mitigate, and consultation between developer and highway authority may be desirable to explore opportunities such as the joint provision of lay-by facilities.

Of the extended off-site impacts, electromagnetic interference must be addressed early during the site selection and layout phase of the project. Visual impact is more problematic, in that it is likely to emerge as a key issue capable of preventing consent from being obtained, yet is notoriously difficult to predict, evaluate and mitigate. These problems are addressed below.

In summary, the distance over which an environmental effect may be projected can be broadly correlated with its potential severity and inversely correlated with its capacity to be mitigated. These relationships should be reflected in the scope and emphasis of the EA.

6. LOCAL COSTS VERSUS GLOBAL BENEFITS

The problem of geographical perspective becomes even more acute when one considers the question of net impact. As mentioned previously, wind energy is with justification promoted for its global benefits. Where consent is withheld, however, this is most likely - and with equal justification - to be defended on the basis of unacceptable local impact. Whilst it is not for the EA to decide between these opposing influences, its presentation of them can be of critical influence on the decision-maker.

The message is one of a need for increased sophistication both in the methodology of EA and in the planning of wind energy projects in response to cost/benefit analysis. The importance of seeking greater community participation and benefit has been acknowledged (see Marcus Trinnick's preceding paper), and innovative approaches to the distribution and pricing of wind energy (e.g. as a substitute for energy derived from fossil fuels amongst local consumers) may be required. Generic statements about the reduction of greenhouse emissions will no longer suffice, and efforts will need to be directed towards a more rigorous analysis of potential costs such as sea-level and climatic change.

7. VISUAL IMPACT

The technique of visual impact assessment is well documented elsewhere, but its application to wind energy projects requires specific consideration, due both to its likely prominence as a planning issue and to the methodological problems it must overcome.

The fundamental conflict between wind energy and landscape is inescapable; the preferred locations for wind farms are invariably in coastal or elevated areas of extensive visibility and relatively high landscape value. The task of the EA is to demonstrate, to an acceptable degree of confidence, the precise visibility of the turbines and their effect on landscape character and on the visual amenity of potential receptors (i.e. observers). Whilst computer modelling techniques have in recent years greatly increased the accuracy and versatility with which future views can be predicted, the difficulty of interpreting these views remains, as any expert witness who has attempted to do so at a planning inquiry will confirm.

The issue of subjectivity is unavoidable; both those responsible for carrying out visual assessment, and those likely to be the recipients of impact, may respond to landscape in very different ways, reflecting a wide range of individual, educational and social influences. Although these influences cannot be built into the assessment process on a scientific basis, the assessment may nevertheless achieve a defensible degree of rigour provided that the assumptions on which it proceeds are reasonable and are clearly explained. It is quite proper, for example, to call upon professional judgement,, in the same way that a practitioner in a conventionally scientific discipline may do; for example, in order to form conclusions about the likely sensitivity of receptors on the basis of their activity at the time of exposure to impact. Such generalisations may be scrutinised closely, but if their premise is sound they can only contribute to the presentation of visual issues in the EA.

The problem of variables is encountered repeatedly in visual assessment. Visibility, for example, may be determined at any time by season, atmospheric conditions, differences in receptor activity and position, and the presence of intervening features. Such factors are likely to affect the distance beyond which turbines may cease to become a significant influence on the character of a view; an issue which has received increasing attention, though not to the extent of producing any universally applicable conclusions. In the absence of ready-made solutions, the practitioner must base the assessment upon intelligent and site-specific scenarios which can be conveyed with relative ease to the decision-maker.

Finally, as wind turbines become more recurrent and familiar features in the landscape, the problem of cumulative impact and receptor sensitivity will require more attention. Whilst research in Cornwall has suggested that public resistance to wind turbines decreases over time, it is possible that this trend could be replaced by one of increased irritation or indifference after repeated exposure to a series of individual wind farms. Moreover, whilst the landscape impact of one windfarm may be mitigated in part by its novelty and its role as a landmark, this advantage would be lost if it were viewed in the context of other wind farms nearby. Different landscapes are able to accommodate different frequencies or scales of wind energy projects causing different degrees of change to their fundamental character. Clearly, as more projects compete for sites within prime areas, greater sophistication will be needed for the evaluation of cumulative effects.

8. CONCLUSIONS

The central theme of this paper is that EA will become an increasingly frequent requirement for wind energy projects, and that a more innovative response is needed to reflect their unique characteristics whilst complying with the Regulations.

Wind farms differ substantially from conventional energy projects, and even from other renewable energy technologies, in terms of their environmental "signature". These differences place particular demands on the scope and analytical depth of the EA process.. Whilst the techniques of assessing individual impacts are in most cases relatively mature, the need to present cumulative effects over different geographical areas and in the light of mitigation of variable effectiveness, will require particular consideration. Specifically, the balancing of global benefits against local effects must in future be achieved with a greater degree of conviction.

Visual impact will continue to be the most controversial issue for most projects, and that most likely to impede the consent process. Here again, traditional assumptions and techniques must be tested and, if found wanting, replaced in order to remain applicable to situations in which wind farms may become recurrent and familiar features.

Like any planning tool, EA must adapt in order to retain its credibility within a context of evolving attitudes, new technologies and changes in policy. There is every prospect that the wind energy sector may lead the way in bringing about such an adaptation.

SELECTED REFERENCES

Department of the Environment/Welsh Office: Environmental Assessment - A Guide to the Procedures, HMSO, 1989.

Department of the Environment: Planning Policy Guidance Note No 22 - Renewable Energy.

Countryside Commission/Land Use Consultants: Environmental Aspects of Wind Farms, 1990.

Countryside Commission: The Treatment of Landscape and Countryside Recreation Issues in Environmental Assessment, 1990.

Noise from wind turbines: activities in the DTI wind energy programme

M. L. LEGERTON, BSc, PhD
ETSU, Harwell, UK

SYNOPSIS Recent activities both within and outside of the DTI's Wind Energy Programme are described which address the issue of noise from wind turbines.

1 INTRODUCTION

Noise generated by wind turbines is an environmental constraint on the exploitation of wind energy. It is a consideration when seeking planning consent for the siting of machines due to the high population density in the UK and low levels of background noise in rural areas. There is, therefore, a need to identify the magnitude and characteristics of noise emitted by wind turbine generators, assess the influences on the propagation of noise through the atmosphere, and provide information to both wind farm developers and planning regulators on likely noise levels. Additionally, wind turbine manufacturers need to understand noise producing mechanisms on wind turbines to develop widely acceptable products.

This paper reviews recently completed work and current projects in the DTI's Wind Energy R,D&D Programme in the above areas. It then discusses more widely ranging activities which relate to wind turbine noise together with plans for work which should increase our understanding of the problems and allow more definitive guidance on noise levels to be given.

2 SOURCES OF NOISE

It is important to reduce the noise emissions from wind turbines to the minimum practicable levels and to be able to predict the level of noise output. A number of projects with these objectives have been initiated in the DTI's programme.

2.1 The Influence of Noise on the Design of Horizontal Axis Wind Turbines
(James Howden & ISVR)

This work was undertaken jointly by James Howden and the Institute of Sound and Vibration Research. The objectives of the work were to identify the principal sources of noise and to make practical recommendations for the reduction of noise from wind turbines The final report (1) is a useful summary of measures available to the wind turbine designer to control noise emissions from wind turbines. It is in four parts: a summary report including a costing of various noise control measures, an experimental investigation of noise emissions from the 1MW Richborough machine and separate studies on the generation and control of aerodynamic and mechanical noise. The report provides a comprehensive review of the aerodynamic and mechanical noise sources of a wind turbine with practical suggestions for their attenuation.

2.2 Sources of Noise on a Wind Turbine
(Windpower & Co)

A comprehensive range of experiments were undertaken to find sources of noise on a 140 kW wind turbine (WP1). In addition to monitoring the usual parameters such as wind speed, power and rpm, the turbine was fitted with strain measurement and a flow visualisation rig which used a rotating video camera focused on flow indicating tufts. Flow visualisation made possible the association of a three times per rev noise with dynamic stall, helped identify an acceptable rotor/tower clearance for the avoidance of upwind thumping and identified the noise associated with fully stalled blades. The flow visualisation technique was used in the development of a new sound measuring method where microphones were suspended close behind the rotor plane to enable blade aerodynamic states to be related to noise generation.

Some of the results reported by the authors are:

- At 25m from the turbine, maximum tone level varied by 5 dB, as a function of direction from the machine. The amplitude of one per rev machinery noise modulation was 4 dB on WP1.
- Slowing tip speed from 60 m/s to 40 m/s reduced rotor noise by about 7 dB.

- Sound power levels varied with wind speed by 0.59 dB per m/s at 59 rpm and 0.46 dB per m/s at 39 rpm. Least scatter occurred in the data when sound power level was related to power, rather than wind speed.

- No evidence was found for noise arising from rotor/tower interaction due to tower acoustic blockage, tower acoustic reflection, or due to the tower's interference with the air flow.

- In winds of over 12 m/s and greater than 10° of yawed inflow, the flow visualisation showed conditions of dynamic stall. This was matched by a 3P, staccato, "chomping" sound which raised spectral levels between 1250 Hz and 3500 Hz by 10 dB, attracted attention and doubled the downwind distance at which the turbine could be heard.

- A continuous "shooshing" sound was heard immediately the flow visualisation showed the blades to have chaotic flow and to have become fully stalled. Spectral frequencies between 450 and 2000 Hz were raised by 1.5 dB but the noise did not attract attention and was not considered a nuisance.

- Mean background levels and to a lesser degree wind turbine sound pressure levels varied over time. This meant that when the long term, mean sound pressure level from the machine and mean background level were nominally equal, they could be intermittently up to 7 dB different when measured concurrently. At times, background noise swamped turbine noise and vice versa. This continuing, apparent modulation made noise from the turbine more noticeable.

- This effect, combined with the "chomping" and modulation of tones from the gearbox due to changes in power combined to draw people's attention to the wind turbine during higher winds. This contradicts the view, often expressed in the literature, that higher winds mask turbine noise.

2.3 Assessment and Prediction of Wind Turbine Noise
(Flow Solutions)

The need to be able to make accurate predictions of noise source levels to assist in the design of wind turbines with low noise levels had been identified. As a result a project was placed with Flow Solutions to develop a prediction method for aerodynamic noise from wind turbines.

A literature review was carried out early on in the programme of work covering aerofoil noise and rotor acoustics, including fans, propellers and helicopter rotors as well as wind turbines. The mechanisms of noise radiation by a wind turbine identified in the study can be separated into the following distinct areas.
Discrete frequency noise at the blade passing frequency and harmonics.

Self induced noise sources.
Trailing edge noise
Separation-stall noise
Tip vortex formation noise
Laminar boundary layer vortex shedding noise
Trailing edge bluntness vortex shedding noise
Noise due to turbulent inflow.

It was found that self noise sources will dominate at low wind speeds, while inflow turbulence sources will be most significant at higher wind speeds at rated power and beyond. Analysis has suggested that the two most important sources are the trailing edge noise due to the passage of the turbulent boundary layer over the trailing edge and inflow turbulence. New prediction models have been developed for both of these sources based upon a combination of theoretical and experimental analysis.

The result of this work was the development of a basic method for predicting aerodynamic noise from wind turbines. The final recommended method was based on a generalised description of the wind turbine and predicted the aerodynamic noise resulting from the passage of atmospheric turbulence over the blades and the convection of boundary layer turbulence past the trailing edge.

A large body of wind turbine noise data from the UK, Europe and the USA was collected. Data from the MOD 2, prototype MS3 and Vestas 39 were compared with predictions from the model. Reasonable agreement has been obtained for these cases.

The theoretical methods developed for the model were also used to investigate parametric trends and formed the basis for some design guidelines. Methods for avoiding the appearance of the two vortex shedding noise sources have been discussed.

The results of the work have been published (2)

2.4 Systematic Comparison of Prediction and Experiment for WT Aerodynamic Noise
(Flow Solutions)

Detailed comparisons of experiment and predictions were undertaken to test the strengths and weaknesses of the prediction method developed in the previous work and to identify any areas requiring improvement. The model was found to under predict the noise by an average of 2.25 dB(A), believed to be due to the presence of mechanical noise and/or excess aerodynamic instability noise in the experimental data. The full report is now available (3).

2.5 Design Prediction Model for Wind Turbine Noise
(Flow Solutions)

This is the third of three linked projects undertaken by Flow Solutions. The model previously developed was based upon a rather generalised description of the wind turbine as it was not possible at that time to use a full detailed description. Whilst the model is useful in demonstrating the parametric trends associated with the key noise generating mechanisms it cannot be used to study the effects of detail design changes on wind turbine noise. Recent advances in aerodynamic prediction methods and greater understanding of the key mechanisms now make it feasible to develop more detailed models which could provide the basis for design of minimum noise turbines. Work is now in progress to extend the model so that the effects of detail design changes on wind turbine noise can be predicted. The objectives of the work are:-

- To develop a prediction model for wind turbine aerodynamic self noise which reflects full details of the wind turbine design.
- To develop new models for stall and tip noise radiation related to existing theory and experimental data.
- To integrate these models in a comprehensive prediction model for wind turbine noise.
- To compare the results with wind turbine noise data.
- To provide suggestions for reduction of wind turbine noise by design.

The work is currently in progress.

2.6 Noise Control Development of the WEG 400
(Wind Energy Group)

This project is primarily concerned with the validation of noise reduction measures already taken and the attainment of further reduction in noise levels. The objectives of the work are to:-

- Verify the accuracy of the noise emission predictions made for the WEG 400, including tonal levels. Noise emission values predicted for the WEG 400 will be verified by taking measurements on the pre-production machine at Cold Northcott. Measurement data from MS-3 production machines will be used to calculate how much allowance must be made for the variation in noise levels between identical examples of the same machine type when establishing sound power level and tonal content that can be guaranteed to clients.

- Assess development of noise control/reduction methods for the purpose of further reducing noise levels of production machines. Areas to be assessed will include as examples:-
The effectiveness of absorption and insulation materials by investigation and machine trials.

The construction and shape optimisation of the cladding.
Anti-vibration mountings.
Air flow and duct design.
Aerodynamic noise (including the effects of tip speed, wind speed, pitch, boundary layer trip and trailing edge fitting)

- Assess the variability of noise emissions that can be expected from a production batch of WEG 400 wind turbines. The correlation between measured vibration levels on gearbox cases and noise level in service will be investigated. This will involve the construction of a load test rig at the gearbox manufacturers together with noise and vibration measurements both on the test rig and repeated on site.

- In conjunction with manufacturers and suppliers, develop more suitable noise emission specifications, with particular emphasis on the practicality of testing for conformance.

- Review currently available techniques of predicting noise radiation and transmission from complex components and structures associated with wind turbines, e.g. gearbox, blades, tower cladding. Determine if such techniques can be used at the concept design stage for optimisation with respect to low noise emissions.

The work is currently in progress.

3 PROPAGATION OF NOISE

By applying a model for the propagation of sound through the air the noise levels at the nearest properties or other noise sensitive areas can be estimated. The simplest and most commonly used method is that outlined in the IEA Recommended Practices (4). This assumes hemispherical spreading of the sound with some additional attenuation due to absorption by the air. Secondary effects such as topography, ground type and wind direction and shear are not included. A number of projects have been commissioned which have assessed and developed noise propagation models.

3.1 The Prediction of Propagation of Noise from Wind Turbines with regard to Community Disturbance
(ISVR)

This study reviewed aspects of outdoor sound propagation relevant to wind turbines, considers the adaptation of semi-empirical propagation models to wind turbine noise, develops an analytical ray-tracing model and compares predictions from this model and others with measured wind turbine noise data.

The ray tracing model was applied to a parametric study on the influences on the propagation of sound. The report concluded that there is an initial zone stretching several hundred meters downwind of the turbine in which the influence of individual parameters is small and gradual, whilst beyond this zone more complex effects could be expected. A report of the work is available (5).

3.2 Noise Propagation Studies at Carland Cross and Coal Clough
(Renewable Energy Systems Ltd)

A study on the propagation of sound over two wind farm sites has been carried out by RES. The sound source was an omni-directional calibrated loud speaker mounted on a 25m mast. Noise levels were recorded over a range of distances from the mast and compared to those predicted by various models. Although the final report has not yet been completed, early indications are that a simple spherical spreading model with air absorption is at least as accurate as some of the more complicated models tested. A report on the work will be available by the end of 1993.

3.3 Noise from Wind Turbines
Joule II project PL920313
(NEL and European partners)

There are two areas of uncertainty associated with the use of propagation models. Firstly, the sound power of the wind turbine used in the calculation is often the result of measurements on one example of its type and may not be representative of all turbines. Secondly, propagation models that are commonly used do not model the effects of changes in the speed of sound with height, ground type and turbulence. This project attempts to quantify the variation in sound power between nominally identical machines and to reduce and evaluate the errors in propagation models by developing and testing a propagation model suitable for wind farm applications.

NEL in conjunction with the Netherlands Energy Research Foundation (ECN), dk-Technik, Danish Acoustical Institute and Deutches Windenergie-Institut, have obtained 50% funding from the CEC Joule II programme for the work which is being topped up by National Governments.

The overall objectives of the Joule II project, managed by ECN are:-

- To provide statistical information on the noise production of 6 different wind turbine types, the noise of each type measured from 5 different machines. Sound pressure levels will be measured on a reflecting board in the downwind position in accordance with the draft IEC standard in preparation. The measurements will be analysed to give the apparent sound power level at a wind speed of 8m/s at 10m height, sound power level as a function of wind speed and narrow band frequency spectra.

- To provide statistical information on the variation in the acoustic noise production of wind turbines with time (ageing effects). Similar measurements will also be made on turbines which have already been monitored giving the variation is noise characteristics over a period of 3-7 years.

- To develop and validate a simplified propagation model suitable for use in wind turbine planning procedures. A sound propagation model suitable for use in wind farm planning procedures is to be developed. This to be based upon the work of Rasmussen of the Technical University of Denmark (6). Rasmussen has developed a model which takes the effect of ground impedance as well as wind and temperature shear into account. Extension of the model to cater for raised source height and a wider range of wind speeds are required. The resulting model will be validated using measurements from an elevated sound source and wind turbines.

The project is of two years duration and started in January 1993.

4 MEASUREMENT OF NOISE FROM WIND FARMS

4.1 Noise Monitoring at Delabole Wind Farm
(Windelectric Ltd)

A comprehensive survey of the characteristics of wind turbine noise is being undertaken at the Delabole wind farm. Topics under investigation include:-

- A comparison of noise levels with and without the turbines operating. Noise levels, wind speed at microphone position, wind speed and direction from the wind farm mast and turbine operational status are being continuously monitored. Noise levels are measured at four locations. Two locations are at a distance of at least 1km from the nearest turbine, one in a sheltered position and one in a more exposed position and so provide measurements of typical background noise levels in a rural area. Spot checks are made to ensure that operation of the wind farm does not affect the noise levels at these locations. Measurements at each site are to be made in two periods, one in Winter and one in Summer. The two other locations are at a distance of approximately 350m from the nearest turbine, one in a sheltered position and one in a more exposed position. Spot checks are made with the turbines switched off to verify that background noise levels at these two locations are similar to those at the 1km locations.

- Measurement of the sound power and spectral analysis of the noise from each turbine.

- Propagation. The propagation of sound from a single turbine is to be assessed by measuring sound levels downwind, crosswind and upwind at distances of 50m, 100m, 200m and 350m from a turbine together with background noise levels. The measurements will be taken at wind speeds as close to 8m/s as possible.

- Variation in turbine noise with wind speed. The noise from a single wind turbine shall be measured over a range of wind speeds at a distance of 50m.

The one year experimental programme commenced in April 1993, the final report is expected in June 1994.

4.2 Noise Monitoring at Cold Northcott and Llangwyryfon
(National Wind Power)

An extensive noise monitoring programme, centred around the Cold Northcott and Llangwryyfon Wind Farms is being undertaken by National Wind Power with 50% funding from the DTI. Areas of investigation include:-

- Measurement of the sound power and spectral analysis of the noise from individual turbines.

- Long term monitoring of noise levels at 50 m from an individual turbine to determine the change in noise levels with time and operating conditions.

- An examination of the suitability of currently available tonal assessment methods to establish which most accurately ranks wind turbine tones.

- Long term monitoring of wind farm noise at nearest residences. Comparisons will be made of background levels prior to installation of the turbines to noise levels now experienced at several nearby locations.

- Noise propagation monitoring. An examination will be made of the factors affecting the propagation of noise over different terrain types and over distances ranging from 50m to 800m. The measured noise levels will be compared to predictions from generally available propagation models.

The project has an expected completion date of June 1994.

5 THE FORWARD PLAN

Some of the requirements for further work are assessed below although the priorities and objectives of work will inevitably change as more is learnt about wind turbine noise.

5.1 Aerodynamic Noise

The aerodynamic noise from wind turbines due to the passage of air over the blades has hitherto been considered to be broadband and unobtrusive in character. Recent experience with operational wind farm has shown that this may not always be the case with anecdotal evidence of 'thumping' and 'swishing', sufficient to draw attention to noise and increase the potential for loss of amenity or annoyance. These effects have been attributed to tower interaction, yaw misalignment, dynamic stall, wind shear and turbulence variations. There is however no definitive study on the causes of the modulation of aerodynamic noise, although the report 'Sources of Noise on a Wind Turbine' identifies some of the mechanisms involved.

It is therefore proposed to commission a study which thoroughly investigates the circumstances leading to the production of the undesirable characteristics of 'thumping' and 'swish' in aerodynamic noise.

Detailed measurements of aerodynamic noise are also required for the development and validation of the Design Prediction Model being developed by Flow Solutions.

5.2 Mechanical Noise

The general principles for the reduction of mechanical noise are well established and their application to the wind turbine are described in the report, "Influence of Noise on the Design of Horizontal Axis Wind Turbines" (1). Manufacturers are well aware of the problems that arise when turbines generate noise that is tonal in character and much work is now in progress to apply best engineering practice to the design of wind turbines. Most work that can be done in this area is the application of existing knowledge to specific turbine designs and configuration and only a limited amount of work of a generic nature can be identified. Therefore the main role of the DTI programme appears to be in assisting UK manufacturers in their development programmes to ensure the noise standards of UK machines are competitive. A number of projects have been identified and it is expected that more will be brought forward in the future

5.3 Noise Measurements in Adverse Conditions

Wind farms are usually sited in wet and windy environments, conditions not ideally suited for the measurement of relatively low noise levels. For example the Bruel and Kjaer literature gives a self induced wind noise level of 54dB(A) at 8m/s for their outdoor noise measurement kit. It is proposed to bring forward an investigation into the limitations of equipment currently in common use leading to suggestions for means of improving the reliability of measurements.

6 WIDER RANGING ACTIVITIES

6.1 IEC TC 88 and BSI Sub Panel PEL 123/-/1, Acoustic Noise Measurement Techniques

The starting point for an assessment of the effect of any noise source on the environment is a knowledge of the characteristics of the source. The characteristics of a source can be described by the sound power, expressed in dB(A) (ref 1pW), narrow band and octave frequency spectra. Due to its size the sound power of a wind turbine is difficult to measure but an 'apparent' sound power is commonly derived from sound power measurements from a microphone mounted on a hard board at ground level at a known distance from the turbine. By assuming that the sound propagates spherically from a point at hub height in the nacelle, the apparent sound power can be calculated from the sound pressure. Planners and developers frequently require information on how the turbine noise level varies with hub height wind speed and also on the tonal content of the noise.

The most frequently used methods at the moment are those described in the IEA Recommended Practice "Acoustics - Measurement of Noise Emission from Wind Turbines" (4), The Danish Standard written by Andersen and Jakobsen (7), and the AWEA Standard 2.1 "Procedure for Measurement of Acoustic Emissions from Wind Turbine Generator Systems" (8).

The current situation is unsatisfactory for a number of reasons:-

1) The measurement methods are all different in detail, which can result in misleading comparisons being made.
2) None of the methods produce information in a form ideally suited for use by planners and developers.
3) The requirements for frequency spectra are poorly specified, if at all.

An international IEC standard is currently under preparation that will provide a standard method for the measurement of acoustic emissions from wind turbines. The UK is represented on the working group responsible for its drafting by Mr R Henderson of the National Engineering Laboratory and Dr J Bass of Renewable Energy Systems Ltd.

In parallel to the IEC standard a revised version of the IEA Recommended Practice on Acoustic Measurement of Noise Emission from Wind Turbines is being prepared by another working group made up largely of members of the IEC panel. The procedures for preparation of the IEA document are simpler than the IEC standard and hence the revised Recommended Practice should be available earlier.

It is unlikely that either document will be able to fully address issues such as the variation in sound power between identical turbines of the same model or site specific factors like the effect of inflow turbulence.

6.2 Planning Policy Guidance Note on Renewable Energy (PPG 22)

The Planning Policy Guidance Note on Renewable Energy, PPG 22 (9), was published by the Department of the Environment and the Welsh Office on 3 February 1993. PPG 22 contains an Annex on Wind Energy which includes some discussion on noise from wind turbines. This annex includes a description of the sources of noise from wind turbines, a discussion on the limitations on the use of BS4142 (10) and advice on noise related information that could usefully accompany a planning application. At the time of writing there was insufficient relevant experience of noise from wind farms and public reaction to the noise, to be able to provide quantitative guidance on suitable noise limits to be set at nearest residences.

6.3 British Standard BS4142

The PPG explains that the use of BS4142 1990, 'Method for rating industrial noise affecting mixed residential and industrial areas', may be inappropriate for the assessment of noise from wind farms for several reasons. These are listed as:-
a) Wind farms are likely to be developed in areas outside of the scope of BS4142 as indicated by the title.
b) The scope precludes situations where background noise levels are below 30dB(A). This level is typical of the background noise level which might be found at wind farm sites.
c) BS4142 states that noise measurements should not generally be made in winds greater than 5m/s average. This restriction guards against the effects of wind noise on the microphone (and influences on sound propagation). Wind farms are likely to be sited in windy areas where the BS4142 conditions may not be satisfied.

A more fundamental problem that may occur using BS4142 for the assessment of wind farm noise lies in the choice of units used to describe the specific noise source and the background noise. BS4142 specifies that the noise source is to be measured as L_{Aeq} and the background noise as L_{A90}. A characteristic of background noise in rural areas is that measurements of background noise measured in these two units can differ by 10dB(A), especially when background noise is wind related or contains relatively loud, intermittent sources. The rating method proposed in BS4142 would therefore indicate the likelihood of complaints even in the complete absence of other noise sources.

6.4 Working Group on Noise from Wind Turbines

With no generally agreed procedure for determining noise levels that are acceptable to nearby residents, planners and developers have been required to use their own experiences to bring forward workable solutions by reference to the particular character and sensitivity of the area. Planners have the benefit of local experience on what the existing noise environment is in their area combined with the public's reaction to new noise making developments, whilst developers have a knowledge of the noise characteristics of wind turbines. Many wind farms, though not all, have had conditions relating to noise levels from the wind farm specified in the planning consents. These have varied in noise level and measurement units (eg L_{90} or L_{50}) from site to site but generally fall in to two classes. Either a flat rate noise level which shall not be exceeded at the nearest residence or a margin above the existing background noise which shall not be exceeded.

It is however recognised within the DTI that there is still a degree of uncertainty among planners and developers. Planners do not have much experience of noise from wind turbines in rural areas. Developers have no noise targets for guidance when selecting sites for wind farms or deciding upon turbine layout.

Therefore the DTI has set up a Working Group largely consisting of outside experts on wind turbine noise. The objectives of the Working Group are:

1) To review recent experience in the field of wind turbine noise. This will include an attempt to relate measured data to complaints and provide an expert assessment of the issues relating to wind turbine noise.

2) Define a framework which can be used to measure and rate the noise from wind turbines. This will include parameters to be measured, measurement methods, units and measurement periods and will fulfil all the necessary criteria required for planning conditions.

3) Provide indicative noise levels thought to offer a reasonable degree of protection to wind farm neighbours and encourage best practice in turbine design and wind farm siting and layout.

4) Encourage the widespread adoption of the Working Group's recommendations.

The Working Group has been asked to address the issues of broadband noise, tonal content and blade swish (the modulation of broadband noise at blade passing frequency). It is intended to produce a report in Spring 1994 which will serve as a working guide to assessing the environmental impact of the noise from wind turbines and establish a framework for associated planning conditions.

6.5   Workshop on Noise from Wind Turbines

A workshop on noise from wind turbines was organised by ETSU and held at Harwell on 30 June 1992 (9). This workshop attracted a wide audience including representatives from local government, industry and environmental bodies and most of the issues were discussed. The views expressed were taken into account in formulating future plans. Several of the papers presented have also appeared in the journal of the BWEA, Wind Engineering.

REFERENCES

1) WATSON, I. The influence of noise on the design of horizontal axis wind turbines. ETSU W/13/00190/REP, 1993.

2) LOWSON, M.V. Assessment and prediction of wind turbine noise. ETSU W/13/00284/REP, 1993.

3) LOWSON, M.V. and LOWSON, J.V. Systematic Comparison of prediction and experiment for wind turbine aerodynamic noise. ETSU W/13/00363/REP, 1993.

4) LJUNGGREN, S. and GUSTAFSSON, A. International energy agency expert group study on recommended practices for wind turbine testing and evaluation, Part 4. Acoustics measurement of noise emission from wind turbines, 2nd edition, 1988.

5) PINDER, J.N., PRICE, M.A. and SMITH, M.G. The prediction and propagation of noise from wind turbines with regard to community disturbance. ETSU WN 5066, 1990.

6) RASMUSSEN, K.B. Outdoor sound propagation under the influence of wind and temperature gradients. JSV, 1986, 104, 321-335.

7) ANDERSEN, B. and JAKOBSEN, J. Noise emission from wind turbine generators, a measurement method. Report No. 109, Danish Acoustical Institute, Lyngby, 1983.

8) AMERICAN WIND ENERGY ASSOCIATION. Procedure for measurement of acoustic emissions from wind turbine generator systems, Volume 1: first tier. AWEA Standard 2.1 - 1989.

9) Planning Policy Guidance Note: Renewable Energy, PPG 22, HMSO 1993.

10) Rating industrial noise affecting mixed residential and industrial areas, BS4142:1990.

11) LEGERTON, M.L. Proceedings of a wind turbine noise workshop, Harwell, UK, 30 June 1992. ETSU-N-123.

Studies of the dynamics of yawed HAWTs in the DTI wind energy programme

J. M. WARD
ETSU, Harwell, UK

SYNOPSIS: Studies in the ETSU/DTI Wind Energy Programme dealing with the effects of yaw on wind turbines are reviewed, and possible applications of their results are discussed.

1 INTRODUCTION

The aim of this paper is to survey an area of R&D in the Wind Energy Programme managed by ETSU on behalf of the UK Government Department of Trade and Industry (DTI). It comprises several projects which were set up to investigate phenomena associated with wind turbines operating at a range of angles of yaw to the wind direction.

2 RATIONALE FOR YAW STUDIES

The motivation and justification for part of the R&D activity funded by the DTI is the reduction of the cost of manufacturing, installing and running wind turbines. It is critically important to reduce the cost and improve the reliability of wind generation of electricity if it is ultimately to become competitive with other forms of electricity supply. The R&D programme aims to achieve these improvements by innovative development of the technology over a range from fundamental aerodynamic studies aimed at increasing energy capture to development of specific components with the object of increasing the commercial viability of wind turbines.

With a great deal of wind plant in operation in Europe and elsewhere from which statistically-based assessments of performance can be made (1), it is possible to identify those elements of a turbine system most likely to contribute to poor performance or availability. Near the top of the priority list (after control systems) come failures associated with yaw

systems, which account for 10-15% of breakdowns in any given year. Examination of the literature and operational experience in this area leads to the following observations:

- There are frequent yaw system failures world-wide on turbines having 3-bladed rigid rotors. This fact emphasises the need to improve the design of yaw systems in order to increase the availability of turbines and reduce their maintenance overheads.
- The behaviour of yawed wind turbines has not been well understood. In general it has not been possible to predict the magnitude of yaw loading accurately.
- Anomalously high yaw loads have been observed by Howden and other manufacturers. These have not hitherto been explained theoretically.

A wind turbine spends much of its time yawed, so it would seem reasonable to expect that designers should have a sufficient understanding of the response of the turbine in that condition to be able to take it properly into account. It is unsatisfactory that turbine components should be designed on grounds not well understood. There is also concern that fatigue in yaw and rotor systems has been underestimated as a result of the lack of knowledge in this area.

Gaining a clear understanding of yaw phenomena poses severe problems. Yaw loads arise from the differential forces experienced by the rotor blades between the upwind and downwind halves of their path. The differences in loading can be caused by mechanical effects (e.g. mass asymmetries between the blades) or aerodynamic effects, since the angle of attack of the blades varies cyclically during a revolution, and a turbulent inflowing wind may influence each blade differently.

Ideally the relevant aerodynamics and structural dynamics, and their interaction, would be sufficiently well understood that the origins of the observed effects could be identified and reproduced in a mathematical model. This would engender confidence that future machines were being designed from a secure position. It is a difficult modelling problem to take account of the effects of unsteady aerodynamics, the structural dynamics of the rotor and tower, gyroscopic effects, and turbulence. It is also difficult to do experiments in wind-tunnels or in the field well enough to obtain accurate and meaningful results.

The Wind Energy Programme has directed effort towards solving yaw problems with the following broad aims:

- Acquisition of well-characterised experimental data at different size scales, both in the field and in wind-tunnels, followed by detailed analysis.
- Theoretical modelling at several levels of sophistication, based on the current understanding of aerodynamics and structural dynamics, to produce predictions to compare with experimental data, resulting in refinement of the models.
- Innovations on the basis of improved understanding: new concepts for advanced design, and improvements to current designs made possible by the ability to predict the effect of modifications.

These approaches are embodied in four specific projects, which are summarised in Table 1.

An additional project, run by the University of Strathclyde, has the primary objective of obtaining data for operation at fixed angles of yaw for a medium-scale (15kW) 3-bladed HAWT, with a rotor diameter of 9m, at the NEL site at Myres Hill, near Glasgow.

3 YAW SYSTEM LOADS OF HAWTS

3.1 Background

The proposal for this first project was prompted by observations and operational experience of wind turbines in the Howden windfarm in California. This work, coupled with limited data analysis and theoretical comparisons, showed that Howden turbines in yaw sometimes performed in a way which was not well understood (2). In particular, measured cyclic

yaw torques occasionally exceeded the estimate values by large margins, and the phenomenon of 'ya banging', a periodic loud noise from the region of th yaw drive system, caused some concern. More seriou was the actual fatigue damage which occurred in th yaw system drive components, such as the slew ring o yaw brake attachment brackets, necessitating repairs o redesign and replacement.

The project was a detailed investigation of ya loads in HAWTs. As can be seen from Table 1, th investigation was to be pursued by analysing existin yaw data from 3-bladed HAWTs, in parallel wi development of appropriate computer models. Th ultimate aim of the project was a complete solution o the problem of predicting yaw system loads.

3.1 Implementation

The aims of the project, though ambitious, have largel been fulfilled. Yaw load and other relevant data fro a Howden HWP330/33 machine in California and th HWP1000/55 at Richborough in Kent has bee analysed, much of it for the first time, and from datasets have been established for the purposes o detailed analysis.

Because of the quantity and high quality of the dat from the HWP330/33, a great deal of effort has bee directed to analysing it. Much of this work has bee subcontracted to Garrad Hassan, who have bee working on HWP330/33 data under a relate ETSU/DTI contract (Section 4). Using finite eleme analysis techniques, and models based on modifie blade element theory and capable of incorporatin modules simulating dynamic inflow and dynamic sta Garrad Hassan have built up an aeroelastic model o the machine. They have used this to obtain prediction of loads to compare with measured data. With the ai of the models, they have been able to analyse the dat in detail to determine the source and nature of th aerodynamic and structural contributions to mean an cyclic yaw torques. In particular they have examine in detail the anomalous yaw torques recorded on on particular data tape (4). They have also conducted a evaluation of the various aerodynamic models o yawed flow, and of dynamic inflow and dynamic sta which have been used in the project.

A 'benchmark' vortex-lattice free-wake mode capable of predicting steady and unsteady yaw load has been implemented and validated as part of thi project by Pesmajoglou and Graham at Imperi College (3). Its predictions have been checked bo against simpler models and against data obtained in wind-tunnel at Imperial College for yawed, sheared, and turbulent flow incident on a small model HAWT (diameter 0.55m). Although simulation runs can take a long time, sometimes of the order of days, the purpose of providing a reference tool for less sophisticated models capable of faster execution has been adequately fulfilled.

Windharvester have made a detailed examination of the yaw loads, yaw activity and yaw error for the HWP1000/55, for which data is more limited. They have also made a detailed finite-element model of the HWP330/33 which they have used to investigate the structural response of the machine to dynamic loading, with the purpose of clarifying the machine's behaviour and validating the method of structural dynamic analysis as a design tool.

The analysis included a study of the magnitude and effect of yaw error on a particular machine, the HWP1000/55 at Richborough. Yaw error, introduced because of the difference between the incident wind direction and the direction measured at a wind vane on the HAWT nacelle behind the rotor, is a potential cause of appreciable loss of revenue from the machine, which has been estimated in this case as approximately 1% or £4000 annually for a mean yaw error of 10°.

3.3 Results

This is an extensive study, and it is not possible to give more than a few of its findings here. The results have either clarified or established for the first time a number of facts, many of which are of general application, while some are specific to the design of the Howden machines but nevertheless exemplify important considerations for design.

The principal source of yaw torque is the blade flap moment differences. Yaw torque synthesised from blade flap signals alone is very similar to the actual torque measured experimentally.

It has been established that cyclic yaw loads on 3-bladed HAWTs are primarily the result of turbulence which is rotationally sampled by the blades. The stochastic loading on the blades around 2P (where P is the passing frequency for a blade), summed at the rotor hub, translates to yaw torque activity mainly at 3P. Yaw torques observed at 1P were deterministic in character, arising from aerodynamic or mass imbalance between the blades; in the case of the Richborough 55m machine, they were caused by misalignment of the tip pitch control surfaces. On the HWP330/33, the 3P random activity was by far the dominant contribution to the cyclic (or fatigue) loads; however, 1P activity was found to affect the yaw stability. The stochastic yaw loads were not sensitive to yaw misalignment, so yaw error is discounted as a reason for excessive yaw loading.

Because of the scale length of turbulence in the wind, cyclic yaw loading is an increasingly serious problem for larger 3-bladed wind turbines, which will experience high yaw moments at 3P.

The mean yaw load is governed by the aerodynamics of the rotor in yawed flow, so that its prediction requires sophisticated aerodynamic modelling. This is in contrast to the cyclic yaw loads, which are well-predicted by modified blade element theory including a turbulent wind input.

The structural dynamic analysis has demonstrated several points of interest which relate to machines with layout similar to the Howden turbines. In the particular case of the Howden HWP330/33, structural dynamics have been shown to play only a small role in determining the magnitude of cyclic yaw loading. Various recommendations have however been derived regarding design considerations for the rotor overhang and tilt, the centre of gravity of the nacelle, the system structural resonances, and the construction of the yaw system.

The study of yaw error for the HWP1000/55 has shown that the constant offset algorithm used to compensate for the wind-vane error is valid only in low winds. Yaw errors of 10°-15° have been occurring regularly. An improved algorithm has been designed on the basis of this work which makes a correction based on the wind-speed or the measured mean yaw error, and should result in less lost revenue.

A general conclusion bearing on future turbine design is that high cyclic yaw loads will inevitably be sustained on large 3-bladed turbines operated at fixed yaw angles. It therefore seems advisable to adopt an appropriate strategy to minimise these loads; one such strategy is limited (damped) free yaw. The turbine is allowed to follow the wind as it sees fit within certain limits, thus relieving the loads automatically; loss of stability resulting in excessive yawing is avoided by damping the yaw motion hydraulically and allowing motion only between end-stops. Other load-relieving strategies, such as cyclic control of the pitch of the blade tips, are possible.

Limited experience of operation of these machines in free yaw has been reported, with similar results for each. In particular, the HWP330/33 was said to be stable in free yaw in low to medium winds, with a marked reduction in tower-head dynamic activity. Yaw stability was poorer in higher winds. Unfortunately no detailed measurements were made for these trials.

A reasonably satisfactory understanding of the dynamic activity of the Howden turbines in yaw, in terms of improved modelling and accurate calibration and interpretation of the data, has now been reached, although further work is indicated in some areas.

4 FURTHER ANALYSIS OF HOWDEN HWP300 AND HWP330 DATA

4.1 Background

This project (Table 1, item 2) came about as a result of the activities of Howden Wind Turbines Ltd. at two locations in California, around 1986. At each location, one at Palm Springs and one at a windfarm in Altamont Pass, a 3-bladed Howden turbine was instrumented with mechanical and wind sensors. There were excellent wind conditions at each site.

4.2 Implementation

40 - 60 channels of analogue data were recorded for durations of real-time data acquisition from 40 - 70 minutes. Around 100 data tapes were acquired in this way, with calibration tests being carried out before and after each period of data acquisition.

4.3 Results

The result was a unique high quality data archive, much of which has now been translated into calibrated digitised datasets, amenable to computer analysis. Some of these have formed the basis for the project on yaw system loads (Section 3) and an analysis of anomalous yaw torques based on them is reported elsewhere in these Proceedings (4).

5 FEASIBILITY OF YAW CONTROL IN HORIZONTAL AXIS WIND TURBINES

5.1 Background

Yawing a windmill, so as to point its rotor out of the wind, was historically one of the earliest means of slowing or stopping it. For modern turbines it is an appealing method of control because of its apparent simplicity. The possibility of dispensing with an expensive and complex pitch control mechanism requiring maintenance, while maintaining more efficiency at lower windspeeds than a stall-regulated system, is undoubtedly attractive.

There are however disadvantages: yaw control is inherently a slower-acting process than direct pitch control, and therefore the machine would have to be able to cope with potential large torques or overspeed until the control action took effect; additionally attempts to yaw a machine rapidly may result in large gyroscopic forces. Safety is a serious consideration: for instance, it is essential that the machine be able to shut down safely in the event of failure of the grid. Nevertheless, the combination of yaw control with a variable-speed generator and a control system which can efficiently limit the possible large fluctuations in torque and rotor speed has its attractions; the Italian Gamma-60 project is an example of its implementation (5).

5.2 Objectives

This project (Table 1, item 3) had the general objective of investigating the engineering feasibility of controlling a wind turbine by yawing out of the wind. In particular the following two modes of yaw control have been investigated for a 300kW 2-bladed teetered machine:

- Power regulation control: active yawing of the machine to control the output power of the turbine (non-stall-regulated rotor). Of concern here were the possible large torque and power excursions (in a fixed-speed system) or large speed excursions (in a variable-speed system) with the associated potential need for a stronger structure.
- Machine shut-down control: standard or emergency shut-downs executed by yawing the machine so as to reduce the rotor torque to the level at which conventional braking can take over.

5.3 Implementation

An advanced aerodynamic model of a 2-bladed teetered rotor at high angles of yaw, incorporating a dynamic inflow treatment, has been developed and incorporated into a turbine structural dynamic simulation model modified to take into account gyroscopic effects. The code has been validated against a prescribed vortex wake model developed as a modification of a helicopter code, and against a standard model based on strip theory within its region of applicability, at low yaw angles.

5.4 Results

Test results from the LS-1 3 MW 60m turbine on Orkney have been compared with the predictions of the model, which has been found to simulate the results much more closely than previous unmodified versions.

Critical load cases (intense wind gusts and shutdowns) have been examined for a notional yaw-controlled version of an existing design of pitch-controlled machine (WEG MS-3). The purpose here is to evaluate predicted power, speed and torque excursions in different circumstances of wind-speed and mode of operation. It has been found that near the rated wind-speed, the turbine needs to be yawed well off the wind in order to get the desired control effect. This implies a high rate of yaw in order to respond quickly, which in turn implies large gyroscopic effects and large teeter excursions, that tend to increase towards high angles of yaw. The strategy has therefore been adopted of yawing at a high rate ($16°s^{-1}$) for the first $10°$, followed by yawing at $8°s^{-1}$ the rest of the way.

Using this approach, the response of fixed and variable speed machines to gusting has been estimated. The results indicate that yaw control combined with the variable speed approach appears feasible, but load and power excursions at fixed speed would be excessive. Even for variable-speed operation, some degree of structural strengthening might be required to cope with the loads produced at the higher speeds.

More importantly, the peak rotation rates would be of order 50-60 r.p.m. or more, resulting in very high tip speeds. The turbine blades would therefore probably produce a lot of noise.

Work continues to use the load case results to assess the implications of yaw control for the cost and reliability of the turbine. The requirement for fail-safe operation to shut down the turbine safely in the event of grid failure has led to an investigation of the design requirements for a system to drive the turbine towards $90°$ yaw, as well as to considerations of the additional requirements on the conventional disc braking system. It was decided that, although complex, a hydraulic system would be the best approach to a fail-safe yaw drive. It is not yet clear whether yaw control has a cost advantage despite the additional yawing and braking demands.

6 YAW BEHAVIOUR OF UPWIND HORIZONTAL-AXIS WIND TURBINES

6.1 Background

The aim of this project (Table 1, item 4) is to improve the understanding of the behaviour of yawed HAWTs by making and analysing detailed measurements of loads on a model upwind horizontal-axis wind turbine, building on earlier work by Noakes and Sharpe at Queen Mary and Westfield College (6). The results, obtained in conditions of smooth, non-turbulent flow, are to be compared with theoretical computer models and field results from full-scale turbines to establish any common characteristics or trends.

6.2 Implementation

The model turbine was designed at RES, built and bladed at the University of Nottingham, and then run in the 1 MW 8m x 4.5m wind-tunnel at the Motor Industries Research Association. A large wind-tunnel was used in order to minimise blockage and interference with the free propagation of the wake by the walls of the tunnel. The turbine was fully-instrumented with strain-gauge and other sensors, and could be configured to operate as either a 3-bladed or a 2-bladed teetered stall-regulated machine.

The detailed data acquired included:
- Blade root flapping moment
- Rotor thrust and torque
- Yaw moment and angle

with operation at angles of yaw up to $\pm60°$, and tests of free yaw, in a range of wind-speeds above and below the rated value.

The data is being analysed by comparison with field data from a full-size machine, the 2 MW Elsamprojekt turbine at Tjaereborg, Denmark, and also with computer models developed by RES and the vortex-lattice free-wake model developed at Imperial College (3).

6.3 Results

One of the main achievements of this project is the establishment of a well-characterised set of small-scale load data of value for computer modelling, for both 2- and 3-bladed configurations. Another is the first detailed study of upwind operation in damped free yaw in a wind-tunnel, and the comparison of the resulting loads with the fixed yaw case, which is reported elsewhere in the proceedings of this conference (7)). The work of comparison with field and modelled data is still in progress.

7 CONCLUSIONS

The programme is producing much useful information and insight into turbine aerodynamics and structural dynamics, with associated implications for improvements in machine design, reduction of the cost of wind-generated energy, and competitiveness of wind energy as a renewable resource. The results of the projects will be published within the next few months as a series of ETSU reports.

Specifically, a large quantity of experimental data has now become available which is either new or has recently been analysed for the first time, and which is being used to produce information of value in the design of current and future advanced machines. In the course of the analysis, new understanding has been gained in parallel with the development of computer tools which will be invaluable aids to turbine design.

The design of new and modified yaw drive systems and control systems should benefit from this work. The new computer models will make it easier to assess ideas such as free yaw. Operation of a large wind turbine at least partially in the free yaw mode, thereby dispensing with all but the simplest yaw drive components and simplifying the control system, is a potentially attractive way of reducing turbine costs by reducing the requirement for expensive components and the loading demands on the system as a whole. However, the realistic evaluation of the concept will require careful modelling taking into account the findings of the projects discussed in Sections 3 and 6.

ACKNOWLEDGEMENTS

The cooperation of Windharvester Ltd, Garrad Hassan, and WEG Ltd, and in particular the staff named in Table 1, in supplying project information is acknowledged. I am grateful to K F McAnulty of ETSU for helpful comments.

REFERENCES

(1) Windstats Newsletter, 1993, **6**, 6.

(2) Measured loads compared to design values on the HWP330/33 wind turbine F3004. *James Howden and Company Ltd.*, Report A420.

(3) PESMAJOGLOU S and GRAHAM J M R. Yaw loads on horizontal axis wind turbines using an unsteady vortex lattice method. *Wind Energy Conversion 1992: Proceedings of the 14th British Wind Energy Association Conference*, 1992, 109-115 (Mechanical Engineering Publications Ltd).

(4) JAMIESON P. Analysis of data from a Howden HWP330/33 wind turbine. *Wind Energy Conversion 1993: Proceedings of the 15th British Wind Energy Association Conference*, 1993 (Mechanical Engineering Publications Ltd).

(5) AVOLIO S et al. Dynamic analysis of GAMMA 60 wind turbine generator and control system: Design and theoretical and experimental validation. *Proceedings of 1993 European Community Wind Energy Conference*, 1993, 465-468.(H S Stephens and Associates, Bedford, U.K.).

(6) NOAKES J and SHARPE D J. The measurement of yaw loads on a model upwind horizontal axis wind turbine. *Wind Energy Conversion 1991: Proceedings of the 13th British Wind Energy Association Conference*, 1991, 161-167 (Mechanical Engineering Publications Ltd).

(7) NOAKES J, SLATER J D T, and ANDERSON M B. The free yaw behaviour of upwind HAWTs. *Wind Energy Conversion 1993: Proceedings of the 15th British Wind Energy Association Conference*, 1993 (Mechanical Engineering Publications Ltd).

TABLE 1 ETSU/DTI PROJECTS ON YAWED HAWTS

Project Description	Objectives	Work
Yaw System Loads of HAWTs. Windharvester Ltd. C G Anderson P R D Agius D H Brown D R R Green M Vartiainen	• Explain yaw loads measured on 3-bladed rotors. • Devise advanced computational models of yaw aerodynamics capable of predicting loads. • Assess the impact of yaw loads on the economics of wind power generation.	• Analyse existing data mainly from Howden HWP1000/55 (Richborough) and HWP330/33 (Altamont Pass). • Develop and validate advanced computer mathematical model of yawed flow. • Develop less sophisticated but faster models for yaw system dynamics and aerodynamics, as design tools. • Make detailed comparisons, draw conclusions, derive implications for design and costs.
Further Analysis of Howden HWP300/26 and HPW330/33 Data. Garrad Hassan Peter Jamieson R I Rawlinson-Smith D C Quarton	• Calibrate and digitise selected analogue real-time data tapes from instrumented Howden turbines. • Perform specified analysis tasks.	• Range of work on data, including yaw results in parallel with project on Yaw System Loads of HAWTs (above).
Feasibility of Yaw Control in HAWTs. Wind Energy Group Ltd. E A Bossanyi B Lineton	• Investigate engineering feasibility of controlling HAWTs by yawing out of the wind, for: 1) Power regulation 2) Shut-down of 2-bladed 300kW machine.	• Develop advanced aerodynamic model for large yaw angles and incorporate into turbine dynamic simulation model. • Obtain test results from full-scale turbine (LS-1) and compare with model predictions. • Examine critical load cases (intense gusts, shut-downs) for notional yaw-controlled machine. • Assess implications for capital cost and reliability.
Yaw Behaviour of Upwind HAWTs. Renewable Energy Systems Ltd. J Noakes J D T Slater M B Anderson	• Obtain high quality yaw load data from a model wind turbine (1.4m diameter) in a wind-tunnel. • Investigate feasibility of free-yaw operation.	• Design and build model wind turbine. • Run in large wind-tunnel to acquire load data at angles of yaw up to ±60°. • Run likewise in free yaw mode. • Analyse data, compare with field data from full-size machines, and with computer modelling.

The free yaw behaviour of upwind HAWTs

J. NOAKES, J. T. D. SLATER and **M. B. ANDERSON**
Renewable Energy Systems Ltd, UK

SYNOPSIS: The tendency for an up-wind horizontal axis wind turbine to align itself to the oncoming wind is a well known phenomenon. However, little has been done to investigate the free yawing behaviour of upwind machines. The work described in this paper aimed to investigate the effect of a wide range of parameters on the free yaw behaviour of a model wind turbine. The work was undertaken by Renewable Energy Systems with financial support from the Department of Trade and Industry.

1.0 INTRODUCTION

The primary causes of yaw error in horizontal axis wind turbines (HAWTs) arise from the need to limit the gyroscopic and unbalanced aerodynamic loads on the rotor blades and prevent excessive yaw drive activity which means that the rotor will only respond to long term changes in wind direction. This combined with the uncertainties in the wind direction measured from the nacelle means that it is inevitable that wind turbines will operate for much of their working life at varying degrees of yaw.

Unsteady aerodynamic effects resulting from the large variations in local angle of attack experienced by a blade operating in yaw can result in high instantaneous blade loads. The effect of unsteady aerodynamics is of particular importance in the design of stall regulated machines which rely solely on the rotor's aerodynamic characteristics to limit the power output and thrust loads. Any modifications to the stalling characteristics may result in extreme instantaneous loads.

A greater understanding of yaw loads will not only help to improve the fatigue life of the blades but will also be the first step in the development of a free yaw wind turbine which may ultimately improve the reliability of future generations of wind turbines and hence reduce the cost of wind energy. This will be of particular importance for offshore wind turbines where, due to the difficulties associated with access, the maintenance cost is likely to be significantly higher than for land based machines.

2.0 OBJECTIVES

The objectives of this project were as follows:-

i) To perform wind tunnel experiments on a model turbine to measure the effect of a wide range of parameters on the yaw loads and behaviour of an upwind HAWT.

ii) To investigate the structural implications of free yaw operation for upwind HAWTs.

iii) To compare the experimental results obtained from the model wind turbine with data from full size machines.

iv) To compare the experimental results with existing theoretical models.

3.0 EXPERIMENTAL SET-UP

3.1 Model Turbine

The wind tunnel model was designed to perform a series of fixed and free yaw experiments to investigate the effect of nacelle tilt, cone angle (3 bladed rotor) rotor overhang and tower blockage on the yaw loads and behaviour of both two and three bladed rotor configurations. Fig 1 shows the model turbine with the two bladed rotor. The model was designed by RES whilst the manufacture of all mechanical components and instrumentation was carried out by Nottingham University.

The two and three bladed rotors had a diameter of 1.4m with blades which were based on the stall

197

regulated LM 17.0 design. The blades were manufactured by Nottingham University using the Resin Transfer Moulding (RTM) technique described in ref (1). The 3-bladed hub assembly permitted cone angles of 0°, 2° and 5° to be investigated. The 2-bladed configuration had 0° coning but ±6° of un-damped teeter motion. The nominal blade tip velocities were 65 m/s and 77 m/s for the 3-bladed and 2-bladed rotors respectively.

Speed control of the rotor was achieved using a d.c. servo motor which enabled a constant rotational speed to be maintained in both the pre-stall and post-stall regions of operation.

The rotor assembly was mounted on a tubular steel tower which located in yaw bearings mounted to the base assembly. A rotary fluid damper was located between the tower and the base plate to provide additional damping during the free yaw experiments.

3.2   Instrumentation

To meet the objectives of the experimental programme, sensors were required to measure the following parameters:

- Rotational speed;
- Azimuth position of the blades;
- Blade teeter angle (for 2-bladed case);
- Blade root flapping moment;
- Rotor torque;
- Rotor thrust;
- Yaw moment;
- Yaw angle.

In general, strain gauges attached to the structural components of the model were used to measure rotor loads. However, a proprietary torque transducer was used to obtain a direct measure of the yaw moment. Blade teeter angle and the yaw angle (during free yaw tests) were measured using precision rotary potentiometers.

The rotor torque was measured using strain gauges on the rotating assembly and transmission of this signal, along with those for the blade flapping moment and teeter angle was via a ten-way slip ring unit.

Calibration of all sensors was carried out prior to entering the MIRA wind tunnel with check calibration being performed during the final test programme.

3.3   Data Acquisition

The PC based data acquisition system comprised a 16 channel A/D card with software written by RES.

A one pulse per revolution disc was used to initiate the data acquisition over the required number of revolutions (typically 100) at a known azimuth position. A 72 pulse per revolution disc was used to trigger the data acquisition from pre-selected channels at a sampling rate of 200 kHz. The high sampling rate enabled data to be collected at each azimuth position within a small angular error of 0.03° per channel.

3.4   Wind Tunnel

The model turbine was tested in the full scale wind tunnel belonging to the Motor Industry Research Association (MIRA). The wind tunnel had an 8 m wide by 4.5 m high working section and provided a low turbulence (typical turbulence intensity quoted as 0.8%) air flow at a number of set wind speeds up to 35 m/s.

The size of the working section in relation to the swept area of rotor resulted in a blockage of 4.3%, however, probably more important was the width of the tunnel which minimised the influence of the tunnel on the sideways deflection of the wake which exists with a yawed rotor.

The results of a traverse of the working section, supplied by MIRA showed that the wind speed variation from the mean level was typically less than 2%.

Nominal wind speeds of 9.2, 13.1, 17.0, 20.2 and 24.4 m/s were available with the four fans running at the same speed setting. Intermediate wind speeds of 4.2 m/s and 18.8 m/s where also achieved by differential fan settings.

4.0   RESULTS

In total, around 250 runs were performed with the model running in fixed and free yaw. The results were analysed to provide information on the time averaged and azimuthal variations in rotor forces and moments for the fixed yaw tests. Time histories of rotor loads and behaviour were used to analyse the free yaw experiments.

The yaw direction is defined for clockwise rotation (as viewed from the front) of the rotor as being positive when the blade travels upwind in the top dead centre position (zero azimuth). The yaw moment is said to be stable if it acts in a direction which tends to reduce any yaw misalignment. The configuration of the torque transducer was such that a negative yaw moment at positive yaw angles was stable and vice versa for negative yaw angles.

The results presented in this paper have been limited to the three bladed rotor configuration.

4.1   Fixed Yaw Performance and Loads

i)   Rotor Power

Fig 2 shows the variation in rotor power as a function of wind speed for a number of yaw angles. The onset of stall at around 18 m/s is clearly evident for the zero yaw case. A reduction in rotor power with increasing yaw angle was observed below the stalling wind speed, however, at yaw angles of ±30° there was a noticeable change in the rotor characteristics in the post stall region. Whilst the peak power remained almost unchanged, the power output in the post stall region was significantly higher for the yawed rotor compared to the un-yawed case.

This result illustrates the sensitivity of a stall regulated rotor to dynamic stall and 3-D aerodynamic effects which could be induced by wind direction changes or imperfect operating parameters causing high yaw tracking errors.

ii)   Rotor Thrust

The rotor thrust was calculated from the bending moments measured on the tower. Fig 3 shows the thrust as a function of yaw angle for specific wind speeds. The measured thrust represents the force acting in a direction normal to the plane of the rotor rather than the force in the direction of the oncoming wind. As expected there was a reduction in thrust with increasing yaw angle.

At wind speeds below stall, the variation in rotor thrust with yaw angle approximated to a cosine relationship. However, in the post stall region, the thrust remained approximately constant up to yaw angles of ±30° and then dropped off rapidly with further increases in yaw angle.

iii)   Yaw Moment

The yaw moments were measured at the centre of the tower axis and included therefore, a component due to the side force acting on the rotor.

The variation in the yaw moment with yaw angle is shown for three wind speeds in Fig 4. The small negative yaw moment present at zero yaw is consistent with the findings given in ref (2) where it was established that tower blockage effects resulted in a small yaw moment with the rotor at zero yaw.

For the higher wind speeds the yaw moment acted in a stabilising direction over a wide range of yaw angles, reaching a maximum at around ±30°, further increases in yaw caused a reduction in the yaw moment.

Asymmetry in the yaw moments between positive and negative yaw angles was observed from the measurements. It can be seen that the yaw moments in the positive yaw region tend to be slightly higher than comparable values at negative yaw.

This result is again consistent with the findings reported in ref (2) where asymmetry was found to be caused by tower blockage effects.

iv)   Blade Root Flapping

The effect of yaw on the cyclic variation in blade root flapping moment is illustrated in Fig 5 which shows a comparison of the ASD between 0° and +30° yaw. The main effect of yaw is found in the increased 1P component with other components up to 6P remaining almost unchanged. It can be seen that there was a 2P component which was identical for both the yawed and un-yawed cases. Tests performed with the model running in still air demonstrated that this component was an inertia load rather than aerodynamic effect and is thought to be the result of small 1P yawing oscillations caused by rotor mass imbalance.

Also of interest is the comparison between positive and negative yaw angles which is shown in Fig 6. The main difference is an increase in the 3P component of flapping moment for a yaw angle of -30°. However, as is clearly illustrated by Fig 7 which shows a comparison of the azimuthal variation in blade flap, the 3P component is small compared to the mean level and 1P fluctuations.

4.2   Free Yaw Test Results

The free yaw experiments were performed by locking the rotor at a predetermined yaw angle and then releasing it when the wind tunnel and model had settled down to a steady running condition. The free yaw tests were performed at three nominal wind speeds of 9.2 m/s, 18.8 m/s and 24.4 m/s with initial release angles in the range ±60°.

Figs 8 and 9 show the progression of yaw angle for the rotor when released from +40° at wind speeds of 18.8 m/s and 24.4 m/s respectively. It can be seen that at both wind speeds the rotor tended to align itself with the tunnel flow. However, there were noticeable differences in the yaw rates and yaw error.

The yaw rate was a maximum at the highest wind speed, -70°/s at 24.4 m/s compared to -35°/s at 18.8 m/s, although the final yaw error in this case was +12° compared to 0° for the lower wind speed. It can be seen that there was little or no overshoot in the yaw angle which was due to the relatively high level of damping offered by the rotary damper unit. The reason for the differences in the final yaw error are not apparent from either the free or fixed yaw results which did not show any significant change with wind speed of the yaw moments measured at small angles of yaw. One possible cause could have been the internal friction of the damper which at small yaw angles would have been greater than the restoring moment being exerted by the rotor.

It was observed from the time averaged results that the rotor power increased with yaw angle when the rotor was running in stall. Fig 10 shows the variation in rotor torque during a free yaw test at 24.4 m/s (the corresponding yaw angle is also shown). It can be seen that as the rotor aligns itself with the tunnel flow, there is a sudden reduction in the rotor torque. The main reduction in torque occurred between yaw angles of 35° and 20°.

Free yaw tests were also performed from higher yaw angles. Fig 11 shows the progression of rotor yaw angle when released from +60°. Although the yaw rate was initially very low, the rotor remained stable. Unlike the previous results, the rotor overshot and ended up at a small negative yaw angle.

The rotor was found to be less responsive when released from negative yaw angles. Fig 11 shows the results of a free yaw test from an initial release angle of -40°. In addition to a lower peak yaw rate of 25°/s, the final yaw error tended to be greater for the negative yaw tests.

The effect of introducing 5° of rotor tilt on the free yaw behaviour of the model is illustrated for positive and negative release angles in Fig 13. The relatively high yaw error for the positive release test may have been due to the tilt angle inducing a component of the rotor torque about the yaw axis.

Further analysis of the experimental data is currently being undertaken to enable further conclusions to be drawn regarding the effect of the various parameters on the yaw behaviour of the model.

5.0 CONCLUSIONS

The experiments performed as part of this project have provided substantial data pertaining to the loads and behaviour of a model wind turbine operated with a number of different configurations in the controlled environment of a wind tunnel.

The results presented in this paper represent a sample of the tests performed to illustrate the effect of yaw on the loads and performance of the model turbine fitted with the 3-bladed rotor.

Based on the analysis performed to date, the following preliminary conclusions can be drawn.

The effect of yaw on the loads and performance of the model were found to differ considerably depending on whether the rotor was stalled. In particular, the rotor power in the post-stall region was found to be sensitive to yaw angle - above 20° large increases in power were observed.

The free yaw experiments demonstrated that an upwind rotor can be stable when released from yaw angles as large as ±60°. Yaw rates were typically in the range 25°/s to 40°/s with extreme values as high as 70°/s during high wind speed experiments.

Typical yaw errors on completion of the free yaw tests were in the range ±10°.

ACKNOWLEDGEMENTS

Financial support for this project was obtained from the Department of Trade and Industry under contract No E/5A/6113/2925.

REFERENCES

(1)   MIDDLETON V et al
        Resin Transfer Moulded Aerogenerator Blades. Paper presented at BWEA 15, York 1993.

(2)   NOAKES J & SHARPE D J
        The Measurement of Yaw Loads on a Model Upwind Horizontal Axis Wind Turbine.
        Wind Energy Conversion 1991. Published by Mechanical Engineering Publications Ltd.

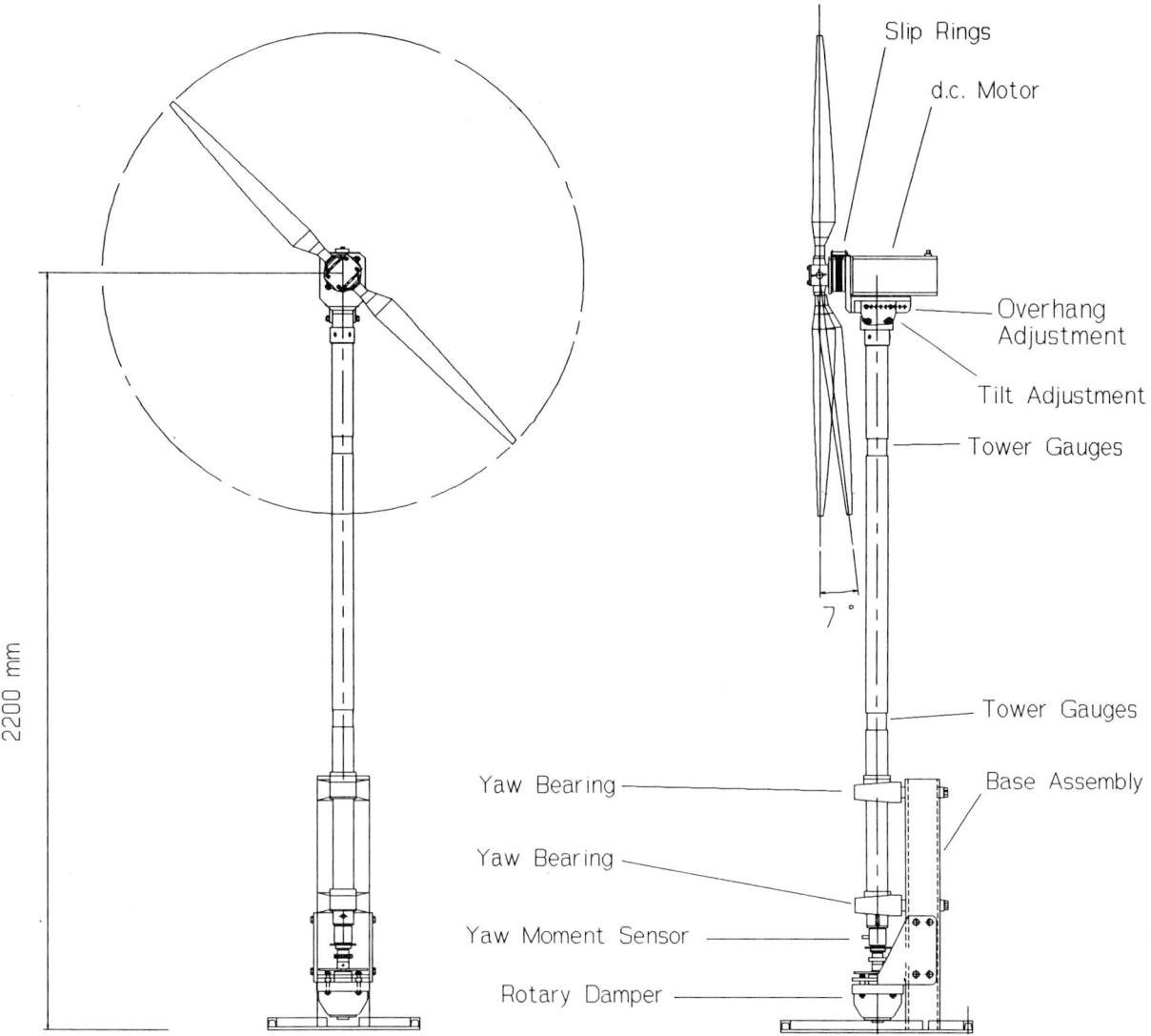

NOTES

1. BLADE TIP CLEARANCE IN NORMAL POSITION = 138 mm

2. BLADE TIP CLEARANCE IN EXTENDED POSITION = 180 mm

3. TOWER DIAMETERS WITH / WITHOUT SHROUD = 104 mm / 60 mm

**Fig 1
Model Wind Turbine shown with
2-Bladed Rotor**

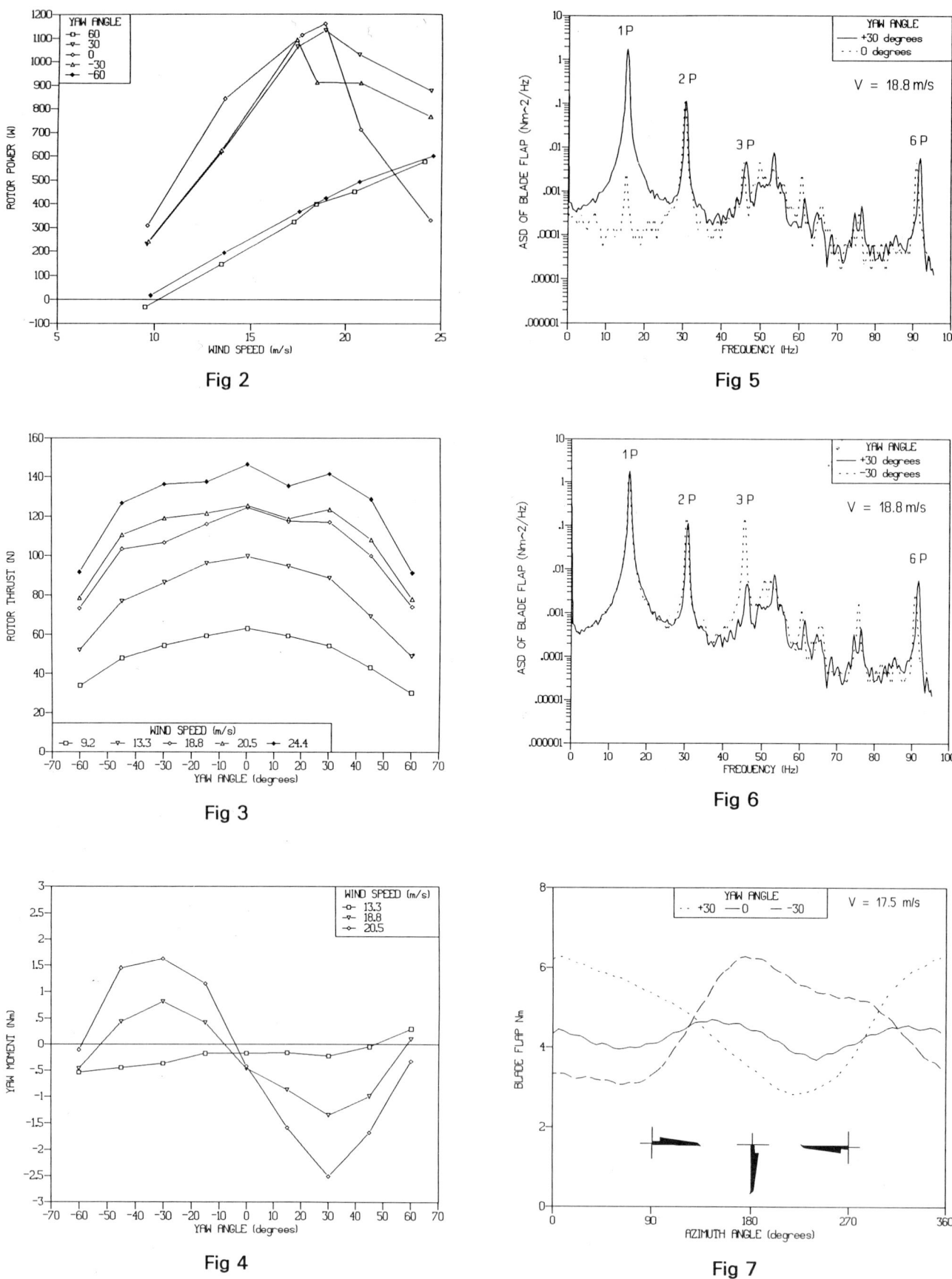

Fig 2

Fig 3

Fig 4

Fig 5

Fig 6

Fig 7

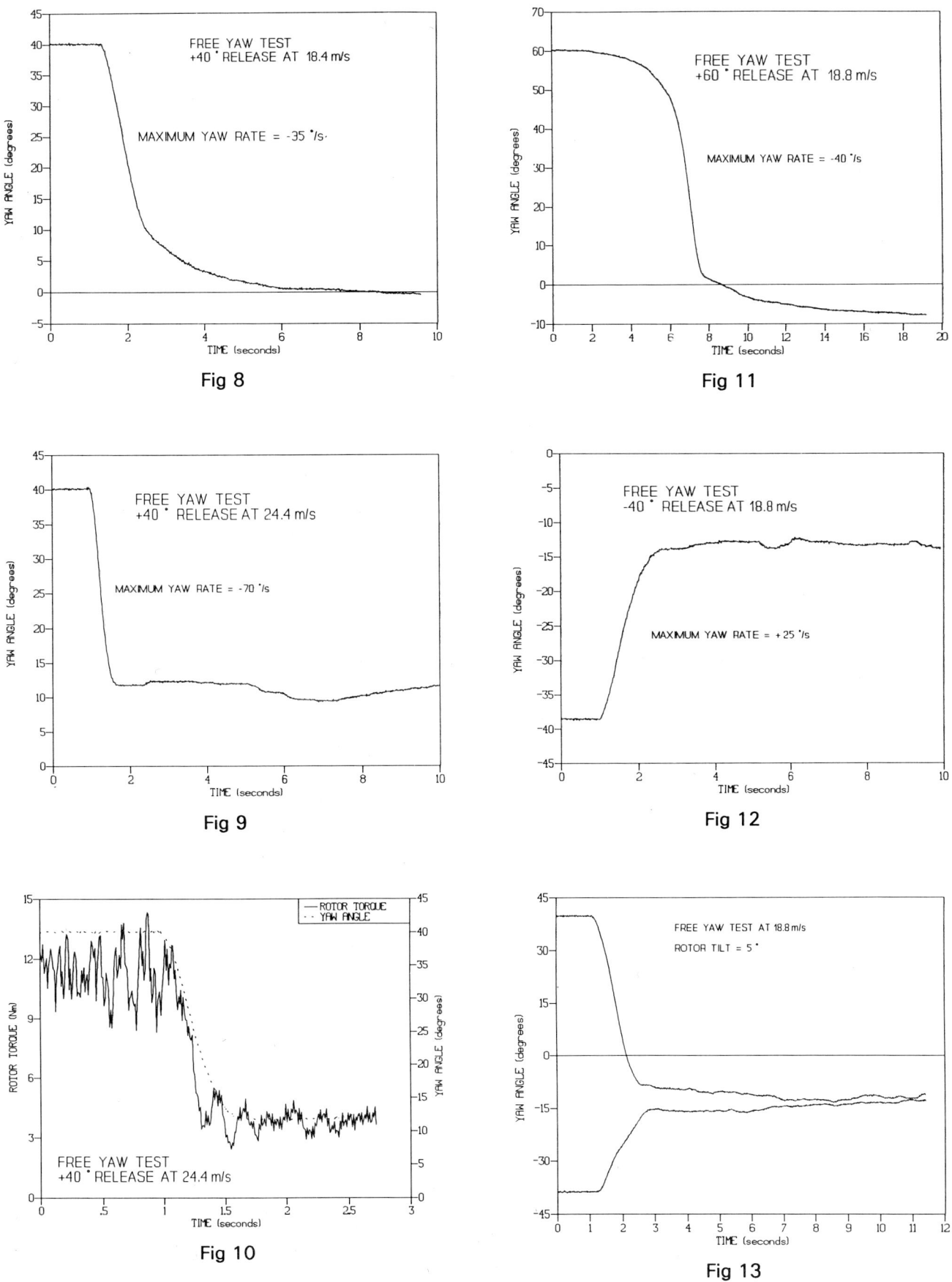

Fig 8

Fig 11

Fig 9

Fig 12

Fig 10

Fig 13

Analysis of data from a Howden HWP 330/33 wind turbine

P. JAMIESON

Garrad Hassan & Partners Ltd, UK

SYNOPSIS Garrad Hassan have a major contract under the project management of ETSU and funded by UK Government to analyse data from a Howden 33m wind turbine sited in Altamont Pass, California. Although the data was collected in 1986, and has been the subject of much previous analysis, the present work shows that further valuable information has been obtained from this data. The data analysis project has been wide ranging, dealing with machine dynamics, control, extreme operational loading, wake effects and 'rapid fatigue' damage evaluation. The present paper concentrates on the issue of yaw system loads where complex, long standing problems have been substantially resolved.

NOTATION

x — an arbitrary variable

$f(x)$ — the distortion function

Y — the yaw torque synthesised from blade flap moments.

a, b, k — constants of the distortion relation

Z — the synthesised yaw torque with distortion applied

1 INTRODUCTION

Howden wind turbines are of an upwind, 3 bladed design with rigid hub connection that is generally prevalent in world markets. Their 26 MW wind farm in Altamont Pass California has operated since 1986 with very high availability after the occurrence of problems (1), which necessitated a complete re-blading of all the machines. The first machine to be recommissioned, was extensively instrumented in order to assess the integrity of the new blade design and its interaction with the wind turbine system. The experimental arrangement, the measurement of power and especially steady state loading has been thoroughly discussed (2), (3). Although the data (collected in 1986) is comparatively old, the extensive nature of it (60 channels per tape and ~ 100, 60 Mega Byte tapes) and the continuing interest in machines of a similar general design type, means that the data has enduring relevance. Because of this the UK government have funded a further major investigation of this data (4) by Garrad Hassan which is the subject of this paper.

The aims of the present project are to form an orderly data base suitable for subsequent analysis and specifically:

- to create and validate an aeroelastic model of the Howden HWP 330/33 wind turbine,

- to critically appraise rules for separation of deterministic and stochastic contribution to loadings,

- to investigate yaw system loads,

- to assess the impact of wakes on machine loading,

- to validate a new method of fatigue load prediction, so called 'rapid fatigue'.

Rapid fatigue is a method of fatigue load prediction being developed in collaboration by Garrad Hassan. It generates equivalent information for fatigue damage calculations but by-passes time series synthesis and the usual direct rainflow counting procedures.

Because of the extensive nature of this project the present paper will focus on one item only, yaw system loads, with some discussion of wake effects in connection with that. It appears that, far from being outdated and exhausted, the data from the Howden wind turbine has enabled further progress to be made in the understanding and prediction of wind turbine loading.

2 YAW SYSTEMS LOADS

The test wind turbine and the data collection system have been discussed (2), (3).

Garrad Hassan has been provided with two data tapes (tape 65 and tape 75) in which a signal is available measuring torsion on the tubular tower at a location just beneath the yaw ring (Fig 1). The yaw torque signal on tape 75 has the normal appearance of a wind turbine vibration but the signal of tape 65 is radically different (Fig 2). It appears to be 'one-sided' with a lower threshold at about − 50 kNm which is not crossed during the whole 71 minute duration of the tape. Closer examination (Fig 3) of the waveform shows the expected central frequency of 3P (3 × blade passing) to be dominant, but that the waveform is much distorted around its minima. The distortion has been established to be real, and not say due to the recording instruments, but the cause of it was not understood.

2.1 Tape 75

The logical starting place, however, in explaining yaw torque behaviour is to assess the 'normal' signal of tape 75. Fig 4 illustrates the forces and moments on the rotor that contribute to yaw torque. It can be shown that by far the most significant contribution to yaw torque (for a rigid bladed rotor system) comes (differentially) from the flapwise bending moments.

Although only 1 blade was extensively strain gauged by Howden, there were gauges measuring flapwise bending on all three limbs of the nodular cast iron hub. Thus by resolving about the yaw axis and summing, yaw torque can be synthesised from these hub flap moments. The similarity between the measured and synthesised signal (Fig 5) is striking and exists for the whole of the operational range of tape 75. As the yaw torque provided by Howden was effectively uncalibrated, the synthesis process links the yaw torque to authoritatively calibrated blade flap signals and hence provides a yaw torque calibration. The validity of the flap moment synthesis process for rigid 3 bladed rotors has also been confirmed by Hansen (5).

Over the past ten years, Garrad Hassan has invested major effort in developing validated models for prediction of wind turbine loads in the presence of wind turbulence (6), (7). These models have realistic wind field representation (usually based on the Von Karman correlation function but now in a general form capable of accepting any accredited spectral model) and allow both frequency domain predictions based on linear modal models of structural dynamics or time domain simulation incorporating appropriate non-linearities. The value of this approach is evident in Fig 6, where most details of the low frequency spectrum are successfully predicted with the exception of the 1P spike (which is due to asymmetry between the blades and therefore not predictable without prior measurement of blade mass distribution, and rotor system geometry). A notable feature is the dominance of the 3P loading which is predominantly stochastic.

Garrad Hassan in various projects in the JOULE programme (8), (9) have been investigating dynamic inflow and stall hysteresis effects. This is another major study and dynamic inflow is of crucial importance in assessing the deterministic 1P component of blade flap moments (usually due to yaw error) and hence the OP (steady) yaw torque and yaw stability. It is evident however that the dominant stochastic 3P loading, which is the major contribution to yaw system fatigue, can be adequately predicted with simple strip theory aerodynamic models, providing there is a good representation of wind turbulence, incorporating the spatial variations in the wind and the rotational sampling of the blades.

Although it is not a major element in the total loading, the 2P content in the auto spectrum Fig 6 is predicted and does not come from the rotor (a symmetric 3 bladed rotor can only feed OP and multiples of 3P into the yaw system). Although rotor asymmetry could be a source of deterministic 2P, the 2P of Fig 6 is mainly stochastic and is due to coupling between

lateral motion of the tower head (which is near to resonance at 2P) and twisting (i.e. yaw motion) of the tower head. The coupling exists because the centre of gravity of tower top mass is substantially forward of the yaw axis due to overhang of the rotor mass. The aeroelastic model of the Howden HWP 330/33 is sufficiently detailed and accurate to be able to predict this. The rise in the auto spectrum (Fig 6) around 4 Hz is due to edgewise blade dynamics and is also predicted.

2.2 Tape 65

Returning to the unusual yaw torque signal of tape 65, the synthesis process (as was illustrated in Fig 5) was repeated. This resulted in a signal that resembled the measured data of tape 65 in its most positive ranges but that naturally did not exhibit the attenuation of the most negative ranges evident in the measured data. Without immediately seeking to explain the underlying physical causes, a "distortion" function was sought that would map the synthesised yaw torque to the measured data. The distortion function is based on the inverse tangent function as follows;

$$f(x) = \frac{1}{\pi} \tan^{-1} \frac{x}{a} + 0.5$$

so that as $x \to \infty$, $f(x) \to 1$ and as $x \to -\infty$ $f(x) \to 0$.

The synthesised yaw torque Y, is then used to generate the distorted yaw torque Z;

$$Z(Y) = k\{(Y+b)\ f(Y+b) - b\}$$

thus as $Y \to \infty$, $Z \to kY$ and as $Y \to -\infty$, $Z \to Z_1$ where the limit Z_1 is given by,

$$Z_1 = -k\ (a/\pi + b)$$

The parameters a, b and the scale factor k allow the distortion to be tuned to reflect the observed position of the threshold Z_1 and also reflect the relative distortion of the upper and lower ranges of the signal.

It may be seen (Fig 7) that there is now a plausible similarity between the measured data and the yaw torque synthesised from hub flap signals with distortion applied.

The physical meaning of the distortion was then considered. It can be shown that this type of distortion is consistent with a non-linear yaw system stiffness and it is thought that this arises due to the distribution of loading on the yaw ring. Fig 8 shows this distribution in the absence of wind where there is a loss of bearing pressure on the side of the tower opposite the rotor, this being due to a large moment from the overhung rotor weight. In light winds, the applied wind thrust helps to restore compression over the whole surface of the ring, and the 'normal' 'linear' behaviour of tape 75 (Fig 5) is evident in the yaw torque. As the wind rises, although the steady wind thrust increases, the machine becomes much more dynamically active and the effective contact area in the yaw ring and hence the effective torsional stiffness of the yaw system becomes variable, with a loss of bearing pressure in one direction only (as the tower head moves into wind). Especially in view of the internal torque loading discussed in the following section, it is also possible that slippage of the yaw brakes (for loading in one direction only) may create the apparent non linear stiffness. Thus the non-linear yaw system response of tape 65 can be accounted for.

2.3 Influence of yaw system design on mean yaw torque

As the steady yaw torque results predominantly from once per revolution (1P) blade flap activity, the synthesis process should be an excellent way of calibrating the mean yaw torque. The gain of the blade strain signals (which can be measured accurately) will establish the mean value and hence appropriate offset for the yaw torque signal.

However, Fig 9 shows a surprising result. The deterministic component of the blade flap signal at 3m radius has been extracted from each minute period of operation, and the 1P component of that (resolved about the yaw axis taking account of its phase angle) is correlated with the associated mean yaw torque. However, the graph splits into two main groups of points, which as it happens, correspond to two distinct nacelle positions. The shift in mean yaw torque between the lines is large (about 40 kNm for this data set from tape 65) and shifts of mean yaw torque associated with nacelle position changes are also observed in data from tape 75.

It should be made clear that these shifts are not simply due to a change in yaw error with change of nacelle position. Fig 10 shows that when the 1P blade flap moment is plotted against yaw error a coherent single line correlation results, as would be expected.

It would therefore appear that the yaw system of this Howden, 330 kW wind turbine is subject to differing amounts of pre-load in yaw torque as nacelle position is altered. This is due to the way in which the hydraulic circuits controlling the 2 yaw motors and 3 yaw brakes operate. In summary, hydraulic pressure can be locked in the system in a way that leaves the motors tending to operate against the fully applied yaw brakes.

Thus the mean yaw torque applied by rotor aerodynamics relates uniquely in the expected way to 1P blade activity (which in turn depends on the subtle aerodynamics of yawed flow), but the total mean yaw torque may include other influences such as preload from the yaw drive system.

3 EFFECT OF WAKES ON YAW TORQUES

Wind farms in Altamont Pass have more or less linear rows of wind turbines, each line at right angles to the prevailing wind direction and often extremely closely spaced (2D is common) along the line. The prevailing winds are totally dominant with a strong diurnal pattern during summer and most of the year but, for a short period in winter, the direction changes and the wind may blow along the line causing severe wake effects.

Tape 44 is a record of 55 minutes of operation in such circumstances. The relative effect of wakes is most pronounced in low winds. Fig 11 shows that the standard deviation of flap moments has risen by a factor of 3 compared to normal operation. The mean windspeeds were low (~ 7m/s) during this period of data collection and in stronger winds the relative effect of wakes would be considerably less. Thus since fatigue damage to blades is not great in low winds, it is argued that such wake effects are not necessarily critically increasing blade fatigue. However, the yaw torque in wake operation (synthesised from hub flap signals because a measured yaw torque signal was not available) exhibits greater variability than in any other non wake operating condition. A time history of yaw torque (synthesised) is shown in Fig 12. The related overturning moment has also been synthesised (Fig 12) and is very similar. The yaw ring of the Howden HWP 330/33 wind turbine is of the cross roller bearing type which means that overturning moment, axial and vertical shears resolve through planes at 45° and all contribute to the variations in contact load distribution.

Thus there is direct evidence that yaw systems may be particularly severely affected by near wake immersion of the rotor.

4 INFLUENCE OF YAW ERROR ON YAW TORQUE

The yaw angle history measured during a five minute extract of tape 65 is shown in Fig 13 and indicates large changes in yaw angle between -30 degrees and +50 degrees. In order to examine the effect of this yaw angle variation on the yaw torque experienced by the machine, the yaw angle history of Fig 13 has been injected into time domain analysis together with the simulated turbulent wind field. Results are presented in Fig 14 and 15 in terms of the auto spectra and periodic wave forms of hub load and yaw torque. Three cases are presented;

Case 1: A turbulent wind field is simulated with a constant yaw misalignment of 11.2 degrees.

Case 2: A turbulent wind field is simulated with the yaw angle time history shown in Fig 13.

Case 3: A coherent turbulent wind field is simulated with the yaw angle time history of Fig 13. The wind field is assumed to be coherent with no spatial variation and with the temporal variation of wind speed represented by a single simulated time history.

Comparing first the results of cases 1 and 2 and considering the hub load the only discernible influence of the yaw angle variation is a slight increase in the low frequency content of the auto spectrum. The yaw torque appears to be almost completely unaffected by the variation in yaw angle. These results are initially somewhat surprising since they indicate that the fluctuating yaw torque experienced by a wind turbine is independent of stochastic variation of the yaw misalignment of the machine. The variation in yaw torque is in fact much more dependent on the characteristics of the longitudinal turbulence. This is confirmed by consideration of the case 3 results. In terms of the hub load the assumption of a coherent wind field results in a considerable increase in load at low frequencies and a reduction at higher frequencies. The yaw torque is reduced substantially at all frequencies.

The periodic components of both the hub load and the yaw torque change only slightly as a consequence of the different assumptions of the three cases. The changes that are observed are a result of the influence of the non-linearities of the wind turbine aeroelastic system on the wave averaging procedure used to extract the periodic components.

It is clear from the results of Figures 14 and 15 that the variation in yaw torque is determined to a very large extent by the spatial properties of the turbulent wind field. By contrast, fluctuation in yaw misalignment has a negligible impact on yaw torque variation. The neglect of the spatial variation of turbulent wind speeds from wind turbine design calculations has far more serious consequences for a yaw system than for a rotor blade. Yaw system loads are determined by differential loading across the rotor and their prediction is therefore particularly sensitive to the realistic modelling of longitudinal turbulent wind speed variations in both time and space. Lateral turbulence giving rise to variations in wind direction and hence variations in yaw error appears to be of negligible importance in the context of yaw loads. Large turbulence structures with significant vertical wind components can however have a major impact on yaw loading (10).

5   CONCLUSIONS

The Howden data has allowed validation of the prediction of yaw system loads on a three bladed wind turbine. Although the issues of mean yaw torque variation and yaw stability depend sensitively on aerodynamic inflow, the dominant fatigue loading in the yaw system is stochastic with a central frequency of 3P and can be accurately predicted using standard strip theory aerodynamics providing a good model of wind turbulence is employed.

An important conclusion from this study is that while most aerodynamic inputs to the yaw system from the rotor can be well predicted (1P blade loading being most problematic), it cannot always be assumed that the wind turbine yaw system, even in locked yaw with brakes applied is a linear structural element. As has emerged in connection with machine dynamics and rotor imbalance effects, the extent to which the rotor weight is overhung may be crucial. This is particularly interesting when the main aerodynamic source of yaw torque (blade flap moments) is actually independent of rotor overhang distance.

Complex non linear behaviour in yaw systems is therefore possible, and much better understood in consequence of the present work.

The data illustrates also the particularly severe effect of wakes on yaw system loading.

ACKNOWLEDGEMENTS

Garrad Hassan are grateful to ETSU for permission to publish this paper.

REFERENCES

1.   JAMIESON, P. Design and development of the Howden 33m diameter wind turbine rotor. *Proceedings of AWEA Conference*, Honolulu, Sept. 1988.

2.   JAMIESON, P. and Anderson, C. G. Power performance measurement on the Howden 33m diameter wind turbine. *Proceedings of 10th BWEA Conference*, London, March 1988.

3.   ANDERSON, C. G. and Jamieson, P. Mean load measurements on the Howden 33m wind turbine. *Proceedings of 10th BWEA Conference*, London, March 1988.

4.   JAMIESON, P. Further analysis of data from the Howden HWP330/33 wind turbine, for DTI, Contract No E/5A/CON/6025/2212. To be published.

5.   HANSEN, A. C. Yaw Dynamics of Horizontal Axis Wind Turbines. National Renewable Energy Laboratory Report for the US Department of Energy, Contract No DE-AC02-83CH10093, May 1992.

6.   QUARTON, D. C., WASTLING, M. A., GARRAD, A. D. and HASSAN, U. The Calculation of Wind Turbine Loads - A Frequency or Time Domain Problem?, *Proceedings BWEA Conference*, Nottingham, March 1992.

7.   MORGAN, C. A., GARRAD, A. D. and HASSAN, U. Measured and predicted wind turbine loading and fatigue. *EWEC '89*, Glasgow, July 1989.

8.   SNEL, H. and SCHEPERS, J. G. Investigation and Modelling of Dynamic Inflow Effects. *ECWEC*, Travemunde, Germany, March 1993.

9.   RASMUSSEN, F., PETERSEN, J. T., WINKELAAR, D. and RAWLINSON-SMITH, R. I. Response of Stall Regulated Wind Turbines - Stall Induced Vibrations. Final report, JOULE I contract, JOUR-0076, for CEC, DG XII, June 1993.

10.   TANGLER, J. Private Communication, NREL Colorado, February 1993.

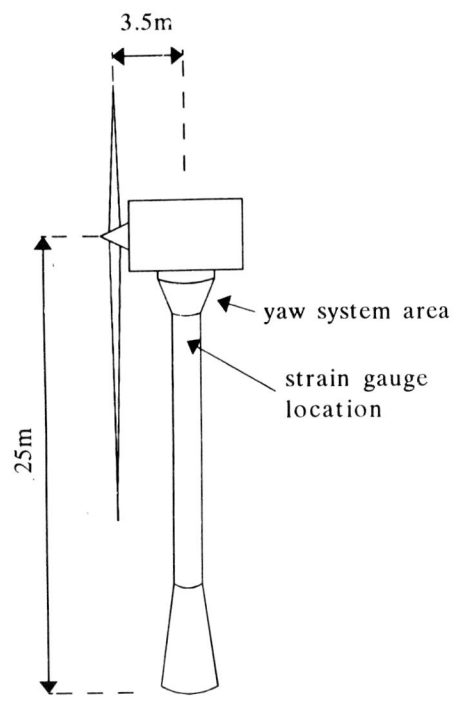

Fig 1   HWP 330/33 wind turbine

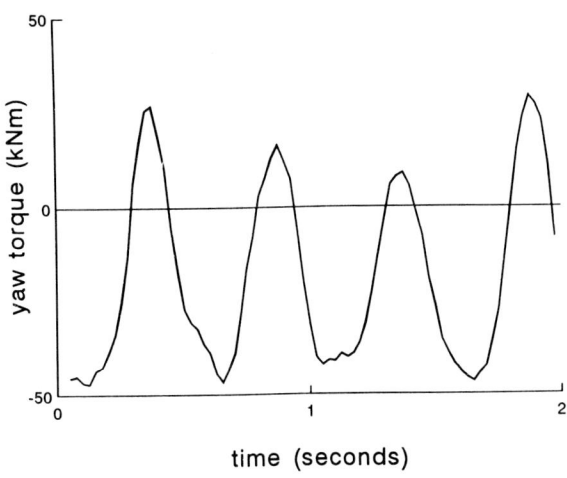

Fig 3   Yaw torque waveform, tape 65

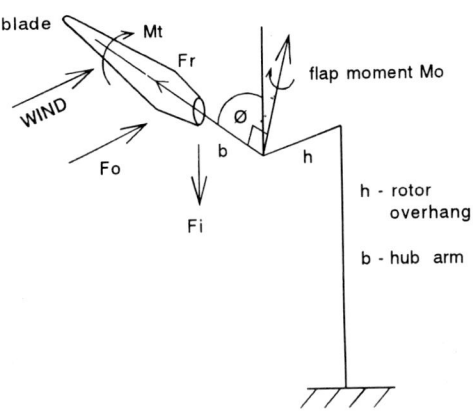

Yaw moment (due to single blade) is;

Y = Fo sin Ø b   (out of plane force)
  + Fi cos Ø h   (in plane force)
  + Fr sin Ø h   (radial force)
  + Mo sin Ø   (out of plane moment)
  − Mt cos Ø   (radial moment ≈ blade torsion
                                     moment)

If the shaft has zero tilt the shaft torque
does not contribute

Fig 4   Components of yaw torque

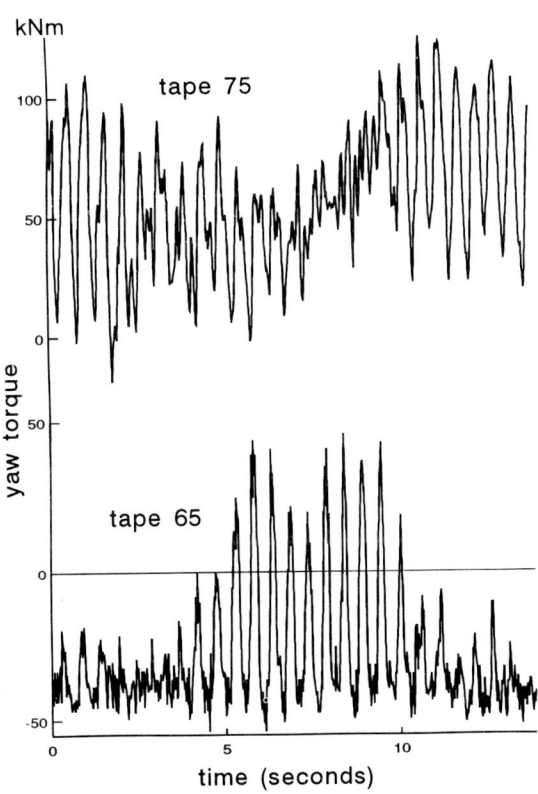

Fig 2   Yaw torque signals

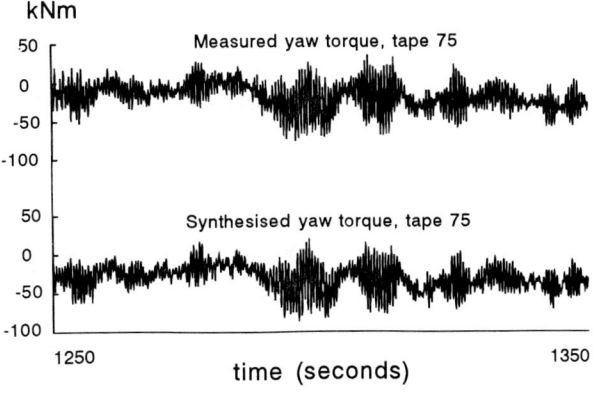

Fig 5   Yaw torque comparisons

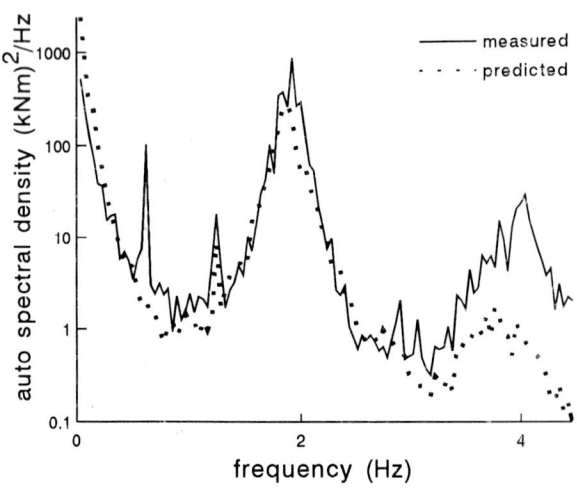

Fig 6 Yaw torque auto spectra, tape 75

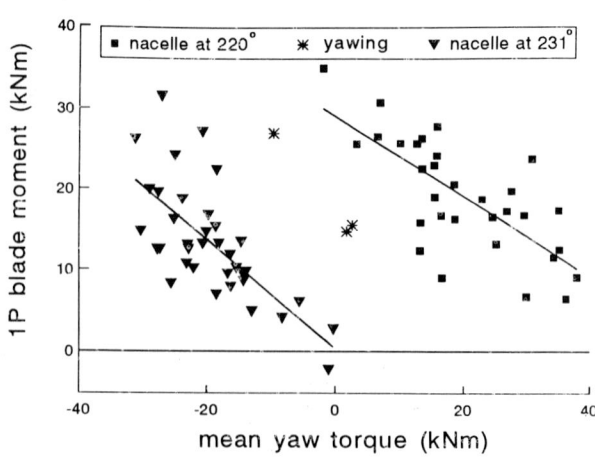

Fig 9   Mean yaw torque correlation

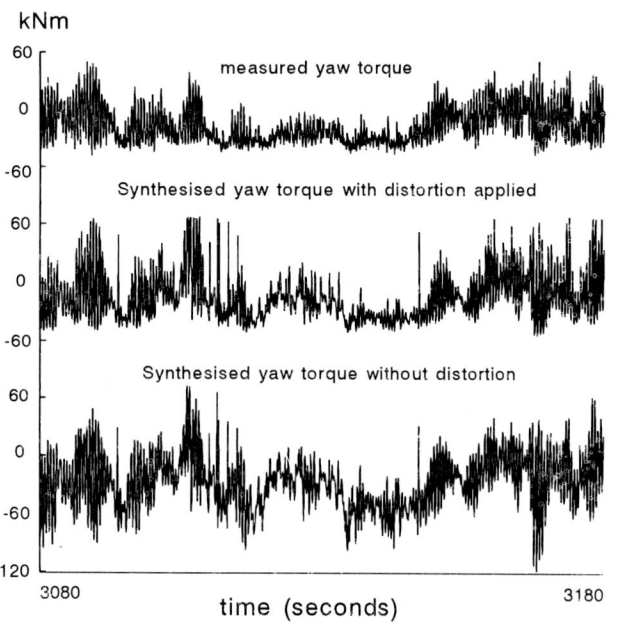

Fig 7   Yaw torque, tape 65

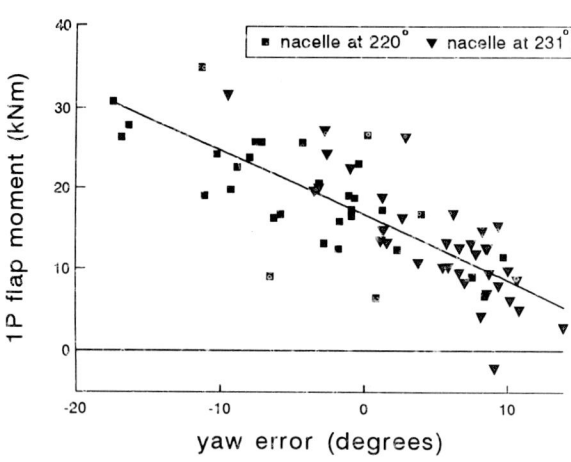

Fig 10   Effect of yaw error on 1P blade response

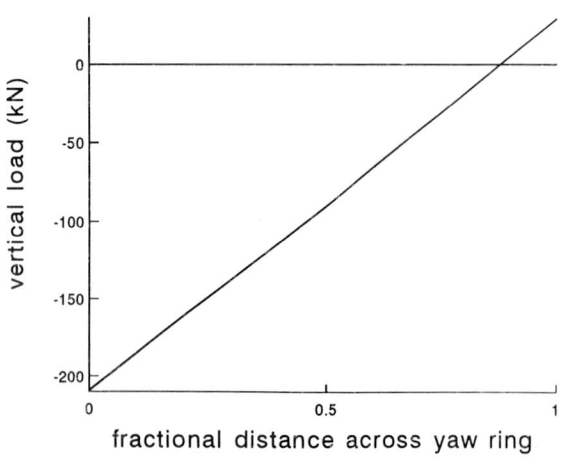

Fig 8   Yaw ring load distribution

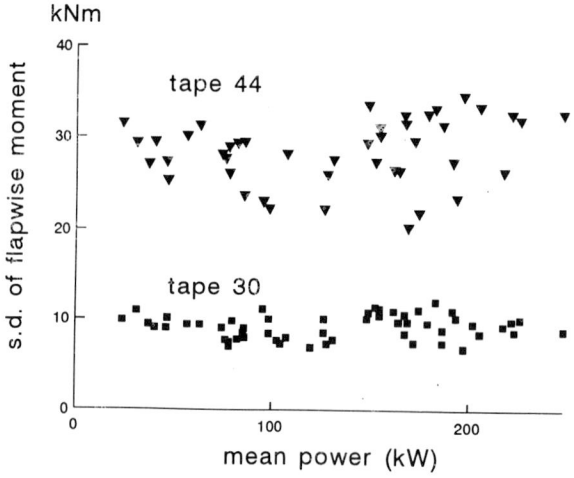

Fig 11   Effect of wakes on blade flap activity

210

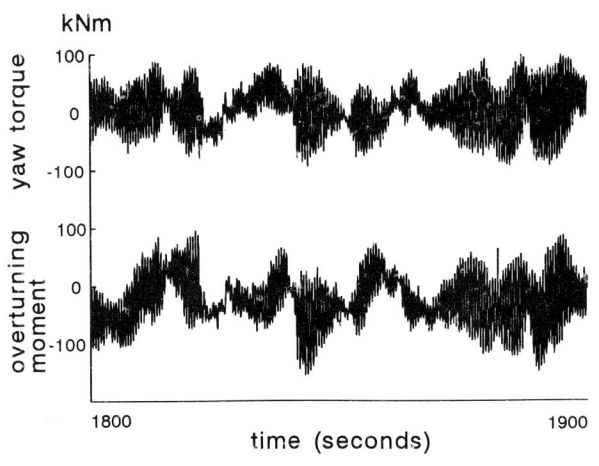

Fig 12　　Effect of wakes on tower top
　　　　　moments

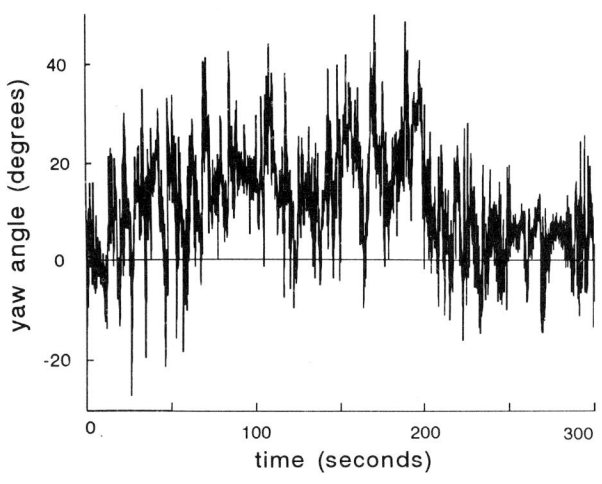

Fig 13　　Yaw angle history, tape 65

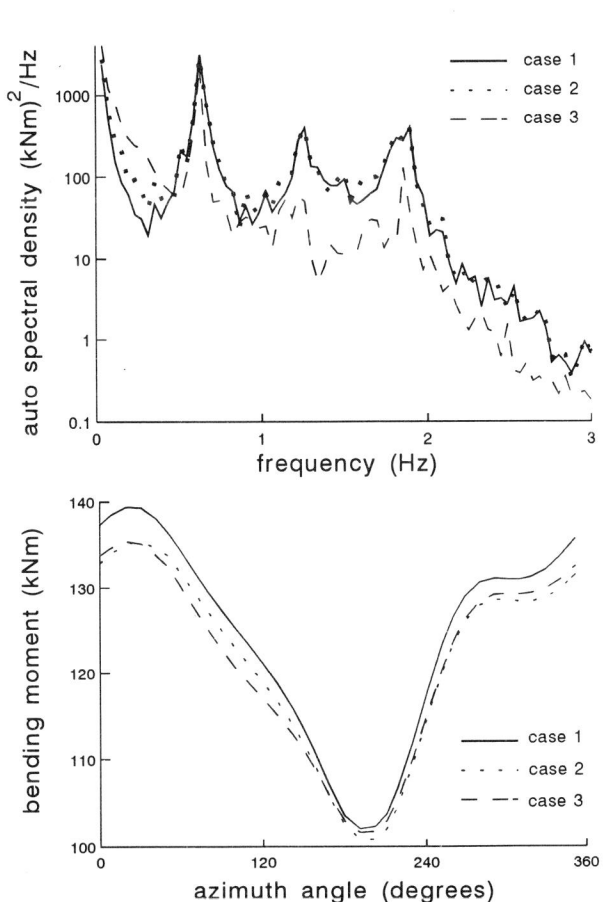

Fig 14　Hub flapwise bending moment
　　　　Influence of yaw angle variation

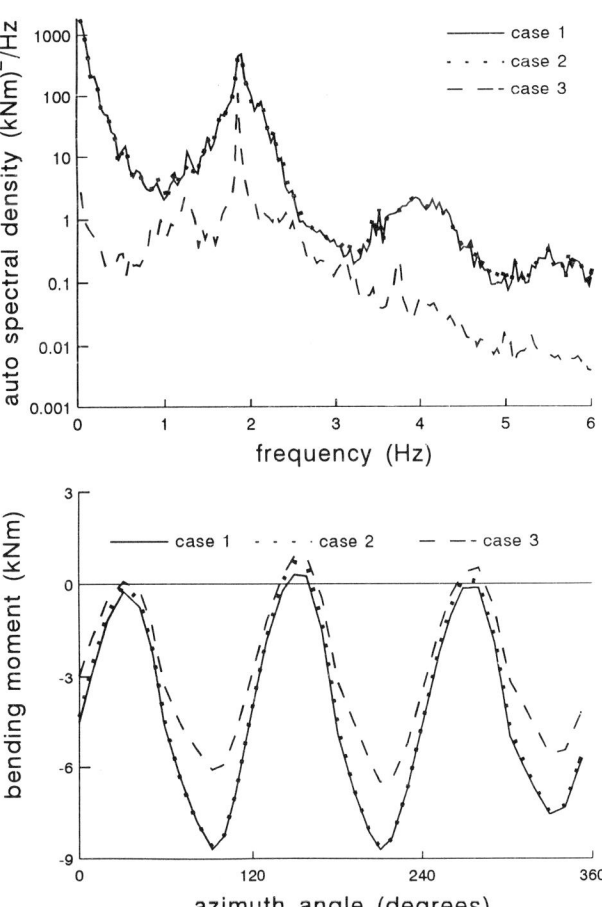

Fig 15　　Yaw torque - influence of yaw
　　　　　angle variation

Relationship of the controllability of power/torque fluctuations in the drive-train to the wind turbine configuration

M. C. M. ROGERS and **W. E. LEITHEAD**
University of Strathclyde, UK

SYNPOSIS : There are some configurations of wind turbine for which pitch regulation is very effective and some for which it is not successful. The purpose of this paper is to quantify the dependence of the effectiveness of the pitch regulation system. For a 300 kW wind turbine, equivalent one, two and three-bladed rotors are designed for both full-span and tip regualtion. The performance is assessed for each choice of rotor for various drive-train characteristics, control system characteristics and windspeeds.

1. INTRODUCTION

Many grid-connected medium scale, constant speed horizontal axis wind turbines have the ability to vary the pitch of the rotor blades or part of the blades. This capability is frequently used to regulate the wind turbine by varying the pitch angle in response to a measurement of generated power. The associated control system has the objective of reducing the transient loads on the drive-train (1). Unfortunately, the performance of the control system is very specific to the configuration of the wind turbine. There are some configurations of wind turbine for which pitch regulation is very effective and some for which it is not successful. The purpose of this paper is to quantify the effectiveness of the pitch regulation system for a variety of 300 kW wind turbine configurations and, thereby, increase awareness of the dependence of the performance of a constant speed HAWT on its configuration.

2. TURBINE CONFIGURATIONS

The aspects of the wind turbine configuration on which the effectiveness of the control system depends are

- the number of blades
- the nature of pitch regulation
- the dynamic characteristics of the drive-train
- the dynamic characteristics of the control algorithm.

The number of rotor blades may be one, two or three. The pitch regulation may be full-span or part-span (tip). The dynamic characteristics of the drive-train are characterised by the frequency of the first-drive train mode. It is an indicator of both how strongly the windspeed turbulence induces transient loads on the drive-train and how highly rated the control actuator need be. The damping factor for the first-drive train mode is chosen to be ideal, i.e. to have a value of 0.7, in all cases. This assumption is reasonable since the value of damping factor for many commercial medium scale wind turbines is typically 0.6 to 1. The dynamic characteristics of the control system are characterised by the crossover frequency of the open-loop system, i.e. the frequency at which the gain is one. It is an indicator of both how active the control system is and how vulnerable it is to high frequency noise.

The spectral density functions for the torques or forces acting on a wind turbine are characterised by two features (1). The first is a low frequency concentration which reflects the turbulence of the wind-field. The second is a series of peaks at integer multiples of the nominal angular velocity of the rotor (P). The latter are caused by the blades sweeping through the wind-field which changes relatively slowly as compared to the speed of rotation of the rotor. The most important of these peaks for an n-bladed rotor is the one at nP. Typical intensities of the nP spectral peaks are known from data monitoring programmes previously conducted on commercial machines (1), (2). To check performance sensitivity to the extent of nP, the 3-bladed wind turbine configurations can have either the nominal intensity of 3P or twice the nominal intensity and the 2-bladed wind turbine configurations can have the nominal intensity of 2P or half the nominal intensity. One further aspect of the wind turbine configuration might be relevant - namely, the angular velocity of the rotor. The sensitivity of the performance to the rotor velocity was investigated and found to be low. Accordingly configurations are not discriminated by rotor velocity in the investigations reported in this paper.

The complete set of configurations for a grid-connected constant speed upwind 300 kW turbine considered here are

1. 1 blade, full span regulation
2. 1 blade, tip regulation
3. 2 blades, full-span regulation, nominal 2P
4. 2 blades, full-span regulation, ½* nominal 2P
5. 2 blades, tip regulation, nominal 2P

6. 2 blades, tip regulation, ½* nominal 2P
7. 3 blades, full-span regulation, nominal 3P
8. 3 blades, full-span regulation, 2* nominal 3P
9. 3 blades, tip regulation, nominal 3P
10. 3 blades, tip regulation, 2* nominal 3p

Equivalent one, two and three-bladed rotors were designed using aerodynamic strip theory, Anderson (3), for both full-span and tip regulation. The rotors are equivalent in that the following are similar for each rotor.

• efficiency below rated windspeed
• rated windspeed
• blade root strees

In addition, for the full-span regulated configurations, the rotors are designed so that the relationships of pitch angle to wind speed are similar to each other and to the relationship typical of commercial machines. For the tip regulated configurations, the lengths of the tips are chosen to provide adequate aerodynamic braking for overspeed protection. Again, similiar to the full-span case, the rotors are designed so that the relationships of pitch angle to wind speed are similar to each other and to the relationship typical of commercial machines. In order to meet these requirements, it is necessary that the rotor have varying radius, solidity, angular velocity and length of tip when tip regulated. The details for the rotors are tabulated in Table 1.

For each configuration the aerodynamic torque coefficient $C_T(\alpha,\lambda)$ is determined where α is the pitch angle and λ is the tip speed ratio - i.e. (rotor radius x angular velocity / windspeed).

Table 1 Details of Rotor Configurations

No of blades	Type of regulation	Radius (m)	Tip length (m)	Solidity	Rotor speed (r/s)
1	Full-span	19.0	-	0.29	6.316
1	Tip	19.0	4	0.29	6.316
2	Full-span	16.5	-	0.67	5.09
2	Tip	16.5	3.5	0.67	5.09
3	Full-span	16.0	-	1	4.125
3	Tip	16.0	2.4	1	4.125

3. PERFORMANCE ASSESSMENT

In these investigations, it is assumed that the drive-train is fairly conventional - that is, it comprises a low-speed shaft, gear-box, high-speed shaft and induction generator with no additional electro-mechanical devices. In these circumstances (1), the most important performance indicator for the drive-train is the torque transients acting on the gearbox and the transients on generated power are good indicators of the loads transients on the drive-train. Hence, above rated windspeed the variance of power is a good measure of performance. It should be noted that it is the extreme loads which are of concern but that these are much greater than three standard deviations above the mean (to perhaps as much as five standard deviations) since

the distributions of the loads transients are not Gaussian.

In the wind turbine control system the actuator is the device which causes the actual pitch angle of the blades to follow the pitch angle demanded by the control algorithm. In any actuator, the pitching of the blades is effected through some internal motive force. Of course there are maximum and minimum limits to the motive force which fundamentally constrain the performance of the actuator and thereby the control system. Because the motive force is related to the second derivative of the displacement induced by the actuator, it is especially vulnerable to high frequency disturbances which can easily cause the actuator to saturate. A major source of high frequency disturbances in the wind turbine are the spectral peaks at integer multiples of the angular velocity of the rotor. The preferred measure of actuator activity is the variance of the acceleration of the pitch angle of the blades since, ignoring any compliance in the mechanical linkage, it is related by a scaling factor, associated with the inertias and gearing ratios of the actuator and pitching mechanism, to the actuator motive force but, unlike the actuator motive force, its value is independent of the specific nature of the actuator.

Because there are a large number of configurations and each is regulated for a range of first drive-train mode frequencies, controller crossover frequencies and windspeeds, it is necessary to employ analytic methods. At an operating point which corresponds to a given mean windspeed, the models of the wind turbine are linearised. The experience of the authors, from both theoretical and practical wind turbine control projects, is exploited to ensure the models are not unrealistic. These models were developed to assist the design of wind turbine control systems (1), (4). They have been thoroughly validated and are known to be a reasonable representation of the various aspects of the wind turbine dynamics (1). It is not possible to be more explicit about these models here, but for further details the interested reader is referred to references (1) and (4). Using linear spectral analysis, the variance of power and variance of pitch angle acceleration are calculated for the linearised wind turbine models.

For each configuration the variance of power is estimated for a range of frequencies of the first drive-train mode, a range of controller crossover frequencies and a range of wind speeds. Of course whether a controller crossover frequency can be achieved for a given configuration depends on the level of actuator activity. Accordingly, the variance of pitch angle acceleration is also estimated in each case.

The performance of configuration 1 to configuration 10 is investigated separately. The performance of each is dependent on the first drive-train frequency, the controller crossover frequency and windspeed. With the windspeed 16 m/s, the dependence is illustrated by Figure 1 which shows the variance of power (on a log scale) plotted

5(b) is for a low windspeed site. For both the low and high windspeed site the 1-bladed turbines have great difficulty in meeting the specified standard deviation on power. For the low windspeed site both the 2-bladed and 3-bladed wind turbines can meet the objective without too much difficulty. However, for the high windspeed site, there is considerable difficulty particularly for the full-span pitch regulated turbines. For a typical first drive-train mode frequency of 6 r/s, the 3-bladed wind turbines meet the goal with a controller of quite low activity (crossover frequency 1r/s) ; the 2-bladed tip regulated wind turbines meet the goal with a controller of quite high activity (crossover frequency 2.25 r/s) which is attainable in practice (3) ; the 2-blade full-span regulated wind turbines meet the goal with a controller of very high activity (crossover frequency 3.5 r/s) which is difficult to achieve in practice.

Of course the controller crossover frequency, which can be attained, is restricted by the actuator capability. The highest controller crossover frequencies which can be attained without the variance of pitch acceleration exceeding 0.01, 0.1 and 1 are tabulated in Table 2, Table 3 and Table 4 respectively for the various configurations and first drive-train mode frequencies. It can be observed that the controller crossover frequency increases slowly as the permitted variance of pitch acceleration increases. In general to achieve the same performance, the actuator rating must be greater for tip regulated rather than full-span regulated machines. In addition, the actuator rating must increase as the number of blades reduces from three to two to one and as the frequency of the first drive train mode decreases. Obviously many of the combinations of wind turbine characteristics, which achieve a standard deviation on generated power of 30 kw or less, indicated in Figure 5 are not feasible since unrealistic actuator capability would be required.

From both Figure 4 and Figure 5, it may be observed that the performance is relatively weakly dependent on the extent of the nP spectral peaks ; that is, there is relatively little difference in performance between Configurations 3 and 4, Configurations 5 and 6, Configurations 7 and 8 or Configurations 9 and 10.

6. CONCLUSIONS

The performance of a variety of configurations of grid-connected, constant speed up-wind medium scale (300 kW) wind turbines have been assessed for a range of first drive-train mode frequencies, controller crossover frequencies and windspeeds. The range of performance is surprisingly large with the variance of generated power differing by up to three orders of magnitude. Such a strong dependence has implications for the design of the complete wind turbines.

Table 2 Controller crossover frequencies with variance of pitch acceleration 0.01 $(rad/s^2)^2$

Config-uration	Frequency of first drive-train mode (r/s)						
	1.0	2.0	3.0	4.0	5.0	6.0	7.0
1	0.49	0.66	0.73	0.73	0.73	0.73	0.73
2	0.49	0.51	0.51	0.51	0.51	0.51	0.51
3	0.5	0.95	1.08	1.1	1.1	1.1	1.1
4	0.5	1.0	1.2	1.3	1.3	1.3	1.3
5	0.5	0.72	0.73	0.73	0.73	0.73	0.73
6	0.5	0.8	0.95	0.96	0.96	0.96	0.96
7	0.5	0.99	1.22	1.5	1.75	1.94	2.02
8	0.5	0.99	1.19	1.49	1.65	1.79	1.85
9	0.5	0.8	1.02	1.22	1.36	1.38	1.39
10	0.5	0.78	1.01	1.13	1.16	1.17	1.17

Table 3 Controller crossover frequencies with variance of pitch acceleration 0.1 $(rad/s^2)^2$

Config-uration	Frequency of first drive-train mode (r/s)						
	1.0	2.0	3.0	4.0	5.0	6.0	7.0
1	0.7	1.14	1.41	1.41	1.41	1.41	1.41
2	0.52	0.96	0.96	0.96	0.96	0.96	0.96
3	0.6	1.1	1.50	1.85	1.94	1.97	2.0
4	0.6	1.15	1.51	1.90	2.05	2.3	2.3
5	0.51	1.02	1.3	1.42	1.43	1.43	1.43
6	0.51	1.03	1.45	1.52	1.61	1.64	1.64
7	0.55	1.1	1.5	1.91	2.09	2.42	2.68
8	0.55	1.05	1.5	1.91	2.08	2.41	2.61
9	0.51	1.0	1.44	1.56	1.93	2.06	2.37
10	0.51	1.0	1.46	1.56	1.92	2.0	2.13

Table 4 Controller crossover frequencies with variance of pitch acceleration 1 $(rad/s^2)^2$

Config-uration	Frequency of first drive-train mode (r/s)						
	1.0	2.0	3.0	4.0	5.0	6.0	7.0
1	0.96	1.46	1.88	2.16	2.25	2.32	2.34
2	0.88	1.33	1.51	1.59	1.59	1.59	1.59
3	1.0	1.46	1.90	2.30	2.50	2.80	3.10
4	1.0	1.46	1.90	2.30	2.50	2.80	3.10
5	0.75	1.30	1.65	1.95	2.25	2.35	2.42
6	0.75	1.30	1.65	1.95	2.30	2.50	2.73
7	0.97	1.47	1.91	2.20	2.50	2.88	3.2
8	0.97	1.47	1.91	2.18	2.50	2.86	3.18
9	0.6	1.14	1.56	1.95	2.34	2.52	2.85
10	0.6	1.14	1.55	1.95	2.34	2.52	2.84

against controller crossover frequency (in r/s) for different first drive-train frequencies (the range is 3r/s to 8r/s in steps of 1r/s) and by Figure 2 which shows the variance of pitch angle acceleration (on a log scale) for the same parameter ranges. Figure 1(a) and Figure 2(a) are for Configuration 7 and Figure 1(b) and Figure 2(b) are for Configuration 9. There is a general tendency for the variance of power to decrease as controller crossover frequency increases and the first drive-train mode frequency decreases. However, when the controller crossover frequency is low the variance of power is not dependent on the first drive-train mode frequency. When the controller crossover frequency is high the variance of power, particularly for the tip regulated wind turbine, is quite strongly dependent on the first drive-train mode frequency. The variance of power of the tip regulated turbine is considerably (an order of magnitude on variance) less than that of the full-span regulated turbine. There is also a general tendency for the variance of pitch acceleration to increase as controller crossover frequency increases and the first drive-train mode decreases. Hence, an improvement in performance as measured by the variance of power alone cannot be achieved without a concurrent increase in actuator activity. However, when the controller crossover frequency is low the variance of pitch acceleration is not dependent on the first drive-train mode frequency. The variance of pitch acceleration of the tip regulated turbine is considerable greater than that of the full-span regulated turbine.

4. COMPARISON OF TWO TYPICAL TURBINES

Two typical but strongly contrasting configurations of wind turbines are compared to illustrate the wide range of performance possible. The first is a three-bladed tip regulated machine, i.e. Configuration 9, with first drive-train mode frequency 7 r/s whilst the second is two-bladed full-span regulated, ie. Configuration 3, with first drive-train mode frequency 6 r/s.

The variance of power (on a log scale) plotted against controller crossover frequency (in r/s) for three different mean windspeeds (12 m/s, 16 m/s and 23 m/s) is shown in Figure 3. The performance is not always improved by increasing the controller crossover frequency (i.e. by upgrading the control system). Several of the plots in Figure 3 have distinct minima. It may be observed from Figure 3(a) that the most testing conditions for the tip regulated turbine is low windspeed conditions which are encountered frequently. Hence, the controller design should be driven by the low windspeed conditions and the variance of power decreases with increasing controller crossover frequency. By contrast, it may be observed from Figure 3(b) that the most testing conditions for the full span regulated turbine is high windspeed conditions which are encountered rather infrequently. Except at high windspeed, the variance of power is only weakly related to the controller crossover frequency. The benefit of increasing the controller crossover frequency would only be perceived in high windspeeds and there is little premium on so doing for a wind turbine

situated on a low windspeed site.

Also indicated on Figure 3 are the controller crossover frequencies at which the variance of pitch acceleration is 0.01, 0.1 and 1 with the windspeed 12 m/s (Since for all configurations the variance of pitch acceleration increases with windspeed, the rating of the actuator is determined primarily by the conditions just above rated windspeed, namely at 12m/s). It may be observed that the actuator rating, required by the tip regulated wind turbine to achieve the same control system capability as the full-span regulated wind turbine, is lower.

The performance of the three bladed tip regulated wind turbine is much better (greater than an order of magnitude) than the two bladed full-span regulated wind turbine when the actuator have the same rating.

5. PERFORMANCE COMPARISON

To enable a direct comparison of performance of the various configurations, the variance of power (on a log scale) is plotted, see Figure 4, against controller crossover frequency (in r/s) for all the configurations with a first drive-train mode frequency of 6 r/s which is typical of commercial medium scale wind turbines. Figure 4(a) is for a high windspeed site and Figure 4(b) is for a low windspeed site with the windspeed rarely exceeding 20 m/s. The range of performance is surprisingly large with three orders of magnitude difference between the best and poorest. The best performance is attained by the 3-bladed tip regulated turbines whilst the poorest performance is attained by the 1-bladed turbines. As the controller crossover frequency increases the difference in performance between full-span regulated and tip regulated turbines reduces. The 3-bladed full-span regulated turbines benefit most from improved controller performance (increased controller crossover frequency). For a high windspeed site, performance for all the 1-bladed turbine and the 2-bladed tip regulated turbines does not monotonically improve as the controller crossover frequency increases. In fact they have a distinct minimum at low frequency. For a low windspeed site, performance for all 1-bladed turbines and all the 2-bladed turbines does not monotonically improve as the controller crossover frequency increases; that is, monotonic improvement is only shown by the 3-bladed turbines. By comparing the Figure 4(b) to Figure 4(a), it may be observed that 2 and 3-bladed full-span regulated wind turbines benefit from the low windspeed site.

Another way in which the performance of the different configurations may be compared is to set a specific design task such as a standard deviation on generated power of 30 kW or less. For each first drive-train frequency there is a controller crossover frequency for which standard deviation of generated power is 30 kW. The loci of these points is shown for each wind turbine configuration in Figure 5. Above and to the left of the loci the design task is achieved, below and to the left of the loci, it is not. Figure 5(a) is for a high windspeed site and Figure

The general conclusion are as follows:

(a) Performance improves as the number of blades increases from one to two to three.

(b) Performance of tip regulated turbines is better than full-span regulated turbines.

(c) Performance is most strongly related to controller ability for full-span regulated machines.

(d) Performance is weakly related to rotor angular velocity.

(e) Performance is weakly related to the intensity of the nP spectral peak where n is the number of blades

(f) The configuration with best performance is three-bladed tip regulated wind turbine.

(g) Increased controller crossover frequency requires greatly increased rating of the actuator.

The results presented in this paper are derived from theoretical studies using simple linear spectral analysis. The estimates of performance therefore, cannot be assumed to be of great accuracy. Nevertheless, the relative performances can be taken as indicative. Some of the wind turbine configurations correspond closely to commercial wind turbines for which monitored data is available. In these cases, the theoretical results are in reasonably good agreement with the measured performance. It is intended to consolidate the work presented here by constructing detailed simulations for a selection of the configurations to obtain better estimates of performance. These will be used to refine the estimates of absolute rather than relative performance.

ACKNOWLEDGEMENTS

The authors wish to thank the DTI (Formerly the Department of Energy) for supporting the work presented here. The advice of Mr. P. Jamieson, Garrad Hassan and Partners, as to what constitutes equivalent rotors is also gratefully acknowledged.

REFERENCES

(1) Leithead, W.E. de al Salle, S.A., Reardon, D., Grimble, M.J., 1992, *Wind Turbine Control System Modelling and Design - Phase I and II*, DTI Report No. ETSU WN 5108.

(2) Leithead, W.E., Agius, P.R.D. 1991, *Application of classical control for the WEG MS-3 wind turbine*, University of Strathclyde report.

(3) Anderson, C.G., 1990, *Wind turbine aerodynamic loading and performance code*, University of Edinburgh

(4) Leithead, W.E. de la Salle, S.A., Reardon, D., 1992, Int. J. of Control, 55, 845-876.

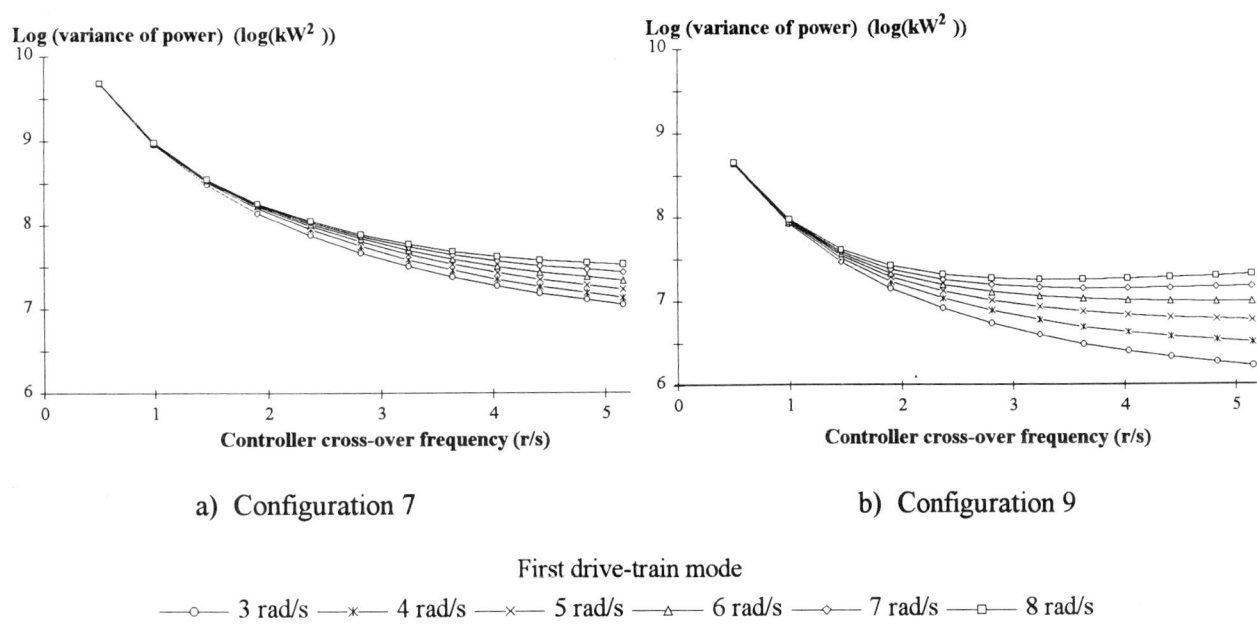

a) Configuration 7   b) Configuration 9

First drive-train mode

—o— 3 rad/s —x— 4 rad/s —×— 5 rad/s —△— 6 rad/s —◇— 7 rad/s —□— 8 rad/s

Figure 1 Performance of various drive-train characteristics

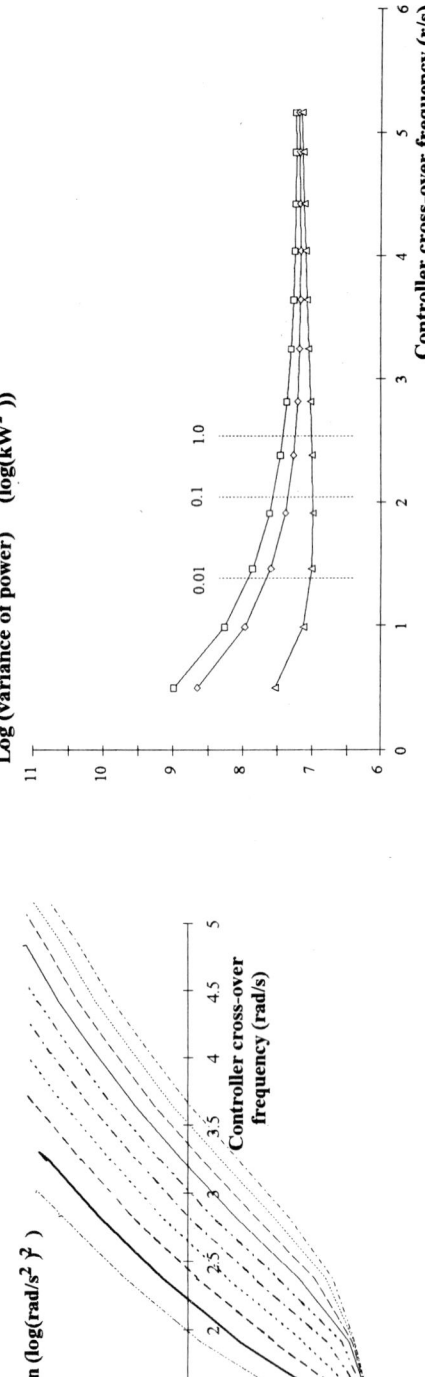

a) Configuration 7

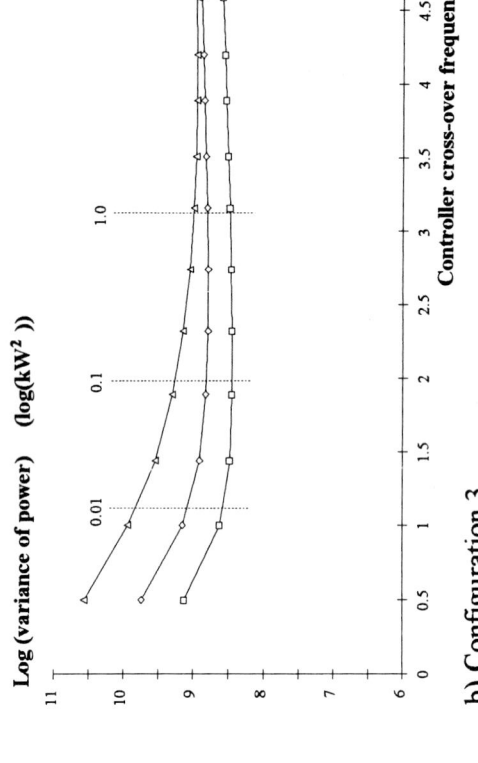

a) Configuration 9

b) Configuration 9

b) Configuration 3

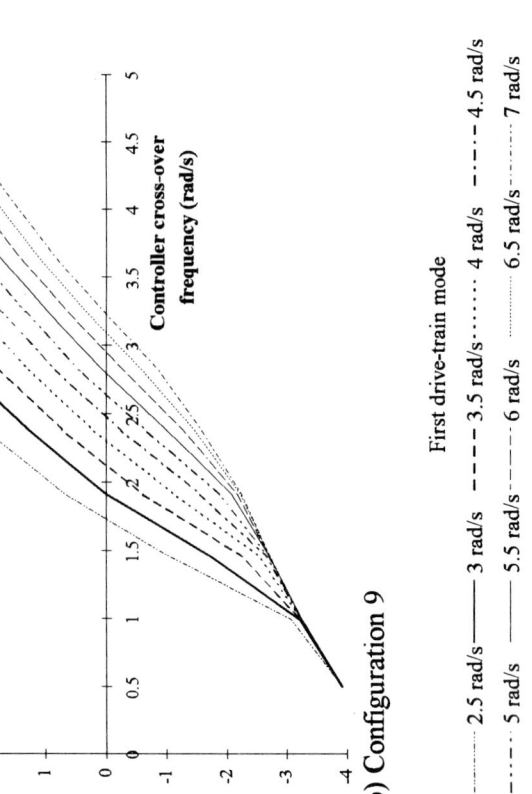

Figure 2 Actuator activity for various drive-train frequencies

Figure 3 Performance for various windspeeds

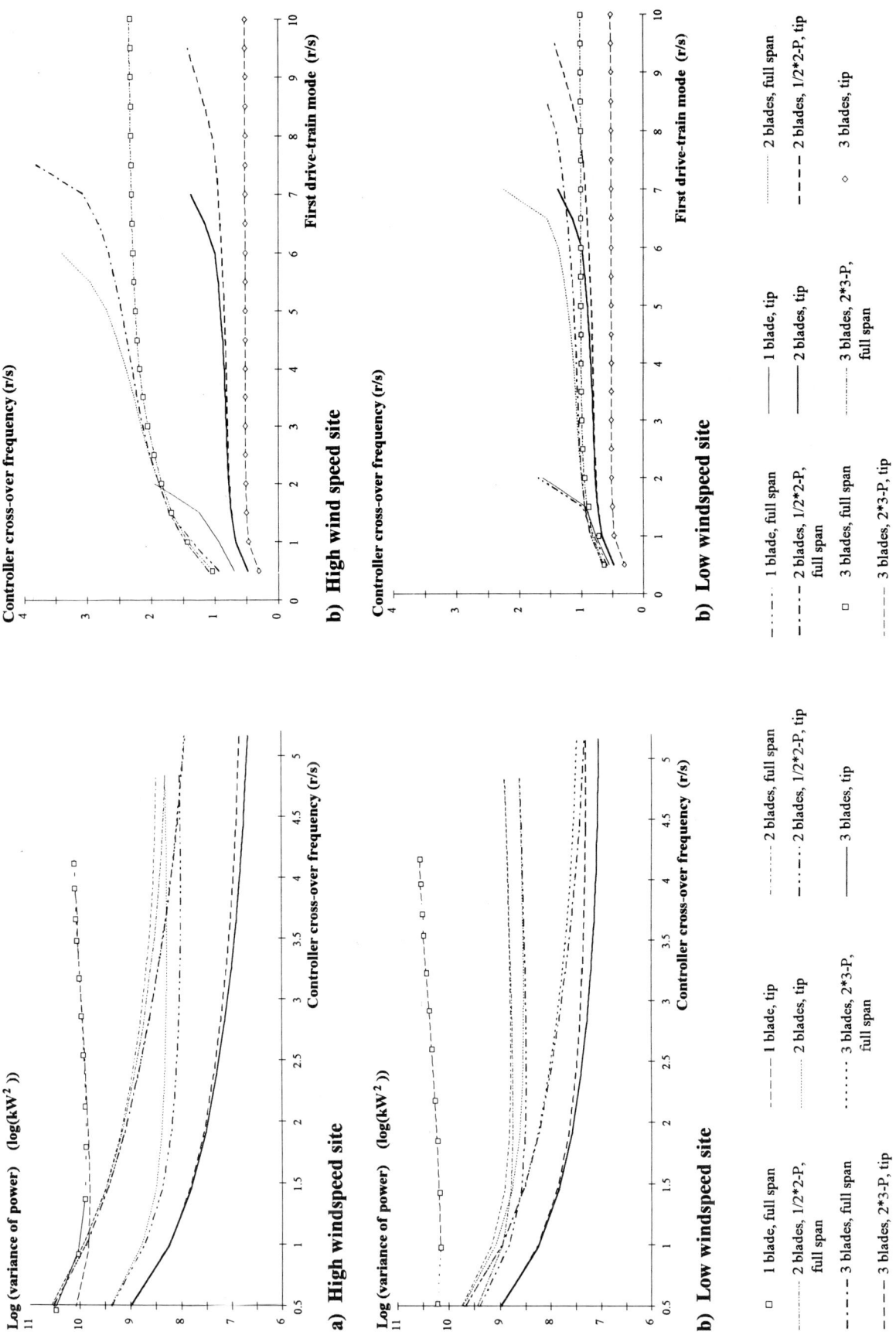

Log (variance of power)   (log(kW²))

a) High windspeed site

Controller cross-over frequency   (r/s)

Log (variance of power)   (log(kW²))

b) Low windspeed site

Controller cross-over frequency   (r/s)

Controller cross-over frequency (r/s)

b) High wind speed site

First drive-train mode (r/s)

Controller cross-over frequency (r/s)

b) Low windspeed site

First drive-train mode (r/s)

□ 1 blade, full span

----- 1 blade, tip

·········· 2 blades, full span

– · – · 2 blades, 1/2*2-P, tip

– · – · 2 blades, 1/2*2-P,
            full span

–––– 2 blades, tip

········ 3 blades, full span

———— 3 blades, tip

– – – 3 blades, 2*3-P,
            tip

            3 blades, 2*3-P,
            full span

Figure 4   Performance of various wind turbine configurations

□ 1 blade, full span

----- 1 blade, tip

– · – · 2 blades, 1/2*2-P,
            full span

········ 3 blades, full span

– – – 3 blades, 2*3-P,
            tip

———— 2 blades, full span

– · – · 2 blades, 1/2*2-P, tip

◇ 3 blades, tip

           3 blades, 2*3-P,
           full span

Figure 5   Conditions for a standard deviation of power of 30 kW or less

Wind turbine tower structural and dynamic analysis using the finite element method

N.EL CHAZLY, PhD, Mech Eng
National research Centre, Egypt

SYNOPSIS The wind rotor hub loads developed in previous works, are transfered to the fixed, or tower coordinate, system and the individual blades summed over the rotor. From a design standpoint, magnitudes and frequencies of the rotor forces and moments are the chief dynamic considerations for the tower and supports. In the present work, a finite element analysis is used to investigate the static and dynamic behaviour of steel towers. A triangular plate element was used to idealize the tower. Strength, strains, deflections; either translational or rotational, eigen values and eigen vectors, are computed for three different types of steel pole tower. Mainly, straight six-sided tower, coned six-sided tower and stepped six-sided tower are compared for best strength and frequencies of free vibrations. For the whole three configurations, the maximum stress occurs at the tower foundation. The tower of maximum strength is the stepped one, and is also the one that satisfies a material saving, proceeding in strength and material saving comes the conical one then the straight tower. The maximum deflection occurs in the wind direction in the three configurations. The deflection is minimum in the stepped tower. Comparing the natural frequencies, it was found that for the 2nd bending mode, the straight tower acquires highest natural frequency.

NOTATION

F_{to}	mean thrust force.
$F_t(\theta)$	thrust force at any angle (θ).
F_{do}	mean force equals to torque (T)/radius (R).
V_r	reference wind velocity, m/sec.
H_r	reference tower height, m
V	wind velocity, m/sec
H	specific tower height, m
ρ	air density, kg/m³ .
C_f	friction coefficient or drag coefficient
A	projected area of one face of the tower under a wind pressure, m²
x,y,z	local coordinates of a plate element
X,Y,Z	global coordinates
u,v,w	nodal linear displacements
$\theta_x,\theta_y,\theta_z$	rotational nodal displacements
$\{q_i\}$	nodal displacements vector
$\{P_i\}$	nodal force vector
$[k^{(e)}]$	element stiffness matrix
$[k]$	combined membrane and bending element stiffness matrix
m	denotes membrane
b	denotes bending
$\{Q\}$	global displacement vector
$\{P\}$	global force vector
$[K]$	global stiffness matrix
$\{\sigma^{(e)}\}$	element stresses vectors
$[M]$	assembly of elemental consistent mass matrices
δ	deflection
γ	eigen value
$\{X\}$	eigen vector

1 INTRODUCTION

Support structures for wind turbines may be of various types and various stiffnesses. The stiffness of a tower reflects its dynamic response to the forcing frequencies of the individual blades or the whole rotor. A further issue is the tower choice which depends on two problems: height, with the trade-off between energy production and the generating cost of that energy; and structural integrity, with the problem of vibration.

The most common structure used is the metal lattice-work tower with a foursided pyramidal shape. Towers may be cantilever tube or guyed pole. Most tubular towers are manufactured in steel and can be uniform in cross-section, step- or full-tapered. Some of the very large turbines have used concrete towers, which look economically attractive for the larger units.

Different approaches have been used either for structural or for modal analysis of wind turbine towers. A detailed finite elemet NASTRAN model of the Mod-0 wind turbine tower (truss type); using bar elements, rod elements and plate elements, was reduced to six beam elements (stick model)(1). The detailed model is too much time consuming while the stick model reduces the computer time, but it does not consider the effect of shear deformation and is very far from accuracy.

The Unarco-Rohn tower sections used for tower design of a 15-KW WIND-SYSTEM (2), were analyzed for structural integrity. Tower sections were modeled as pinjointed truss sections. The wind machine/tower system was modeled as a lumped-mass and the first and second tower bending frequencies were calculated, assuming that shear deflections are neglected.

The present work is based on a finite element program used for structural and vibrational analyses of wind turbine towers of pole type. A simple and time reducing triangular plate element defining both in-plane and bending actions is used.

2 NUMERICAL ANALYSIS

The tower is the support structure of wind turbines, so it has to withstand different loads transfered from rotor hub; mainly thrust and torque and the weight of the aloft system. The tower must also resist wind shear distributed over its height. Eigen value analysis must also be performed to avoid tower resonance. So, the strength of the tower and supports must be sufficient to resist the maximum expected transient loading, and the tower stiffness must be determined to avoid tower resonance at the rotor frequencies.

2.1 Tower structural model

Three different tower configurations have been analysed, having in common the material used which is steel, the height; five meters and a six-sided cross-section. The configurations used are:
1. Straight pole tower.
2. Tapered or coned tower.
3. Stepped tower.

Finite element idealization: The tower body is discreteized into triangular plate elements, of eighteen degrees of freedom each, allowing for in-plane and bending actions. Three idealized configurations are shown in Fig. 1.

External boundary forces: The structural integrity of the tower is dependent on the strength of the tower. The tower must be designed to with stand the forces imposed to it.

The supporting tower of a wind generator is subjected to various forces, the dominant ones being the steady loads aroused from rotor thrust, torque and the weight of the aloft system, plus the horizontal force of the wind (wind shear), Fig. 2. Also, because the wind is usually fluctuating in time, the tower is subjected to vibratory forces and hence fatigue forces, these forces are beyond the scope of this work.

The detailed boundary forces are as follows:

1. Integrated thrust force over blades length:

$$F_t(\theta) = F_{to} \sin \theta \qquad (1)$$

2. Torque due to drag over the blades:

$$F_d(\theta) = F_{do} \sin \theta \qquad (2)$$

3. Horizontal Wind Shear
   The wind velocity increases with height according to a power law where the velocity is proportional to the height raised to a power, (α) say 0.2 (the exponent depends on the ground roughness, among other factors), the relation is given by:

$$\frac{V}{V_r} = (\frac{H}{H_r})^\alpha \qquad (3,4) \qquad (3)$$

We can then compute the wind load P or drag force at each tower height using the wind speed "V", (5), obtained from the equation:

$$P = \tfrac{1}{2} \rho V^2 C_f A \sin \theta \qquad (4)$$

From tower solidity and Reynolds numbers, the drag coefficient was calcuated (6).

Finite element stress analysis: The triangular plate element used for the analysis of three dimensional structures, provides the aspect of coupling the membrane and bending actions. The basic approach is outlined in a previous work of the author (7), and will be mentioned in brief in the present work, Fig. 3.

Defining the combined nodal displacements as:

$$\{q_i\} = \begin{Bmatrix} u_i \\ v_i \\ w_i \\ \Theta_{xi} \\ \Theta_{yi} \\ \Theta_{zi} \end{Bmatrix} \qquad (5)$$

and the appropriate forces as:

$$\{P_i\} = \begin{Bmatrix} P_{xi} \\ P_{yi} \\ P_{zi} \\ M_{yi} \\ M_{xi} \\ M_{zi} \end{Bmatrix} \qquad (6)$$

we can write

$$\begin{Bmatrix} P_i \\ \vdots \end{Bmatrix} = [K] \begin{Bmatrix} q_i \\ \vdots \end{Bmatrix} \qquad (7)$$

or

$$\{P^{(e)}\} = [k^{(e)}] \{q^{(e)}\} \qquad (8)$$

The stiffness matrix is made of the following submatrices:

$$[k_{rs}] = \begin{bmatrix} [k_{rs}^m] & 0 & 0 & 0 & 0 \\ & 0 & 0 & 0 & 0 \\ \hline 0 & 0 & & & \\ 0 & 0 & & [k_{rs}^b] & \\ 0 & 0 & & & \\ 0 & 0 & & & \end{bmatrix} \qquad (9)$$

The combined stiffness matrix is given by Nihad (7). Transforming the local stiffness matrices to a common set of global coordinates, together with the element forces vector $\{P^{(e)}\}$, (8), we get:

$$\{P\} = [K] \{Q\} \qquad (10)$$

The element stresses are given by:

$$\{\sigma^{(e)}\} = [D] [B] \{q^{(e)}\} \qquad (11)$$

where [D], and [B] are given in (8).

2.2 Eigen value analysis

The modal analysis used to predict tower natural frequencies and mode shapes is based on Rayleigh-Ritz subspace iteration (9) to find the solutions for the eigen vectors $\{X\}$ associated with the lowest eigen values "γ", given by:

$$[[K] - \gamma [M]] \{X\} = 0 \qquad (12)$$

where [M] is the assembly of all elemental consistent mass matrices, representing both membrane and bending ones, and given in details by the author (7).

3 RESULTS AND DISCUSSION

3.1 Stresses and deflections

The results will focus on the structural behaviour of the tower; either stresses, deflections, ...etc., and on the modal analysis.

Fig. 4 shows the stress distribution along the tower height for three different configurations. The stress is maximum at tower foundation for the three shapes. The stepped tower acquires maximum strength and maximum material saving, then comes the conical one and at last the straight pole tower.

Fig. 5a gives the lateral deflection "δx", or deflection in the wind pressure direction, for the three configurations. The maximum deflection occurs in the straight tower and the minimum in the stepped one. The same trend is found in the "z" or tower height direction, it is obvious that "δz" is very small compared to "δx", Fig. 5b.

Fig. 6a illustrates the deflection "δx" of tower cross sections at different tower stations. The deflection decreases going from the tower top to the base. Minimum deflection occurs in the stepped tower. The "δy" deflection together with "δz" for the top section are given in Fig. 6b.

3.2 Natural frequencies and mode shapes

Table 1 compares the natural frequencies for the three configurations. As obvious, the straight tower acquires the highest natural frequency for the second bending mode.

Figs. 7-9 show the mode shapes of the bending modes given in Table 1.

Table 1 Tower natural frequencies, rad/sec.

Mode	Straight tower	Coned tower	Stepped tower
1st bending-x	w_2=1.254	w_1=1.331 w_3=1.396	
1st bending-y	w_1=1.253	w_2=1.336	
2nd bending-x	w_3=7.031		
2nd bending-y			w_1=0.117 w_2=0.118 w_3=0.123

4 CONCLUSIONS

1. The combined plate element is a best substitute for the shell complicated element to treat and analyze different wind turbine tower configuration of small thickness, taking into account the stresses and deformations in the three dimensions x, y and z.

2. For all the configurations, maximum stress occurs at the base of the tower, while maximum deflection occurs at its top.

3. The stepped tower proved to have maximum strength, maximum material saving and minimum eigen values.

REFERENCES

(1) TIMOTHY, L.S. Simplified modeling for wind turbine modal analysis using NASTRAN. NASA Conference Publication 2034 DOE Publication CONF-771148, "Wind Turbine Structural Dynamics", Nov. 1977.

(2) ENERTECH, COR. Enertech 15-KW wind-system development. REP-3341/2, UC-60, 1981.

(3) DANIEL, M.S. Wind power. Noyes Data Corp., Park Ridge, 1975.

(4) FRERRIS, L.L. Wind energy conversion systems. Prentice-Hall, Inc., Englewood, Cliffs, N.J.I 1990.

(5) JOHN, J.B. and MICHAEL, L.S. Aerodynamic for engineering. Prentice-Hall, Inc., Englewood Cliffs, N.J.I, 1980.

(6) SACHS and PETER, Wind forces in engineering, Pergamon Press, 1975.

(7) NIHAD, M. EL-CHAZLY, Static and dynamic analysis of wind turbine blades using the finite element method. Accepted for Publication in J. Renewable Energy, 1992 and J. Numerical Methods in Eng., 1993.

(8) RAO, S.S. The finite element method in engineering. Pergamon International Library,1982.

(9) BATHE, K-J. Finite element procedures in engineering analysis. Prentice-Hall, Inc., Englewood Cliffs, N.J., 1982.

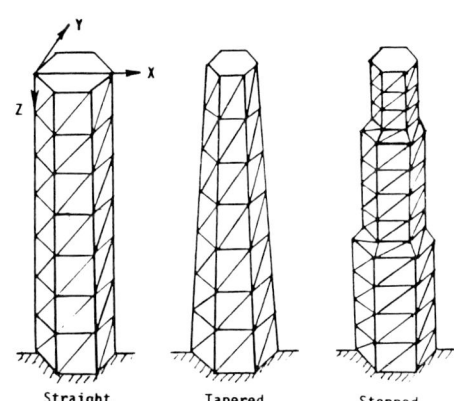

Fig. 1 Finite Element Idealization.

Straight   Tapered   Stepped

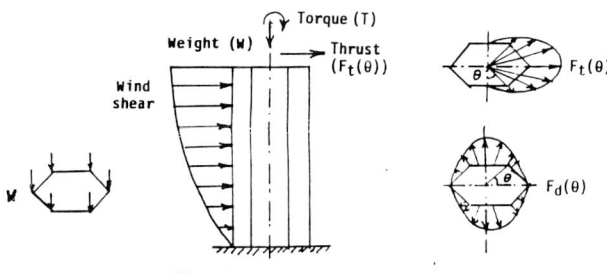

Fig. 2 Boundary Conditions.

223

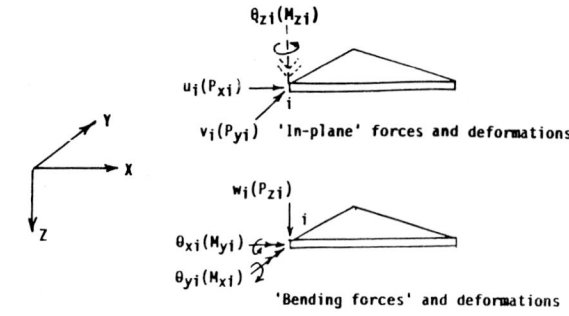

'In-plane' forces and deformations

'Bending forces' and deformations

Fig. 3 A flat element subject to "in-plane" and bending actions.

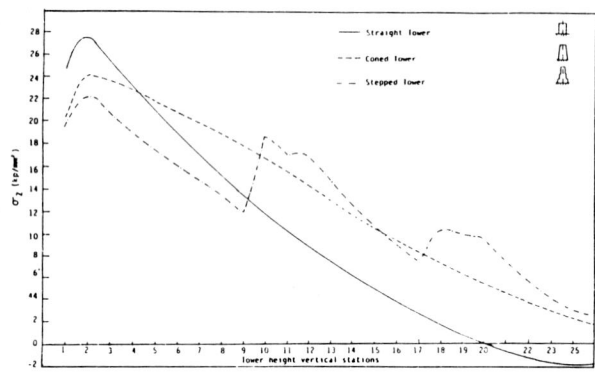

Fig. 4 Stress distribution σ_z along tower height.

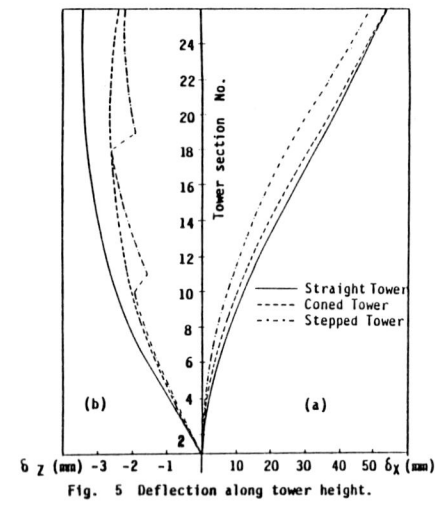

Fig. 5 Deflection along tower height.

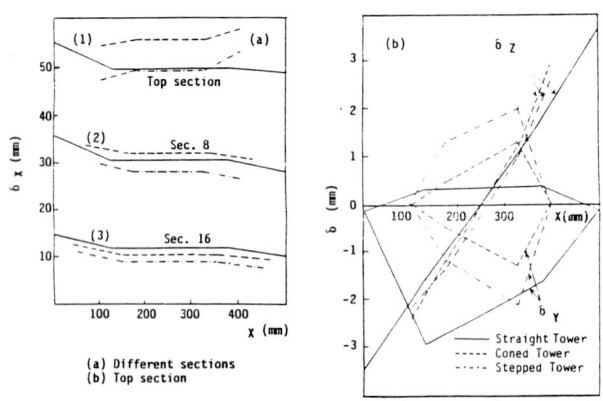

(a) Different sections
(b) Top section

Fig. 6 Tower cross sectional deflection.

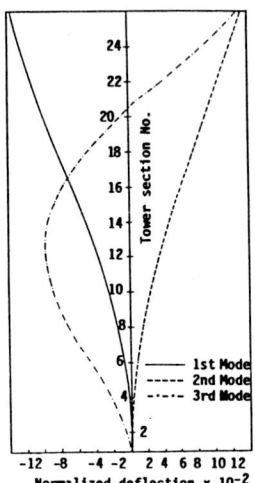

Fig. 7 Straight tower modal shapes.

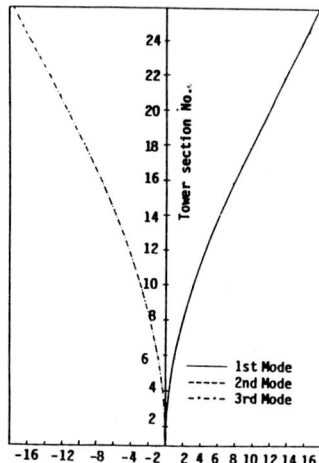

Fig. 8 Coned tower modal shapes.

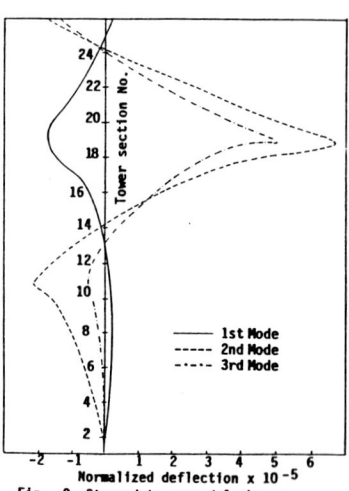

Fig. 9 Stepped tower modal shapes.

An analytical approach to wind farm design

A. D. GARRAD, A. S. MERCER and **B. M. ADAMS**
Garrad Hassan and Partners Ltd, UK
N. R. JENKINS
UMIST, UK

1 Introduction

This paper addresses the problems, and some of the solutions, which arise in the design and analysis of wind farms. The areas which are considered may be divided into two separate sections: resource and constraints. The challenge for the designer is to maximise the exploitation of the resource within the bounds placed on it by the environmental constraints. With the exception of the visual effect of the wind turbines all the other aspects of a wind farm are amenable to computational treatment. The process will be illustrated in this paper by an example. The site for this example has been chosen at random. Its only virtue is that the data describing it is available free as a demonstration from the Ordnance Survey. A map of the site is shown in Figure 1 together with an initial layout of the wind farm.

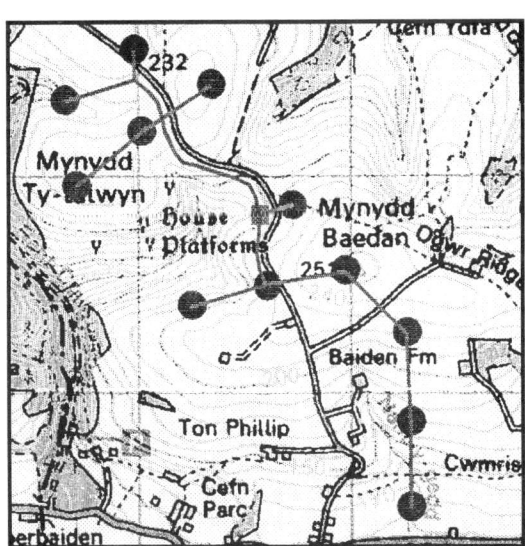

Figure 1   Site layout and cable routes

2   Wind Resource: Measure, correlate, predict

The hub height of a typical modern commercial machine is 40m. In order to reduce uncertainty in estimates of wind speeds on the site, it is usual to use a meteorological mast of a similar height to the hub height of the machine and to make measurements for about one year.

It should be stressed that these estimates of the long term wind speed statistics are based entirely on the correlation process and historical precedent. Quite large year on year variations can be expected, easily within the order of 15% either side of the mean and the error in the correlation may give rise to a further uncertainty of 10%. The statistical estimate of the wind speed is an estimate of the long term average and not the expected value for the next year. Proper appreciation of the statistical nature of these estimates is vital in formulating the cash flow projections and looking at the commercial down side for the farm.

An exercise such as described above will result in the central estimate for the mean wind speed and also an estimate of the long term wind rose.

3   Topography

A calculation of wind flow over a hill, or a series of hills, is a very complicated matter and much fundamental work on computational fluid dynamic tools is being devoted to this very subject. These types of calculation are not amenable to design and evaluation, which needs to be carried out in a rapid and cost effective fashion.

The wind speeds recorded on the site are compared with the Met Office wind speeds at a nearby station. From such a correlation, it is possible to calculate the speed up of the wind speed between the Met Office Station and the local site measurements. These figures are dependent on the wind direction, hence speed up ratios are calculated on directional basis, usually using a 30° sector. The purpose of the correlation is to allow the limited period of local measurements to gain access to the long term Met Office Station data. A common misconception in this process is that the absolute accuracy of the Met Office Station is of importance. In fact, provided the measurements made at the Met Station are consistent, the absolute accuracy is not important. The resulting wind rose therefore contains both wind speed and wind direction information which is a vital ingredient for the future analysis of the site.

There are various tools available for simpler calculations, which include the effects of roughness and topography, but do not include the possibility of viscous separation. If viscous separation effects are likely to occur on the site, then it may well be that the site is not suitable for development as a wind farm. The two most commonly used tools are both based on the theory developed by Mason and Sykes. One has been developed by the Canadian Met Office and is called MS3DJH/3R and the other, known as WAsP, has been developed by Risø National Laboratories in Denmark. The authors have used the two models back to back for certain applications and found them to agree very well. In theory WAsP will provide an absolute spot value for a point on the site, initiated from a local, approved, Met Office Station. In practice, in undulating terrain, these estimates are not sufficiently reliable to decide whether or not to proceed with the wind farm development. They are, however, a useful tool for "negative screening" of sites: if a WAsP estimate comes out to be particularly low, it is probably unwise to consider any further work on the site. GH uses MS3DJH to provide a complete map of the site, giving wind speed contours. It also provides information, as does WAsP, about the shape of the boundary layer on the site which can be useful in extrapolating from the height at which the measurements were made to another one.

GH often uses MS3DJH in conjunction with on-site measurements to extrapolate from the spot measurements of the meteorological mast to other points on the site where windmills might be placed. A typical set of wind speed contours on a wind farm site are shown in Figure 2. These contours have been normalised to show "speed-up ratios" rather than absolute values of wind speed. Such a plot enables the wind farm designer to determine the best place, from an ideal point of view, to place the wind turbines. The choice of optimum positions is not, however, simply determined by the location of the highest wind speeds.

4   Wake Interaction

A wind turbine removes energy from the wind and creates a wake. The wake has two important characteristics: reduced mean wind speed and increased turbulent energy. The reduced mean wind speed indicates a reduction in power production from machines operating in the wakes of others. The increased turbulence intensity results in an increased dynamic loading on the downstream machines. Both of these considerations are important. GH has developed a description of the development of a wind turbine wake, both in terms of mean and turbulent characteristics. The calculation of these parameters is performed using an eddy viscosity model, EVFARM, which has been validated against various measurements from wind farms in flat terrain.

A lot of detailed work in this respect has been performed as part of a European Community Joule Project, using a Dutch wind farm at Sexbierum, [1]. This wind farm contains eighteen identical machines with variable spacing as shown in Figure 3. A lot of detailed validation calculations can be performed looking at individual machines but, from the commercial wind farming point of view, the most important feature of a wind farm's behaviour is its energy production. The basic parameters which determine the wind farm's production are the wind resource covered above, and the wind farm power curve. Figure 4 shows such a power curve as a function of wind speed and direction. The values predicted using EVFARM are compared with those measured on the site. This power curve has been developed for the wind farm shown in the previous figure, Figure 3. It is quite clear, from this graph, that there are substantial drops in the power produced from the farm for specific wind directions, when the wind is blowing along a line of machines. This trend is, of course, no surprise, since the upwind machines are removing wind from the downwind machines. The ability to calculate this power curve is, however, very important since the total loss integrated over the whole wind rose may well be of the order of 5%-10%. The precise values depend on the size of the machines and their relative spacing.

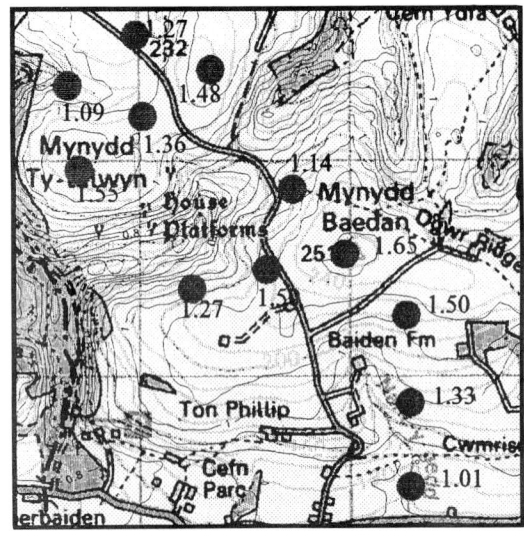

Figure 2   Wind speed contours

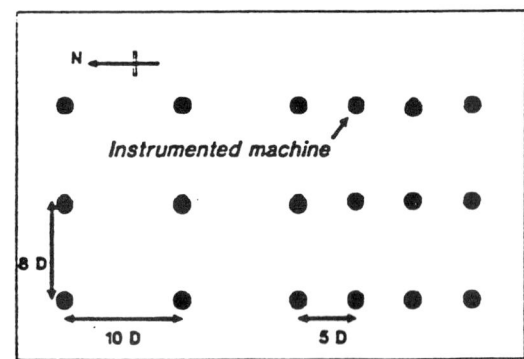

Figure 3   Layout of Sexbierum

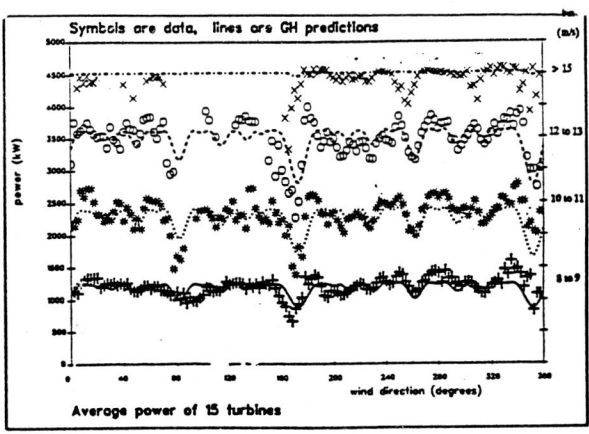

Figure 4   Wind farm power curve

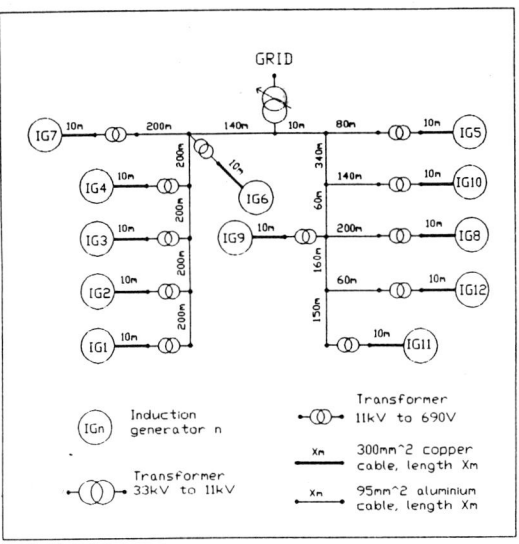

Figure 5   Typical electrical layout

In UK applications, the wake losses occur not in flat terrain but in more complicated topography. It is therefore necessary to combine the two calculation approaches: one which produces the wake interaction and the other which looks at the wind flow across the hills. To do this in a rigorous sense is very difficult. It is, however, possible to use a super-position approach whereby the mean wind speed calculations, produced by topographical models, are combined with the wake interaction models, by superposing the mean wind speed field. There is, to date, very little in the way of high quality measured data for wind farms in undulating topography. A CEC Joule II project entitled "Dynamic Loads in Wind Farms II" [2] started about a year ago in which this specific issue is addressed, in conjunction with that of the increasing dynamic loads resulting from wake interaction.

The evaluation of energy production for which the wind farmer is paid has not yet been addressed. It is a relatively simple matter to combine the wind speed distribution on the site with the power curve of the machine or, when a farm is considered, the power curve of the farm. The convolution of the wind speed distribution and the wind farm power curve over a year gives the annual energy production for the farm. The energy production for each of the machines in the example farm, ignoring electrical losses, is shown in Figure 2.

5   Electrical Losses

A wind farm is really a power station. It is a power station with particular characteristics, since it has a large number of small generating units dispersed over quite a wide area. These generators must be connected through transformers, usually to 11 kV cable network, and then the power collected at one substation usually working at 33 kV. There is therefore a potential for significant energy loss on the wind farm side of the Regional Electricity Company's meter, which is generally placed on the high voltage side of the main sub-station. A typical electrical layout is shown in Figure 5.

Electrical losses occur at each transformer and are dictated by the design and loading of that transformer. There may be 5km-6km of cable within a wind farm and so the electrical losses of the cables may also be significant. Finally the substation itself has electrical losses associated with it. A comprehensive analysis of the electrical behaviour of the wind farm can be performed to calculate real and reactive power flows and voltage drops in the network. This is normally carried out using the fast de-coupled load flow technique. Relatively simple loss calculations can, however, be performed without the necessity of undertaking a detailed load flow analysis. Garrad Hassan has developed such a code, known as ELOSS, which conforms to the Danish Standard DEFU77, and which has been calibrated back-to-back with a detailed load flow analysis undertaken by UMIST. These calibration calculations have demonstrated that the simple approach which is very rapid, agrees in loss terms to within half a percent of the more comprehensive analysis for typical wind farm conditions of circuit impedance and loading. Garrad Hassan would normally undertake a full load flow analysis when the design had been finalised but use the faster parametric technique for design calculations.

The electrical loss calculation incorporates all the items described above. Spot values of power are calculated for each machine for each of 36 wind direction sectors for a range of wind speeds. The power produced by each individual machine is then fed into the electrical loss calculation and the net power produced by the farm is calculated. This process is repeated for all the points on the wind rose and may result in, perhaps, 400 spot calculations performed for each machine. The net energy produced by the wind farm as a whole can then be compared with the energy which that farm would have produced in the absence of any electrical losses. For a well designed wind farm electrical losses usually account for between 2%-3% of the total gross energy. By making detailed measurements within a wind farm, GH is in the process of calibrating these loss calculations against full scale measurements.

6 Visual considerations

There is increasing concern about the environmental effect of wind farms. The main issues addressed and those which are amenable to analytical consideration are visual impact and noise. It is a vital part of an environmental assessment undertaken as part of a planning application to be able to provide an objective assessment of these two essential characteristics of a wind farm.

Topographical information is required in order to perform the energy calculations described above. That same information, usually purchased from the Ordnance Survey, can be used to calculate the visual envelope of a farm. Normal procedure is to purchase OS data for terrain surrounding a wind farm to a distance of, perhaps, 10km. The layout of the farm can then be used to calculate the zones of visual influence, ZVI shown in Figure 6. The shaded part of the terrain indicates all the points from which any part of the wind farm can be seen. The variation in density of shading, indicates the number of machines which can be seen from any individual position. Such a calculation is useful for determining the locations at which photo-montages would later be assembled to give an idea to the planning authorities of the likely visual impact of the farm. The ZVI is a means whereby important or sensitive views can be located. Using the ZVI approach, combined with a suitable analytical or numerical model, it is possible to move some elements of the wind farm to minimise the visual impact of the development. This procedure is not, of course, peculiar to wind farms, but is a standard procedure used for many new developments in sensitive terrain.

7 Noise

Noise has become an increasingly sensitive issue in the development of wind farms. It has become a standard procedure for manufacturers to quote the noise source power of their machines together with the associated octave band analysis. This is an approximate representation of the noise produced by the machine. Using this approach, the machine is represented by a single source on top of the tower. Although approximate it appears to give reasonably good agreement with measured data. Use of this simple representation allows the noise envelope around a wind farm to be computed rapidly. Noise contours, in dB(A), for the example wind farm, are shown in Figure 7. A potential wind farm site is usually chosen to be distant from any sensitive dwellings or other important landmarks, where it is important that there is no noise nuisance.

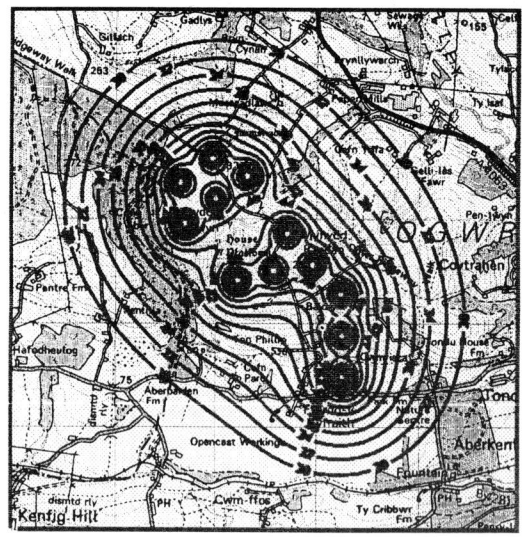

Figure 7   Noise contours

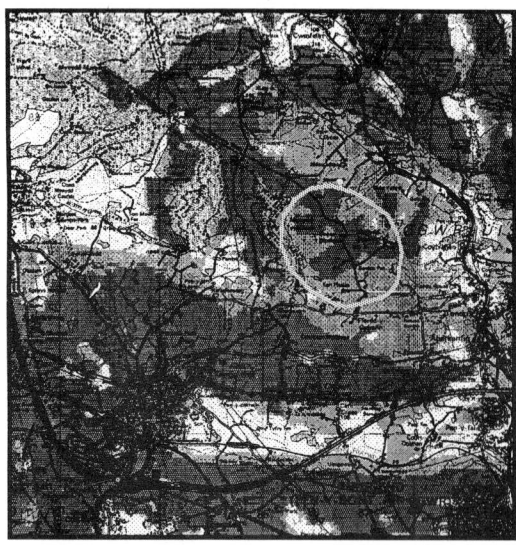

    ▒ = up to 33%
        of WTGs

    ▓ = 33 to 66%
        of WTGs

    █ = 66 to 100%
        of WTGs

       = Site

Figure 6   Zones of visual influence

Again, this approach is rapid and simple and can be used effectively for wind farm design. Unfortunately, in the case of noise, the precise noise figures at any location are dependent, not only on the noise source power, but also on the local topography, ground cover and air quality. It is, however, unsafe to assume that any particular point, which might not be visible from the wind farm, will not be affected by noise. It is therefore necessary to adopt a conservative estimate of the noise envelope. The assessment of the noise levels at a site has to be made using the predicted levels described above and measured background levels. A standard means of assessment of the noise from wind farms has not yet been developed in the UK and this area remains one in which considerable uncertainty still exists. However, by any means of assessment, the example wind farm would clearly be unacceptable!

8 Conclusion

This paper has described a series of computer programs which allow all the salient features of a wind farm to be assessed rapidly and with a high level of reliability. These programs allow not only the energy production and resource but also the environmental impact of the farm. For ease of explanation the programs have been described in their "analysis" mode rather than as a means of optimisation. Clearly they represent powerful tools for the optimisation of a wind farm within the external constraints.

The environmental calculations are as important as the financial calculations, since there is little to be gained by designing and erecting an efficient and therefore profitable wind farm, if it runs into planning and public acceptability difficulties. In the UK we are now entering a period in which it will be important to assess the efficacy of these different approaches. It is likely that all of these tools, which have been used to design and assess potential wind farms, will undergo improvement as a result of field trials.

REFERENCES

1   Tindal A, "Dynamic Loads in Wind Farms", Report of JOULE contract number JOUR-0084-c, August 1993.

2   "Dynamic Loads in Wind Farms II", JOULE contract number JOUR-CT92-0094

A method for the aerodynamically optimal design of wind parks

S. G. VOUTSINAS and **K. G. RADOS**
National Technical University of Athens, Greece

SYNOPSIS A method for designing aerodynamically optimal wind parks is proposed. The method consists of positioning the wind turbines so as to maximize the energy they absorb or equivalently minimize the loss of energy due to wake effects. To this end, the optimal design problem is formulated as a constrained minimisation problem for the functional that represents the loss of energy due to wake effects, the latter being modelled by means of an improved kinematic wake model. The minimization problem thus obtained, is solved by a discrete version of the gradient method. Numerical results are presented for three sites in Greece with different wind roses so as to cover a wide range of possible wind data. In all cases, the efficiency of the optimal configurations was found greater than that of the configurations of uniform spacing.

1 INTRODUCTION

The promotion of wind energy as an alternative resource, is closely related to the design of reliable and cost effective wind parks. To this end, the design of a wind park or equivalently the lay-out of its wind turbines in space, should maximize the energy output as well as the life-time of the machines. For flat terrains, both issues are basically affected by the development of wakes downstream of every turbine, i.e. regions exhibiting velocity deficit and high turbulence levels. Consequently inner wind turbines will produce less energy as compared to the production of a stand alone machine, as a result of their reduced inflow. Moreover the additional fluctuating wind loads due to the higher levels of turbulence, would probably diminish their fatigue life-time. Therefore the design and particularly the optimal design of wind parks is based on the modelling of wake effects. In this connection we must be able to predict: (1) The velocity field in the wake of a stand alone turbine (the wake model), (2) The velocity field established when the wakes of several turbines overlap (the interaction model), and finally (3) The turbulence intensities and the length scales that are needed for fatigue analysis (the turbulence model).

As regards wake models, several ones exist, of varying complexity. They can be classified in two general categories: (a) the explicit or kinematic wake models[1÷3], and (b) the implicit wake models[4÷6]. Explicit models are the most widely used, because of their simplicity and low computational cost, two indispensable features of all wake models intended for analysis and design of wind parks. The main advantage of explicit models is due to the closed-form expressions that are used for calculating the velocity field. This is a direct consequence of the self-similarity of the velocity profiles, an assumption that can be made only for the far-wake region (i.e. the region the flow has fully developed). Following the literature on the subject, in [3]

we have proposed SWAKE: an improved explicit wake model that was validated with respect to the Nibe experiments. The results reported therein suggested that, in spite of its simplicity, this explicit wake model can provide reasonably accurate velocity predictions. This is particularly true in the far-wake that happens to be the region of interest for wind park analysis. That is why we have introduced this model to most of our developments on wind park analysis. Concerning implicit models, they were developed as elaborate alternatives of the explicit models, aiming at providing a more detailed account on the formation and development of the wake of a wind turbine. They were based on approximations of either the Navier Stokes [4,5] or the unsteady vorticity equations [6]. Due to their complexity and their considerable computational cost, these models were basically used for the cross-checking and calibration of the explicit models.

As regards interaction models, instead of a linear superposition, an energy interpretation was adopted that led to a quadratic interaction relation [3]. This relation together with a depth averaging technique for taking into account ground effects complete the "kinematic" modelling of wake effects. Therefore, in terms of energy production, both the performance and the design problem of a wind park can be formulated. We have specified the corresponding optimization problem as the "aerodynamically optimal design problem" because of the close connection between energy production and aerodynamics. Fatigue analysis and consequently turbulence modelling was not included in the present work (for details see [7,8]).

The Aerodynamically Optimal Design Problem was formulated in [9] as a constrained minimisation problem for the functional of the energy loss (cost function). The constraints introduced comprised geometrical inequality constraints as well as operational constraints for the wind turbines of the park. Also in [9] a localised gradient method was proposed, for solving this design problem. Basis of the method was the decomposition of the full problem into a sequence of intermediate ones. In each intermediate problem-step the configuration of the wind park was determined so that the Annual Energy Production (AEP) of all turbines was greater than a prescribed value. Because the lower bound for AEP of every turbine is closely connected to the number of turbines of the park, this sequence of intermediate problems served as a tool (*search function* of the algorithm) for converging towards the desired final number of turbines. The results presented in [9,10], showed that at least for the tests considered therein, a considerable energy gain could be accomplished. However, definite conclusions could not be made, since the wind data of the tests had only two wind directions of equal probability. Therefore a systematic study was initiated, for realistic wind data. It was found that as the number of wind directions increased, convergence of the scheme became difficult. Moreover, in some cases the iterations had an oscillating behaviour. Thus the method was reconsidered leading to an improved calculation of the velocity performed at intermediate stages of the iterative procedure. Moreover a second step was added. More specifically as soon as the desired number of turbines is reached, a non-local gradient method is applied to give the final configuration.

Application of this improved method, showed that not only convergence is ensured but also an even greater gain in energy production is accomplished. Regarding the mathematical characteristics of the scheme, one could argue that the configurations obtained are optimal since the final solution is accomplished by means of a nonlocal gradient method. Nevertheless, the fact that the output of the optimal configurations obtained is in all cases greater than that of the configurations of uniform spacing, indicates such an assertion.

2 FORMULATION OF THE DESIGN PROBLEM

Let us consider the problem of installing N turbines in a site with planeform Ω. Given the wind potential of the site as a probability density function $p_u = p_u(U_w, \Theta_w)$ and the power curve of the turbines to be used $P(U)$, the annual energy production of the wind park E as well as the energy losses due to wake effects E_L are obtained from:

$$E = 8760 \int_0^{2\pi} d\Theta_w \int_0^\infty dU_w p_u(U_w, \Theta_w) \sum_{i=1}^N P(U_i) \qquad (1)$$

$$E_L = 8760 \int_0^{2\pi} d\Theta_w \int_0^\infty dU_w p_u(U_w, \Theta_w) \sum_{i=1}^N [P(U_w) - P(U_i)] \qquad (2)$$

$p_u(U_w, \Theta_w)$ gives the number of hours per year that the wind blows from the $\Theta_w \in [0, 2\pi)$ direction with mean velocity $U_w \in [0, \infty)$. U_i denotes the mean inflow to the i-th

turbine that is a function of Θ_w and U_w. Clearly $\{U\} = \{U_i, i=1(1)N\}$ constitutes the basic unknown. In order to determine U, the wake modelling described in [3] is used. In brief, the inflow velocity U_i is determined by setting the total energy loss on the i-th turbine equal to the sum of the independent losses from all the overlapping wakes at:

$$(U_w - U_i)^2 = \sum_{j=1, j\neq i}^N \left(U_j - (u_j)_i\right)^2 + \Delta u_g^2 \qquad (3)$$

In (3), $(u_j)_i$ denotes the velocity that the j-th turbine induces to the i-th turbine. According to theory[3], the velocity $(u_j)(x,y)$ at a point (x,y) in the wake of a turbine, e.g. the j-th, is given in closed form and depends on: its mean inflow U_j, its initial deficit ΔU_j and the ambient turbulence intensity α. If U_j and α are assumed known, ΔU_j is determined by means of a modified Bernoulli equation that uses the power curve of the wind turbine. Therefore, ΔU_j also depends on U_j. Consequently we conclude that $u_j = u_j(x,y \mid U_j)$ or else that $(u_j)_i = u_j(X_i, Y_i \mid U_j)$, where (X_j, Y_j) denote the co-ordinates of the j-th wind turbine. The last term in (3), is added to account for ground effects and depends on U_w (see [3] for details). By means of (3), the incident velocities U_i are determined and thus E and E_L can be evaluated.

3 THE NUMERICAL SCHEME

Usually the constraints imposed to the designer are of economic nature. In most cases these constraints can be translated to a given nominal power of the park or equivalently to a given number of turbines for every type of turbine among those considered as candidates for installation. By specifying the type of the turbine, the power curve is also specified. Therefore, for given wind rose, the optimal design problem is formulated as follows:

Find $\{Z\} = \{Z_i = (X_i, Y_i), i=1(1)N\}$ that minimize E_L under the geometrical inequality constraints $Z_i \in \Omega$.

It is simple to observe that with no additional constraints, non-admissible configurations from the engineering point of view are possible. This is expected, because minimization of the energy losses due to wake effects is equivalent to the maximization of inter-machine spacing. In order to ensure admissible configurations, a minimum for the spacing of the turbines is imposed. Computationally this is accomplished by means of a cartesian grid. The spacing of the grid is chosen so that only one turbine can be placed in every of its cells. Note that in order to account for all the wind directions, the cells of the grid are constructed to be of almost equal area. This is an additional constrain of geometrical nature. Based on the above assumptions, the optimization problem is reformulated as follows:

Let C denote the set of the centroids of all the cells of the grid. Then the positions of the wind turbines are given by:

$$Z_i = Z_{io} + \Delta Z_i, \quad Z_{io} \in C, \quad |\Delta Z_i| < \epsilon \qquad (4)$$

where ϵ denotes the upper bound of the spacing. Z_{io} defines the cell within which the i-th turbine is set. Computationally, this is declared by assigning to each turbine the number of the cell to which it belongs. The exact position of the i-th turbine is defined by ΔZ_i that can be interpreted as a virtual displacement. A rather direct method for solving the minimisation problem can be formulated by searching all possible combinations defined in C. However we do not dispose a search (or selection) function. As the number of these configurations can be large, such an unguided method will converge slowly. In order to construct a more efficient algorithm, we introduce the following auxiliary problem:

Given a spacing or equivalently a grid C, find the number N_k and the positions of the wind turbines $\{Z\}_k$ that all operate with annual efficiency greater than a prescribed minimum value $\{\eta_{eff}\}_k$.

Clearly this is an intermediate problem because its solution, i.e. the lay-out of the wind turbines obtained, depends on $\{\eta_{eff}\}_k$. Clearly the value of $\{\eta_{eff}\}_k$ that corresponds to the solution of the optimal design problem $\{\eta_{eff}\}_{opt}$, is unknown. Assuming that $\{\eta_{eff}\}_{opt}$ is uniquely determined by the desired number of turbines N, it can be determined by an iterative scheme. In every iteration k an intermediate problem is solved and then the lower bound $\{\eta_{eff}\}_k$ is modified accordingly. Convergence of the iterative scheme is assumed when N_k reaches the desired number of turbines. Note that in the general case where the whole of the wind rose (i.e. all the wind directions) is taken into account, every intermediate problem is a non-linear elliptic one, because, on the performance of a turbine on an annual base, is influenced by all the other machines. In order to overcome this difficulty, the following procedure is introduced.

Let (X_i^k, Y_i^k), $i=1(1)N^k$ denote the configuration obtained after the k-th iteration has been completed. If the annual energy efficiency of all the turbines is greater than $\{\eta_{eff}\}_k$ the procedure is terminated. In the opposite case all the cells of the grid are checked ($k+1$ iteration). For every cell the gradient of the cost function is considered in order to determine the best position. This constitutes a localised version of the gradient method. Clearly if the efficiency constraint is not satisfied, the current cell is excepted. In this case the number of turbines is decreased by one. With a specific cell as starting point, all the cells are checked sequentially. The sequence is defined so that at any stage of the procedure, the next cell to be checked is adjacent to the subregion defined by the cells already checked. This means that the checking of the cells is performed in spiral strips. For every strip, the velocity at a point is defined by:

the contributions of the turbines that were placed in the inner strips during the $(k+1)$-iteration, and (b) the contribution of the turbines of the k-iteration contained in the current and outer strips. As soon as a strip is completed, the new turbines take the place of the old ones that were included in this strip. Then the velocity field within the entire domain is corrected. This procedure can be regarded as a semi-elliptic scheme. Fully elliptic schemes have been tested as well. However the rate of convergence is low, leading to exceedingly high computational cost. The scheme can be initialized either by filling all cells with turbines and choosing a low $\{\eta_{eff}\}_1$, or by placing just one turbine with a low $\{\eta_{eff}\}_1$. This constitutes the first step of the whole scheme.

The result of the first step, is considered as an initial approximation of the optimal configuration. In order to obtain the final solution, the minimisation of E_L is performed by a non-local gradient method [11]. Let $\nabla_i E_L$ denote the gradient of E_L with respect to the positions of the wind turbines Z_i. Then a new approximation of the optimal configuration can be obtained by moving every turbine along the direction of the corresponding component of $\nabla_i E_L$, by minimising the cost function. Since the spacing constraint should not be violated, the displacement of every turbine can be made only to those cells of the grid that are free. Moreover in order to ensure minimisation of the cost function it suffices that at the new position the turbine operates at a greater efficiency. This constitutes the second and final step of the scheme.

4 RESULTS

In the sequel results are presented for the optimal design of a wind park within a 1000x1000 (m^2) region, for three different wind roses, denoted as cases A, B and C, corresponding to specific sites in Greece where the installation of wind parks is planned:

Case A: Ano Moulia, in the island of Crete (Fig. 1 $\div$ 3)
Case B: Kalivari, in the island of Andros (Fig. 4 $\div$ 6)
Case C: Volimes, in the island of Zakynthos (Fig. 7 $\div$ 9)

In all cases, a wind turbine with 34m diameter and 400kW rating power, was used. The validation of the method is based on the absolute and relative gains G_a and G_r obtained by the optimal configurations with respect to those of equal inter-machine spacing.

$$G_a = \frac{E_o - E_u}{E_u} \qquad G_r = \frac{E_o - E_u}{1 - E_u} \qquad (5)$$

where E_o and E_u denote the annual energy production of the optimal and uniform configuration respectively.

The wind rose of case A exhibits three dominant wind directions: N, NNE and WSW. Consequently, the method places the wind turbines approximately along a line normal to the diagonal defined by the N-NNE and the WSW directions (Figure 3). For 20 machines this is quite clear whereas for 40 and 60 most of the turbines are placed in the Northern region while a second group of machines is positioned in the West-South-Western region. This behaviour is expected, since the method tends to maximize the distances between the turbines of the park along directions normal to the dominant ones. This fact assures the minimization of the energy loss due to wake effects of the upstream turbines.

In the second wind data set (Figure 6) the wing energy is concentrated into two wider and almost opposite regions: the NNW-NNNE and the SW-SSW-S. Again as expected, the method places the turbines into the Northern and Southern region of the perimeter of the park. As the number increases, the density of these two groups of turbines also increases. Moreover, the initial gap that separated these two groups is decreased.

Finally, in case C, the wind rose (Figure 9) is characterized by a distribution of the wind energy in almost all the wind directions. Therefore a more distributed and rather structured configurations are expected. It is clear from the configurations obtained, that as the number of the turbines increases the method places the turbines in approximately equidistant rows.

Regarding the quantitative features of the results, the following remarks can be made:

- As the number of turbines increases the park efficiency of the uniform configuration must decrease since the energy loss due to the formation of wakes is pronounced. This means that the efficiency of an equally spaced configuration decreases. Consequently, for a method designed to give optimal solutions, the margin of energy loss to be recovered is greater. This should result the increase of the absolute gain with the number of turbines (Figures 1, 4, 7). Clearly this cannot characterise the variation of the relative gain, that decreases with respect to the number of turbines. This is due to the fact that the loss of energy due to wake effects increase as the number of turbines increases since more intensive wake interactions are expected. Thus, the upper bound of the park efficiency, that can be reached by the optimization method, decreases.

- From Figures 2, 5, and 8 we can observe that the trend of the difference between the efficiencies of the uniform and the optimal configurations is approximately the same for all the cases considered. This is related to some extend to the method used. As already pointed out the main guideline of the method is of geometric nature, i.e. the method places the turbines in order to maximize their mutual distances. Therefore the method is expected to behave independently from the wind data.

Regarding the computational cost of the method, it is noted that the code was designed to run on a PC computer. The computer time required was found to be proportional to the number of wind directions considered. For example for a PC running at 33MHz the CPU time required for 60 turbines is approximately 15 min per direction. Finally, as regards the size of the working area needed, the limit of 2Mbytes was never exceeded.

5 CONCLUSIONS

A simple and computationally cost effective method for the optimal design of wind parks was presented.

Although we have not proved that the configurations numerically obtained are strictly optical, compared to the best uniformly spaced configurations a portion greater than the 30% of the energy losses due to wake effects is recovered.

REFERENCES

(1)   LISSAMAN, P.B.S. Energy effectiveness of arbitrary arrays of wind turbines. *Journal of Energy*, 1979, **3**, 323.

(2)   KATIC, I., HOJSTRUP, J., JENSEN, N.O. A simple model for cluster efficiency. *Proc. of EWEC'86*, 1986, Rome.

(3)   VOUTSINAS, S.G., RADOS, K.G., ZERVOS, A. On the analysis of wake effects in wind parks. *Wind Engineering*, 1990, **14**, 205-218.

(4)   SMITH, D., TAYLOR, G.J. Further analysis of turbine wake development and interaction data. *Proc. of 13th BWEA Conference*, 1991, Swansea.

(5)   CRESPO, A., MANUEL F., HERNANDEZ, J. Numerical modelling of wind turbine wakes. *Proc. of EWEC'90*, 1990, Madrid, 166-170.

(6)   ZERVOS, A., HUBERSON, S., HEMON, A. Three dimensional free wake calculation of wind turbine wakes. *Journal of Wind Engineering and Industrial Aerodynamics*, 1988, **27**, 65.

(7)   VOUTSINAS, S.G., RADOS, K.G., ZERVOS, A. Wake effects in wind parks. A new modelling approach. *Proc. of ECWEC'93*, 1993, Travemunde, 444-447.

(8)   CRESPO, A., HERNANDEZ, J. Analytical correlations for turbulence characteristics in the wakes of wind turbines. *Proc. of ECWEC'93*, 1993, Travemunde, 436-439.

(9)   VOUTSINAS, S.G., RADOS, K.G., ZERVOS, A. Prediction of the energy effectiveness and optimal design of wind parks. *Proc. of ECWEC'89*, 1989, Glasgow, 527-531.

(10)   VOUTSINAS, S.G., RADOS, K.G., ZERVOS, A. The effect of the nonuniformity of the wind velocity field in the optimal design of wind parks. *Proc. of EWEC'90*, 1990, Madrid, 181-185.

(11)   POLAK, E. Computation Methods in optimization. *Academic Press*, 1971, New York.

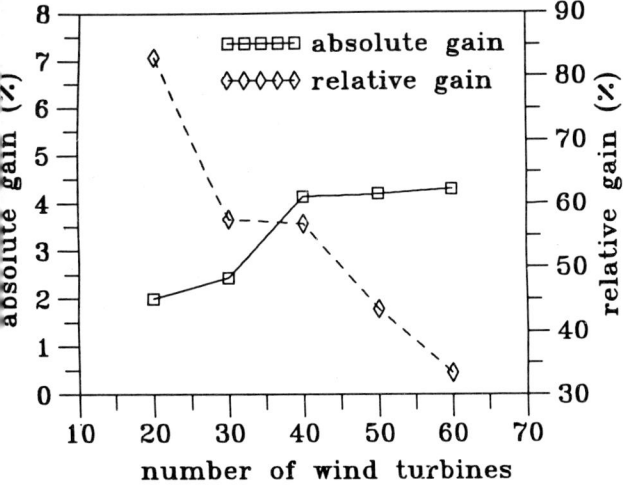

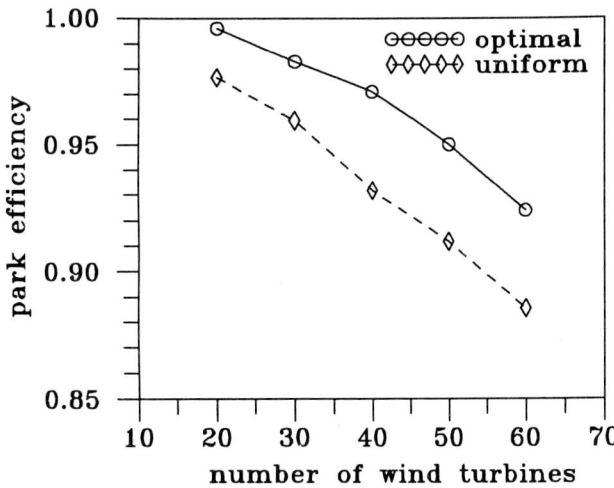

Figure 1 Variation of the absolute and relative gain with respect to the number of turbines (case A)

Figure 2 Variation of the efficiencies of the optimal and the uniform configurations with respect to the number of turbines (case A)

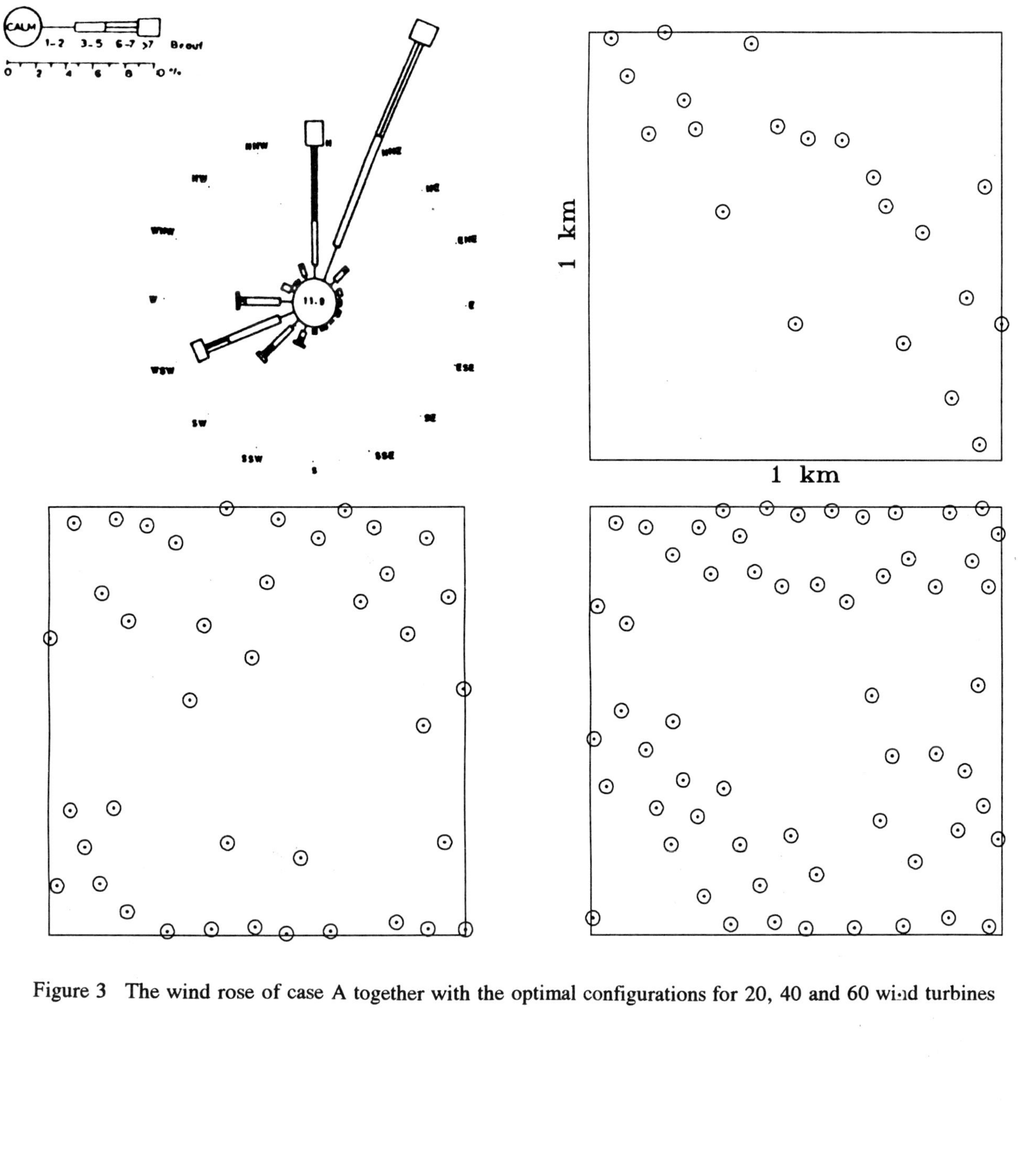

Figure 3 The wind rose of case A together with the optimal configurations for 20, 40 and 60 wind turbines

235

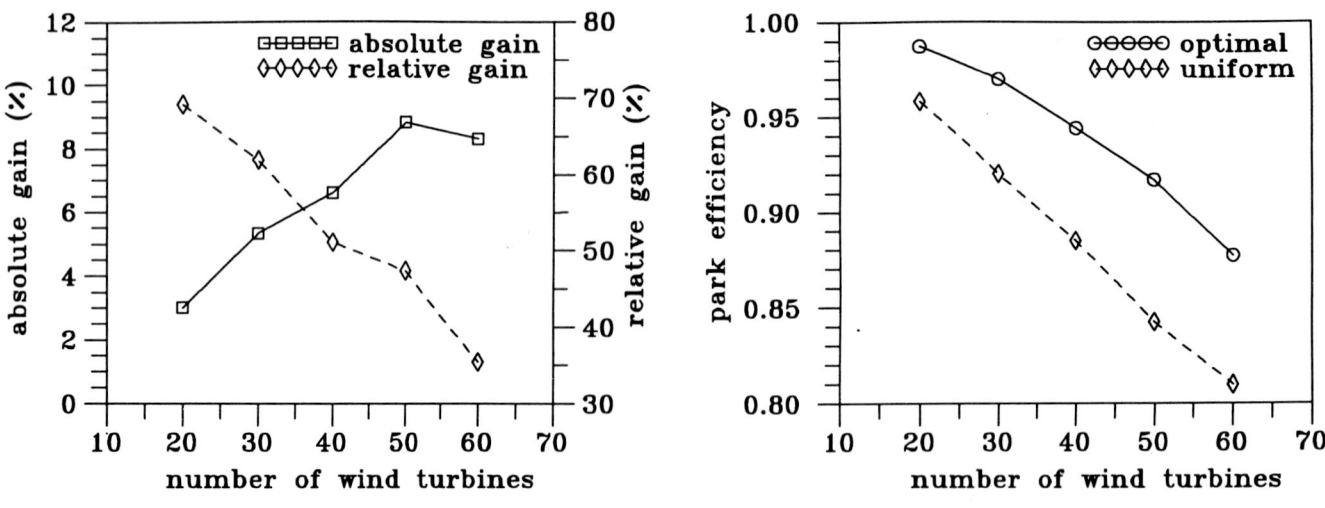

Figure 4 Variation of the absolute and relative gain with respect to the number of turbines (case B)

Figure 5 Variation of the efficiencies of the optimal and the uniform configurations with respect to the number of turbines (case B)

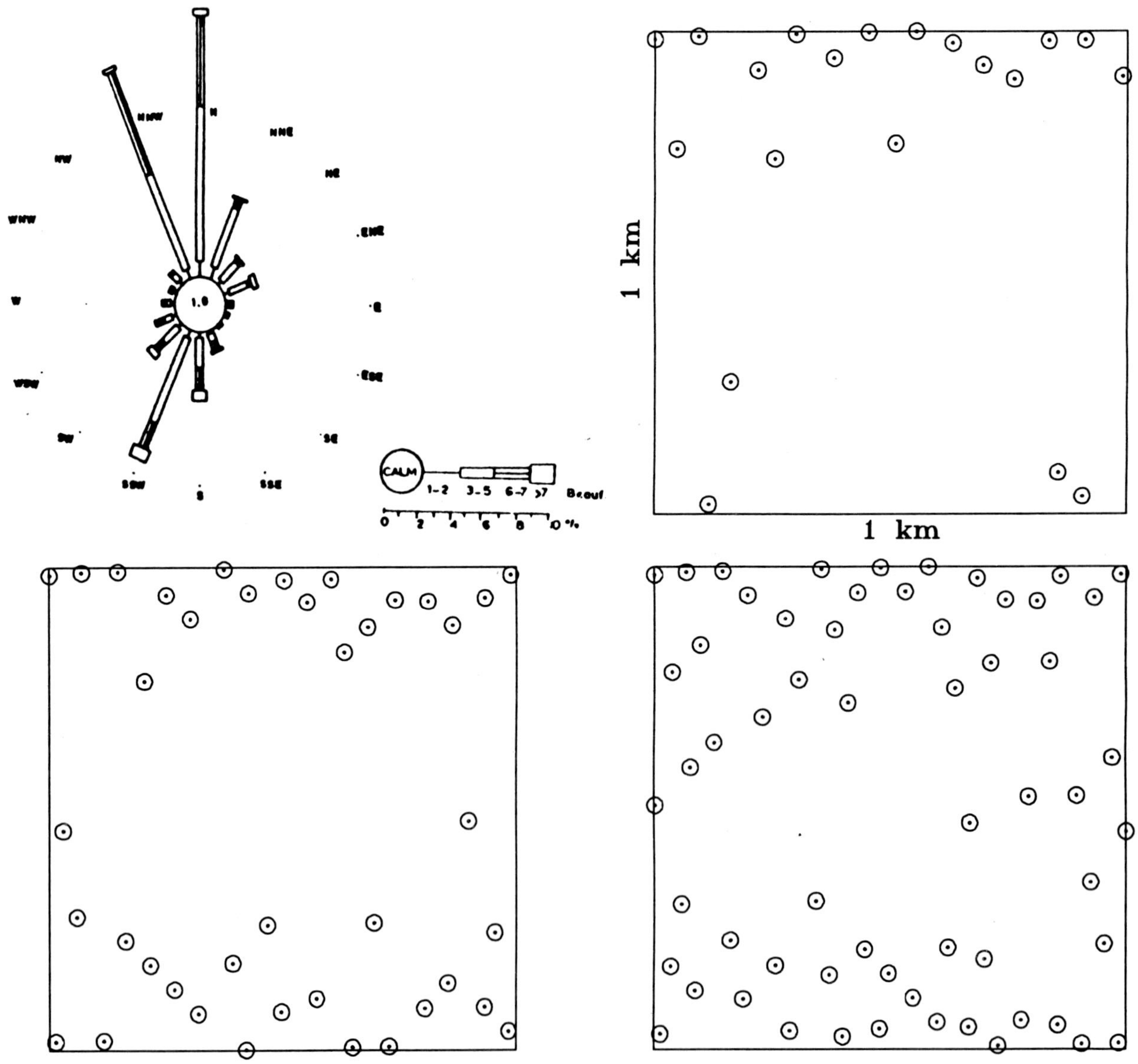

Figure 6 The wind rose of case B together with the optimal configurations for 20, 40 and 60 wind turbines

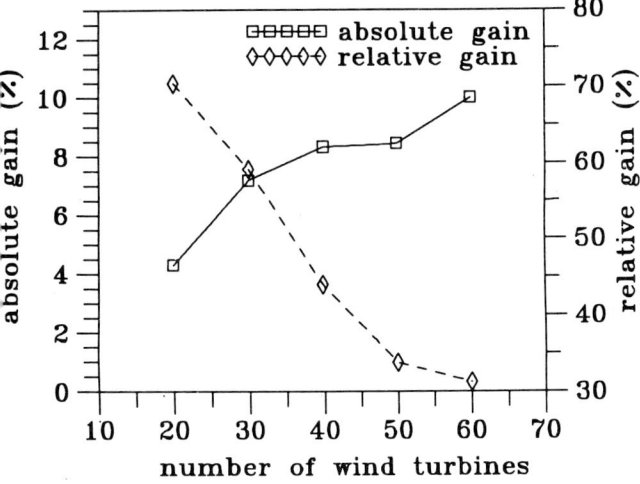

Figure 7 Variation of the absolute and relative gain with respect to the number of turbines (case C)

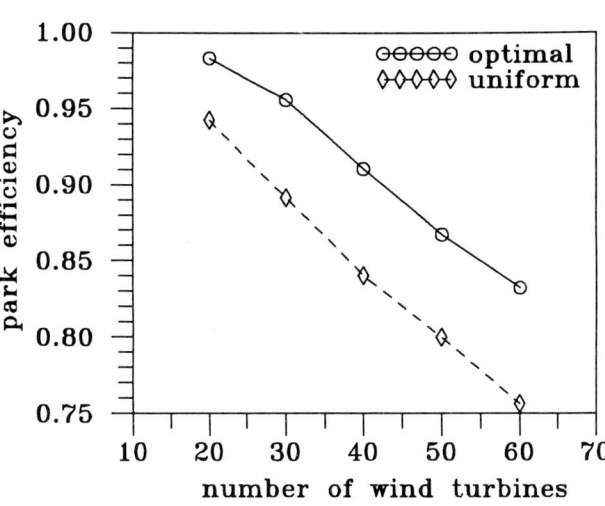

Figure 8 Variation of the efficiencies of the optimal and the uniform configurations with respect to the number of turbines (case C)

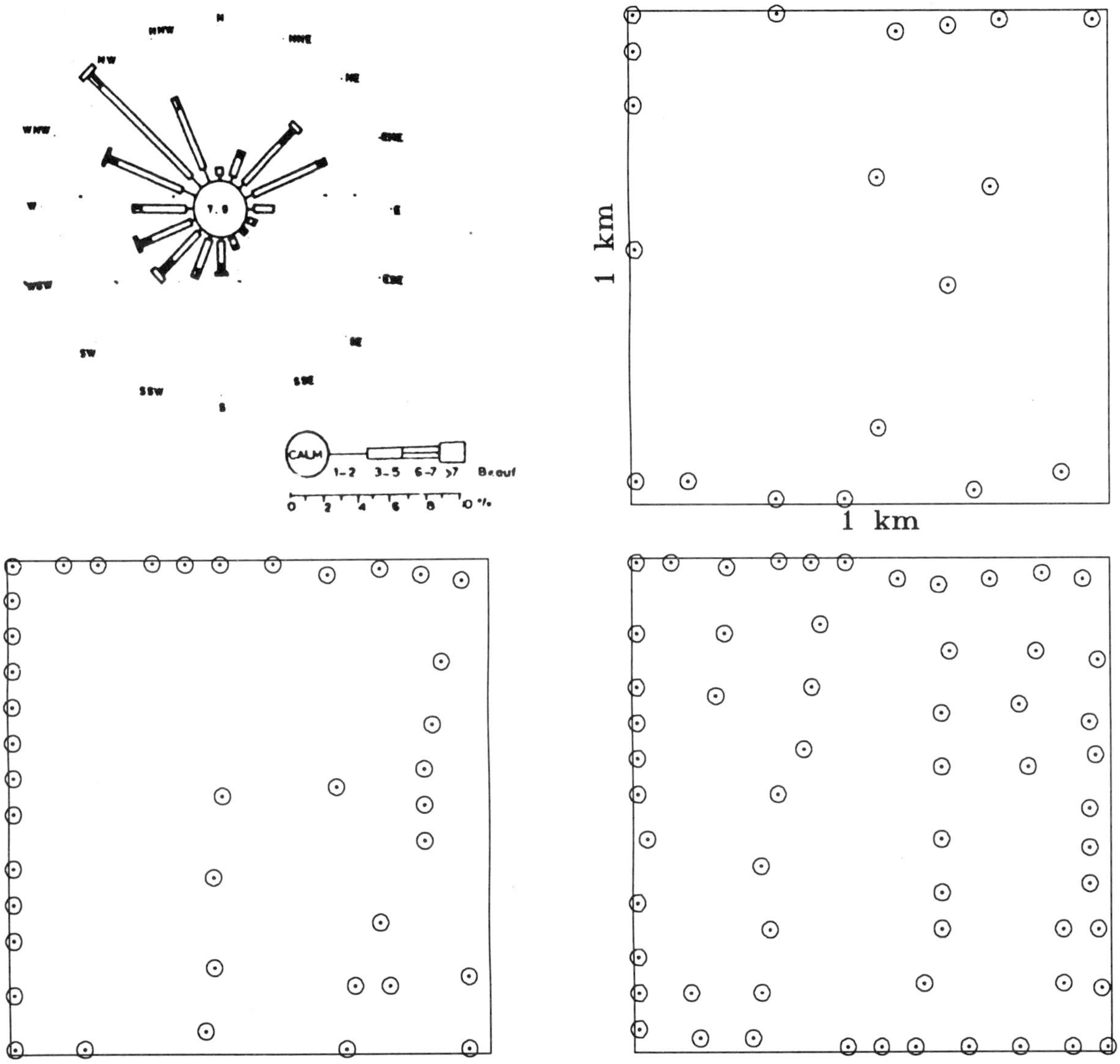

Figure 9 The wind rose of case C together with the optimal configurations for 20, 40 and 60 wind turbines

Economically optimized design of the electrical layout in wind farms

J. KRISTIANSEN, BSc. Elec. Eng.
Elsamprojekt A/S, Denmark

SYNOPSIS Development of windfarms implicate several phases and activities from idea to operation of the wind farm. When evaluating the project economy some of the keypoints are: optimization of the power output and the capital investment, completion of the project in time and within budget in order to obtain the minimum total lifetime costs of the project. This paper deals with considerations regarding the optimization of the electrical layout in order to achieve the minimum lifetime costs for the power collecting system. The electrical layout of the power collecting system in wind farms has been investigated with regard to the value of the electrical losses. Experience gained by development and design of wind farms has shown that, by means of marginal efforts, a considerable reduction in lifetime costs for a wind farm project is obtainable if the electrical layout is designed on basis of capitalized losses. For that purpose the WIFAGO design tool has shown to be very suitable in order to optimize the layout and to reduce the required time consumption in the design phase.

Furthermore, the investigations has shown that the electrical layout for a wind farm on a specific site is highly dependable on local conditions such as prices for selected equipment, interest rate, energy price, soil compound regarding excavation etc. and wind speed distribution. The amount of electrical losses has an important influence in choice of layout. The alternative with the lowest electrical losses is not necessarily the optimum choice when the layouts are evaluated with regard to lifetime costs comprising equipment cost and capitalized electrical losses.

1. INTRODUCTION.

The development of projects for generating power and energy in general increases the demand for an environmentally sound solution. New schemes are now beeing evaluated based on technical, economic and environmental criteria. One of the technologies satisfying future requirements is wind energy as the cost of generating wind power is becoming more competitive. Furthermore, fine wind conditions and favourable energy prices makes it profitable to establish wind farms in Great Britain.

Development of windfarms implicate several phases and activities from idea to operation of the wind farm. Some of the key points for a sustainable project are:

- Optimization of the power output regarding wind farm layout and electrical layout.
- Minimize the capital investment.
- Completion of the project in time and within budget.
- Secure agreements and approvals for works and necessary permissions.

When evaluating the project economy the optimization of the power output and and capital investment is essential in order to obtain the minimum total lifetime cost of the project. This paper deals with considerations regarding the optimization of the electrical layout in order to achieve the minimum lifetime costs for the power collecting system.

The increasing size of wind turbines and windfarms makes it profitable to evaluate the layout of the electrical power collecting system, not only based upon voltage variations, but also with regard to the electrical losses. The value of losses together with costs for the installed equipment are investigated in order to achieve the minimum lifetime costs for a wind farm.

Experience gained, by development and design of wind farms have shown that, by means of marginal efforts, a considerable reduction in lifetime costs for a wind farm project is obtainable if the electrical layout is designed on basis of capitalized losses.

For that purpose a design tool has been developed. WIFAGO (WInd FArm Grid Optimizing) is a design tool for optimizing the electrical power collecting system in the wind farm. The WIFAGO design tool has shown to be very suitable for optimizing the layout and to reduce the required time consumption in the design phase.

On the basis of general information about the wind farm and wind statistics etc., the tool gives proposals for the economically optimum selection of equipment. To determine the overall layout the considered layout proposals for the electrical power collecting system are evaluated during iterations making it possible to compare different configurations. When the overall electrical layout is determined each piece of equipment is optimized on basis of equipment costs and electrical losses.

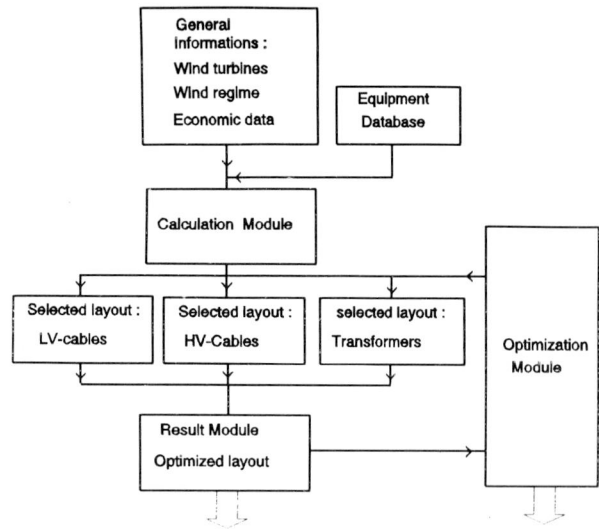

Fig. 1 Wifago optimization tool.

The paper presents investigations made and considerations, regarding the above. The investigations are based upon evaluations and comparisons of several proposed layouts for Taff Ely Wind farm in Southern Wales.

The wind farm consists of 20 450 kW Nordtank wind turbines equipped with induction generators The induction generators are partly compensated for their reactive power consumption.

2. OPTIMIZATION.

2.1 Optimization method for windfarm layout.

Prior to the design of the electrical layout in the windfarm a resource assessment has been performed. The wind turbines are located on basis of terrain, wake effects and the prevailing wind directions in order to assess and optimize the windfarm production.

For that purpose the well known resource assessment tools WAsP and PARK developed by RISØ National Laboratory are used. Further to these tools special design tools has been developed to assist in both resource assessment and optimization of windfarm layout. These special design tools makes it possible to obtain a detailed map of the wind energy potential in the selected area and to optimize the windfarm layout by moving the wind turbines around on a digitized map.

Especially in mountainous terrain the resource assessment is complicated and the economy of the project is much more sensitive to changes in layout than for a windfarm in flat terrain. Therefore, a reliable resource assessment is essential. This is obtained by using a measure - correlate - predict method by means of the the above tools.

Experience gained from using the above tools for resource assessment and optimization of the layout of Taff Ely Windfarm in Southern Wales show good correlation between measured wind data on the site and long term statistic wind data from measurement stations located with a cosiderable distance to the site, if these data are extrapolated by WAsP to the site. Using the special design tools to optimize the windfarm layout increased the energy output considerable compared to the initial layout.

2.2 Optimization method for electrical layout.

Based on the location of the wind turbines, access roads etc., the electrical layout of the wind farm is optimized during several iterations making it possible to compare different configurations.

If the wind farm layout consists of identical rows or matrixes the wind farm may be divided into groups represented by an equal number of wind turbines, identical lines and transformers. Each transformer or cable within the group is separately optimized. This procedure simplify the calculations to be performed.

With a given wind speed distribution for the area and the power curve for the wind turbines, a load duration curve is calculated, allowing the energy losses in the power collecting system to be calculated.

2.3 Input parameters for optimizing the electrical layout.

In order to design and evaluate the different layout alternatives, it is essential to identify the parameters influencing the layout. For the optimization tool data has to be collected and evaluated regarding their influence on the layout. Examples of topics to be considered are mentioned below:

General information on the windfarm:

Characteristics for wind turbines:
- Power curve for active and reactive power.
- Nominal voltage.
- Reactive power compensation system, size and steps.

Economic informations on equipment and works:
- Equipment costs: Cables, transformers, switchgear etc.
- Local wages regarding the electrical works.
- Soil compound regarding costs of excavation for cable trenches and foundations.
- Topography and roads on site regarding cable routing.

To determine the capitalized lifetime costs of the losses in the power collecting system, the following information is needed:

- Development in energy price
- Interest rate
- Wind speed distribution
- Project lifetime
- Equipment characteristics

The losses in all components are calculated and capitalized using the NPV-method (Net Present Value). This gives a figure for comparison of the future costs. With a given interest rate, energy price and lifetime, the NPV of the losses is calculated. The total lifetime costs are the sum of equipment costs and NPV losses in the wind farm power collecting system during the project lifetime.

2.4 Optimization procedures.

The optimization is made by finding the equipment, used in the layout, with the lowest combined cost of equipment and of the capitalized losses during the project period.

Calculating the load distribution, the basic data for the generator and the wind conditions must be defined. The generator is described by the nominal voltage, maximum active and reactive power output and the power curve for the wind turbine.

The local wind conditions are described by the Weibul A and C parameters which are used to calculate the wind speed distribution by means of the accumulated probability density function W_f:

$$W_f = \frac{C}{A} e^{((C-1)\ln(\frac{WS}{A}))} e^{(-e^{(C\ln(\frac{WS}{A}))})}$$

WS -   Wind speed
A -   Weibul A scale factor
C -   Weibul C shape.

The annual number of hours during which a certain wind speed occur is:

$$W_t = W_f \, 8760 \text{ hours}$$

In figure 2, the wind distribution for the specific site is shown. The Weibul parameters A and C are 8.15 and 2 respectively.

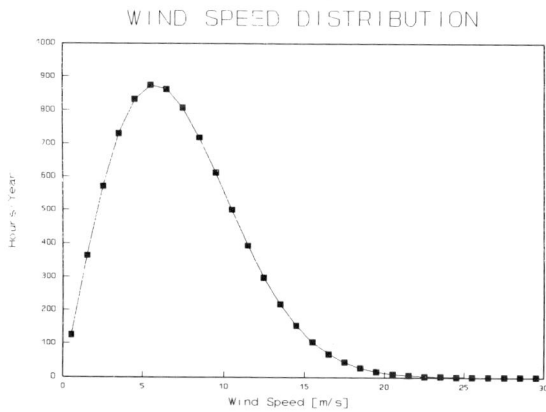

Figure 2.   Wind speed distribution for site.

Combining the wind turbine power curve with the wind speed distribution curve, giving the generator load curve, the power production and the losses can be calculated for each wind speed bin as:

$$\Delta P = \frac{r}{U_n^2} (P^2 + Q^2)$$

where P and Q are the power transmitted, r is the ohmic resistance in the equipment and U_n is the voltage level.

In the NPV-calculations the losses are multiplied with the energy price to determine the cost of annual losses. The energy price can be fixed or change over the years.

The NPV-function calculates the Net Present Value with the formula:

$$NPV = \sum_{i=1}^{n} \frac{c_i}{(1+interest)^i}$$

n   expected wind farm lifetime in years or duration with fixed energy price.
interest   interest rate
c_i cost for annual losses.

When the Net Present Value of the losses is known, the lifetime costs are calculated as the sum of costs for NPV-losses and the equipment costs. Foundation costs are included in transformer costs, and costs for excavation, reconditioning underground installations and backfilling are included in the cable costs.

The voltage increase in cables and transformers are calculated using the approximate formula for evaluating the voltage variations in the wind farm. These calculations are only performed to observe the voltage increase at the generators at full production. If more accurate calculations are needed, load flow calculations have to be performed. The voltage increase are calculated using the approximate formula:

$$\Delta U = \frac{-P_{max} r + Q_{max} x}{U_n^2}$$

U_n Nominal voltage
P_{max} Maximum active power transmitted
Q_{max} Maximum reactive power transmitted
r   Resistance
x   Reactance.

3.   DESIGN OF WIND FARM.

3.1   Wind farm data.

Examples of design considerations are given to illustrate some of the aspects mentioned above. The investigations are based on Taff Ely Wind farm located in Southern Wales. The Taff Ely Windfarm consists of 20 450 kW stall controlled wind turbines equipped with induction generators with a nominal voltage of 690 V. The voltage level of the power collecting system is 11 kV. The wind farm is connected to the grid via a 11/33 kV transformer. Based on the resource assessment and optimization of the wind farm layout the wind turbines are located in 4 rows with 2, 6, 6 and 6 wind turbines, with an approx. mutual distance of 150 m.

3.2   Reactive power compensation.

To reduce the reactive power consumption from the grid the wind farm is compensated for the reactive power demand at idle. Each wind turbine is equipped with a capacitor bank with a capacity equal to the reactive power demand on the wind turbine generator at idle. The capacitor bank is connected to the grid when the wind turbine is producing. This secures a cos φ better than 0.97 for all load cases and no production of reactive power out of the wind farm. Only alternatives with this compensation are considered.

In figure 3, the power curve for the 450 kW wind turbine is shown. Both active, reactive power consumption and compensated reactive power consumption are shown.

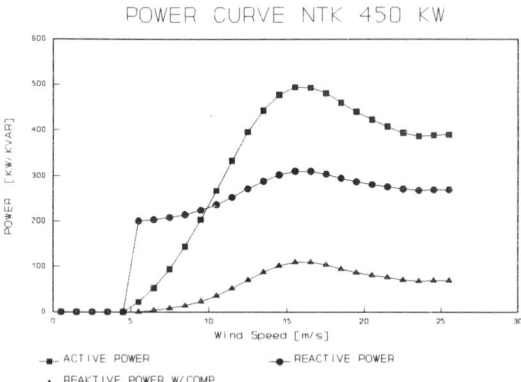

Figure 3. Power curve characteristics for Nordtank 450 KW wind turbine.

3.3 Layout alternatives.

If the wind farm layout consists of identical rows or matrixes the wind farm may be divided into groups represented by an equal number of wind turbines and identical lines and transformers. This procedure simplify the calculations to be performed. As the Taff Ely wind farm consists of only 20 wind turbines and the rows are not identical all equipment in the wind farm are evaluated.

For the Taff Ely wind farm a LV/HV-layout and a HV-layout are investigated to determine the most economical solution and to emphasize some of the parameters influencing the layout.
The costs used to evaluate the layouts of the electrical power collecting systems are based on prices for cables inclusive of excavating, laying and backfilling, transformers inclusive substation transformer, switchgear, housings, foundations and losses calculated and collected for this specific study. The capitalized losses are calculated on basis of 9.25 % interest rate, 4 % inflation, a project lifetime of 20 years and an energy price of 0.11 £/kWh (basis 1993) in the period 1993 until 1999, thereafter a price of 3.49 (basis 1999). Only costs influencing the choice of overall layout and used for optimization of the equipment are included. Costs for additional civil and electrical works and equipment have to be included in the total wind farm costs.

In the LV/HV-layout the wind turbines are connected to a common step-up transformer in groups of 2, 3 and 4 wind turbines via LV-cables and a LV-switchgear. On the HV-side of the transformer a ring main unit is used for looping to other groups or to the main panel in control building. The transformer, switchgear and ring main unit are located in a common housing located at a turbine in the group. The wind farm is connected to the grid via a 10 MVA 33/11 kV transformer.

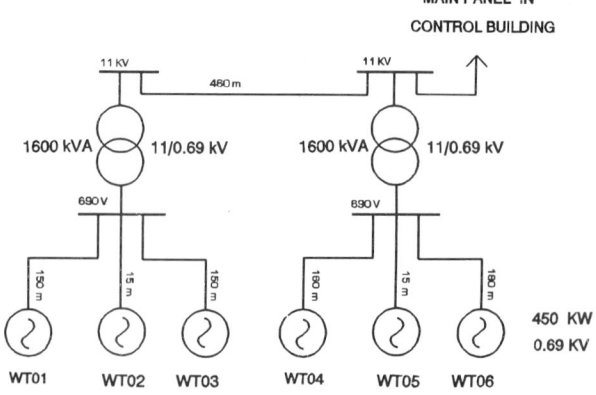

Figure 4. Section of wind farm with LV/HV- layout.

In the HV-layout a step-up transformer and a HV-ring main unit is located at each wind turbine. In this layout the LV-cables are as short as possible in order to reduce the losses.

This layout is evaluated with regard to 2 different preconditions: In the low cost-layout the equipment are selected on basis of voltage variations and loads, e.g. current carrying capacity, SC-ratio etc. to maintain low equipment costs.
In the optimized layout selected transformers and cables are optimized in order to obtain the lowest total lifetime costs i.e. the beneficial project economy. In figure 6 a section of a HV-layout is shown.

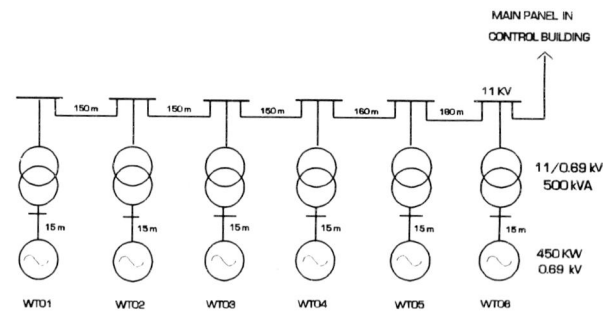

Figure 5. Section of wind farm with HV-layout with transformer at each wind turbine.

In table 1 below, a calculation of the total lifetime costs for a LV/HV-layout is shown, including costs for equipment and capitalized losses calculated by the NPV-method.

All figures in £	Equipment	NPV Losses	Total
LV-cables	86,531	79,754	166,285
HV-cables	90,456	42,497	132,953
WT transformers	68,600	246,374	314,974
33/11 KV trans	65,000	125,865	190,865
HV-switchgear	25,400		25,400
LV-switchgear	29.400		29,400
Housing/found.	56.000		56,000
	421,387	494,489	915,877

Table 1. Lifetime costs for LV/HV layout.

The total lifetime costs for a HV-layout based on low equipment costs gives:

All figures in £	Equipment	NPV Losses	Total
LV-cables	7,100	14,780	21,880
HV-cables	98,006	62,341	160,347
WT transformers	90,000	254,641	344,641
33/11 KV trans	60,000	149,163	209,163
HV-switchgear	88,000		88,000
LV-switchgear			
Housing/found.	80.000		80,000
	423,106	480,925	904,031

Table 2. Lifetime costs for a HV-layout based on low equipment cost.

The total lifetime costs for a HV-layout based on optimization of transformers and cables in order to obtain the lowest lifetime costs gives:

All figures in £	Equipment	NPV Losses	Total
LV-cables	9,880	9,100	18,980
HV-cables	108,941	45,359	154,300
WT transformers	96,000	214,416	310,416
33/11 KV trans.	65,000	125,865	190,865
HV-switchgear	88,000		88,000
LV-switchgear			
Housing/found.	80,000		80,000
	447,821	394,740	842,561

Table 3. Lifetime costs for a HV-layout optimized with regard to equipment costs and capitalized losses.

In Figure 6 The costs of the layout alternatives are divided into equipment costs, costs of capitalized losses and total lifetime costs comprising equipment costs and capitalized losses.

LIFETIME COSTS FOR LAYOUTS

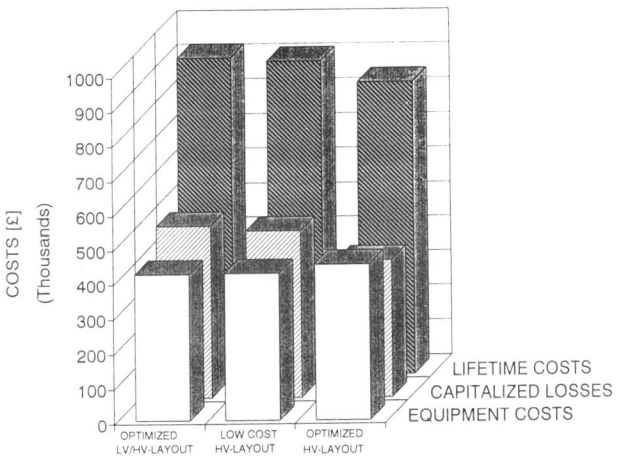

Figure 6   Lifetime costs of layout alternatives.

In the initial layout evaluation the overall layout is decided: The figure show that the HV/LV-alternative has the highest total lifetime costs compared to the HV-alternatives. The HV/LV-alternative has relatively low equipment costs and considerably higher costs of capitalized losses. For this specific site a HV-layout is therefore chosen.

By optimizing the HV-layout with regard to equipment cost and capitalized losses a considerable reduction in total lifetime costs for the power collecting system in the wind farm is obtainable. This cause a minor increase in equipment costs due to the use of low loss transformers and cables with other dimensions which reduces the capitalized losses considerably.

If some of the parameters affecting the evaluation of the layout deviate from the values used in the above calculations, the result may be different.

3.4 Parameters influencing the WIFAGO model and the electrical layout.

In the preliminary phases of a project some of the parameters used in the selection of the electrical power collecting system layout are unknown or uncertain.

Deviations and uncertainties in the informations have highly varied influence on the layout. To minimize the total uncertainty, the topics of major importance have to be as accurate as possible.

To illustrate the sensitivity of the layout alternatives, some of the above-mentioned parameters and their influence on the modelled layouts are shown in figure 7. The values in figure 7 are relative and show the change in total lifetime costs for the evaluated layouts and are based upon a 10% change in the parameter specified.

LAYOUT SENSITIVITY

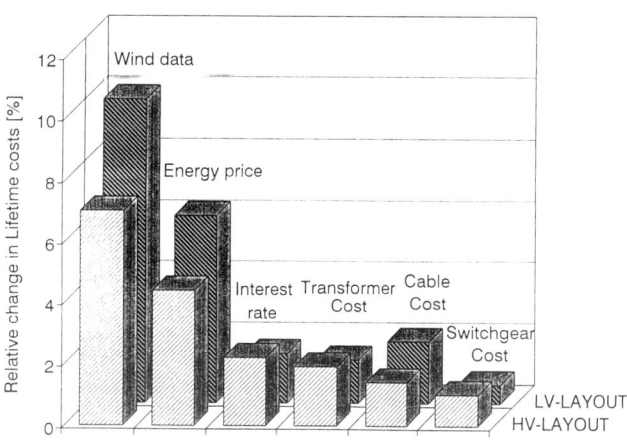

Figure 7.   The design parameters' relative influence on costs for the modelled layout.

As seen in figure 7 a LV/HV-layout is sensitive to deviations in data for wind speed distribution. Therefore an accurate ressource assessment and wind regime for the site is essential in the evaluation of a LV/HV-layout. This is due to the influence which the wind data has on calculating the capitalized costs of the losses. In general, the data used for calculating the capitalized costs of the losses are important for evaluating a LV/HV-layout. Evaluating the HV-layout, the equipment costs, such as costs of transformers, cables and switchgear, are parameters of major importance to the total lifetime costs as these costs amounts to a large part of the total lifetime costs. The data used for calculating the costs of capitalized losses are of minor importance, compared to a HV/LV-layout.

One of the most important parameters for evaluating the electrical layout with regard to lifetime costs based on capitalized losses is the energy price and the development of energy price in the project period.

In figure 8 is shown the influence of energy price on equipment costs and capitalized losses and in figure 9 is shown the influence on the optimum size of losses in the power collecting system. The figures are based on a HV-layout optimized on basis of 9.25% interest rate, 4% inflation and a project lifetime of 20 years. The layout is optimized using various energy prices of resp. 0.05, 0.07, 0.09, 0.11, 0.13, 0.15 and 0.17 £ (basis 1993) for the period 1993 to 1999 and 0.016, 0.022, 0.029, 0.0349, 0,041, 0.048 and 0.054 £ (basis 1999) for the period 1999 to 2013.

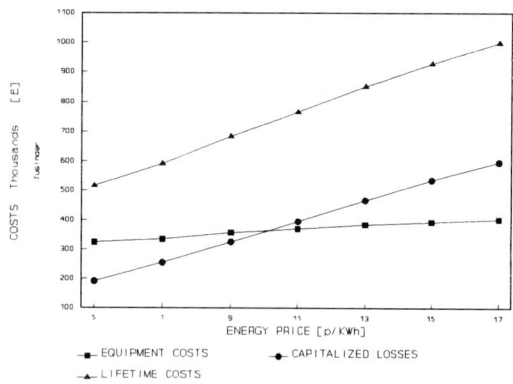

Figure 8   The equipment costs, capitalized losses and total lifetime costs with various energy prices.

Figure 8 show that the optimum choice of equipment demands a larger equipment cost by increasing energy price. Due to the increased value of the capitalized losses the optimum layout requires less losses by increasing energy price. Figure 9 show that the optimum size of losses in the power collecting system are depending on the energy price.

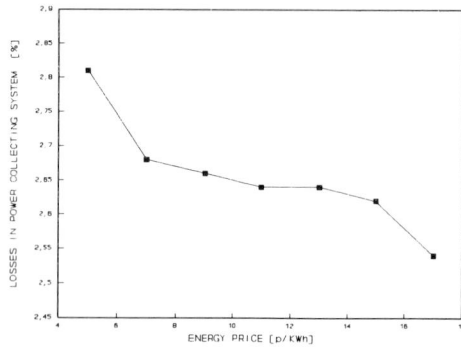

Figure 9   Optimum electrical losses in power collecting system with various energy prices.

3.5   Optimizing a cable.

All cables in the power collecting system are optimized with regard to capitalized losses and cable costs. As an example, a cable is chosen in the wind farm. The cable has a length of 1,330 m and loops from a row with a total of 6 wind turbines to the main panel in the central control building. In the database in the WIFAGO tool, 5 different HV-cables are selected for the optimization: 50, 95, 150, 185 and 240 mm^2 aluminium XLPE, all cables with 3 sector-formed conductors and a copper screen. With the load duration curve calculated on the basis of wind speed duration, power curve for the wind turbines connected and availability factor etc. the total lifetime costs are calculated.

The total lifetime costs for the cables are stated in table 3.

Dimension mm^2	NPV Life-time losses £	Cable costs £	Total costs £
3*50+16	53022	21081	74103
3*95+25	26470	25536	52006
3*150+25	17205	31056	48261
3*185+35	13979	35511	49490
3*240+35	10588	40964	51552

Table 3.   Lifetime costs for a 1,330 m cable with a load of 6 wind turbines.

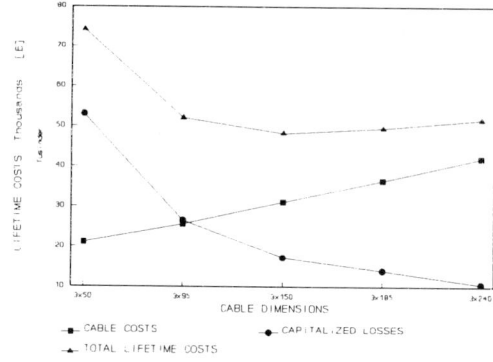

Figure 10   Lifetime costs for a 1,330 m cable with a load of 6 wind turbines.

As seen in table 3 and figure 10 the best choice of cable regarding lifetime costs, including capitalized losses is a 3*150+25 mm^2 cable, showing that the alternative with the lowest losses or the lowest equipment costs is not necessarily the best choice. In Figure 11 the maximum voltage variations in the cable is shown. The optimum voltage variaton for the specific cable is approx. 0.7 %.

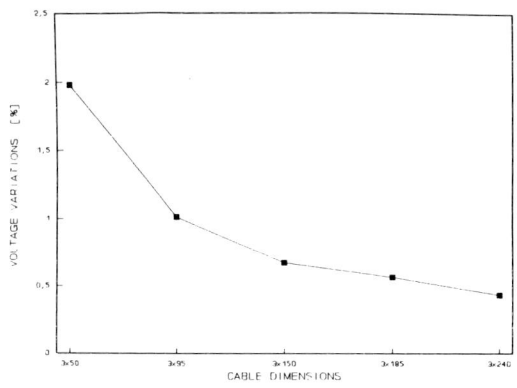

Figure 11   Voltage variations in cable with varying cross sections.

CONCLUSIONS.

The electrical layout of the power collecting system in wind farms has been investigated with regard to electrical losses. The losses have been evaluated to obtain the economical optimum of the total lifetime costs, comprising capitalized losses and equipment costs.
Experience gained by development and design of wind farms have shown that, by means of marginal efforts, a considerable reduction in lifetime costs for a wind farm project is obtainable if the electrical layout is designed on basis of capitalized losses.

For that purpose the WIFAGO design tool has shown to be very suitable in order to optimize the layout and to reduce the required time consumption in the design phase.
Furthermore, the investigations show that the electrical layout for a wind farm on a specific site is highly dependable on local conditions, such as prices for selected equipment, interest rate, development in energy price, soil compound regarding excavation, foundation and the wind regime on the site.
The amount of losses has an important influence on the choice of layout. The alternative with the lowest losses is not necessarily the best choice when the layouts are evaluated with regard to lifetime costs. The parameters of major importance for the calculation of the capitalized losses are established to be cable length of LV-cables, the energy price and the wind regime for the specific site.

Parameters which are in favour of a HV-layout are: large mutual distance between (large) wind turbines, high energy price and low transformer prices. In favour of a LV/HV-layout are: short mutual distance between (small) wind turbines, low energy price and high transformer prices. A general conclusion is that it is normally an economically sound solution to locate transformers as close as possible to the wind turbines. Whether a low or a high voltage layout is the economical optimum is highly dependable on the wind farm layout, the parameters used in the evaluation and the size of wind turbines. Therefore, it has to be evaluated for each individual case.

No specific criteria can be given for cabling due to the influence from local conditions, especially in mountaneous areas. The general rule should be to make the power flow (cabling) in the general direction of the grid connecting point. This should be incorporated in the routing of access roads, as the cables are usually following these.

REFERENCES

[1]   Jensen, U.S. a.o.
        European Wind Power Study
        EEC DGXII, JUR-0041-DK

[2]   Kristiansen, J. Hansen, J.C.H.
        Economically optimized design of wind farms.
        ECWEC 1993.

A wind park linearized model

RUI M. G. CASTRO MSc, **J. M. FERREIRA DE JESUS** PhD
Instituto da Energia/Instituto Superior Técnico, Portugal

SYNOPSIS: In this paper the development of a wind park linearized model from the detailed non-linear model is presented. Furthermore, some of the methodologies currently used in power systems to reduce the order of the models have been applied to the case of the wind park linearized model, in order to build up dynamic equivalents. These studies have allowed a comparison between the performance of each method, thus enabling the choice of the one that is best suited for the specific purpose of retaining the relevant dynamics of a wind park.

1 INTRODUCTION

The development of clean renewable energies for future large scale applications in electric utility systems has been selected as one of the main objectives for research and technological development in the field of renewable energy sources. As a consequence, both the USA and Europe have issued specific legislation with the purpose of promoting the multiplication of small dispersed generation sources, which is becoming a widespread and profitable way of producing electrical energy.

Wind Energy Conversion Systems (WECS) have been identified as promising renewable energy sources due to its negligible impact on the environment, and the advances in technology in this field in recent years has confirmed the expectations placed in this form of energy conversion.

The production of a significant amount of power from the wind using wind generators has required both the development of larger and more efficient and reliable wind turbines and the use of more than one machine at each site constituting the popular wind farms (or wind parks). Currently, the most used configuration of wind parks consists of a set of wind turbines each one driving a squirrel cage induction machine, which, in turn, are assembled in group(s) and directly coupled to the existing a.c. system.

In order to assess the interaction between the wind farms and the utility grid to where they are connected, some computational tools have been developed with the aim of assisting both wind park and distribution system planners and designers.

These computational tools are based on models that are able to accurately simulate the behavior of wind parks under transient situations. Two types of models have been produced:

A. Models that are able to simulate the transient behavior of each WECS belonging to a park as well as the transient behavior of the park with respect to the utility system, henceforth denoted by detailed wind park models.

B. Models that are able to simulate the relevant dynamics of the park with respect to the utility system, henceforth denoted by wind park reduced order models.

These models address different problems and will be able to assist designers and planners in different but complementary issues. Whereas the detailed wind park models will assist in the design of the configuration of the wind park, the reduced order models will assist in the interconnection between the wind park and the utility grid.

This paper is concerned with the development and application of wind parks reduced order models above denoted as type B. models, type A. models being the subject of a companion paper "The influence of transmission system disturbances on the dynamic behavior of a wind park" (1).

In this paper the development of a wind park linearized model from the detailed non-linear model is presented. Furthermore, some of the methodologies currently used in power systems to reduce the order of the models have been applied to the case of the wind park linearized model, in order to build up dynamic equivalents.

These studies have allowed a comparison between the performance of each method, thus enabling the choice of the one that is best suited for the specific purpose of retaining the relevant dynamics of a wind park.

The linearization of the detailed wind park model was achieved by applying *Taylor* expansion about an operating point (2). The resulting set of equations describes the dynamic behavior of small disturbances about this operating point. This is exactly the case of a wind park running in steady-state conditions: both the wind speed and terminal voltage changes constitute small perturbations about a steady-state operating point.

Due to its crucial importance, a dedicated algorithm was developed in order to obtain the steady-state operating point, i.e., the initial conditions for the set of differential equations. This technique appeals for classical power flow concepts modified in a suitable way, to calculate the relevant quantities, based only on the system parameters and on the mean wind speed.

Several methods currently used in power systems to reduce the order of the models and, therefore, to determine dynamic equivalents are published in the literature.

Modal truncation is one the first reduction schemes that has been applied to electric power systems (3). This technique is based on the pole location of the linear system. The state variables are transformed in modal variables and the fast decay poles and/or those associated with high frequencies are neglected, thus enabling a reduction in the order of the system.

Balanced reduction techniques take a slightly different approach, because they are based in the input/output behavior of the system (4). Actually, the original state-space system is transformed into a new representation that has the property of each state-space variable being as controllable as observable. In order to achieve a reduced order model, states that are strongly influenced by the inputs and strongly connected to the outputs are retained, whereas states that are weakly controllable and observable are truncated.

Another method used in power systems order reduction is the so called Optimal Hankel-Norm approximation (5). This criterion will try to achieve a compromise between a small worst case error and a small energy error.

The technique known in the literature as Singular Perturbations decomposes the system according to its fast and slow dynamics and then lowers the model order by first neglecting the fast dynamics phenomena (6). The effect of fast dynamics are then reintroduced as "boundary layer" corrections calculated in separated time scales, which leads to correct static gains.

Finally, another approach is the coherency based technique in which clusters of generating units that tend to swing together at close frequencies are identified and aggregated in order to form one or more dynamic equivalents (7). This approach is currently under investigation in order to determine whether it is appropriate to apply to the case of a wind park or not.

Among these power systems reduction order techniques, Modal analysis, Balanced reduction, Optimal Hankel-Norm approximation and Singular Perturbations technique seem to be the most prominent. Therefore, they have been applied to the case of the wind park linearized model, thus enabling to obtain different reduced order models.

In order to be able to draw some conclusions regarding the comparative behavior of all the detailed non-linear, full order linearized and reduced order wind park models some computer simulations were performed. A test system composed by a wind park, connected to an a.c. system through a feeder, operating in steady-state conditions has been used for these purposes (Fig. 1).

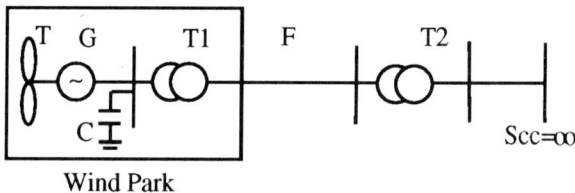

Fig. 1 - System studied.

The results obtained were compared in order to draw some conclusions on the ability of the different reduced order models to accurately simulate the steady-state behavior of a wind park.

2 FULL ORDER LINEARIZED WIND PARK MODEL

The set of differential equations that describe the behavior of a wind park are non-linear. However, the behavior of the system regarding small disturbances may be analyzed using the set of linearized differential equations.

This is the case of a wind park operating under steady-state equations: both wind speed and terminal voltage changes act as small disturbances about a steady-state operating point.

So, let us consider a wind park composed by n turbine/generator groups each one made up of a wind turbine driving a squirrel cage induction generator equipped with a reactive power compensation system. All turbine/generator groups are assembled in parallel and connected to a step-up transformer. The non-linear detailed model that describes this system, which is fully presented in the companion paper (1), is composed of $5n+4$ differential equations.

By applying the classic *Taylor* expansion about an operating point in steady-state conditions one obtains a linearized version of the model in the state-space representation.

$$\frac{d}{dt}[\Delta x] = [A][\Delta x] + [B][\Delta u]$$
$$[\Delta y] = [C][\Delta x] + [D][\Delta u]$$

where:

    $[\Delta x]$ - state space vector (selected state variables)

    $[\Delta u]$ - input vector (wind speed, terminal voltage)

    $[\Delta y]$ - output vector (active, reactive output power)

Matrices A, B, C and D depend upon the selected operating point, namely the wind speed, the mean wind speed at the wind park location being chosen for obvious reasons.

Some tests were carried in order to find out if the linearized model could be used to describe with a reasonable degree of accuracy the steady-state behavior of the wind park. In Figs. 2 it is shown a comparison between the results obtained for both the wind park active and reactive power output using the detailed non-linear model and the linearized model. It was assumed that all the wind turbines were "seeing" the same mean wind speed. The simulations were performed using an available wind time series (8) and a specific network configuration.

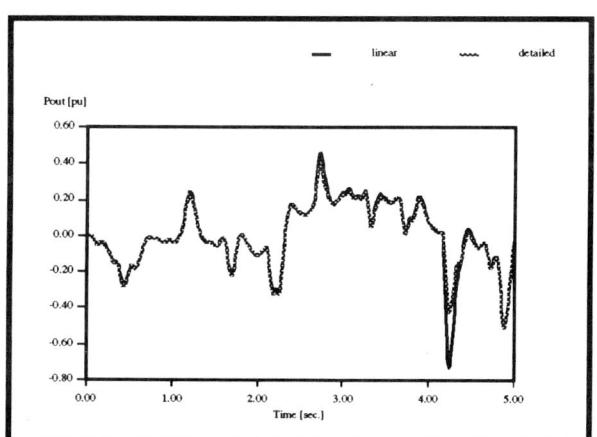

Fig. 2a - Active power (detailed and linearized models).

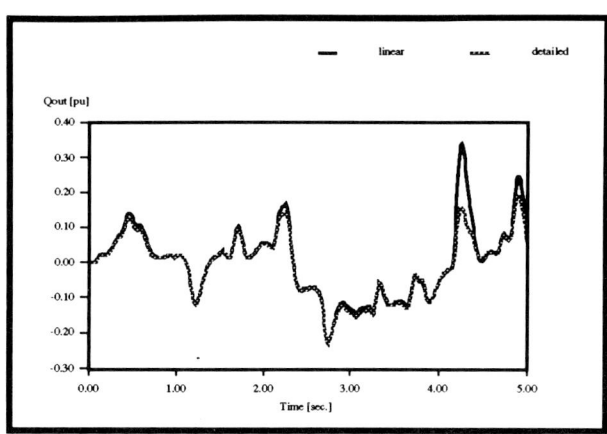

Fig. 2b - Reactive power (detailed and linearized models).

The results obtained showed good agreement except for strong wind gusts, resulting in an overall medium error of 0.2%. However, the computer time saves reached more than 75% which may be of the highest importance when the issue of predicting the steady-state behavior of a park composed by a great number of units is to be assessed.

3 REDUCED ORDER WIND PARK MODELS

The methodologies currently used in power system to reduce the order of the systems were applied to the case of the wind park linearized model.

From the application of those methods a common conclusion could be drawn, which comes straightforward from minimal realizations theory: as long as all the machines within the wind park "see" the same mean wind speed, the park can be reduced to one single equivalent turbine/generator group, at least for steady-state purposes.

The studies performed within the scope of this paper are strictly restricted to steady-state behavior and to the case of equal mean wind speed inside the park, the case of transient behavior and different mean wind speeds being currently under investigation.

Both Hankel-Norm and Balanced reduction methods, although produce very good approximation of transient response, have the drawback of high reduction errors at low frequencies, due to the intrinsic mismatch between the *DC* gains of the full and reduced order models. Therefore, these methods are not suitable to steady-state applications. In order to overcome this difficulty, some authors have proposed frequency-weighted balanced reductions that tend to shape the reduction error both at high and low frequencies. The applicability of these techniques to wind park models is being currently investigated (5).

The applied modal based reduction techniques make the discarded dynamic behavior static, by updating the *D-matrix* so that the *DC* gain of the reduced model is equal to the original model one. Therefore, this method, which actually neglects the fast dynamics phenomena, is thought to be the best suited to steady-state applications. However, due to its inherent lack of a clear criterion for determining the proper order of the reduced-order model, it becomes strongly dependent on the specific system to which it is applied. In fact, when the issue of neglecting the so called fast poles is to be assessed one has to be extremely cautious and a pre-analysis of the characteristics of the system is required.

The general rule of neglecting the poles having a large absolute value and/or a large real part may lead to erroneous conclusions about the order reduction, if a previous knowledge of the dominant modes is not taken into account. However, if this precaution is undertaken very good approximations of the behavior of the system may be achieved as it can be seen from Figs. 3. These figures show the results obtained for a case study in which it has been considered that only the wind speed input is actually changing, the terminal voltage input changes being neglected. The figures displays the simulation of the active power output (Fig. 3a) and the reactive power output (Fig. 3b) for a two wind turbine wind park using both the full and the reduced order models. In this situation, the application of Modal reduction technique to the *14th* order full linearized system lead to a *3rd* order reduced system. This reduction has been achieved after a prior analysis of the system which has recommended that the poles having a large absolute value should be neglected in order to retain only the ones strongly associated with the wind speed input changes. This procedure has resulted in an excellent agreement between full and reduced order models.

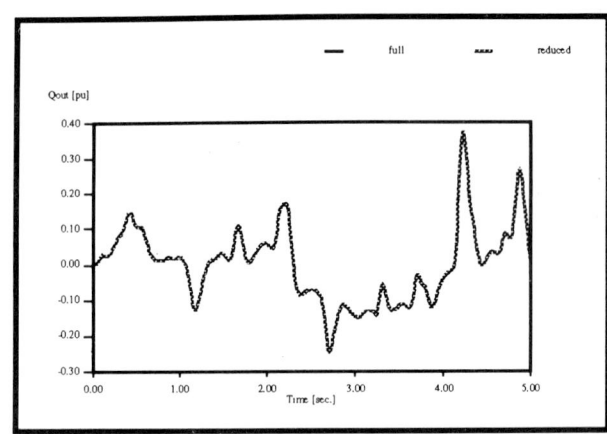

Fig. 3b - Reactive power (full and reduced order models).

The above shown results indicate that for this specific study, a *3rd* order Modal reduced model reproduces very accurately the results obtained with the *14th* full order model. However, conclusions about the application of Modal reduction technique to all studies involving wind parks can not be drawn from them.

From the authors point of view a more prominent reduction technique is Singular Perturbations. Due to the fact that the effects of fast dynamics are reintroduced in the reduced system, this technique seems to be more general, namely in what concerns the prediction of both steady-state and transient behavior of the systems. However, to the particular steady-state case-study performed in the scope of this paper, the application of this technique lead to a more inaccurate approximation.

Figs. 4 show the simulation of the active power output (Fig. 4a) and reactive power output (Fig. 4b) of a wind park operating under the same conditions of Figs. 3, but with the difference that a *4th* order reduced model was obtained through the application of Singular Perturbations technique.

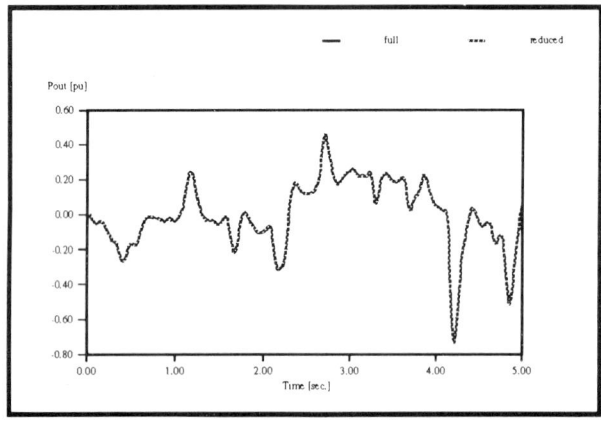

Fig. 3a - Active power (full and reduced order models).

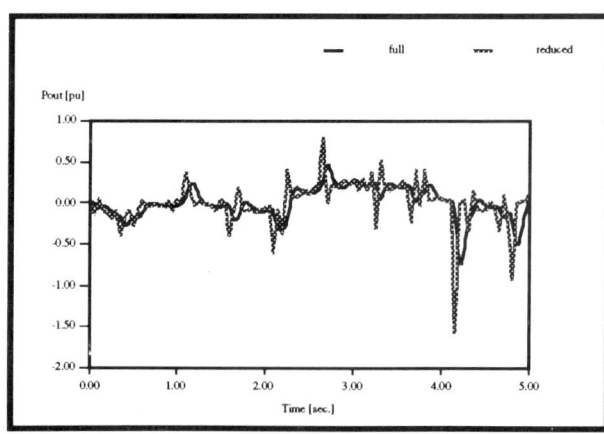

Fig. 4a - Active power (full and reduced order models).

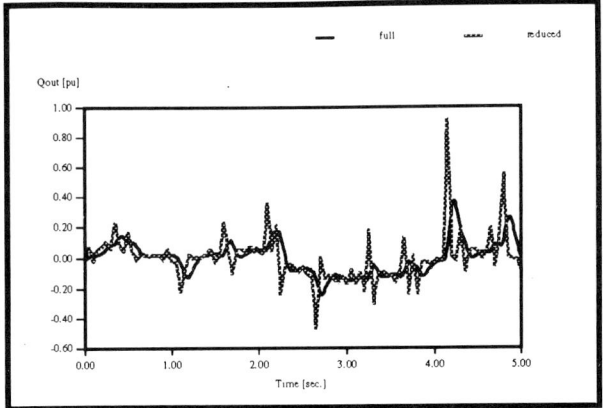

Fig. 4b - Reactive power (full and reduced order models).

These results clearly show that for this specific case-study a worse approximation has been reached. However, it remains to be sorted out what would be the results if the transient behavior of the wind park was to be assessed. It is the authors belief that under these circumstances Singular Perturbations would behave better than Modal reduction. This issue is currently under investigation.

4 CONCLUSIONS

A wind park linearized model obtained from the classical *Taylor* expansion about an operating point was presented. The detailed non-linear model which served as the basis for the development of this linearized model is fully presented in the companion paper (1). Some computer simulations were performed using both descriptions, the results obtained showing a good performance of the linearized model in predicting the steady-state behavior of the park.

In order to alleviate the computer charges resulting from the generalization of the individual components models to a wind park composed by n units, some of the procedures currently used in power systems to reduce the order of the models were applied to the case of a wind park.

For the time being only steady-state behavior was targeted, transient behavior being currently under investigation. In this conditions, Modal reduction and Singular Perturbations technique were found to be the best suited methods. Once more, some computer simulations were performed in order to allow a performance comparison between the above mentioned techniques. These studies allowed the conclusion that although Modal reduction gives the best results in the prediction of steady-state behavior, Singular Perturbation should behave better in forecasting the overall steady-state and transient behavior of the wind park.

5 ACKNOWLEDGMENTS

The authors wish to gratefully acknowledge the active contribution of *Dr. Peter Bongers* from *Delft University of Technology* to this paper, namely by making available the software package *MR* (9) which allowed that some of the studies presented here could be performed.

6 REFERENCES

(1) Rui M.G. Castro, Ana I.L. Estanqueiro, Ricardo Aguiar, Jorge A.G. Saraiva, J.M. Ferreira de Jesus, "The Influence of Transmission System Disturbances on the Dynamic Behavior of a Wind Park", BWEA15, British Wind Energy Association Annual Conference, York, Oct. 1993.

(2) P.C. Krause, "Analysis of Electrical Machinery", McGraw-Hill Book Company, 1986.

(3) J.M. Undrill, A.E. Turner, "Construction of Power System Electromechanical Equivalents by Modal Analysis", IEEE Transactions, PAS-90, 1971.

(4) B.C. Moore, "Principal Component Analysis in Linear Systems: Controllability, Observability and Model Reduction", IEEE Transactions on Automatic Control, Vol. AC-26, Feb. 1981.

(5) M. Bettayeb, U.M. Al-Saggaf, "Practical Model Reduction Techniques for Power Systems", Electric Power Systems Research, 25, 1992.

(6) P.V. Kokotovic, R.E. O'Malley, P. Sannuti, "Singular Perturbations and Order Reduction in Control Theory - An overview", Automatica, Vol.12, 1976.

(7) R. Podmore, "Identification of Coherent Generators for Dynamic Equivalents", IEEE Transactions, PAS-97, 1978.

(8) Ana I.L. Estanqueiro, Ricardo Aguiar, Jorge A.G. Saraiva, Rui M.G. Castro, J.M. Ferreira de Jesus, "The Development and Application of a Model for Power Output Fluctuations in a Wind Park", 1993 European Community Wind Energy Conference and Exhibition, Travemunde, Germany, Mar. 1993.

(9) Peter M.M. Bongers, "Factorizational Approach to Robust Control: Application to a Flexible Wind Turbine", PhD Thesis, Delft University of Technology, The Netherlands, 1993.

The implications of fatigue on the cost of HAWTs

N. D. P. BARLTROP, I. P. WARD and **D. J. DAW**
WS Atkins Consultants Limited, UK

SYNOPSIS This paper describes a fatigue costing methodology, which can be used to assess the relative importance of fatigue for any type of wind turbine and its components. Typical calculation results are presented for a 180 kW stall-regulated turbine and a 2000 kW pitch-regulated turbine. General fatigue and cost trends are demonstrated, relative fatigue costs are shown to be about 5-10% of the total cost of the turbine, with the greatest changes in fatigue due to different control techniques.

1   INTRODUCTION

Wind turbines are sensitive to fatigue damage since they are subjected to fluctuating loads, from both wind turbulence and deterministic effects associated with rotor rotation. The dynamic loads and the design detailing will determine the fatigue life. Studies have been carried out on the influence of control action on fatigue and the fatigue properties of wind turbine blades.   However, since many uncertainties still exist in the design methodologies and since there has been little full-scale verification, the real implications of fatigue on design and turbine cost are not well understood.

A study was initiated so that progress could be made in determining the cost implications of fatigue. By conducting detailed dynamic analyses and other sensitivity studies on two typical HAWT's, a 180 kW and a 2000 kW turbine, and associating fatigue with relative extra capital costs, a methodology has been developed and some trends identified. (1)

2   METHODOLOGY

The fatigue costing analysis methodology is based on time history simulations using full aeroelastic turbine models (Figure 1), fatigue analyses and fatigue-costing relationships. The general assumption is that the original capital cost of the machine is based on design for one of: fatigue, strength, deflection, frequency separation or buckling. Then, if fatigue is found to be governing, a corresponding extra capital cost can be allocated, based on increased component size to meet fatigue design criteria.

For a selected turbine design and wind environment, a dynamic analysis is conducted for representative mean wind speeds and yaw error positions. The rainflow counting method is used to count equivalent stress (or load or moment) ranges and associated cycles, for each component of interest. Then, fatigue S-N curves are selected and the collection of stress ranges and cycles are combined, using a formula consistent with Miner's rule, to produce the Weighted-Mean-Range. For i ranges of stress R_i and n_i cycles, with m as the slope of the S-N curve, the Weighted-Mean-Range is,

$$\text{WMR} = \sqrt[m]{\frac{\sum\limits_i R_i^m n_i}{\sum\limits_i n_i}} \qquad (1)$$

The calculated WMR is interpolated/extrapolated to give effective values at all wind speeds (selected discrete values) within the wind speed probability distribution, and all probable yaw error angles.   Using joint probabilities, the number of cycles are counted for each specified WMR, and the fatigue damage is summed.

A unity check is then performed on the predicted fatigue life of the component by calculating the Fatigue-Usage-Factor, defined as the ratio of fatiguing stress to permissible fatiguing stress. This is related to fatigue life by the slope of the S-N curve, and can be directly compared to stress or strength criteria. The Fatigue-Usage-Factor is:

$$FUF = \left(\frac{design\ life\ required}{design\ life\ provided}\right)^{1/m} \quad (2)$$

Other stress unity checks are calculated based on strength in compression, tension or buckling. The strength check is performed, in accordance with European wind turbine design guidelines, for extreme wind events (usually 50 year return period), with the turbine at rest. The strength unity checks, or usage factors, are modified by any appropriate load factors and partial material factors.

The usage factors are then compared in order to ascertain the importance of fatigue over other possible design drivers. The Usage-Factor-Ratio is:

$$UFR = \left[\frac{FUF}{highest\ other\ usage\ factor}\right] \quad (3)$$

Fatigue costs are allocated, for each turbine component, based on the Usage-Factor-Ratio (UFR), percentage cost of turbine, and cost-weight relationships. In general, if the UFR>1, then an extra cost is associated with fatigue, and if UFR<1 fatigue does not govern the design and therefore does not give rise to an increase in cost.

To determine the cost-weight relationship, preliminary calculations are performed to compare effects of an increase in component weight (or cross-sectional area A) to the reduction in fatigue, or Fatigue-Usage-Factor (FUF). Assuming a simple power-law relationship a weight-fatigue relationship is developed. Then, the cost-exponent 'a' in the power-law relationship can be calculated as,

$$a = \frac{\log(A_1/A_2)}{\log(FUF_1/FUF_2)} \quad (4)$$

The resulting cost-exponent parameter 'a' is sensitive to the section properties, scale, the amount of area increase and the amount of additional work required for the larger component.

For a fatigue dominated component , the percentage extra cost needed as a result of fatigue can be related to the percentage increase in cost by using the following relationship (1):

$$\Delta_{cost} = k \times (1 - UFR^a) \quad (5)$$

where k is a constant which reflects the balance of fixed cost and material cost for the component. The extra costs for each component can then be converted into a percentage increase in cost for the complete machine. The cost benefits of changing the design to increase the fatigue life of a particular turbine component, or change overall turbine design, can also be assessed.

3 TURBINE ANALYSES

Time history analyses were conducted on two base case turbines, a 180 kW 3-bladed stall-regulated turbine and, a 2000 kW 3-bladed pitch-regulated (full span) turbine using software developed by RISØ National Laboratory in Denmark (2). Detailed fatigue calculations were performed at the blade root, blade mid-span, hub blade, hub, main shaft, 2 main bearings, gearbox input shaft gearbox gear teeth, tower base and tower top. The base environmental condition was an annual mean wind of U_{30}=6.6 m/s and turbulence intensity I=20%.

The analyses were based on actual turbine designs; the Danwin 23 (3) and the Tjæreborg prototype turbine at Esbjerg, Denmark (4). However, the precise construction details were not used for the study so the results herein should be considered to be only indicative of their fatigue sensitivity.

The analyses show that there is a significant extra capital cost (of the order of 5-8%) which can be attributed to fatigue. The results for the two base case turbines are presented in Figures 2 and 3. The fatigue cost trends are similar for both the small and large turbine:

- fatigue governs the design of the hub, main shaft gearbox input shaft and gear teeth;
- in some of these cases, the additional component weights are large (say 50-70%), but the effect on overall turbine cost is less, due to the non-linear relationship between component weight and cost, and the fact that the component cost may be a small percentage of the overall cost;
- generally for GRP blades, fatigue does not govern hence either no or only marginal fatigue costs result. However, this may not be the case for steel root details;

- fatigue governs the design of the cast steel hub on both the blade and the shaft side, dominated by deterministic (gravity) loads due to rotor rotation, rather than stochastic loads;
- the main shaft fatigue is dominated by deterministic bending due to rotor overhang, and to a lesser extent, torsional loading;
- the gearbox input shaft fatigue occurs primarily due to fluctuating torque from aerodynamic and control effects;
- gearbox fatigue and cost is dominated by gear tooth root bending, rather than tooth contact stress; the relative importance of fatigue to strength requirements depends on the design or extreme braking torque; the internal shafts will also tend to be fatigue dominated.

The fatigue costs of the larger pitch-regulated turbine are higher than the smaller stall-regulated turbine, predominantly due to:

- pitch-regulation of the 2000 kW machine, which causes significantly more fatigue damage than the stall regulation system of the 180 kW machine;
- use of mechanical brakes on the high speed shaft of the stall-regulated turbine, and corresponding higher rated torque, resulting in relatively less fatigue damage;
- the rotor overhang from the front bearing, which is proportionally larger on the 2000 kW turbine, resulting in greater shaft bending moments, hence greater shaft fatigue;
- other differences including: different hub height, wind speed, turbulence correlation and wind shear characteristics, different blade mass and stiffnesses (due partly to pitch control system requirements), planetary gearbox on larger turbine and parallel type on smaller one, also different drive train torsional stiffnesses, different relative importances of component weights.

4 SENSITIVITY STUDIES

Additional studies were conducted to assess the influence of design changes and operating environment on turbine fatigue cost. The base case analyses were repeated for different wind turbulence intensities, wind resource levels, blade materials and gearbox design. The studies show that changes to the design or operating environment may change the additional machine cost by up to 7%, and that similar trends are shown for turbines of different sizes.

Overall design changes from 2 to 3 blades, or from fixed to teeter hub, has little effect on total turbine fatigue cost, as shown in Figure 4, although some individual components may be affected differently by fatigue. However, changing the blade control system from stall to PI controlled pitch-regulation increases fatigue cost significantly for any turbine design.

The comparison between 2- and 3-bladed turbines was made using simple momentum/blade-element theory calculations. It was found that:

- the 2-bladed rotor should be 10% larger in all dimensions to give the same power as the 3-bladed rotor;
- each blade of the 2-bladed rotor has a resulting increase in the extreme blade bending moment, section modulus and mass (by a factor of $1.1^3=1.33$);
- the 2-bladed rotor mass is 0.9 times that of the 3-bladed rotor; however, rotor mass moment of inertia increases by $1.1^5(2/3)=1.07$, which affects braking torque;
- for the 2-bladed rotor, the maximum rotating flapwise moment is 1.44 times higher, and edgewise, 1.5 times higher;
- for the 2-bladed rotor, fatigue thrust reduces to 0.92 times, while fatigue torque increases by 1.09 times.

These results compare well in general with those from a similar exercise carried out by Rasmussen and Kretz (5).

A comparison between teeter versus fixed hub illustrates that:

- teeter hinge reduces effects on hub of antisymmetric loads (wind shear, yaw error, turbulence), but not of symmetric loads (mean wind, turbulence (symmetric component), gravity loads);
- the teeter hub will reduce blade root flapwise bending moments by about 35%. effectiveness of reduction of blade flapping moments away from the hinge depends on the balance between loading distribution and dynamic structural and aerodynamic resistance along the length of the blade;
- other moments and forces, except for main shaft bending, are hardly affected by teeter;
- extreme flapwise moment and thrust loads are not affected, since teeter is in a locked position;
- braking and operating torque, thrust and gravity induced main shaft bending, are unaffected by teeter.
- main shaft bending moments due to aerodynamic effects (wind turbulence, wind shear, yaw error) are reduced to zero by the teeter hinge.

Overall the fatigue cost saving offered by teeter (primarily at the blade root and main shaft) is small.

The cost of manufacture for fatigue increased from roughly 5% to 10% for the turbines with pitch-regulated control. For the pitch versus stall regulation comparison, calculations, based on Leithead et al's work (6) on control, show that PI controller controls excessive power production, but it does not produce a smooth torque, and causes greater stress cycling in the turbine components. Furthermore, as demonstrated by Quarton (7), control by blade feathering inherently produces large flapwise bending moments on the blades. Physically, when the wind speed increases so that less power is required, the blade is feathered to produce less lift. However, the remaining lift is now applied at an angle which provides more torque, so that an even greater change in pitch angle is required in order to obtain the required power reduction. This study's results indicate that:

- feathering of the blade results in a proportional change in blade bending stress which is 2-3 times greater, which is potentially significant in fatigue terms;
- stall regulation produces smoother power and lower stress fluctuations, particularly in the blades (this was verified by measurements at RISØ (8));
- the stall-regulated coefficient of variation (COV) of torque is similar in magnitude to the turbulence intensity value, but may be 2-3 times greater for a pitch-regulated turbine;
- most other moments and forces are not directly affected by control, since gravity effects dominate, or because the extreme event is not affected by the control system;
- more advanced control systems avoid much of the unnecessary stress fluctuation and benefit the fatigue life of nearly all the wind turbine components, and also improve the quality of the power output;
- the best compromise for a pitch regulated machine may be an advanced control system in conjunction with assisted stall pitch control; aerodynamic devices which control power by increasing drag should also cause less fatigue damage to the blades.

The influence of component design changes can be potentially significant both on overall turbine fatigue cost, and on fatigue life (and cost) of that particular item. For instance, the use of lighter, more fatigue resistant materials (such as GRP or wood-epoxy blades over steel ones) has potential fatigue and cost benefits. Improved designs, such as planetary vs. parallel gearboxes, may have increased relative fatigue costs, but may still be more practical (lower weight, capital cost).

The following points are worthy of note from the planetary (combined epicyclic) versus parallel gearbox comparison:

- planetary gearbox is lighter and smaller than parallel gearbox;
- planetary type has a greater relative fatigue cost than parallel, up to 1% more turbine cost;
- gear tooth root bending fatigue tends to govern the gear designs of either type;
- both types require rigid casings to limit out-of-parallelism etc;
- gearbox fatigue is significantly greater for pitch controlled systems, due to increases in fluctuating torque, such that up to a 20% increase in gearbox power rating may be required;
- for any turbine, a 20% reduction in torque COV results in a 5-7% reduction in gearbox power rating and a corresponding 0.5-1% reduction in turbine cost.

Changes in operating environment have a limited impact on overall turbine fatigue cost, hence capital cost, although specific component designs may be affected due to different relative importances of fatigue and extreme loadings. For the gearbox, the importance of fatigue increases as turbulence intensity or annual wind speed increases, since changes to mean wind speed and turbulence primarily affect gearbox fatigue loading. In contrast, for the hub and main shaft, increasing annual wind speed or turbulence predominantly affects only the extreme loading, because the fatigue loading is dominated by gravity effects.

Similarly, changes in design parameters such as cut-out wind speed or design life have a limited impact on fatigue life and fatigue cost. Increasing the cut-out wind speed from 25 to 30 m/s has virtually no impact on fatigue cost. For an increase in fatigue design life from 20 to 50 years, the additional fatigue cost is less than 1%.

5 SUMMARY OF RESULTS

a) Designing for fatigue increases the cost of the wind turbines studied by about 5-10%, depending on turbine design, size, and control technique.

b) The type of control system has the largest overall effect on the relative importance of fatigue in wind turbine designs of any size. A Proportional Integral (PI) controller with feathering pitch control may increase the fatigue cost by 5% of total machine cost.

c) Three blades are less fatigue prone than two blades, and a teeter hub reduces blade root, hub and shaft fatigue loading. However, these effects only amount to a 1-2% change in the fatigue cost.

d) Increasing turbine size tends to make fatigue more important; especially for the blades and main shaft. However, the significance of fatigue in relative cost terms is less for the large machine than for the small machine. This is caused in part by a change in the distribution of component costs.

e) Design features can significantly affect the extra cost of fatigue for individual components, such as long rotor overhang, stress concentrations, type of gearbox and mechanical braking technique.

f) The most fatigue sensitive turbine components are the hub, main shaft, gearbox input shaft and gearbox parts.

g) The material of construction is important due to weight, strength and fatigue properties: GRP and wood epoxy are more tolerant of fatigue loading than steel, welded steel has a poorer fatigue performance than cast steel; a reinforced concrete tower is much less fatigue sensitive than a welded or lattice steel tower; gear tooth hardening techniques change the relative tooth fatigue sensitivity.

h) The local environmental conditions and wind resource level have an effect on component fatigue, although total turbine fatigue cost change is marginal.

i) A change in other design parameters, such as increased cut-out wind speed, and increased design life (50 years), have little or no impact on fatigue and fatigue costing.

j) Fatigue problems are more easily built into a structure than an extreme resistance shortfall because fatigue is very dependent on local details, whereas strength tends to be much more dependent on the overall amount of material provided to resist the loading. At the design stage fatigue lives can be increased at small additional cost by considering design modifications. Once a wind turbine is built, fatigue problems can be very difficult to put right.

6 CONCLUSIONS

a) The study has shown that fatigue increases the capital cost of a wind turbine by 5-10%.

b) The design of the control system and the physical control method significantly affects the fatigue damage and cost of the turbine overall.

c) Although the fatigue related capital cost increase is relatively small, fatigue typically governs the cost of the components from the blade hub through the drive train to the generator.

d) Changes in turbine design details, environmental conditions, and design parameters have little impact on overall turbine fatigue cost, but the fatigue life of certain components can be affected.

e) Minor errors in design, detailing or fabrication are much more likely to cause additional fatigue problems than overall strength problems. Therefore fatigue is an important consideration at the design, detailing and fabrication stages.

7 RECOMMENDATIONS

a) The fatigue costing methodology should be applied to commercial HAWT designs, to identify components which are sensitive to fatigue damage, assess fatigue-cost effects. The method could also be extended to consider the power quality, and actual capital, operational and maintenance costs.

b) Further work should be carried out which compares the different methods of regulating power and assesses their significance on fatigue and cost for various machine concepts, in order to determine the optimum combination of control method and control algorithm.

c) Further fatigue related research may help to reduce the additional fatigue cost. The maximum benefits would occur with components tested with realistic load spectra.

ACKNOWLEDGEMENTS

The authors gratefully acknowledge the support of this project by the Department of Trade and Industry (Energy Technology Support Unit) and for the very helpful project guidance given by Mr. K. McAnulty. The project was managed by Mr. R.A. Lyons. RISØ National Laboratories, Denmark provided data and conducted some of the analytical work.

REFERENCES

1. Barltrop, N.D.P., Ward, I.P. and Daw, D.J. `A Fatigue Costing Study of Horizontal Axis Wind Turbines'* WS Atkins Consultants Limited report for Department of Trade and Industry (ETSU), UK. 1993.

2. Petersen, J.T. `Kinematically Nonlinear Finite Element Model of a Horizontal Axis Wind Turbine'. RISØ National Laboratory, Denmark. 1990.

3. Pedersen, T.F., Petersen, S.N., Themsen. K., Madsen, P.H. and Højstrup, J. `Loads for Wind Turbines in Inhomogeneous Terrain Measurement Report: Danwin 23 Study'. RISØ National Laboratory, Report RISØ-M-2922, Denmark. 1991.

4 van Grol, H.J. and Bulder, B.H. `Procedures to Determine the Fatigue Life for Large Size Wind Turbines'. EWEC, Travemunde, Germany. 1993.

5 Rasmussen, F. and Kretz, A. `Dynamics and Potential for the Two-Bladed Teetering Rotor Concept'. RISØ National Laboratory, Report, Denmark. 1992.

6. Leithead W.E., de la Salle S., Reardon D., 1991. `Assessment of Control Design Methods for Constant Speed HAWT's', 13th BWEA Conference.

7. Quarton D.C. `The Impact of Pitch Control on Wind Turbine Fatigue', 12th BWEA Conference. 1990.

8 Thomsen K. `Sammnlgning af Bladlaster pa Stall - og Pitchregulerede Vindmoller', RISØ National Laboratory, Report Riso-M-2905, Denmark. 1991.

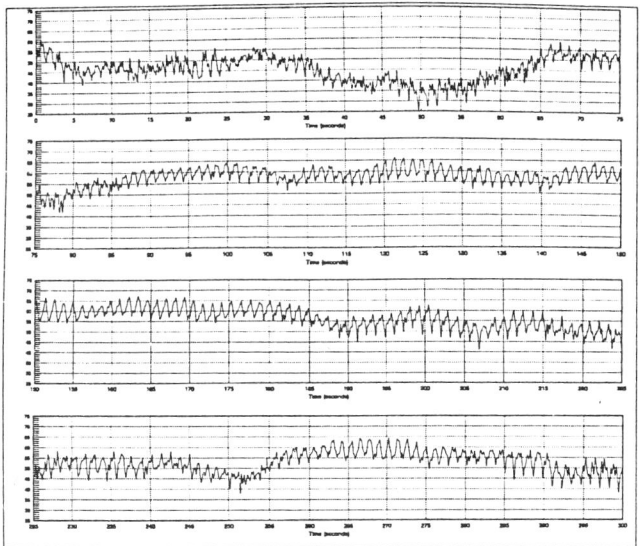

Figure 1   Example time history simulation (blade flapwise bending moment, deterministic + stochastic loads).

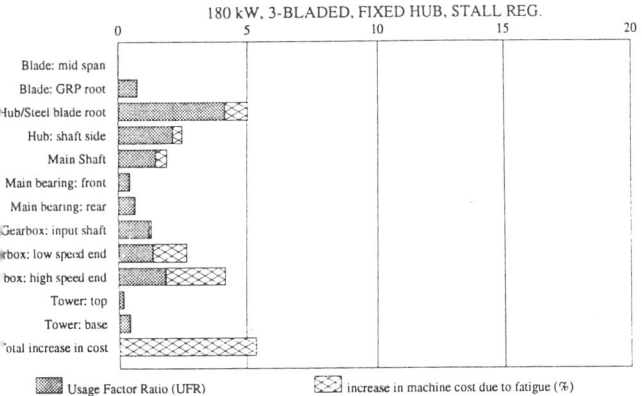

Figure 2   Fatigue cost for 180 kW, 3-bladed, stall-regulated turbine.

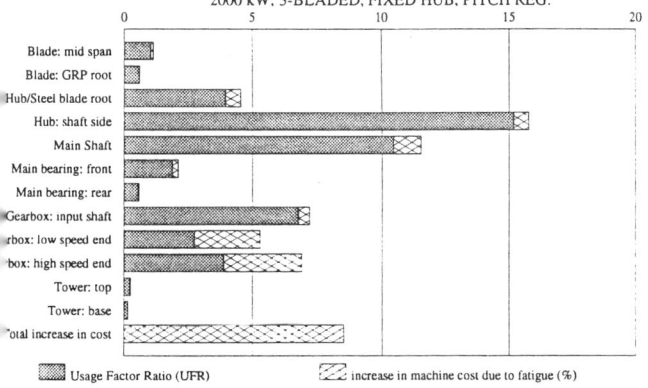

Figure 3 Fatigue cost for 2000 kW, 3-bladed, pitch-regulated turbine.

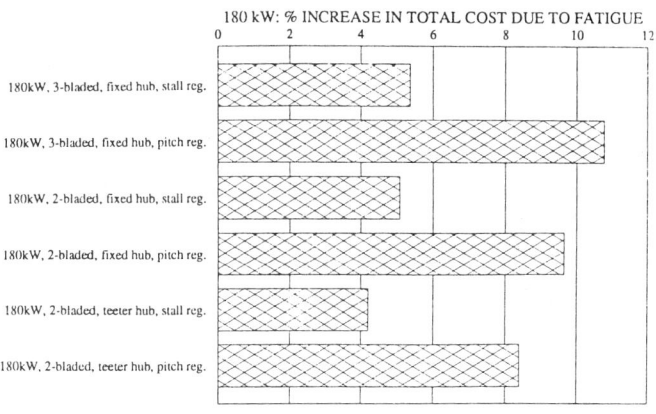

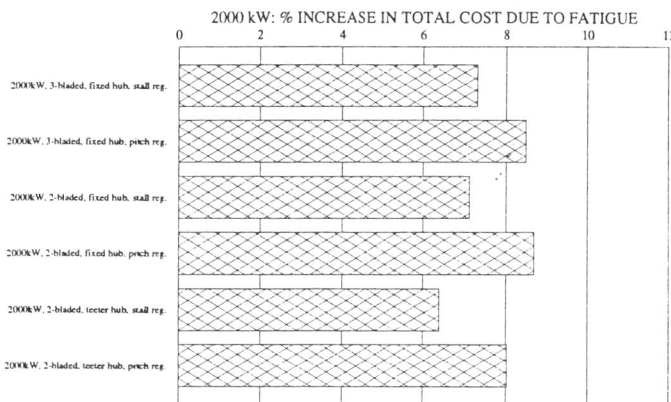

Figure 4   Comparison of fatigue costs for base case turbines with overall design changes.

259

Methods for the rapid evaluation of fatigue damage on the Howden HWP330 wind turbine

N. W. M. BISHOP, Z. HU, R. WANG
University College London, UK
D. QUARTON
Garrad Hassan & Partners Ltd, UK

Abstract

Spectral methods for predicting fatigue damage of wind turbine blades have been shown to be efficient and accurate for the analysis of loading situations which are reasonably stationary and Gaussian. The aim of this paper is to assess the adequacy of frequency domain fatigue techniques for analysing wind turbine blade loads using measured responses from a Howden HWP330 wind turbine. Two most promising frequency domain approaches, Dirlik's empirical solution and Bishop's theoretical solution, are applied. Mean stress effect is discussed. As expected, this data was found to have significant non-stationary, non-Gaussian and deterministic components and so the statistical results obtained were compared with the results obtained earlier for the MS1 machine. Further work is currently being carried out to make the spectral methods more applicable to such data and so the paper will also present the progress made so far with this.

1. INTRODUCTION

The frequency domain approach to fatigue analysis is a particularly useful tool for the design of wind turbine blades because of its computational simplicity and speed. Furthermore, the use of frequency domain based techniques for the analysis of structural systems is now widely accepted in many engineering fields. The offshore industry, for instance, relies very heavily on the use of PSD (Power Spectral Density) techniques for the dynamic analysis of deepwater platforms [1]. Dynamically sensitive structures are efficiently analysed in this way. Statically responsive structures can also be more efficiently analysed this way when the loading conditions are complex.

However, the application of PSD based analysis techniques for predicting fatigue damage assumes that the response is stationary and Gaussian (normal). Many structural systems do not conform rigidly to these requirements. In particular, wind turbine blade loads are likely to have significant non-stationary, non-Gaussian and deterministic components.

During the mid 1980's, Howden Wind Turbines Ltd recorded a considerable volume of measured data from 2 comprehensively instrumented 3 bladed machines in California. Working under contract to ETSU, Garrad Hassan have recently been involved in a detailed analysis of the data. The work reported in this paper has been carried out at University College London under subcontract to Garrad Hassan and has made use of measured data from 1 of the 2 instrumented machines, a 33m diameter HWP330 located within a large windfarm in the complex terrain of Altamont Pass.

The aim of this paper is therefore to assess the adequacy of frequency domain fatigue techniques for analysing wind turbine blade loads using measured responses from a Howden HWP330 wind turbine. Four monitored results files were analysed, each containing three flapwise and three edgewise signals at 3.00m, 8.09m and 13.04m radius locations. A summary of the four basic load cases is given in Table 1.1.

Table 1.1. Description of Howden HWP330 data.

tape(no)	wind speed (m/s)	turbulence intensity (%)	mean yaw (degrees)	duration (sec)	data length
18	10.68	19.6	-11.7	2560	102000
26	14.07	9.2	-6.5	3260	130400
27	16.86	10.7	-11.8	3863	154518
30	8.51	15.3	2.4	3512	140398

As expected, this data was found to have significant non-stationary, non-Gaussian and deterministic components and so the (statistical) results obtained were compared with results obtained earlier [2,3] from a Wind Energy Group MS1 machine. For this machine, six monitored results files were used corresponding to the flapwise load cases at 1.35m, 3.28m, 4.94m and 7.24m along the blade respectively. A summary of the six basic load cases is given in Table 1.2.

Table 1.2. Description of Wind Energy Group MS1 data.

case	wind speed (m/s)	yaw (degrees)	turbulence intensity (%)	hub conf	duration (sec)	data length
A	18.4	7.1	9.7	fixed	300	37500
B	23.7	3.0	11.1	fixed	240	30028
C	11.1	-12.5	8.9	fixed	300	37500
D	16.5	2.5	10.1	fixed	300	37500
E	15.6	-3.1	6.6	teet	102	12823
F	11.3	12.3	15.0	teet	300	37500

2. BASIC THEORY

The methods used for this project rely on the assumption that a standard *S-N* curve can be used to model the material properties of the components being analysed. This simply shows that under constant amplitude cyclic loading, a linear relationship exists (with slope, *m*) between cycles to failure *N* and applied stress range *S* when plotted on log-log paper. Because 'real' signals rarely conform to this ideal constant amplitude situation, an empirical approach is used for calculating the damage caused by stress signals of variable amplitude. Despite its limitations, Miner's rule is generally used for this purpose. The rule states that at failure;

$$\sum \frac{n}{N} = 1.0$$

Rainflow ranges have been widely used for estimating fatigue damage from random signals since Matsuishi and Endo first introduced the concept to the scientific community over twenty years ago [5]. When the loading sequence is specified as a time history the procedure for calculating rainflow ranges is then relatively simple.

For structural systems analysed in the frequency domain a method is required for extracting the pdf of rainflow ranges directly from the PSD of stress. The characteristics of the PSD which are used to obtain this information are the *n*th moments of the PSD function:

$$m_n = \int_0^{\infty} f^n G(f) df$$

Some very important statistical parameters can be computed from these moments such as the expected number of zeros and peaks per second;

$$E[0] = [\frac{m_2}{m_0}]^{1/2}, \qquad E[P] = [\frac{m_4}{m_2}]^{1/2}$$

from which the irregularity factor γ can be computed, which gives an indication of the spread of frequencies present in the signal;

$$\gamma = \frac{E[0]}{E[P]}$$

γ varies between 1.0 and 0.0. A value of 1.0 corresponds to a narrow band signal which, as the term implies, means that the signal contains only one predominant frequency. A

value of 0.0 implies that the signal contains an equal amount of energy at all frequencies.

3. ESTIMATING FATIGUE DAMAGE FROM PSD'S

A number of ways now exist for estimating fatigue damage from PSD's based on rainflow cycle counting. They can be classified into two groups: empirical and theoretical approaches. A summary of the various methods is given elsewhere [6][12].

4. INCLUSION OF MEAN STRESS EFFECTS

Most *S-N* curves apply to constant amplitude cycles with zero mean stress. In order to use stress cycles with a non-zero mean stress it is usual to convert the stress range into an equivalent stress cycle range at zero mean stress using, for instance, the *Goodman* formula;

$$S_e = \frac{S_a}{(1 - S_m / S_{ult})}$$

In which, S_a is a stress cycle range with mean stress, S_m is the mean stress, S_e is a stress cycle range when $S_m = 0$ and S_{ult} is the ultimate tensile stress of the material.

In the absence of desired stress measurements all analysis was carried out on the basis of bending moment data. Therefore, the ultimate stress of the material was converted to an ultimate bending moment for each cross sections of the blade. These values are given in Table 4.1.

Table 4.1. Ultimate bending moment information for the Howden data.

radial station	m=12 flapwise	m=12 edgewise	m=8 flapwise	m=8 edgewise
3.0	325.5	452.9	538.1	902.9
8.09	170.3	166.1	277.3	329.5
13.04	40.52	31.97	64.92	62.47

The ultimate bending moments given in Table 4.1 were obtained by ignoring the actual structure of the blade, but instead by performing an approximate prediction of the strength that would be required to design against fatigue. This is of course less realistic than using the sectional properties of the blade but provides a consistent means of comparing fatigue results for different slopes of *S-N* curve .

An additional assumption was made that mean stresses can be used with rainflow cycles which, by their very definition, are made up of sections of signal which may be separated by a large time interval. The concept of mean stress then becomes rather abstract. However, as with many engineering situations the use of mean stress in conjunction with rainflow ranges would appear to be the best option available and one which has some experimental backing. Wood epoxy wind turbine blades have been shown to be very fatigue sensitive to mean stress [13] and so it is important that this parameter is included in any design tool. However, only the Theoretical Solution is capable of being adapted in this way since the Dirlik solution is a closed form empirical expression which is impossible to separate and rework.

To consider the influence of mean stress in fatigue damage, the pdf of stress cycles is calculated together with the mean stress level for each cycle. In the time domain analysis, the mean stress associated with the cycle is easily obtained when the cycle is counted. In a frequency domain analysis, the Theoretical method offers a way to distinguish the relative mean stress level of the cycles when their probability densities are calculated.

To assess the influence of mean stress on fatigue damage, the pdf is converted to a corresponding pdf with zero mean stress. This is done using the Goodman relationship described previously. This means that, a cycle S_a with mean stress S_m will be enlarged to range S_e.

For a time domain analysis, the zero mean can easily be obtained from the signal. For a frequency domain analysis, however, there must be a way to set the zero mean stress level because PSD's have lost the information about mean stress. For application of the Theoretical solution in this project the mean value is taken from the corresponding time domain analysis. However, in practice the use of frequency domain methods is likely to be for fast calculations at the design stage. At this point the appropriate mean value to apply with each PSD should be easily established.

5. STATISTICAL ANALYSIS

A statistical analysis was carried out for all the data. For convenience of analysis, each signal was broken up into a number of blocks which were then analysed as independent records. Arithmetic averages were then used to compute a single value for the particular variable of interest .

The *Mean, Root Mean Square (RMS)* and *Irregularity Factor* were calculated directly from the appropriate number of blocks of signal and then summed to get the arithmetic average.

Stationarity and *Trend Tests* were carried out using the *Reverse Arrangement Test (RAT)*, i.e., after calculating the mean/rms of every block, a sequence of N (number of blocks) observations (denoted as $x_i=1,2...N$) of a random variable x is obtained. The number of times that $x_i>x_j$ for $i<j$ is then counted (This is called the *number of reverse arrangements*). If the sequence of N observations are independent observations of the same random variable, then the number of reverse arrangements is a random variable A, with the mean and variance as follows[14]

$$\mu_A = \frac{N(N-1)}{4}$$

$$\sigma_A^2 = \frac{N(N-1)(2N+5)}{72}$$

For the significant level $\alpha=0.05$, the accepted regions for the hypothesis that the observations are independent observations of a random variable depend on the number of blocks and are listed in Table 5.1

Table 5.1 95% confidence region

N	μ_A	σ_A	lower bound	upper bound
31	232.5	29.4	174	290
34	280.5	33.7	214	346
37	333.0	38.2	258	407
63	976.5	84.3	811	1142
68	1139.0	94.4	953	1324
75	1387.5	109.3	1173	1601
78	1501.5	115.9	1274	1728
91	2047.5	145.8	1761	2333

A *Normality Test* was carried out using two approaches. The χ^2 *(Chi-square) Goodness-of-Fit Test* and also by using the *Kurtosis Test*. The χ^2 Goodness-of-fit Test was performed using K.Pearson's statistics [14],

$$\chi^2 = \sum_{i=1}^{k} \frac{(f_i - F_i)^2}{F_i}$$

where f_i is the observed frequency, i.e., the number of observations falling within the ith class interval; F_i is the expected frequency, i.e., the number of observations that would be expected to fall within the ith class interval if the true probability density function of x is Gaussian. According to the *Pearson Theorem*, χ^2 obeys the χ^2 distribution of Degree of Freedom $k-1$. For practical problems, the statistical parameters (i.e., mean and root mean square) are not known, and have to be estimated. The actual Degree of Freedom is then $k-3$. The results for $k=43$ (i.e., D.O.F=40) are listed. For reducing error, the intervals are chosen so that all $f_i>3$. For the significant level of 0.01, the upper bound of the accepted regions for χ^2 (40) is *63.69*.

Without losing generality, we assume the signal is stationary with a zero mean value, when carrying out the *Kurtosis (Coefficient of Excess) Test* [15,16] . Denote μ_i (i=1,2,4) as the ith central moment, σ^2 as the variance. Kurtosis (or Coefficient of Excess) is then defined as

$$\kappa = \frac{\mu_4}{\sigma^4} - 3$$

κ can be used as a measure of the degree of flattening of a frequency curve near its centre. Kurtosis therefore gives an

263

indication of the drift of the signal from a Gaussian distribution. The minimum value of κ is -2, and this occurs only when x is a symmetric binary random variable ($|x|$ =constant). At the other extreme, κ may be infinite for a probability density function with slowly decaying tails. For a normal distribution, since $E[x^4]=3\sigma^4$ [17], then $\kappa=0$.

For real signals, the signal is transformed to a zero-mean random variable. σ^2 and $E[x^4]$ are then estimated and used to get an estimation of kurtosis. If kurtosis is around zero, the signal can be regarded as Gaussian. When taking an arithmetic average, some blocks of the signal give a positive kurtosis and some give a negative kurtosis. A very small kurtosis can then be obtained. An improvement of this may be obtained by averaging the absolute value of kurtosises for every block. κ_1 is therefore the average of the absolute value and κ_2 is the arithmetic average.

6. RESULTS

6.1 Basic fatigue calculations (without mean stress)

Ignoring the influence of mean stress, both the Dirlik and the Theoretical spectral methods were applied to the data. Because these spectral methods are very sensitive to some parameters such as the FFT window size and cutoff frequency of the PSD, predefined techniques were used to set these variables. The window size of the FFT was set at 4096 data points (in the time domain). The clipping ratio was set at 6 times the rms value. The frequency cutoff point for the PSD was set at the frequency corresponding to

99.5% of m_0^{100}, where m_0^{100} is defined as the zeroth moment obtained by integration up to the maximum frequency limit (see section 6.4).

Tables 6.1, 6.2 and 6.3 list the Howden data results for S-N slopes of 4, 8 and 12. In these tables the numbers given are the ratios between the frequency and time domain estimates of fatigue damage rates, for instance the damage obtained using Dirlik's method divided by the damage obtained using a rainflow count of the original signal.

Table 6.4 lists the results (from [3]) for the WEG MS1 machine with an S-N slope of 5.

Figures 6.1 shows the pdf curves for the Howden data. Figures 6.1(a) is an edgewise signals. Figure 6.1(b) is a flapwise signal. Fig. 6.2 shows the pdf of a WEG data sample.

Table 6.1 Fatigue damage ratios for the Howden data without mean stress m=4

Chnl	tape 18 Dirlik	tape 18 Th'cal	tape 26 Dirlik	tape 26 Th'cal	tape 27 Dirlik	tape 27 Th'cal	tape 30 Dirlik	tape 30 Th'cal
3m flapwise	0.825	0.905	0.858	0.852	0.854	0.971	1.012	0.927
3m edgewise	1.702	1.965	1.616	0.832	1.581	0.519	1.758	2.324
8m flapwise	0.740	0.740	0.841	0.840	0.830	0.860	0.811	0.844
8m edgewise	1.625	1.410	1.548	0.493	1.466	0.397	1.693	1.822
13m flapwise	0.773	0.323	1.108	0.163	1.113	0.290	0.839	0.357
13m edgewise	1.495	0.660	1.334	0.294	1.281	0.306	1.483	0.896

Table 6.2 Fatigue damage ratios for the Howden data without mean stress m=8

Chnl	tape 18 time	tape 18 Th'cal	tape 26 time	tape 26 Th'cal	tape 27 time	tape 27 Th'cal	tape 30 time	tape 30 Th'cal
3m flapwise	0.416	0.581	0.366	0.446	0.575	0.730	0.808	0.838
3m edgewise	16.508	18.147	13.959	8.377	12.594	5.107	18.149	23.338
8m flapwise	0.278	0.332	0.334	0.452	0.488	0.596	0.407	0.478
8m edgewise	14.604	12.665	11.691	4.620	9.477	3.341	16.098	16.301
13m flapwise	0.299	0.161	0.697	0.145	0.500	0.177	0.170	0.092
13m edgewise	11.897	6.225	4.845	1.430	6.405	2.053	11.753	8.107

Table 6.3 Fatigue damage ratios for the Howden data without mean stress m=12

Chnl	tape 18 th'cal	tape 18 Dirlik	tape 26 th'cal	tape 26 Dirlik	tape 27 th'cal	tape 27 Dirlik	tape 30 th'cal	tape 30 Dirlik
3m flapwise	0.136	0.215	0.126	0.174	0.256	0.335	0.255	0.290
3m edgewise	387.302	406.441	309.107	193.658	260.951	114.635	448.297	559.042
8m flapwise	0.085	0.113	0.097	0.162	0.177	0.237	0.083	0.104
8m edgewise	316.215	274.071	209.757	90.025	146.376	63.549	367.660	352.934
13m flapwise	0.086	0.059	0.424	0.128	0.169	0.084	0.009	0.006
13m edgewise	245.986	132.810	7.540	2.906	62.961	26.879	221.880	165.038

Table 6.4. Basic fatigue results for WEG MS1 data m=5.

load case	Dirlik	Th'cal	load case	Dirlik	Th'cal
a(1.35m)	1.03	1.07	d(1.35m)	0.84	0.81
a(3.28m)	1.00	1.09	d(3.28m)	0.83	0.83
a(4.94m)	1.59	1.23	d(4.94m)	1.01	0.81
a(7.24m)	2.34	1.37	d(7.24m)	1.12	0.87
b(1.35m)	0.77	0.79	e(1.35m)	0.99	0.90
b(3.28m)	0.81	0.86	e(3.28m)	1.01	1.02
b(4.94m)	1.07	0.91	e(4.94m)	1.03	1.08
b(7.24m)	1.48	1.30	e(7.24m)	1.11	1.41
c(1.35m)	0.76	0.61	f(1.35m)	0.98	1.01
c(3.28m)	0.73	0.61	f(3.28m)	1.00	1.13
c(4.94m)	0.74	0.51	f(4.94m)	1.01	1.11
c(7.24m)	0.76	0.46	f(7.24m)	0.98	1.17

Several points are clear from the pdf curves and fatigue damage tables. Firstly, agreement between the true (time domain) and frequency domain (either Dirlik or Theoretical) methods is excellent for the WEG data. For the Howden data the agreement is much poorer with the worst agreement being for the edgewise data where significant deterministic components are present. The worst damage ratio obtained for $m=12$ is 559%. When converted to an equivalent stress range parameter this represents an overprediction of 70%. Sources of error for this data are discussed later.

6.2 Mean stress fatigue calculations

For the Howden data the probability density functions including mean stress levels obtained from both the time domain and frequency domain analysis methods are shown in Figs 6.3. Results for the Howden data are listed in Tables 6.5 and 6.6 for S-N slopes of 8 and 12. The damage ratios given are the ratios between the estimates. with and without mean stress included.

Table 6.5. Fatigue damage ratios for the Howden data with mean stress m=8

Chnl	tape 18 time	tape 18 Th'cal	tape 26 time	tape 18 Th'cal	tape 27 time	tape 27 Th'cal	tape 30 time	tape 30 The'cal
3m flapwise	3.910	4.394	4.680	4.554	3.781	3.712	3.029	3.090
3m edgewise	1.585	1.486	1.103	1.093	1.129	1.128	1.076	0.873
8m flapwise	2.234	2.412	2.281	2.583	2.059	1.937	2.155	2.248
8m edgewise	2.791	2.751	1.053	1.040	1.068	1.068	1.031	1.017
13m flapwise	1.394	1.443	1.080	1.225	1.010	0.973	1.314	1.496
13m edgewise	1.639	1.609	1.018	1.000	1.000	0.973	1.045	1.025

Table 6.6. Fatigue damage ratios for the Howden data with mean stress m=12

Chnl	tape 18 time	tape 18 Th'cal	tape 26 time	tape 26 Th'cal	tape 27 time	tape 27 Th'cal	tape 30 time	tape 30 Th'cal
3m flapwise	35.887	52.163	65.208	57.928	43.748	32.368	17.247	19.368
3m edgewise	4.129	3.944	1.348	1.354	1.442	1.493	1.247	1.049
8m flapwise	7.533	9.477	7.383	11.388	7.227	5.357	6.632	7.863
8m edgewise	26.456	26.202	1.266	1.156	1.218	1.258	1.094	1.116
13m flapwise	2.182	2.476	1.162	1.660	1.017	0.945	1.719	2.707
13m edgewise	4.491	4.393	1.007	1.017	1.000	0.946	1.154	1.099

It is clear that the Theoretical spectral method provides an accurate method for assessing the influence of mean stress for the Howden data for any *S-N* slope value. The level of agreement for the Howden edgewise data is surprising because of the discrepancies between the relevant pdf curves (see Figs 6.3(c) and 6.3(d)). However, there is a relatively lower mean stress level present for these edgewise load cases and so it is the flapwise load cases, where the agreement is better, which are of most importance.

6.3 Statistical results

The statistical analysis results for the Howden data are given in Table 6.7.

Table 6.7. Howden Data Statistics
(Head & Tail of signal removed: 4096 pts/blk)

Data	Blks	Mean	Ratm	RMS	Ratr	$\chi^2(40)$	γ	$\kappa(1)$	$\kappa(2)$
h18ch5	24	91.08	93	9.34	131	44.53	0.239	0.346	0.044
h18ch6	24	50.41	147	29.86	142	604.15	0.408	1.459	-1.459
h18ch7	24	29.17	92	4.02	131	45.67	0.263	0.475	0.332
h18ch8	24	39.66	184	6.90	144	532.79	0.321	1.444	-1.444
h18ch9	24	3.04	84	0.67	139	34.49	0.291	0.548	0.48
h18ch10	24	3.62	16	2.26	134	417.45	0.134	1.412	-1.413
h26ch5	31	93.80	316	11.37	168	33.87	0.28	0.31	0.028
h26ch6	31	10.31	118	35.31	125	451.81	0.35	1.42	-1.421
h26ch7	31	31.61	335	5.65	171	52.50	0.24	0.53	0.179
h26ch8	31	1.38	127	12.40	135	369.86	0.24	1.38	-1.385
h26ch9	31	1.79	334	1.49	170	189.19	0.18	1.11	0.374
h26ch10	31	-0.21	291	2.25	220	219.98	0.13	1.21	-1.216
h27ch5	37	82.23	566	13.95	246	32.14	0.397	0.348	0.257
h27ch6	37	14.03	384	35.35	445	387.88	0.26	1.39	-1.393
h27ch7	37	2.87	344	12.49	267	322.00	0.20	1.35	-1.349
h27ch8	37	22.63	571	7.15	357	32.77	0.34	0.43	0.371
h27ch9	37	-2.29	580	1.77	458	71.21	0.22	0.52	0.407
h27ch10	37	-3.08	304	2.24	464	206.44	0.12	1.23	-1.227
h30ch5	34	72.10	451	9.69	364	42.49	0.27	0.37	-0.101
h30ch6	34	7.99	48	35.34	297	654.41	0.44	1.47	-1.467
h30ch7	34	27.44	459	3.76	362	42.74	0.28	0.54	0.437
h30ch8	34	1.09	57	12.26	276	581.80	0.37	1.46	-1.455
h30ch9	34	3.38	445	0.60	338	41.30	0.30	0.98	0.981
h30ch10	34	0.15	536	2.16	260	430.85	0.15	1.42	-1.417

It is likely that there are three main sources of error in the fatigue damage estimate for the Howden data. Clearly for the edgewise signals there is a significant, and dominant, edgewise (deterministic) load. However, some of the flapwise signals also have quite large damage ratio discrepancies. The error for these load cases is likely to be from two further sources, non-stationarity and a non-Gaussian tendency. In order to improve the accuracy of the spectral fatigue methods progress will probably be required to deal with data containing deterministic component, non-Gaussian tendancy and non-stationarity..

6.4 Computational considerations

An important part of the Theoretical solution involves a numerical solution to a Markov Chain model. Although a more efficient means of solution almost certainly exists it

has so far not been obtained. It is therefore necessary to discretise the stress signal range into a manageable number of divisions. Generally the value used for this is 32. As the interval width increases, with reducing interval number, this discretisation error can increase to a point where the Theoretical solution becomes unstable. The Theoretical solution is particularly sensitive to this problem when the irregularity factor is low because in this situation the peak to trough transition matrix becomes confined to the band along its diagonal associated with low cycle range transitions.

In order to deal with this problem firstly the clipping ratio was fixed at 6 by experience. Both the Theoretical and Dirlik solutions were tested in this way although Dirlik's solution is not susceptible to the instability. Then, the moments and damage values (Theoretical and Dirlik) were plotted as a function of PSD cutoff frequency .

For these data sets the stable region for the Theoretical solution is for a cutoff frequency below 6Hz. The complication here is that the cutoff frequency cannot be less than a point below which most of the energy in the PSD is present. It was for this reason that the 99.5% of m_0^{100} was chosen as the means of establishing a cutoff frequency. Furthermore, even with Dirliks solution, the fourth moment is particularly sensitive to energy in the PSD at high frequencies. Since the fourth moment has a profound effect on all frequency domain methods for predicting fatigue this is also an important reason why the cutoff frequency needs to be properly established. An investigation of this problem for the WEG data was carried out in [6].

Another problem for both time domain and frequency domain methods is the choice of sample length used. In order to investigate this, each load case was analysed from its beginning and the damage ratios plotted as a function of increasing sample length. The results for the typical Howden and WEG data files are shown in Figs. 6.4(a) and (b).

The results show the degree of non-stationarity present in the Howden flapwise signal. Furthermore this is not just a frequency domain problem, it is just as important when working in the time domain.

7. CONCLUSIONS

Two frequency domain methods (Dirlik and Theoretical) have been used to analyse the Howden wind turbine blade loads. Both methods show the same tendency for calculations which ignore mean stress. This is reassuring because Dirlik's method is considerably easier to implement and is much more stable when being computed.

The results for fatigue damage ratio which ignore mean stress show large discrepancies for a high S-N slope value of 12. However when these ratios are converted to equivalent stresses the errors are less than 70%.

Dirlik's method cannot deal with mean stress calculations although the Theoretical method can. The Theoretical method has been tested on the Howden data and has been shown to give very good agreement with the result obtained from a time domain analysis. For a design tool, it will be necessary to produce a Dirlik type solution including mean stresses which is easy to compute and stable. Work on this is currently being carried out with a reasonable degree of optimism that a solution can be obtained.

In order to improve the spectral estimates for data such as the Howden blade load data, progress will be required to deal with the following (in order of priority).
1. Deterministic components.
2. A non-Gaussian tendency
3. Non-stationarity.

As mentioned above, work is currently underway on (1) to produce a Dirlik type solution which will deal with 1 significant deterministic component such as that which is present in the Howden data. Work on (2) and (3) is also being carried out although this is still very much at the research stage. An improved theoretical solution is being proposed which is simpler to apply to engineering data.

ACKNOWLEDGEMENT

The authors of this paper are most grateful to Garrad Hassan & Partners Ltd for the provision of data from a Howden HWP330 wind turbine as a part of work done under ETSU/DTI contract no. W/24/00198, and to SERC for their financial support under grant number GR/G 53903

REFERENCES

[1] N W M Bishop, Dynamic fatigue response of deepwater offshore structures subjected to random loading, Structural Engineering Review, SER 76/11, Aug 1991.

[2] N.W.M.Bishop and Hu Zhihua, The Analysis of Non-Gaussian Loadings from Wind Turbine Blades Using Frequency Domain Techniques, British Wind Energy Conference 13, Swansea, April 1991.

[3] N.W.M.Bishop and Hu Zhihua,The Fatigue Analysis of Wind Turbine Blades Using Frequency Domain Techniques,EWEC '91, Amsterdam, Oct 1991.

[4] NWM Bishop, Fatigue life prediction from power spectral density data.Part 1, traditional approaches and Part 2, recent developments.Environmental Engineering, Vol.2, Nos. 1 and 2, 1989.

[5] M.Matsuishi and T.Endo, Fatigue of metals subject to varying stress, paper presented to Japan Soc Mech Engrs, Jukvoka, Japan, 1968.

[6] N.W.M.Bishop, Hu Zhihua and Ruhuai Wang, Fast Frequency Domain fatigue life assessment of wind turbine blades, BWEC 14, Nottingham 1992.

[7] G.Lindgren and I.Rychlik, Rainflow cycle distributions for fatigue life prediction under Gaussian load processes, Fatigue Fract. Engng. Mater. Struct. 10, No 3, 1987, pp 251-260.

[8] P.H.Wirsching and M.C.Light, Fatigue under wide band random loading, J Struct. Div., ASCE, pp1593-1607, July 1980.

[9] G.K.Chaudhury and W.D.Dover, Fatigue Analysis of Offshore Platforms Subject to Sea Wave Loading, Int J Fatigue 7, Jan 1985.

[10] J.C.P.Kam and W.D.Dover, Fast fatigue assessment procedure for offshore structures under random stress history, Proc. Instn. Civ. Engrs., Part 2, 85, Dec, 689-700, 1988.

[11] T.Dirlik, Application of computers in Fatigue Analysis, University of Warwick Thesis, Jan 1985.

[12] N.W.M.Bishop and F.Sherratt, A theoretical solution for the estimation of rainflow ranges from power spectral density data. Fat. Fract. Engng. Mater. Struct., 13 no.4, 1990.

[13] P Bonfield and MP Ansell, Fatigue testing of wood composites for aerogenerator rotor blades. Part V. Life Prediction and Hysteresis. Wind Energy Conversion 1990, eds TD Davies et al.

[14] Bendat, J.S. and Piersol, A.G., Random Data: Analysis and Measurement Procedures (2nd ed.), John-Wiley & Son, 1986. (Chap.4, sec.6)

[15] Cramer, H., Mathematical Methods of Statistics, Princeton Univ. Press, 1946, (13th printing, 1974). (chap.15, sec.8,10.)

[16] Bickel, P.J., and Doksum, K.A., Mathematical Statistics, Holden-Day Inc., 1977.(chap.9, sec.6)

[17] Middleton, D., An Introduction to Statistical Communication Theory, McGraw-Hill Book Company, 1960. (eq.(7.29a))

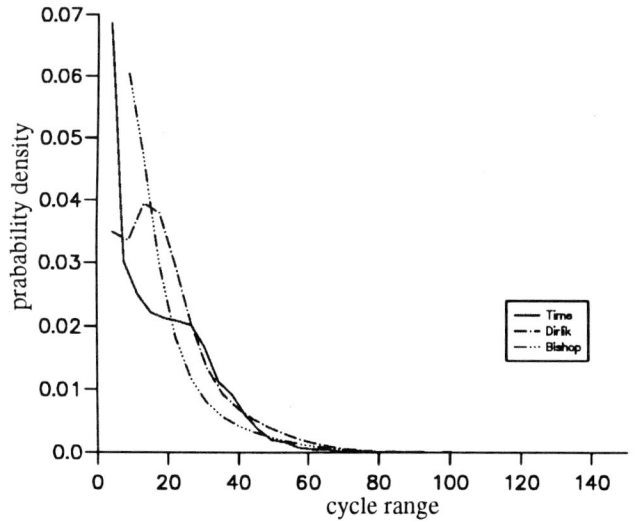

Fig 6.1(a) Howden data pdf curves for tape 26, 3m flapewise

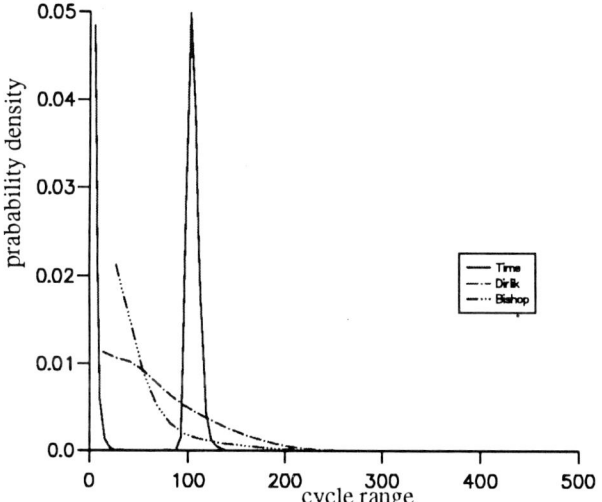

Fig 6.1(b) Howden data pdf curves for tape 26, 3m edgewise

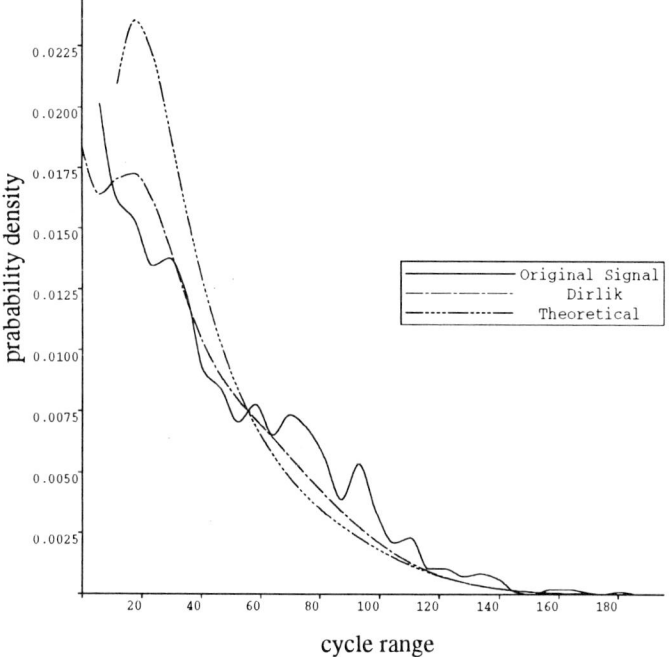

Fig 6.2 WEG data pdf curves for load case a (1.35m)

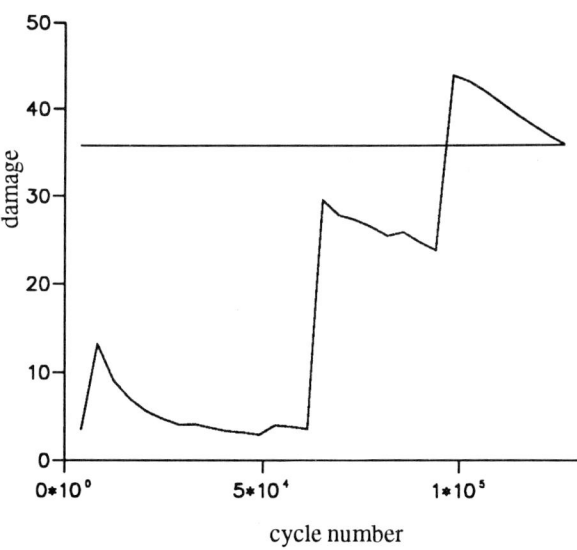

Fig 6.4(a). The effect of sample length on fatigue damage for Howden load case 26, 3m flapwise.

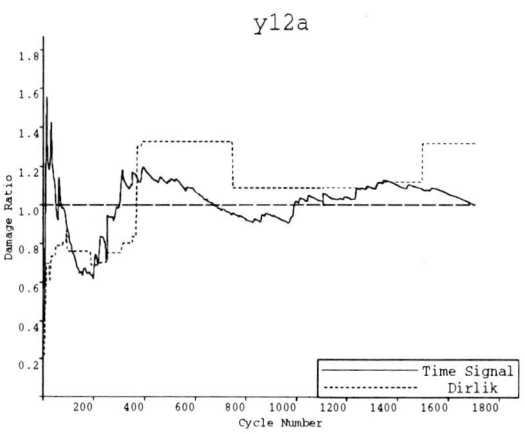

Fig 6.4(b). The effect of sample length on fatigue damage for WEG load case a (1.35m).

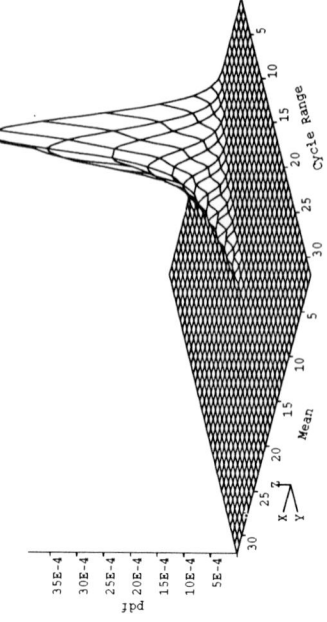

Fig 6.3(c). The joint pdf of rainflow ranges and means computed from the time sample of tape 26 3m edgewise (Howden data)

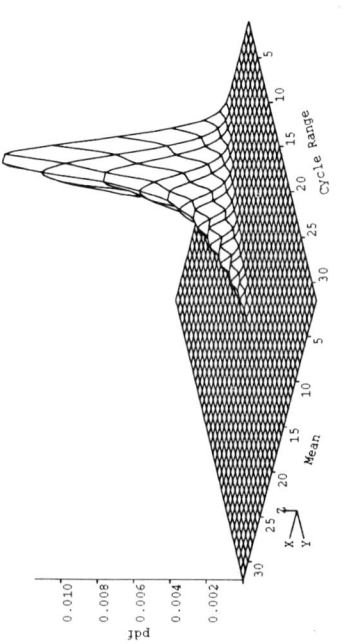

Fig 6.3(d). The joint pdf of rainflow ranges and means computed using the Theoretical Solution applied to tape 26 3m edgewise (Howden data)

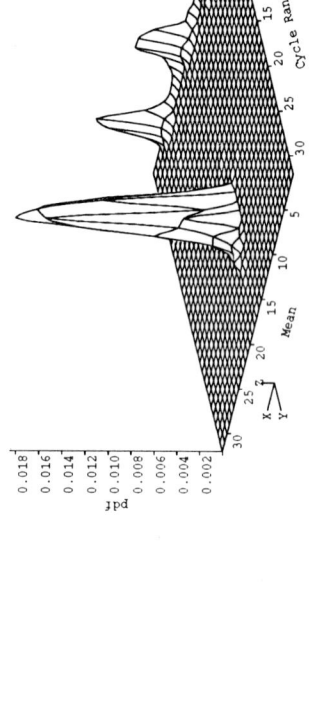

Fig 6.3(a). The joint pdf of rainflow ranges and means computed from the time sample of tape 26 3m flapwise (Howden data)

Fig 6.3(b). The joint pdf of rainflow ranges and means computed using the Theoretical Solution applied to tape 26 3m flapwise (Howden data)

Fatigue testing of wood composites for aerogenerator blades, part IX. Alternative adhesives

C. L. HACKER, I. P. BOND and **M. P. ANSELL**
The University of Bath, UK

In order to reduce the cost and facilitate the manufacture of wood laminate wind turbine blades a range of cheaper adhesives have been investigated. Block shear tests were carried out on various room temperature curing amino, phenolic and vinyl acetate adhesives. Fully laminated samples were produced using phenol-resorcinol formaldehyde (PRF), melamine urea formaldehyde (MUF) and poly vinyl acetate (PVA) adhesives. These were tested statically and in fatigue at $R=-1$ ($R=\sigma_{min}/\sigma_{max}$). Although the epoxy performed better than the other adhesives the fatigue performance of all adhesives converged at around 10^7 cycles. This suggests that, in low stress, long life applications such as wind turbine blades, the adhesive will not significantly affect the blade's fatigue performance. Consequently, it should be possible to select the cheapest, safest and easiest to use adhesive, namely PVA, for the manufacture of wind turbine blades unless other factors such as its creep behaviour make it unsuitable.

1. INTRODUCTION

The UK wind energy industry is now established and is in the more commercial phase of wind farm construction and management. Research is being carried out to meet some of its current needs under the third of a series of grants from the Energy Technology Support Unit (ETSU) [formerly part of the Department of Energy but now under the control of the Department of Trade and Industry]. The thrust of the work is three-fold: firstly alternative adhesive systems, alternative woods (1) (2) and joint systems (1) (3) are being investigated. Significant cost savings are possible if cheaper resin systems and woods have the necessary fatigue performance to replace the current materials. Secondly the areas of fatigue life prediction (4) and damage accumulation in wood laminates (3) are being explored in more depth. The effect of complex load-time histories relating to different wind farm sites and wind turbulence levels on damage build-up in wood will also be determined. Thirdly, the lack of design codes for wood composite blades is to be remedied by the production of design rules (5) which are directly related to the composite's fatigue performance.

This paper, the ninth in the series 'Fatigue testing of composites for aerogenerator blades', concentrates on one particular aspect of our work, namely the assessment of alternative resin systems for use in the production of wood composite wind turbine blades.

2. ALTERNATIVE ADHESIVES

The range of adhesives which may be used with wood is extensive and includes elastomers, thermoplastics and thermosetting resin systems (6). Of all the available adhesives the ones which are used most commonly with wood are the amino (UF) or phenolic (PF) resins, their derivatives and PVA. Slightly less common ones include isocyanate and epoxy. Other adhesives have various wide ranging applications but in terms of the volumes consumed per year are much less significant.

It is important that with such a wide choice of adhesives available the correct one is used in a particular application. The bond must meet the strength requirements and survive the environmental conditions it will experience during service without weakening. It must also be compatible with the manufacturing process. Epoxy has been used successfully in the manufacture of wind turbine blades both in the United States and Europe and at present WEG Ltd. use about 200kg of two part, room temperature curing structural epoxy adhesive in an 18m blade costing approximately £1000.

Epoxies have excellent cohesive strength and will adhere well to most materials including wood. Their moisture resistance makes them suitable for the harshest of outdoor applications. However, they are not a widely used wood glue because of their high cost in comparison to other glues which have adequate properties. Table 1 shows the approximate prices of one mixed kilogram of adhesive as supplied by the manufacturer for a ten kilogram quantity (7).

It cannot be denied that Khaya-epoxy laminates perform extremely well in fatigue (1). However, the price of epoxy in the UK. is double that in the U.S.A. (8), so when the technology was imported to the UK. the cost of a blade increased by some £500 on account of the adhesive alone. A recent test of a 12m Khaya-epoxy blade at City University (9) showed that the blades had considerable design margins. After having withstood 3×10^6 cycles of fatigue loading the blade was statically loaded to failure and the calculated stress in the composite at failure was 75Nmm^{-2}. Since the design ultimate stress of the laminate was only 30Nmm^{-2} there is clearly scope for reducing costs by redesigning some features of the blade.

3. EXPERIMENTAL TECHNIQUE

Samples of phenol resorcinol formaldehyde (PRF), melamine urea formaldehyde (MUF), urea formaldehyde (UF), cross-linked polyvinyl acetate (PVA) and phenol formaldehyde (PF) were obtained from adhesive manufacturers for testing. The two-part, room temperature curing epoxy resin which is currently used by WEG was also tested.

The testing program fell into three sections; firstly, static shear testing of single gluelines; secondly, tensile testing of fully laminated samples and thirdly, fatigue testing of fully laminated samples. Stage one served to eliminate adhesives from the investigation which were of low strength or incompatible with the manufacturing process. The method of block shear testing was adopted because it avoids large stress concentrations and peel forces which are present in simple lap shear testing.

Laminated blocks of Khaya were bonded together and a shear force applied to the glueline using a jig as specified in BS 4169 (10), figure 1. The complete range of adhesives tested is given in table 1. An Instron 1195 was used to apply a load with a cross-head speed of 1mm min^{-1} and a chart speed of 5mm min^{-1}. The maximum load applied to the sample was recorded by the Instron and from this the shear force exerted on the joint was calculated. Four or eight samples were tested for each adhesive and the mean shear strengths were calculated. All the surfaces of failed block shear samples were characterised by inspecting them under an optical microscope.

As a result of the initial block shear tests poly vinyl acetate, PVA (Britannia), phenol resorcinol formaldehyde, PRF (Dyno) and melamine urea formaldehyde, MUF (Dyno), were identified as possible candidates for replacing the epoxy adhesive currently used in blade manufacture. Factors such as high cost and incompatibility with the manufacturing process eliminated several adhesives from the investigation. Conversely, PVA despite its inferior strength was selected on account of its low cost and ease of use. Fully laminated samples were produced using each adhesive and comprised 5 pre-scarf jointed 4mm thick Khaya veneers with a glass - epoxy outer coating. In addition to the all PVA samples, a set was produced containing epoxy scarf joints and PVA laminations.

Static and fatigue testing were carried out using a Mayes 20 tonne servohydraulic fatigue test rig. Four or five samples of each type were tested and the mean tensile strength and

standard deviation calculated. Once the static testing was complete fatigue tests were set up and SN data obtained for each adhesive system at R=-1 (R=$\sigma_{min}/\sigma_{max}$). A constant loading rate of 400MPa/s was used and the moisture content of the sample kept constant at 65%RH. The number of data points obtained was limited by the small number of samples and the time available.

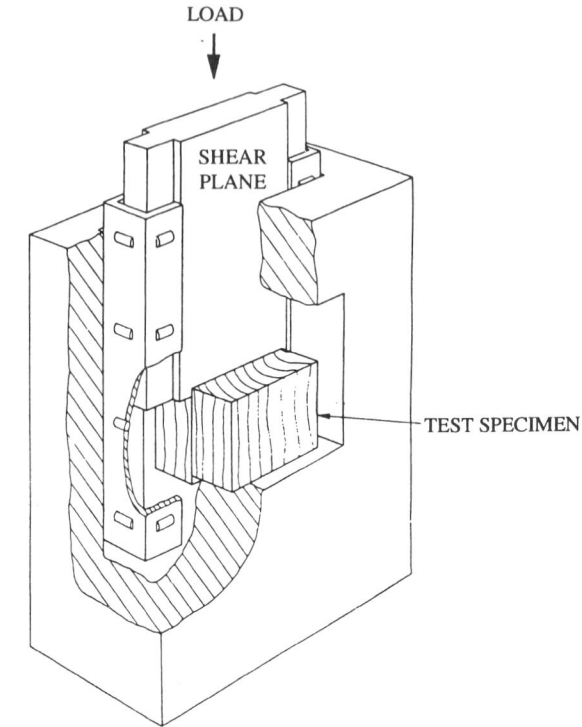

FIGURE 1. BLOCK SHEARING JIG. BS4169, 1988.

4. BLOCK SHEAR TESTING

The shear strengths of the different adhesive joints measured by the block shear method are presented in table 1, with values ranging from 6.2MPa for PRF (Borden) to 13.0MPa for Cascophen. It was observed that the samples failed by wood shear away from the glueline (deep wood shear), wood shear adjacent to the glueline (thin wood shear) or by shearing of the adhesive itself, figure 2. In some of the adhesives tested a mixture of the above failure types was observed but usually one particular type predominated.

The adhesives which failed predominantly by deep wood shear, namely Epoxy, PRF (Dyno) and Cascophen, tended to have the highest shear strengths. The failure path in such samples passed through both adherends up to 3mm away from the glueline. In the epoxy samples regions of adhesive failure were present and were usually associated with voids in the glueline. It is clear that in all these samples the strength of the joint is higher than that of the wood alone (9.6MPa). This is attributed to infiltration of the wood by the adhesive which reinforces the wood in the vicinity of the joint (11).

TABLE 1. ADHESIVES INFORMATION AND TEST DATA FOR KHAYA

Sample (Supplier)	Adhesive cost per mixed kg £ 10kg quantity	Moisture resistance (12) (13)	Block shear strength (Std. dev.) MPa	Tensile strength of fully laminated samples (Std. dev.) MPa	SN equation
Epoxy	5	WBP	12.5 (1.2)	67.2 (10.1)	S=50-4.8logN
PRF (Borden)	—	WBP	6.2 (1.6)	—	—
PRF (Dyno)	2.6	WBP	9.1 (0.9)	48.6 (1.8)	S=37-3.1logN
MUF (Dyno)	1.5	BR	9.9 (1.7)	54.5 (8.4)	S=40-3.8logN
PVA (Britannia)	1.0	Intermediate	7.6 (0.8)	64.4 (8.7) 56.2 (9.1) (Samples containing PVA scarf joints)	S=32-2.4logN S=31-2.1logN
PF (British Petroleum)	N/a	WBP	10.7 (1.3)	—	—
U.F.-Cascophen (Humbrol)	4.5	MR	13.0 (2.1)	—	—

WBP = Weather-proof and boil-proof   BR =Boil resistant
MR = Moisture resistant and moderately weather resistant   Intermediate = Properties between WBP and MR

Cascophen could prove to be a suitable candidate for laminating wind turbine blades since it is safe to use, its open and closed times are acceptable and the joint strength is high enough. If it could be obtained for a low price of around £1 per kg then it would warrant further investigation. However, it is important to remember that UFs have a lower moisture resistance than PRF or MUF, table 1. The level of environmental stability necessary for use in a wood laminate blade is yet to be established although it is not thought to be particularly high because the glass-epoxy outer layer on the blade acts as an effective barrier to the elements.

Other adhesives failed mainly by thin wood shear. Such failure surfaces were smooth with a thin layer of wood cells adhering to them and included MUF (Dyno), PVA (Britannia) and PRF (Borden). It is possible that failure by thin wood shear was caused by damaged surface wood on the substrates. Slatts (11) suggested that damaged wood acts as a weak boundary layer between the sound wood and the adhesive and also prevents the adhesive from penetrating into the good wood. The combination of these two effects would cause serious weakening of the bond which could not be compensated for by using higher performance adhesives. This view is supported by the fact that in the block shear tests it was observed that the samples which failed in the wood close to the glueline were usually weaker than those exhibiting predominantly deep wood shear.

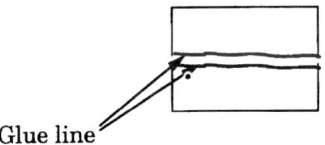

Adhesive shear failure

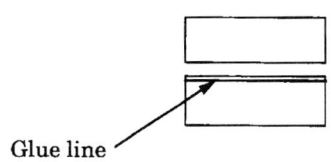

Thin wood shear failure

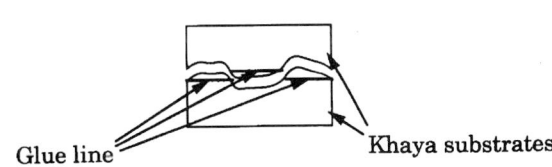

Deep wood shear failure

FIGURE 2. BLOCK SHEAR FAILURE TYPES

The only adhesive to fail entirely in the glueline was British Petroleum's cold curing PF, with a shear strength of 10.7MPa. Failure had occurred in the resin itself or at the wood-adhesive interface. However, with the Phencat 10 catalyst both the pot life and release time cannot be balanced appropriately for the time schedule used in wind turbine blade manufacture so this adhesive does not merit further investigation.

5. STATIC AND FATIGUE TESTING OF FULLY LAMINATED SAMPLES

The static tensile strengths and standard deviations of the fully laminated samples and the best fit linear equations fitting the fatigue data are given in table 1. The SN data are presented graphically in figure 3.

The mean tensile strength of the samples laid up using alternative adhesives are all lower than the mean tensile strength of the epoxy samples. Samples made with epoxy scarf joints and PVA laminations are the strongest of the alternative systems, followed by the all PVA, MUF and PRF samples.

In figure 3 the data for all the different adhesive systems have been plotted. The error bar on the static tensile strengths is equal to the maximum possible range in scatter for all the data. It is clear from this figure that the samples produced with epoxy adhesive have the highest fatigue performance. However, it is difficult to establish more than that and to differentiate between the performance of the other samples produced with alternative adhesives.

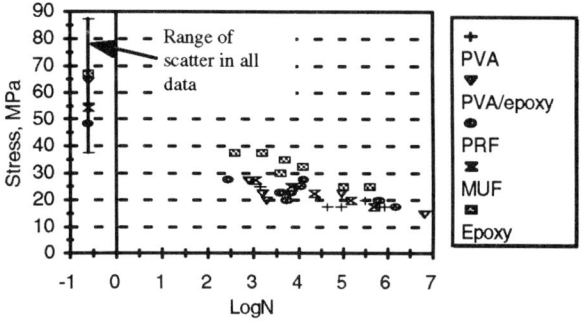

FIGURE 3. SN DATA FOR ALL THE ADHESIVES TESTED. R=-1. KHAYA VENEERS

The equations of the lines fitted by a linear regression to the fatigue data alone (the lines do not pass through the static strengths), however, reveal the differences in performance of the samples produced with alternative adhesives more clearly. These equations are summarised in table 1 and also plotted in figure 4. The equation which fits the data of the epoxy samples, which have the highest fatigue performance, has the highest intercept value and steepest gradient. The PVA equation has the lowest intercept and gradient, closely followed by that for the PVA samples with epoxy scarf joints. The gradient and

intercept for the PRF samples are higher still, but lower than the values obtained for the MUF samples. The sequence in which the intercept and gradient values increase is the same as the order in which the shear strengths of the different adhesive gluelines also increase.

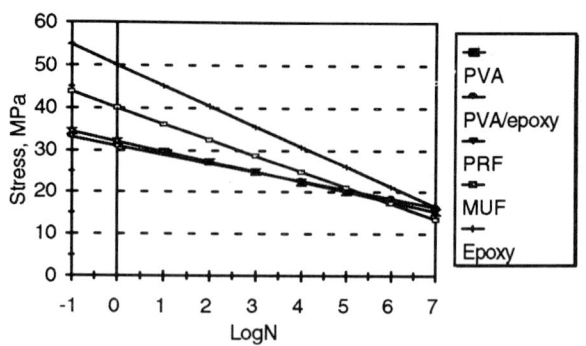

FIGURE 4. THE BEST FIT LINES FOR EACH ADHESIVE SYSTEM PLOTTED WITHOUT THE DATA POINTS

If the equations of the lines are plotted they all converge towards around 10^7 cycles and consequently the differences in the fatigue performance of each adhesive system are much less distinct. As wind turbine blades are designed to last over 10^8 cycles it would appear that for such low stress and long life applications the choice of adhesive does not significantly affect the composite's performance and consequently costs might be reduced by switching to a cheaper adhesive system. However, there is a range of other factors which must be investigated before such a significant change in blade material can be made. In addition to determining the composite's creep performance and the importance of its environmental stability, manufacturing issues must be addressed. It will be necessary to determine the adhesive's compatibility with the current blade production process and the viability of any modifications which might be required for its successful use. For example the adhesive must be compatible with the other materials in the blade, such as the epoxy used to form the scarf joints, and the blade production rate should also be maintained.

6. CONCLUSIONS

1. The fatigue performance of samples produced with the epoxy adhesive system, currently used in blade production, is superior to that of the samples laminated with possible alternative adhesives (MUF, PRF and PVA).

2. Although the fatigue performance of the samples laminated with MUF, PRF or PVA is, on the whole, fairly similar the resin systems can be distinguished from each other and listed in order of decreasing fatigue performance: Epoxy, MUF, PRF, PVA with epoxy scarf joints and PVA. This is the same as the order in which the shear strengths of each adhesive decrease.

3. The fatigue performance of all the adhesive systems converges at around 10^7 cycles and consequently are

indistinguishable at this point. In a low stress, long life application such as a wind turbine blade which is designed to last over 10^8 cycles it would appear that the choice of adhesive is not going to affect the performance of the blade. However, other factors such as the adhesive's creep behaviour might become significant in limiting a blade's life.

4. Once any production issues have been addressed, potential cost savings could be made by selecting a cheap, safe and easy to use adhesive system such as PVA for the laminating adhesive in wood composite wind turbine blades.

7. ACKNOWLEDGEMENTS

I would like to acknowledge the Energy Technology Support Unit (Department of Trade and Industry) at Harwell for awarding their contract No. E/5A/6111/2818 to the University of Bath which has enabled this research to be carried out. Thanks is also due to Mark Hancock of Wind Energy Group Ltd. for his practical advice on this work.

8. REFERENCES

(1) Bonfield, P.W., Bond, I.P., Hacker, C.L. and Ansell, M.P. Fatigue testing of wood composites for aerogenerator rotor blades. Part VII. alternative wood species and joints. In Wind Energy Conversion. Proceedings of the Fourteenth BWEA Wind Energy Conference, Nottingham 25-27th March 1992. Ed. Clayton, B.R. pp.243-250 (Mechanical Engineering Publications Ltd.).

(2) Bond, I.P., Hacker, C.L. and Ansell, M.P. First progress report on "Optimizing and standardising the fatigue design of commercial wood composite turbine blades". 1992. For the Department of Energy, Contract No. E/5A/6111/2818.

(3) Bonfield, P.W. (1991). Fatigue evaluation of wood laminates for the design of wind turbine blades. PhD thesis, University of Bath.

(4) Bonfield, P.W. and Ansell, M.P. Fatigue testing of wood composites for aerogenerator rotor blades. Part VI: spectrum fatigue loading, life prediction and damage rates. In Wind Energy Conversion. Proceedings of the Thirteenth BWEA Wind Energy Conference. Swansea 10-12th April 1991. Ed. Quarton, D.C.and Fenton, V.C.. pp.311-316 (Mechanical Engineering Publications Ltd.).

(5) Ould, M.J. and Ansell, M.P.. Development of norms and standards for wood composite materials used in the construction of wind turbine blades. Proceedings of the Fourteenth BWEA Wind Energy Conference. Nottingham 25-27th March 1992. Ed. Clayton, B.R.. pp.251-258 (Mechanical Engineering Publications Ltd.).

(6) Pizzi, A., Ed., Wood adhesives. Chemistry and technology. Vol.2. 1983 (Marcel Decker Inc., New York, U.S.A.)

(7) Hacker, C.L.. "An investigation into alternative adhesive systems suitable for use in the production of wood laminate wind turbine blades. Static testing of the adhesive wood joint shear strength". 1992. For the Department of Energy, Contract No. E/5A/CON/6013/2115.

(8) Corbet, D.. Investigation of materials and manufacturing methods for wind turbine blades. 1993. Contract report for the Energy Technology Support Unit, no. ETSU W/44/00261/REP.

(9) Hancock, M., Richmond, B.N., Morris, M.J. and Jones, W.A.. Fatigue testing of 12m wind turbine blades. In Wind Energy Conversion. Proceedings of the Fourteenth BWEA Wind Energy Conference. Nottingham 25-27th March 1992. Ed. Clayton, B.R.. pp.229-236 (Mechanical Engineering Publications Ltd.).

(10) BS 4169 (1988). Manufacture of glued-laminated timber structural members.

(11) Slatts, M.A.. Eliminating glueline failure in bonding hardwood. Adhesives Age, 1979 22, 6, pp.18-22.

(12) BS 1204: Part 1: 1979. Synthetic resin adhesives (phenolic and aminoplastic) for wood. Specification for gap filling adhesives.

(13) BS 5442: Part 3: 1979. Adhesives for construction. Adhesives for use with wood.

Fatigue design curves of fibreglass blade material compared to test data

P. A. JOOSSE
Stork Product Engineering BV, The Netherlands
D. R. V. VAN DELFT
Delft University of Technology, The Netherlands

SYNOPSIS A vast amount of fatigue tests on fibreglass blade material has been carried out. The results of these tests are compiled in the FACT-database. The information of the draft version of the FACT-database is used to draw some preliminary conclusions. From the data it becomes clear that the influence on the fatigue life depends strongly on the stress ratio and is independent on the matrix material. Present design fatigue curves as used for certification purposes are compared to the test data. A design curve is presented with a good fit to the available data.

1 INTRODUCTION

The wind turbine industry is shifting towards turbines with an installed power of 1 MWatt and more. For this type of product the risks have to be minimised because the investment costs are very high. The risks have to do with a lack of knowledge on one hand and a strong urge for low-cost turbines on the other hand.

The loading spectrum of a turbine is fatigue dominated. To design a cost effective turbine it is therefore necessary to know the fatigue behaviour of the used materials.

Most of the wind turbine rotor blades are made of fibreglass reinforced plastics (GRP). In several countries fatigue curves for GRP have been established for certification purposes.

Because these fatigue design curves are formulated on the basis of on-going research, they reflect the knowledge at the time of formulating. From time to time the fatigue design curves must be compared with the test data available and must be adapted, if necessary. This paper will comment on some of the fatigue design curves and proposes a fatigue formulation which is applicable for a variety of glass-matrix materials.

2 DUTCH FATIGUE RESEARCH ON GRP

2.1 Objectives

The fatigue behaviour of GRP has been the subject of both national and C.E.C. funded research projects for several years. The Dutch national research program works in close cooperation with the C.E.C./Joule II fatigue research.

The purpose of the program is to give a complete and practical formulation for the fatigue behaviour of glass fibre reinforced polyester (GRP) in laboratory conditions and, if possible, the effect of parameters like humidity, matrix etc. As this formulation will be used in design and certification, only materials and details as used by the manufacturers are tested.

At the start of the program all manufacturers used glass/polyester, at this moment there is a shift towards other fibre - and matrix materials. This paper will therefore include other matrix materials too.

The main task for the present phase of the research program is to establish a draft version of the fatigue formulation. The activities are focused on achieving a reliable database and statistically correct interpretation of the test data. The statistical interpretation of the available test results is carried out at the moment. Results will be available mid 1994.

Parallel to these computer-based activities tests are carried out to get more data in the (ultra) high cycle region for both constant amplitude and variable amplitude testing.

An overview of the test data of the program is already presented in [1]. The aim is to achieve the fatigue formulation within two years.

2.2 The FACT database

All relevant information on fatigue tests of glass/matrix material is compiled in a large database. This database is called 'FAtigue of Composites for Wind Turbines' and is compiled by ECN. The database comprises test results from 11 different laboratories, among them are CIEMAT, DLR, ECN and Sandia National Laboratories. The major part of the tests are carried out on glass/polyester, but glass/epoxy, glass/vinylester and glass-carbon/polyester are investigated too.

Results are available from constant amplitude (c.a.) tests and variable amplitude (v.a.) tests. The former tests are carried out at different R-ratios, mostly R=-1, R=0.1 and R=10; the v.a. tests are according to WISPER and its variants.

The fibre volume fraction of the tested specimen range from .30 to .69; the latter being very high compared to industry practice.

The information stored in the FACT database comprises of the following items (with a few examples):
- the specimen (e.g. dimensions, shape, stiffness, ultimate strength),
- the testing method (shape of fatigue cycle, control method),
- the test parameters (maximum load, R-ratio, frequency),
- test results (cycles at failure, failure mechanism).

Not all of these items are available for every test and of some the quality of the data is questionable; this complicates the evaluation of the database.

The database will be updated frequently; a first issue will become available mid 1994.

3 EVALUATION OF THE TEST DATA

3.1 Introduction

At this moment a thorough statistical evaluation is carried out on the available test data. In this paper a preliminary evaluation will be carried out, concerning the mean fatigue behaviour.

The test data are to be compared to present fatigue formulations and should improve the fatigue formulation. In [1] a number of requirements is formulated to achieve a generic fatigue formulation.

In short the formulation should describe the fatigue behaviour of a variety of stress combinations and material uses. The different stress combinations are commonly defined by the R-ratio (minimum to maximum ratio) and a load value (extreme or range).

A generic formulation can use the strain or a normalised stress amplitude as the relevant variable.

As has been illustrated in the past (e.g. in [1]) the use of the normalised stress gives relative low scatter. The test data will therefore be evaluated on a strain basis and a normalised stress basis.

For the purpose of this paper all run-out results are omitted; only the failure results are used.

The test results of FDL and CIEMAT are clearly outliers, they all fall in the lower region. FDL has tested glass/epoxy laminates with a very high fibre volume fraction. CIEMAT has tested glass/polyester laminates for different R-ratios. Although the CIEMAT-data are relatively low, the data are mutual consistent.

3.2 Important variables

3.2.1 Normalisation of the stress

For almost all tests the applying strain is given, even if the tests are load controlled. For a number of tests the static strength and strain are not available. For these tests the data can not be normalised: therefore more data are available for strain-cycle presentation.

In this paper the stress amplitudes are normalised by the ultimate strength. For tension-tension cycling the tensile strength is used, while the compressive strength is used for compression-compression. For tension-compression (R=-1) testing the lowest of the ultimate strengths is used.

3.2.2 Influence of the lay-up

The test data show a significant difference between lay-ups with a certain amount of 0° layers and lay-ups with different orientations (e.g. ±10°). The structural laminate of most rotor blades has some fraction of 0° layers in it. Therefore laminates not having 0° layers will not be considered here.

3.2.3 Influence of the matrix

From the view point of data reduction the matrix material is not an interesting variable, as is illustrated in figure 4. The test data of glass/epoxy and glass/vinyl spread randomly between the glass/polyester data.

3.2.4 Influence of the R-ratio

The R-ration is clearly an important variable. As figure 6 illustrates, the R=-1 test data are significantly higher than the other data (mostly R=0.1 and R=10).

In figure 5 the results and a linear curve-fit are plotted for three different R-ratios. The linear curve fit of the R=-1 data gives a high correlation coefficient (0.91), whereas the correlation of R=0.1 and R=10 is low (resp. 0.68 and 0.65).

As is expected the slope of R=-1 data is the steepest one, the intersection with the y-axis lies above 1. This is higher than expected and is probably caused by differences in failure mechanism and/or the influence of stress concentrations between the static and dynamic loading.
The data of R=0.1 and R=10 were expected to give almost the same curve, since only the sign of the loading is different. Figure 5 gives separate curve fits for R=0.1 and R=10, but due to the large scatter this difference may not be significant.

3.2.5 Choosing strain or stress

Comparing figure 4 and 5 the stress-cycle presentation shows less scatter in y-direction compared to the strain-cycle presentation. The scatter in cycle-direction is comparable. The influence of the R-ratio is larger for the stress-cycle graph.
The stress amplitude - cycle presentation gives a smaller scatter band.

4 COMPARISON WITH DESIGN CURVES

4.1 Comparison with existing curves

Both in the Netherlands and Germany a design fatigue curve is formulated in the certification regulations, resp. by CIWI [2] and Germanischer Lloyd (GL) [3].
These certification codes have the same basis: a linear relationship is given for the log of the normalised stress amplitude and the log of the number of cycles. The effect of the mean stress is taken into account by a Goodman line.
In both codes proposed curves give the mean of the population, whereas safety factors define the lower band.

The CIWI code uses a slope of 1:10 on a log-log scale. If the absolute values of the ultimate tensile and compressive strength are unequal, the top of the constant life lines (forming a 'Goodman triangle', figuur 1) should be situated at the mean of these strengths.
The slope of 1:10 is plotted in figure 2, together with the test data and a linear regression. For this R-ratio the CIWI-curve is conservative, although the

conservatism is small in the region above 10^5 cycles. When using the regression lines from figure 5 there is no proof for a shift of the top of the Goodman-triangle. Evaluation of the test data on this subject is difficult because the ratio of static compressive to tensile strength ranges from 0.6 to 1.5.

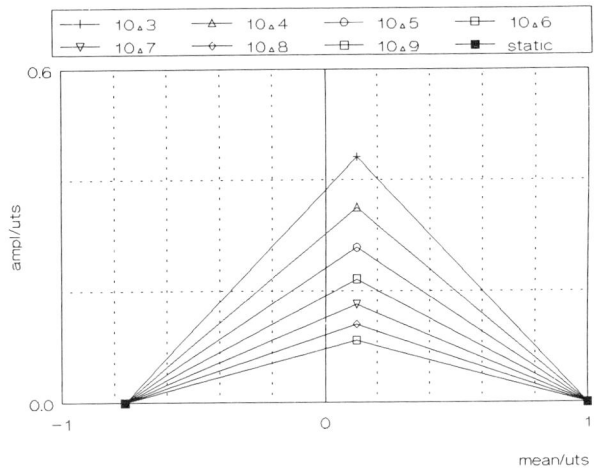

Figure 1: Goodman diagram according to CIWI, for UCS=0.79*UTS

GL makes a distinction between the matrix material: epoxy is expected to be less fatigue sensitive than polyester. The slope on a log-log scale is 1:9 for polyester and 1:12 for epoxy.
As is illustrated by figure 4 there is no proof for the statement that epoxy is less fatigue sensitive. Figure 3 shows the GL-curve for epoxy to be too optimistic for R=-1 tests above 10^6 cycles.

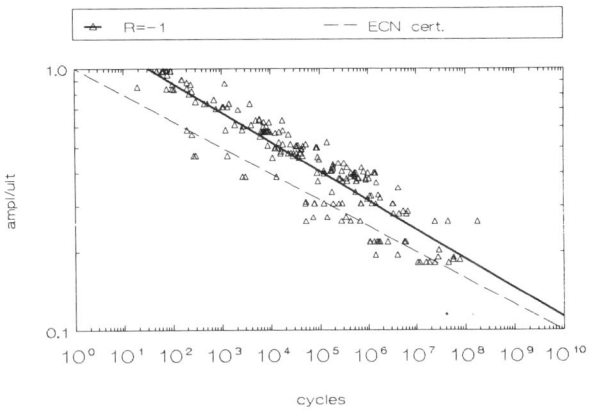

Figure 2: The design curve for R=-1 according to the ECN criteria, compared to test results of various matrix materials.

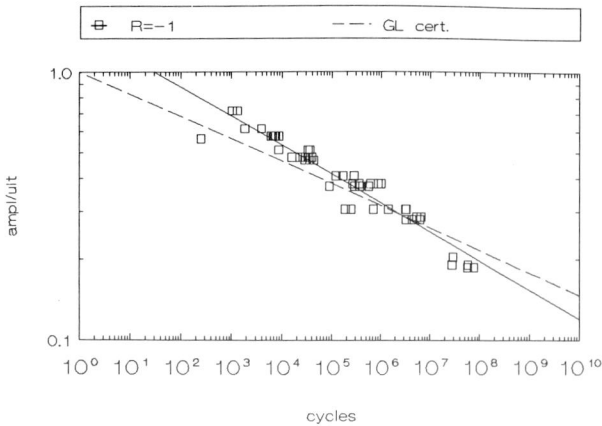

Figure 3: The design curve for glass/epoxy (R=-1) according to the Germanischer Lloyd criteria, compared to test results of DLR on glass/epoxy.

4.2 Proposal for a new curve

The test data suggest a linear relationship for R=-1. For practical reasons this line intersects the y-axis at the static strength. Bearing in mind the large scatter a slope of 1:10 is selected. A Goodman triangle is used, top at R=-1, when the mean stress is unequal to zero. The normalisation should be carried out as given in 3.1.

5 CONCLUSIONS

The test data from several fatigue research programs on glass/matrix material are stored in the FACT-database. In this paper a preliminary evaluation is carried out.

Comparison of the test results show a significant difference between laminates with and without 0° layers. For design purposes laminates containing a significant fraction of 0° layers are of more interest, for this reason only this type of laminate is evaluated.

The matrix material has no significant influence on the fatigue behaviour, whereas the R-ratio has a marked, significant influence.

Presentation of the test data on a normalised stress-cycle graph offers less scatter compared to a strain-cycle graph. To improve the usefulness of the European fatigue research it is recommended to give the normalised stress as load-variable.

Linear regression analysis of the test data suggest that the R=-1 test data can be interpreted as a linear relationship between the normalised stress amplitude and the number of cycles (both on a log scale).

The test data give no proof for a shift of the top of the Goodman-triangle.

The influence of the matrix material is shown to be negligible

ACKNOWLEDGEMENT

Most of the data presented in this paper are extracted from the FACT-database which is compiled at ECN by mr. Bach and de Smet. Although the paper presents the view of the authors they are indebted to the on-going discussions with the Dutch colleagues in the research on fatigue of GRP (granted by NOVEM, the Dutch agency for energy and environment).

REFERENCES

[1] P.A. Joosse, D.R.V. van Delft
Fatigue behaviour of fibre glass wind turbine blade material
ECWEC 93, Lubeck-Travemunde, Germany

[2] L.W.M.M. Rademakers e.a.
Proposal to change the technical criteria with respect to the fatigue properties of fibre reinforced composites (in dutch)
ECN, DE-Memo-91-58, 10 Dec. 1991

[3] Regulation for the Certification of Wind Energy Conversion Systems
IV - Non-Marine Technology, part 1 - Wind energy
Germanischer Lloyd, 1993

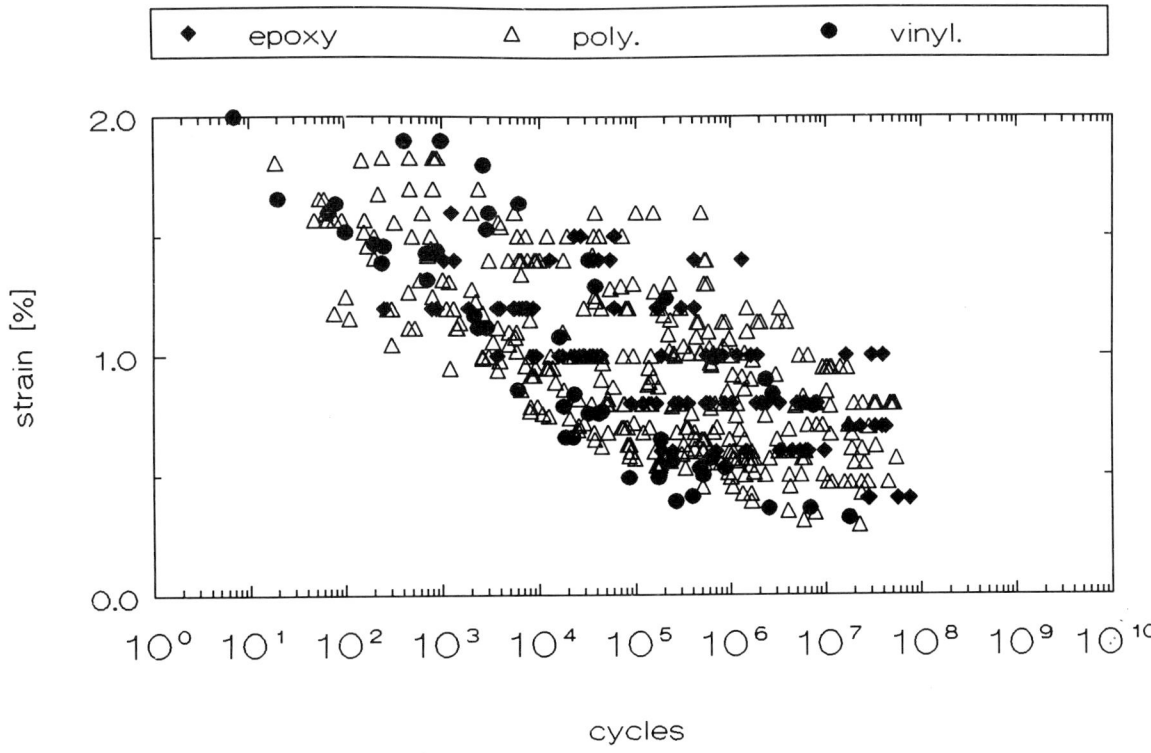

Figure 4: The test results of GRP (failed specimen, lay-up includes 0° orientation) for three matrix materials.

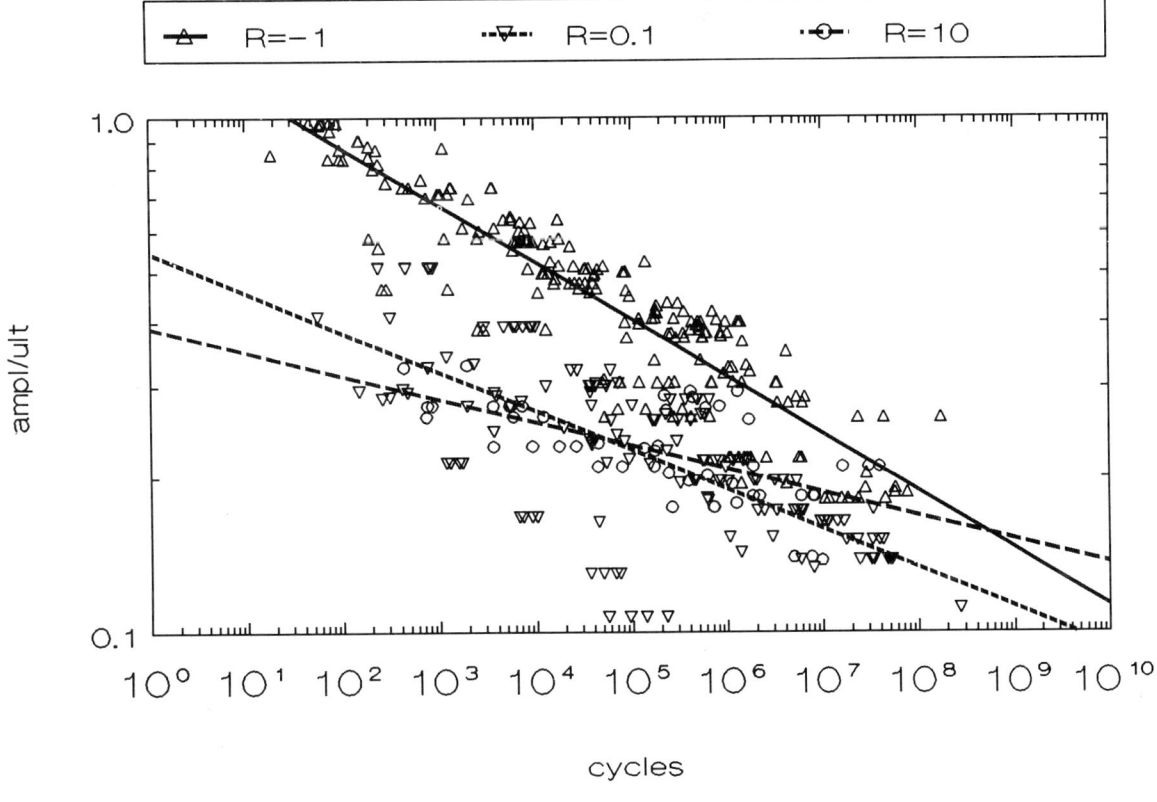

Figure 5: Linear curve-fits for different R-ratios, various matrix materials.

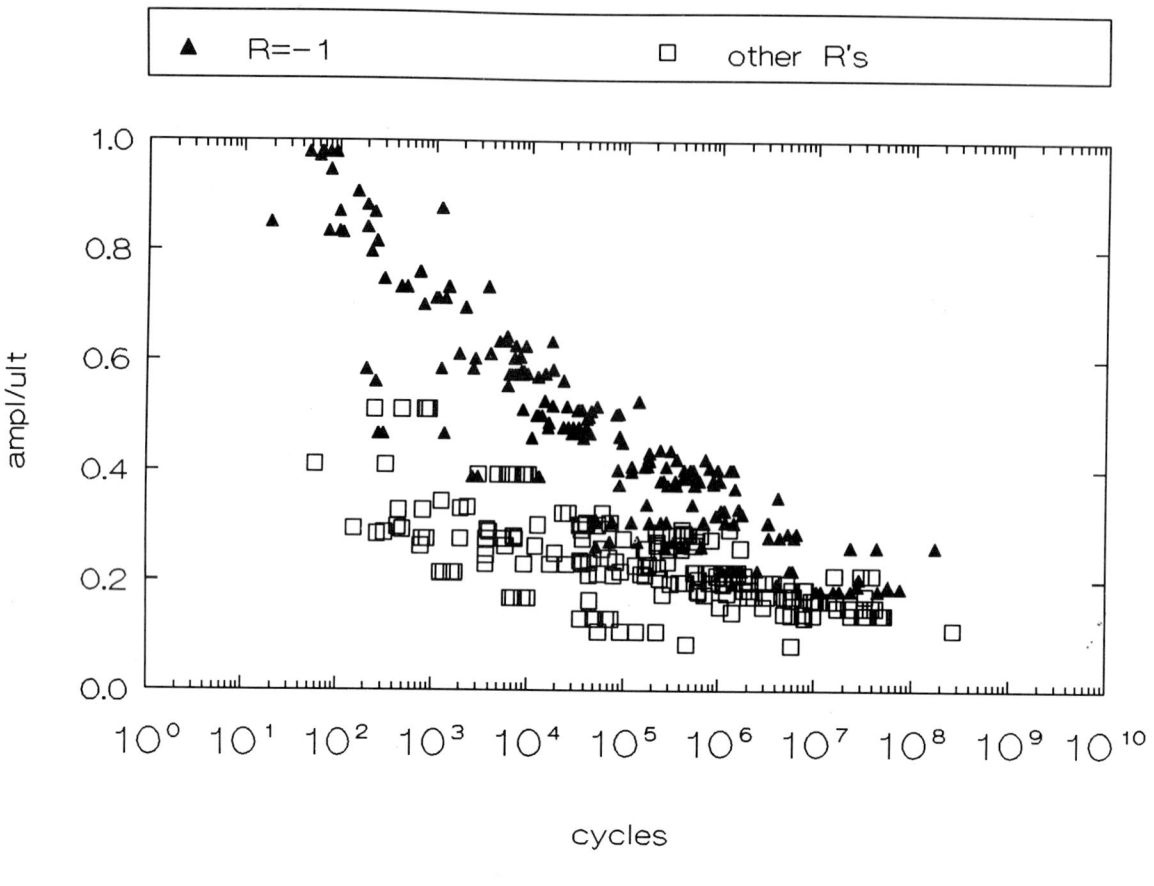

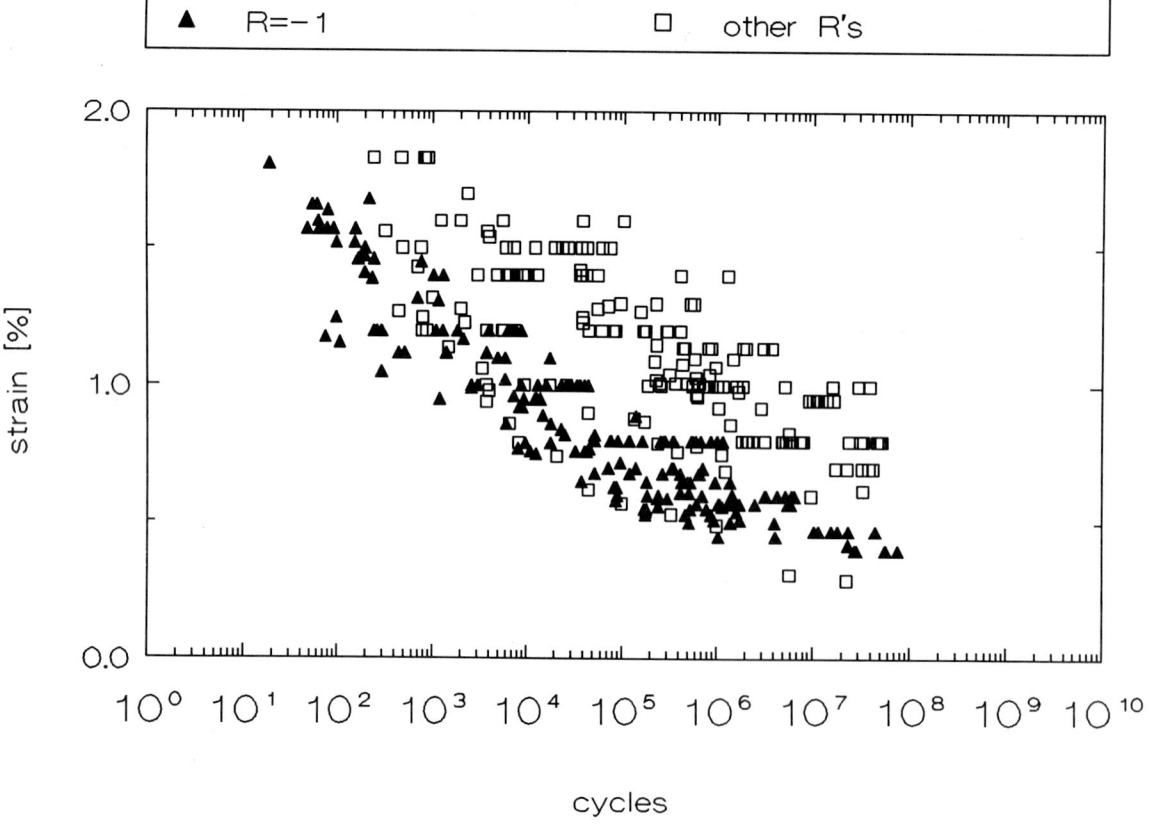

Figure 6: Influence of the R-ratio on the test results of GRP for the same set of data (failed specimen, lay-up includes 0° orientation, known ultimate strength).

Fatigue behaviour of fibreglass wind turbine blade material in the very high cycle range

D. R. V. VAN DELFT and **H. D. RINK**
Delft University of Technology, The Netherlands
P. A. JOOSSE
Stork Product Engineering, The Netherlands

SYNOPSIS Fatigue tests are carried out on fibreglass wind turbine blade material. The tests are performed at low stress levels in order to obtain results at a number of cycles in the order of 500 millions. This is a typical number of cycles for material in a wind turbine rotor blade. The test results until now indicate that the relation between applied stress amplitude and number of cycles is better represented by a log-log relation than a linear-log relation. Furthermore there is no indication for a fatigue limit at this number of cycles.

NOTATION

a, b	Constants
N	Number of cycles
N_f	Number of cycles at failure
R	Ration between minimum and maximum stress during a cycle.
UCS	Ultimate compressive strength
UTS	Ultimate tensile strength
$\Delta\sigma$	Stress range during a cycle

1 INTRODUCTION

In the design of a wind turbine rotor blade the structural detailing is mainly based on the fatigue analysis. In the design operational life the number of cycles in the fatigue loading is in the order of 500 millions.

The fatigue behaviour of fibreglass material in this very high cycle range is not known yet. Although much fatigue data on fibreglass material as used for wind turbine blades is becoming available now, most data is limited to the cycle range up to 10 million cycles. This is due to the costs and required testing time needed for a larger number of cycles.

The different ways of extrapolating the available test results into the very high cycle range can lead to substantial differences in blade mass [1].
To get a better understanding of the fatigue behaviour in the range of cycles as experienced by blade material a research project has been started to carry out fatigue tests in the very high cycle range (10 to 1000 million cycles).

2 THE FATIGUE FORMULATIONS

The results of fatigue tests on Glass fibre Reinforced Plastics (GRP) material are normally presented by plotting the applied stress range level against the number of cycles at failure. The number of cycles is always plotted on a logarithmic scale whereas the applied stress level is plotted on either a linear scale or a logarithmic scale.

Although the results will exhibit some scatter, in both cases these results will normally indicate a linear relation between the applied stress range level and the number of cycles as plotted. This is even true when the same data are plotted in either way [1].
When the stresses are plotted on a linear scale this relation can be expressed by:

$$\log(N_f) = a + b.\Delta\sigma \tag{1}$$

When stresses are plotted on a logarithmic scale the relation can be expressed by:

$$\log(N_f) = a + b.\log(\Delta\sigma) \tag{2}$$

In many cases both assumptions can be applied to the data since there appears to be no significant difference. This is due to the fact that in the cycles range of available fatigue data which ranges from about 100 to 10 million cycles the difference between the relation is limited with respect to the scatter. This is shown in Figure 1 where the same data taken from ref [2] are plotted with the applied stress range on a linear and logarithmic scale respectively. In both figures the regression lines for the mean and for two times the standard deviation are given.

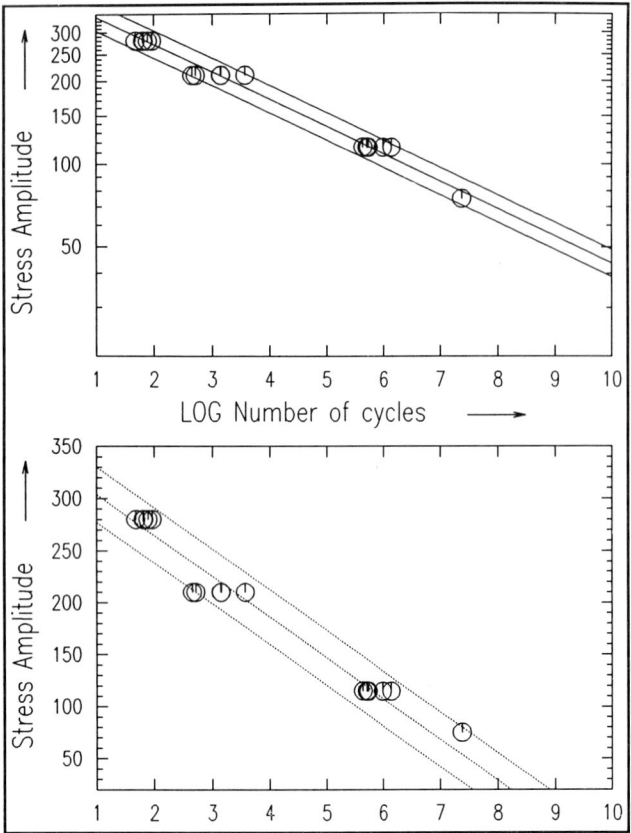

Fig 1   Same set of test results [2] plotted with the stress linearly and logarithmical.

When these two sets of regression lines of both graph are plotted in one graph (see Figure 5) the overlap becomes clear.

Although in the area were the data is generated the lines do not differ significantly, when extrapolated beyond 10 million cycles a substantial difference occurs. Since the number of cycles of the GRP material in a rotor blade is in the order of 500 millions, the design can be affected by which of the two lines is applied. The difference in blade mass will depend on other design considerations as well but can be in the order of 30% [1].

3 EXPERIMENTAL

3.1 General

The investigation of the fatigue behaviour at the very high cycle range is part of the dutch national research programme on GRP material for wind turbines. In this programme fatigue test have been carried at ECN [2] in the range up to 10 million cycles. These test results from the same specimens and material are used here as a reference.

3.2 Test programme

The tests are carried out at the R-value 0.1 and -1. The aim was to have at least four specimens tested for each R-ratio in the very high cycles range. The target number of cycles was 500 millions.

With the estimated test frequency of 12 Hz the testing time is in the order of 1.5 years for each specimen. Therefore it was decided to have 3 test rigs in which up to four specimens were tested simultaneously according to the scheme in Table I. In one rig four specimens were tested simultaneously at R=0.1 and in the others one or two specimens were tested simultaneously at R=-1. At R=-1 the number of specimen that can be tested simultaneously is restricted due to buckling.

Table I   Scheme of test program

test rig	R-value of tests	Number of test simultaneously
I	-1	2
II	-1	2
III	0.1	4

3.3 Material and Specimen

The dimensions of the specimens are given in Figure 2. The specimen have a dog-bone shape in order to avoid failure at the clamping area.

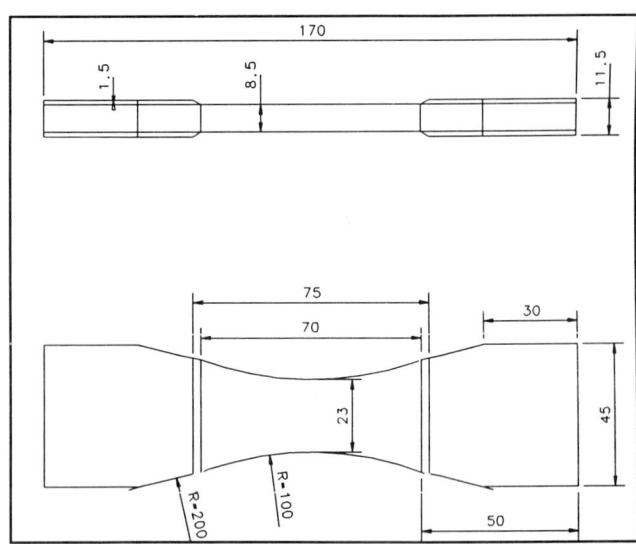

Fig 2   Geometry of specimen

The specimens were cut out from a glass fibre reinforced plate. The plate contains 7 layers of

500 g/m² unidirectional and 8 layers of 480 g/m² +/-45° glass fibre in a symmetrical hand lay-up. More detailed information is given in ref [2].

3.4 Test rigs

In order to keep the cost of the tests low three simple dedicated test rigs and control systems were developed and build. In the test rig more than one specimens can be placed in line and tested simultaneously (Figure 3). The actuator control in the test rig is of the normal closed loop servohydraulic type. During the tests the actuators are load controlled.

Fig 3   Photo of test rig III with 3 specimens tested at R=0.1 (the fourth specimen failed and is replaced by a dummy).

For monitoring and processing the data of the various transducers a PC was connected to each control system of the test rigs enabling communication with the control system.

The test rigs were provided with ventilators for cooling of the specimens.

3.5 Clamping of the specimen

In a test rig the specimens are connected to each other by clamping them between two steel plates with the bolted connection. The uppermost specimen is connected to the load cell (almost directly) by a bolted connection with two angle sections (see Figure 4) whereas the most lower specimen is in the same way directly connected to the piston rod of the hydraulic actuator. At each connection the two bolts (type M12) at each are strain gauged, which makes it possible to obtain a controlled preload in the bolts. Originally this preload was set to 20 kN but in a later stage increased to 30 kN.

After mounting the specimens in the test rig the alignment is measured by the four strain gauges placed on each specimen at low load level. If necesary adjustments are made until a satisfactory alignment of the specimens is obtained. Although clamping the specimens in the rig was time consuming due to the attention that was paid to the alignment of the specimens it was negligible with respect to the testing time for these type of tests.

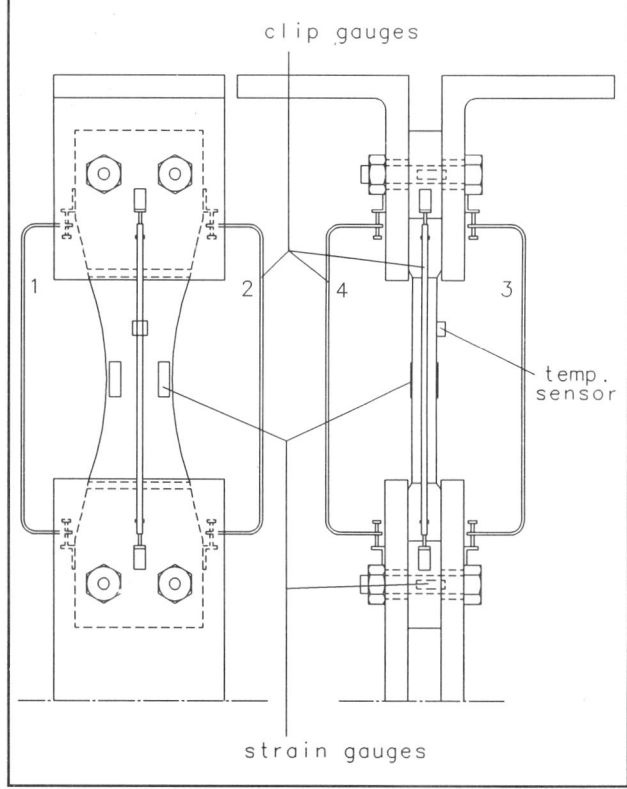

Fig 4   Clamping of uppermost specimen and location of measuring devices.

3.6 Measurements

Each specimen was provided with the following measuring devices.
- four strain gauges.
- four clip gauges measuring the elongations on all four sides of the specimen.
- one temperature sensor.
- four strain gauged bolts.

All these are continuously measured by the PC's together with the actuator force and displacement.
At each fatigue cycle the average and range of the above reading are calculated. These values are checked against predefined tolerances and periodically (in general each hour) written to disk.
The strain gauges, which are basically used for the alignment, mostly fail due to fatigue after a short period.

The values of the four elongations measured at each specimen are used to detect which of the specimen is possibly going to fail. It also indicates whether a specimen starts bending due to asymmetric damage. In that case it can be removed to avoid eccentric load transfer to the other specimens.

4 RESULTS

4.1 Test data

The results of the tests are given in Table II

Table II   Test results

Test rig	R	$\Delta\sigma$ [MPa]	N_f, N [Mc]	remarks
I	-1	120	276	taken out
I	-1	120	276	taken out
I	-1	230	1.3	failure
I	-1	230	1.4	failure
I	-1	130	40	running
I	-1	130	40	running
II	-1	150	174	failure
II	-1	150	44	failure
III	0.1	82	268	failure
III	0.1	82	315	running
III	0.1	82	315	running
III	0.1	82	315	running

These results are plotted in Figure 5 and Figure 6 for the R=-1 and R=0.1 tests respectively together with earlier results from ECN. The number of cycles is at failure for failed specimens. The stress amplitude is plotted again linearly and logarithmical.

The lines given in the figures are linear regression lines for the mean and plus and minus two standard deviations based on the ECN results. There are two sets of these lines in these figures. One is based on a linear scale and the other on the logarithmic scale of the stresses. The set of dotted lines is based on a linear scale of the stresses and therefore according to equation (1) whereas the others are based on a logarithmic scale of the stresses and therefore according to equation (2). For the results of R=-1 (Figure 5) these lines are the same as given in Figure 1.

Some specimens are taken out after a period of testing due to overloading as a result of power failure. These specimens are indicated in the plots by the arrow pointing upwards. Specimens marked with an arrow pointing to the upper right are still running.

4.2 Test frequency and temperature

The test frequency was basically limited by the maximum allowed temperature of the specimen. The specimens are each cooled by a ventilator. The air blown by the ventilators was of normal room temperature. As a result the temperature of the specimen varied during seasons and day and night. The maximum allowable temperature measured at the sensor was set to 35°C. With this temperature the test could be run at a frequency between 12 and 16 Hz depending on load level and R-value.

5 Discussion of results

5.1 Results of R=-1

Two specimens had to be taken out after running for 276 million cycles due to overload as a results of power failure. There was no indication that these specimen would have failed soon. These two results are marked by the right most point in both graphs of Figure 5.

Although their failure would have occurred at a higher number of cycles they are already beyond the scatter band predicted by a lin-log relation based on the results of ECN (dotted lines). The results are close to the mean line of the scatter band based on the log-log relation.

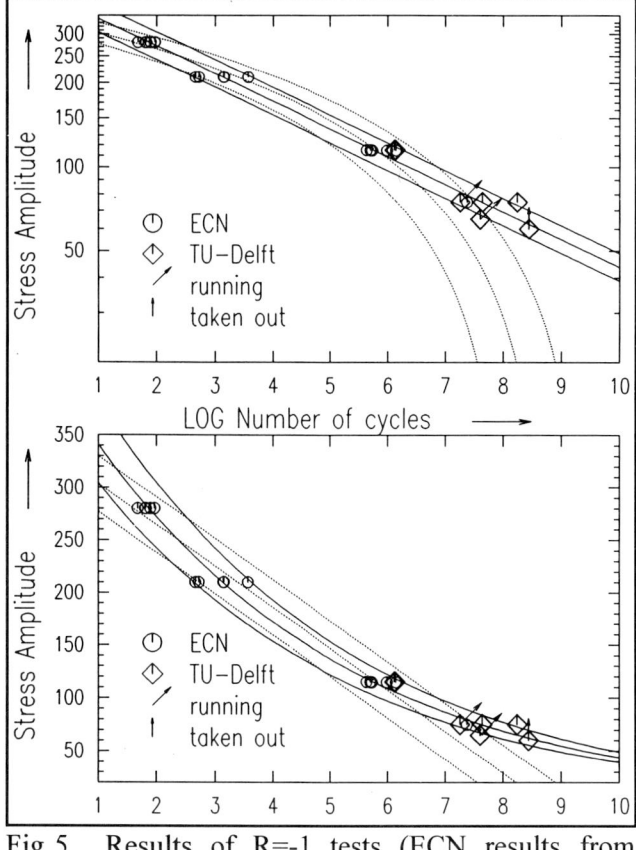

Fig 5   Results of R=-1 tests (ECN results from ref [2]).

The two specimens which failed at 44 and 174 million cycles respectively are also far beyond the lin-log scatter band. One is very close to the mean of the log-log prediction whereas the other is even slightly beyond the scatter band of the log-log prediction.

These four results discussed above seem to be well covered by the log-log relation as can be seen from Figure 5. When results are plotted with the stress level on a linear scale (lower graph in Figure 5) it is also clear that the data does not fit on straight line but follows the curve of the log-log relation.

It can be concluded that these four results indicate that the log-log relation according to equation (2) represents the fatigue behaviour better than the lin-log relation according to equation (1).

The three specimen which are currently running are still in the area where both scatter bands overlap. Therefore no conclusions can be drawn from those yet.

5.2 Results of R=0.1

One of the four specimen tested simultaneously failed after 268 million cycles. This specimen was therefore

replaced by a steel dummy specimen in order to continue testing for the other three specimens. These specimens have now been subjected to 315 million cycles. The failed specimen is outside the lin-log scatter band (see Figure 6) and close to the mean line of the log-log relation. The same applies for the specimens still running (all three indicated by one marker) which are even closer to this mean line at this moment.

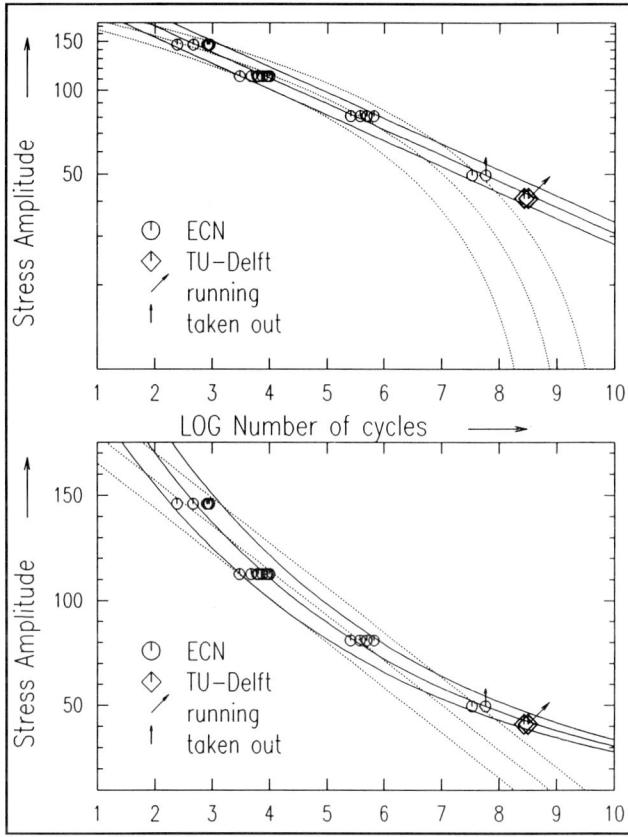

Fig 6   Results of R=0.1 tests (ECN results from ref [2]).

Similar to the R=-1 results these four results indicate that the log-log relation according to equation (2) represents the fatigue behaviour better than the lin-log relation according to equation (1).

5.3 Fatigue limit

As the results are now the lin-log relation does not represent the fatigue behaviour of the specimens very well. The results seem to follow the log-log relation. Assuming a lin-log relation (which is probably not correct as discussed above) one might conclude that the SN-curve is levelling off indicating that a fatigue limit is nearby.

However if the log-log relation is used (which seems to be the correct one) then the fact that specimens failed after 174 and 268 million cycles might indicate that there is no fatigue limit in this range of cycles.

5.4 Stiffness degradation

Figure 7 shows the stiffness degradation of the specimen tested at $R = 0.1$ which failed after 268 millions cycles. The stiffness plotted in this figure is based on the average value of the four clip gauges measuring the elongation of the specimen.

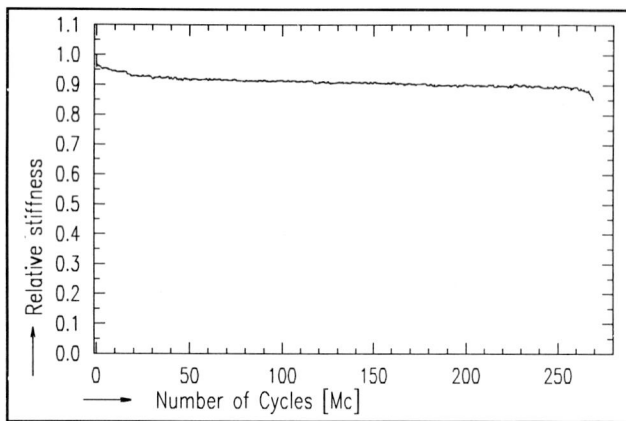

Fig 7   Stiffness degradation of the specimen.

Three stages can be discriminated in the curve. In the early stage of the test the stiffness decreases rapidly. Secondly there is a long stage where the reduction of the stiffness is moderate and linear with the number of cycles. Then at the last stage the decrease in stiffness increases announcing the failure of the specimen. The other specimens have similar curves for the stiffness degradation.

6 CONCLUSIONS

Tests have been carried out in the very high cycle fatigue range at R-values of -1 and 0.1. Although the test programme is not finished the following prelimenary conclusions are given.

- The test results until now indicate that the relation between applied stress amplitude and number of cycles is better represented by a log-log relation than a linear-log relation.

- If the log-log relation is applied there is no indication for a fatigue limit at cycle range investigated.

- The stiffness of the specimen degrades rapidly at the beginning and at the failure stage. However for the larger part of the fatigue life in between the stiffness reduction is moderate and linear with the number of cycles.

7 ACKNOWLEDGEMENTS

The research presented here is a part of the Dutch national research programme and sponsored by NOVEM.

8 REFERENCES

[1] D.R.V. van DELFT, F. HAGG, P.A. JOOSSE
The influence of fatigue design line criteria on the rotor blade design.
ECWEC conf., Madrid, 1990, pp395-9

[2] P.W. Bach
Fatigue properties of glass- and glass/carbon polyester composites for wind turbines.
ECN-C-92-072, November 1992

Resin transfer moulded aero generator blades

V. MIDDLETON BSc, PhD, CEng, MIMechE
I. A. JONES BSc, BEng, CEng, MIMechE
B. R. CLAYTON BSc(Eng), PhD, CEng, FIMechE, FRINA
The University of Nottingham, UK

SYNOPIS The resin transfer moulding (RTM) process is described and various applications of the process are listed. A description is given of the design and manufacture by the RTM process of the blades for a small wind turbine. A computer program used for the stress analysis and design optimisation of the blade is introduced together with verification results from the ABAQUS finite element package. This program relies upon simple beam theory and takes account of geometrical nonlinearities to model centrifugal stiffening of the blade. This simplified model is extended to consider coupling between twist, tension and bending in orthotropic laminated blades, with a view to performing aeroelastic design of "smart" composite blades.

INTRODUCTION

This paper describes the design and production of blades for a small research turbine manufactured by the resin transfer moulding (RTM) process. The blades were part of a 1.4 metre diameter model to test the performance of Vestas 400 type wind turbines. They were instrumentated to measure the flat wise bending under operational loads. Although only a model turbine design, build was quite critical in order that the instrumentation would respond to normal loads and a blade would be safe operating in free running mode in the event of a load failure.

2 THE RTM PROCESS

In this process dry fibrous material is placed within a closed mould and the resin is injected, either under vacuum or with pressure assistance. Advantages of the process are that the directional material can be laid in the dry state so that the laminated structure is accurate and by the use of a double-sided mould, the fibre to resin ratio is also closely controlled. The process therefore delivers composites with a consistently high quality which are not dependent upon operator processing. The use of closed moulds also addresses several safety issues in that there are no heavy vapours in the work space, and the component requires no trimming after injection, thus eliminating dust hazards. Variations of the process can be used to manufacture car bodies (Lotus Cars) and very high performance aero propellers (Dowty Aerospace). It is now being actively developed for the volume production of automotive parts and also as a replacement manufacturing process for the prepreg/ autoclave route for carbon fibre products in the aerospace industry.

The pressure injection process requires matched moulds and since small pressures over large mould areas produce large forces and hence deflections, moulds for this process must be substantially stiffened. Depending upon the mould life required, process temperature and component size, mould materials may consist of GRP backed with ply-wood formers, resin and sand mixture, electroplated shells or machined from the solid. Pressure processing ranges from filling under gravity up to approximately 5 bar.

For larger parts the Vacuum-Assisted Resin Injection process (VARI) is available from Crystic Systems, a joint company formed by Lotus Cars and Scott Bader. The system uses two GRP moulds with patented edge seal. The mould surfaces can be gel coated or even primer painted before placing the reinforcement. After closure of the mould halves, the action of drawing a partial vacuum closes the mould tightly and compresses the reinforcement. An advantage for the process is that flexible moulds can be used which are supported by the reinforcement. The compression of the reinforcement gives a high glass content and uniform thickness to the laminate. Air is withdrawn from the mould in several places and resin is allowed to flow into the mould until it appears at the vacuum connections. The resins are specially formulated to provide the right viscosity for the process and flow rate is critical to avoid bypassing areas of fibre, whilst allowing flow to be complete before the resin gels. In the field of boats one production hull requiring 800 kg of resin to be injected has been made on a regular basis. The process produces better laminations than hand laminating in the sense that it has a uniform thickness, consistent glass content and two fair surfaces. Consistent thickness and uniform weight are obviously important for blade production in order to minimise blade balancing and vibrational problems.

3 THE MODEL AEROGENERATOR

The blades were models of LM17 production components. As reported elsewhere in this conference[1] the model was to be used to assess the yaw performance of such a turbine in both two blade form with teeter hinge and three blade fixed configuration. Instrumentation was required to determine the bending stress within the blade the teeter angle, torque and speed of the rotating system and mast load and yaw angle of the support.

Figure 1 Wind tunnel with turbine installed

A photograph of the wind turbine installed in the M 8 metre truck wind tunnel is shown in Figure 1. T turbine hub was directly mounted on a DC servo mo which itself was attached on to the mast which rota in plummer bearings on a short stand. The mast w strain gauged at two sections along orthogonal directic and yaw angle measured at the base. A torsio vibration damper driven through a torque metre was a located at the foot of the mast, together with a di which enabled the assembly to be locked predetermined yaw angles. Instrumentation wiring w led through multiple leads to a break out box in wind tunnel control room where connections w regrouped and coupled to a data acquisition card ir personal computer through a 50-way ribbon cable.

4 BLADE DESIGN

For dynamic similarity, the operating speed of t turbine was 1100 rpm in two-blade configuration a 873 rpm in three-blade form. Maximum overspeed the event of the load being dropped was 4550 rpm a 4000 rpm for the two- and three-blade configuratio respectively. The tests required that the blades sufficiently flexible to enable aerodynamic loading to determined at the normal operating speeds, yet be stro enough to withstand the very high loading should runaway occur. Thus blade design and structure we quite critical. The profile of the blade was supplied computer readable form by Renewable Energy Syster at 18 radial stations. It was decided to make the bla as a single RTM moulding in a manner similar to t used by Dowty Aerospace to fabricate aero propelle [2]. The basic structure (Figure 2(a)) is a thin skin +/- 45° inside which there are unidirectional spa attached to the front and rear skins which effective take the majority of the centrifugal and bending loads the blade. The core is injected polyurethane foam whi limits the curing temperature which can be applied to t epoxy resin system. The University has considerab experience in manufacturing automotive parts using similar procedure [3].

A computer program was developed which, workir from the profile of the blade, determined the mass of t various laminations working inwards towards the cor Table 1 shows the properties of the laminate used in t design. The program was then extended to calcula deflections and stresses under the aerodynamic ar centrifugal loading, and this program is described Section 6.

The final structure consisted of two skin preforms ar three spar preforms. The inner skin was dropped off a

Table 1 Material Properties

Element	Skin	Spar	Core	Membrane
Supplier	Cotech EBX 318	Cotech ELPb 567	Baxenden	Dowty
Material	Glass +/- 45°	Glass 0° (95%) 90° (5%)	Polyurethane foam	Glass/ film adhesive
Density	0.318 kg/m²	0.567 kg/m²	55 kg/m³	0.633kg/m²
Vol. fraction	0.50	0.60	-	-
E_1 MN/m²	-	40,000	-	-

Figure 2
(a) True structure of blade (thickness exaggerated)
(b) Simple beam model
(c) Laminated model

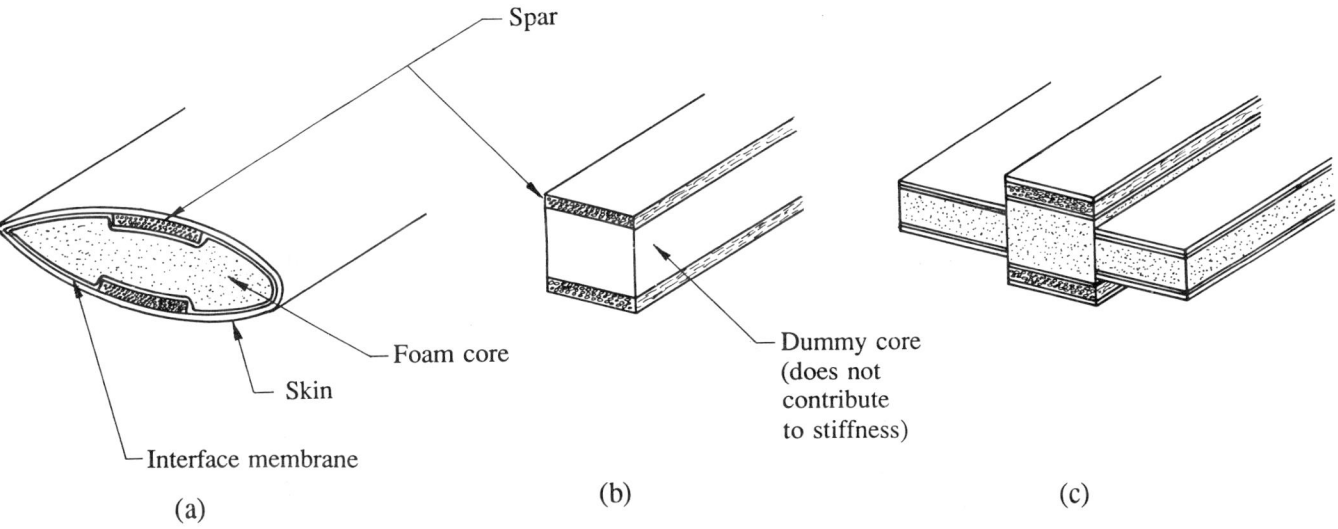

428 mm radius and the two inner spars were dropped at 428 and 621 mm radius respectively. The blade was terminated in a circular root into which a steel attachment bolt was bonded. Templates were made for all the preforms to be cut. Front and back skins were made as a single piece unfolding about the leading edge. The preforms were laid up in a simple slate epoxy copy of the mould. Initially, the layers were bonded together with Neoxil powder but in the interests of a more consistent product, this was changed to Tigermesh at a later date. The final layer in the preform was an impervious membrane, subject to patent protection by Dowty, which prevented the foam core from permeating into the unfilled preform. The shell was then folded along the leading edge and the membrane sealed along the trailing edge and at the root. This was then placed in the injection mould and polyurethane foam injected into the core, thus compressing the preform against the mould face. This method of manufacture ensures a much better volume fraction for the composite than the alternative method of premoulding the core, since in the latter case generous volumes must be left for the skin and spar structure otherwise pressure on the fibre would be too great and the mould will not fill. The fibrous preform was then injected with epoxy resin. In order to obtain full wet out, this had to be done with vacuum assistance and with the flow of resin controlled so that the mould was filled very slowly, the process taking approximately 30 minutes. The mould was then placed in an oven and the

epoxy cured. Because of the very small blade section, significant difficulties were experienced, both in filling the core and the composite sections of the blade, which from our work with larger components would not be experienced should the scale be increased.

5 TOOLING

Owing to the small size of the blade it was desirable to produce the tools directly from the numerical data by an NC machining process rather than the traditional method of manual pattern making. Quotes received for doing this phase of the work were unacceptably high so an intermediate route was taken. Profiles of the blade were cut from 1.5 mm thick steel sheet using an extremely fine numerically controlled laser. These were keyed together in egg box construction with similarly machined longitudinals. The cavities were then filled with an epoxy marble and the blade surface faired to produce top and bottom half moulds. Epoxy slate shadows were then taken and the mould and preforming tools taken as a copy from these. The moulds were then mounted on steel backing plates with ancillary clamping bars. After foaming, the blades were removed from the mould. Prior to resin injection, wires were inserted for the strain gauge connections between the root of the blade to approximately one-third up the aerofoil section where bending stresses were maximum. Only a short lead time was available so no attempt was made to do an in-depth finite element analysis or to study the aeroelastic behaviour of the blades before they were finally manufactured. A simple finite element verification of the design was undertaken subsequently (Section 7). The preform shapes can be readily modified in order to accommodate the findings from this study since no hardware in the form of mould tools for core production are required by this manufacturing route.

The blades performed admirably in the wind tunnel. As expected they were subjected to overspeed and remained intact even though the paint on them had cracked showing they had been subject to considerable strain.

6 STRESS ANALYSIS TECHNIQUES

Consistent with the short time available for the design process, a simplified model of the blade was encoded as a QuickBasic program. This model used simple engineering beam theory and took account of geometric nonlinearities due to the centrifugal stiffening effect. The blade was discretised into short elemental beams whose flexural rigidity was calculated by assuming that only the spar contributed to the stiffness of the blade (Figures 2(a) and (b)). The following loads on the blade were considered:

(a)   the centrifugal loading
(b)   the wind loading
(c)   the "straightening moments" induced into the blade by the centrifugal loading on the deflected shape giving the effect of centrifugal stiffening.

An iterative solution was employed to take account of the centrifugal stiffening effect. The deflected shape was calculated taking account of existing estimates of deflections (initially assumed to be zero). The straightening moments were then calculated and the deflected shape re-estimated using the moment-area method. Direct use of this estimated value of deflection gave instability in the iteration process and it was found necessary to apply an under-relaxation factor to the estimated deflections when re-calculating the straightening moments. When the calculated deflected shape of the blade was within a specified tolerance of the previous estimate of the deflected shape, the iteration was assumed to have converged.

Figure 3(a) Normal running conditions (873 rpm)

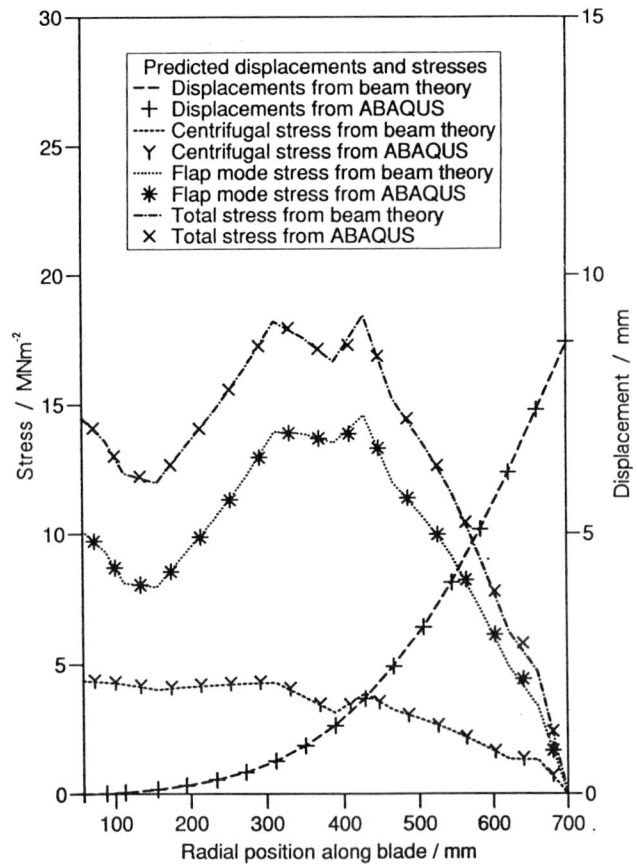

Figure 3(b) Typical overspeed conditions (4500 rpm)

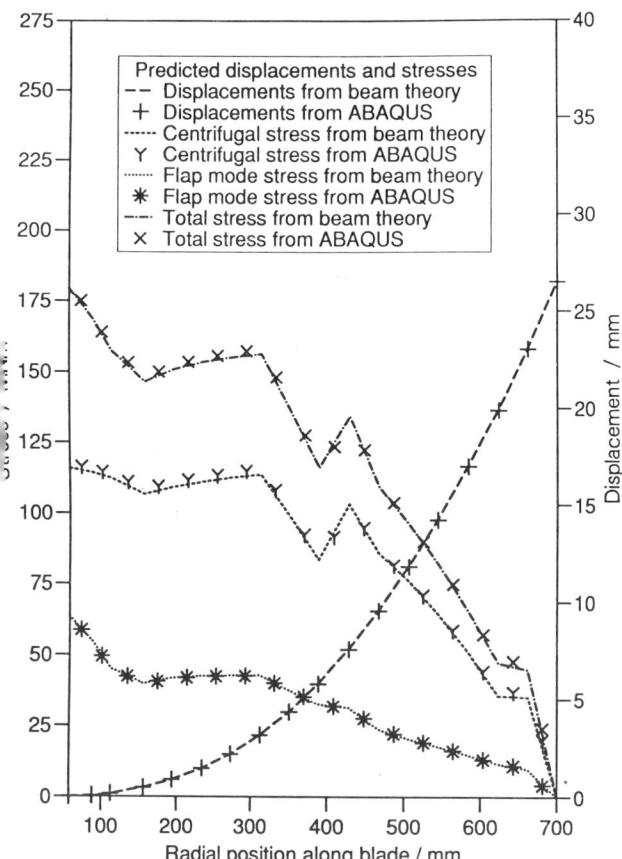

rate of twist $\frac{1}{2}\kappa_{xy}$ are obtained from the tensile force F_x and bending moment M_x using the following equation:

$$\begin{Bmatrix} \kappa_x \\ \kappa_{xy} \end{Bmatrix} = \begin{bmatrix} C_{31} & C_{33} \\ C_{41} & C_{43} \end{bmatrix} \begin{Bmatrix} F_x \\ M_x \end{Bmatrix} \quad \text{where}$$

$$[C] = \left[\sum_{i=1}^{3} b_i ([U_i] - [V_i][W_i]^{-1}[V_i]^{\mathrm{T}}) \right]^{-1} \quad \text{and where}$$

$$\begin{bmatrix} U_{11} & U_{12} & U_{13} & U_{14} & V_{11} & V_{12} \\ U_{12} & U_{22} & U_{23} & U_{24} & V_{21} & V_{22} \\ U_{13} & U_{23} & U_{33} & U_{34} & V_{31} & V_{32} \\ U_{14} & U_{24} & U_{34} & U_{44} & V_{41} & V_{42} \\ V_{11} & V_{21} & V_{31} & V_{41} & W_{11} & W_{12} \\ V_{12} & V_{22} & V_{32} & V_{42} & W_{12} & W_{22} \end{bmatrix}_i = \begin{bmatrix} A_{11} & A_{16} & B_{11} & B_{16} & A_{12} & B_{12} \\ A_{16} & A_{66} & B_{16} & B_{66} & A_{26} & B_{26} \\ B_{11} & B_{16} & D_{11} & D_{16} & B_{12} & D_{12} \\ B_{16} & B_{66} & D_{16} & D_{66} & B_{26} & D_{26} \\ A_{12} & A_{26} & B_{12} & B_{26} & A_{22} & B_{22} \\ B_{12} & B_{26} & D_{12} & D_{26} & B_{22} & D_{22} \end{bmatrix}_i$$

where $[U]_i$, $[V]_i$ and $[W]_i$ are partitions of the laminate stiffness matrix

$$\begin{bmatrix} [A] & [B] \\ [B] & [D] \end{bmatrix}_i$$

for each region i across the width of the blade and b_i is the width of region i; $[A]$, $[B]$ and $[D]$ for a general laminate are defined in reference 4.

7 VERIFICATION & RESULTS

Although commercial finite element (FE) codes do not generally include the facility to model beams with coupling effects between bending and twisting, it is straightforward to use a large-deflection FE analysis to verify the iterative approach used in the models based upon simple beam and lamination theories. ABAQUS level 5.2 was the FE code used for this verification. Elements B23 (cubic beam elements for 2D problems) were used within a geometrically nonlinear static analysis. For typical loadcases of normal duty and overspeed, the results from the simple beam model are compared with corresponding FE results. The results from the two sets of results are shown in Figures 3(a) and (b) to agree very closely. (The values of blade properties and overspeed angular velocity used for this verification are marginally different from the design specification). Although the values of aerodynamic load on the blade are very high under overspeed conditions (typically 1000N per metre of blade length towards the blade tip), it may be observed that the centrifugal stiffening effect results in the deflections being quite small (approximately 26mm) and hence the centrifugal stresses are the predominant effect. Under normal running conditions, the aerodynamic loads are much smaller but the reduced effects of centrifugal stiffening and stress mean that the bending (flap mode) stresses are the predominant ones.

n modelling the blade for the purposes of the design process, no attempt was made to utilise the opportunity for aeroelastic effects or to model such effects. However, the possibility exists that "smart" blades could be designed aeroelastically to twist and stall under conditions of overspeed. This would require careful choice of materials, fibre directions and lamination sequence used for the skin and possibly the spar. In order to investigate this phenomenon the simple model of the blade was extended to consider the stiffness contributions of the skin, membrane and foam core and to extend the calculation of flexural rigidity to include the calculation of coupling terms linking twisting to tension and bending. The blade cross-section is simplified to that shown in Figure 2(c) and classical lamination theory [4] is used to evaluate the properties of each component of the cross-section. This extended solution reduces to the simple beam theory model if the contributions to stiffness of the skin, core and membrane are neglected by setting their moduli to low values. The moment-area method is easily extended to this new application by the inclusion of a coupling term $F_x C_{31}$ in the calculation of blade curvature κ_x. Twist of the blade is obtained by trapezium-rule integration of the varying rate of twist along the length of the blade. The curvature κ_x and

The usefulness of the extended (lamination theory) model is apparent in evaluating the kind of twisting effects that may realistically be achieved using orthotropic materials in manufacturing the blade. Although the aeroelastic design of a blade for this particular application would be a purely academic exercise, some simple numerical experiments have shown that blade end twist under conditions of overspeed of 1°-2.5° may be achieved using a directional skin with a high value of Young's modulus. A more radical re-design of the blade would be necessary to achieve larger values of twist.

It is suggested that an integrated approach to aerodynamics, materials selection, design and manufacture could be used in conjunction with the analysis method described to obtain an optimal blade structure. A rigorous three-dimensional finite element analysis of the final structure could then be carried out to verify the results obtained from the simple laminated model.

8 CONCLUSIONS

A manufacturing route has been developed for wind turbine blades based upon the RTM process. After considerable experimentation and process development, a set of blades has been successfully manufactured by this route and these blades have performed well under test. A simple design analysis program used during the development of the lamination sequence of the blades has been verified using the ABAQUS finite element system and has been extended to model twisting effects due to the chosen lamination sequence. Simple variations upon the actual design of blade give only small values of twist even in the case of overspeed; a more integrated aeroelastic design approach would be required if a "smart" wind turbine blade were required, although the laminated blade model would form a sound basis for the design optimisation.

9 ACKNOWLEDGEMENTS

The authors gratefully acknowledge the financial support of Renewable Energy Systems Limited. They also wish to thank Mr S L Waddell for investigating the use of ABAQUS for centrifugal stiffening problems of this kind as part of the work forming the project for his BEng degree course.

10 REFERENCES

1. NOAKES, J *et al*, The free yaw behaviour of up-wind horizontal axis wind turbines. *British Wind Energy Association Conference*, University of York, October 1993.

2. McCARTHY, R, Composite propellor blades for commuter aircraft and hovercraft. *Progress in Rubber and Plastics*, 4(4), 1988.

3. OWEN, M J, MIDDLETON, V, HUTCHEON, K F, SCOTT, F N and RUDD, C D, The development of resin transfer moulding (RTM) for volume manufacture. *Proc IMechE*, 1989-2 *Design in Composite Materials Conference*, 7-8 March 1989, 107-114.

4. JONES, R M, Mechanics of Composite Materials. New York: Hemisphere, 1975.

Measurement of stall delay on a model of a stall controlled rotor

J. M. R. GRAHAM and **C. J. BROWN**
Imperial College, London, UK

SUMMARY This paper describes a wind-tunnel investigation of the stalling behaviour of a horizontal axis wind turbine rotor. The rotor was based on the geometry of a prototype stall controlled rotor. Measurements of mean and fluctuating pressures on the surfaces of the blades were carried out. The flow in the region of the blade was visualised using smoke and laser sheet illumination. Some measurements of surface temperature from which surface skin friction could be inferred were carried out using temperature sensitive liquid crystal paint.

NOTATION

c	Blade chord
C_D	Drag coefficient
C_L	Lift coefficient
C_{Lmax}	Maximum value of lift coefficient
r	Radius of measurement station
a	Incidence
a_{max}	Incidence for CLmax

1. INTRODUCTION

It has been found in previous work that large suction pressures and greater than predicted torques can occur on the inboard sections of rotors operating at low tip speed ratios (Brown and Graham (1), (2) and Barnsley and Wellicome (3)). The results strongly suggest that stall delay with an accompanying increase in maximum lift coefficient above that predicted by two-dimensional aerofoil data occurs on inboard sections of turbine rotors similar to that found on propellers by Himmelskamp (4). Depending on how far outboard this effect persists, and whether an equivalent reduction in maximum lift coefficient occurs near the tip, as has also been suggested, dictates whether a higher than predicted power output will occur on the rotor.

The present work was carried out to try to understand the reason for the stall delay phenomenon. It was hoped that by carrying out flow visualisation experiments together with further surface pressure measurements, insight into the flow mechanism might be obtained. A secondary aim was to make the measurements on a rotor representative of a practical turbine design, thus providing three-dimensional blade characteristic data at model scale

Reynolds numbers and information on how far along the blade span towards the tip the stall delay was significant.

Three possible explanations were considered for the high suction peaks at high incidence and the large lift coefficients found on rotor blades at low tip speed ratios: (a) the effect of the radial gradient of dynamic pressure, arising from the varying relative velocity along the blade, acting on the fairly stagnant fluid in the separation region after the onset of stall; (b) a time dependent process whereby, rather like the case of dynamic stall, a vortex grows in the separation region and is then shed leading to a short period of reattached flow followed by the build up of another vortex and so on; and (c) the occurrence after separation of a permanently structured leading edge vortex with strong edgewise convection of vorticity as on a slender wing. In order to assess these visualisation of the flow both off and on the surface of the blade was undertaken. Smoke visualisation was used for the former and liquid crystal paint for the latter. This surface flow visualisation used a technique insensitive to centrifugal and gravitational force, namely temperature sensitive liquid crystal paint. These crystals are encapsulated in resin and hence remain fixed on the blade surface and do not respond to moderate forces. By preheating the blade the surface temperature contours could be recorded by observing the colour of the crystals as the flow cooled the surface. Since the heat transfer rate from a surface is proportional to the skin friction on that surface, the surface temperature patterns can give strong indications of the locations of regions of high skin friction (attached flow, high heat transfer and hence low surface temperature) and low skin friction (separated flow, low heat transfer and hence high surface temperature).

The surface pressure measurements were undertaken using the technique described previously (reference 1). A Scanivalve was used to sample each pressure tapping on the blade in turn. The pressure transducer output was transmitted from the hub to an external data acquisition system by a fibre optic link. A rapid response pressure transducer was used for the unsteady pressure measurements.

The 2m. diameter rotor had two blades designed from the coordinates of a prototype WEG stall regulated blade. This blade has a NACA632xx section of varying thickness from the root to the tip and is twisted and tapered with a parabolic pointed tip. The full scale blade is designed to deform under load, with an out-of-plane curvature when unloaded. The model blade was made of carbon fibre reinforced plastic and designed to be very stiff with no out-of-plane curvature, thus modelling the heavily loaded case. The instrumentation in the hub necessitated a hub diameter of 135 mm at the blade root followed by a wider transition piece behind the rotor of 250 mm diameter. Figure 1 shows the rotor assembled, without the fairing over the shaft and d.c motor.

The experimental procedure followed that used in previous work. The rotor rig was mounted in the large settling chamber of the department's 4.1/2' x 4' wind tunnel and run at 200 RPM for most cases over a range of wind speeds in the settling chamber from approximately 2 m/s to 8m/s, giving Reynolds numbers between 3×10^4 and 10^5. Mean pressures were sampled over several rotations and the average taken as the output. Unsteady pressures were sampled at 20 Hz. which was a compromise between the aim of capturing the significant fluctuation frequencies and the tube lengths in the pressure system.

2. RESULTS

The rotor was run for a range of tip speed ratios and hence local blade angles of incidence. The angles of incidence shown have been corrected, as far as possible, for the effect of the induced velocity field of the rotor by using measurements of axial velocity made a half chord length in front of the rotor plane and one half of the values of swirl velocity measured behind the rotor plane. The incidences shown are all measured from each section chord line thus taking into account the twist of the blade. The radial coordinates of the measurement stations and the blade chords are given in table 1.

Table 1. Planform Geometry.

Section	Radius(m)	Chord(m)
A	0.1925	0.115
B	0.2675	0.118
C	0.3425	0.113
D	0.4875	0.097
E	0.6375	0.079
F	0.7825	0.060

2.1 Mean surface pressures and loads.

Figures 2 and 3 show examples of mean pressure distributions measured at different radial stations r and at different effective local section incidences a. Figure 4 shows the section lift data. The lift coefficient for the inboard most section rises to a maximum value of about 2.2 at approximately 30° incidence. This figure also shows the results for the 2-dimensional aerofoil section for comparison. It is clear that at high incidences the inboard sections A, B and C in particular are generating much higher lift than would be indicated from the two-dimensional aerofoil data. When the blade starts to stall the pressure distributions show more pronounced suction towards the leading edge than in 2-dimensional flow. Within the experimental scatter of the measurements the lift curves all follow approximately the same curve until close to C_{Lmax}. The incidence a_{max} at which C_{Lmax} is reached gets progressively larger as r/c decreases.

For the cases measured here and in reference 1 a very approximate fit to the data is given by: $a_{max} = a_{max.2D} (1 + 4 c^2 / r^2)$
where $a_{max.2D}$ is the value of a_{max} for the same aerofoil section in 2-dimensional flow and the incidences are all measured relative to the zero lift incidence. This result may provide some general guidance for other blades but it should be noted that these results have been obtained for low Reynolds numbers at which 2-dimensional stall occurs at moderately low incidences. There is also an indication of a reduction in C_{Lmax} at the outermost station (F) which is at just under 80% of the tip radius. Part of this reduction may be caused by the low Reynolds number of the small tip sections and part by the outward drift of the boundary layer on the blade causing boundary layer thickening towards the tip which contributes to early stall. Figure 5 shows the pressure drag measurements which show consistency between all the stations and no significant effect of the parameter r/c.

2.2 Fluctuating pressure measurements

The time histories of fluctuating pressure from pressure tappings near the leading edge at the innermost station of the rotor blade only showed significant increases in the level of fluctuating pressure at incidences above 35° indicating development of a large scale separation region. The general level of fluctuation was low until this critical incidence was reached above which the fluctuating pressure started to increase rapidly. This critical incidence is higher the further inboard the station, and the incidences are well above the two-dimensional stall incidence for the section. The inference from this is that large scale (open) separation does not occur until very high incidences are reached. Below these incidences the flow may be separated but the indications are that the separation is limited.

2.3 Flow visualisation

Smoke flow visualisation indicated that at the lowest incidences the flow was attached. As the incidence was increased limited separation was observed to occur on the suction surface from a separation point some way back from the leading edge. Further increase of incidence caused the separation to move forward and widen into fully developed stall. At the highest incidences the separation point had reached the leading edge. Under conditions prior to fully developed stall the separation region above the suction surface was observed to be moderately thin.

Figures 6 shows an example of surface temperature patterns measured on the rotor blade under low tip speed ratio conditions. To obtain these the blade was first heated to a temperature at which the liquid crystal paint became uniformly black. The rotor and air stream were then set into motion and the blade surface photographed as it cooled down. Areas of high temperature indicative of low heat transfer and therefore low skin friction are seen, coloured black, near the root of the blade, whereas lower temperature, lighter colour, indicating the opposite are seen in the outer region and also in the root leading edge region indicating higher heat transfer and hence skin friction there. These results agree qualitatively with the smoke flow visualisation that the separation in the root region starts further back and is less strong than it is at intermediate stations further out along the blade. Static cooling tests, however, suggest that the results should only be considered as qualitative evidence since the internal heat transfer rates on this blade are not uniform owing to the blade structure. Also it should be noted that the blade rotation causes the skin friction generally to rise toward the blade tip.

3. CONCLUSIONS

The mean pressure measurements taken on a two bladed rotor, whose blade geometry was based on a prototype stall controlled rotor, show the same effects of stall delay in the inboard region as have been observed in previous work on other rotors. The effect appears to depend on the parameter chord/radius and is therefore less pronounced than for some previously tested geometries but is still significant.

Flow visualisation shows that the flow over the inboard sections starts to separate at a comparable incidence to that at which it separates in two-dimensional flow. But the separation starts some way back from the leading edge developing initially as a fairly thin reattaching bubble. As the incidence is increased the separation develops but does not appear to attain the condition of a full stall with failure to reattach before the trailing edge and hence open separation until very high incidences have been reached. This flow picture is supported by the measurements of fluctuating pressures and the qualitative study of surface cooling of a heated blade. These results are compatible with the first (a) of the three suggested physical mechanisms for stall delay and do not support the second or third ones (b, c).

There is some indication of reduction in the maximum lift coefficient in the outermost region of the blade. However this conclusion is complicated by the low Reynolds number of the sections there. The pressure drag coefficient does not appear to be influenced by radial station effects, in agreement with previous findings.

ACKNOWLEDGEMENT

This work was supported by ETSU under the Department of Trade and Industry Wind Programme.

REFERENCES

1. Brown C.J. and Graham J.M.R. Proceedings XIIth BWEA/EWEC Conf. Glasgow. 1989, p117.

2. Brown C.J. and Graham J.M.R. Final Report to ETSU on contract agreement number E/5A/CON/6041/2347. 1991

3. Barnsley M. and Wellicome J., Proceedings XIIth BWEA/EWEC Conf. Glasgow. 1989

4. Himmelskamp. H., Max Planck Inst. For Fluid Dynamics, Göttingen. 1950 Rept. No. 2.

Figure 1. Test rig showing rotor blades, hub and two part nose cone.

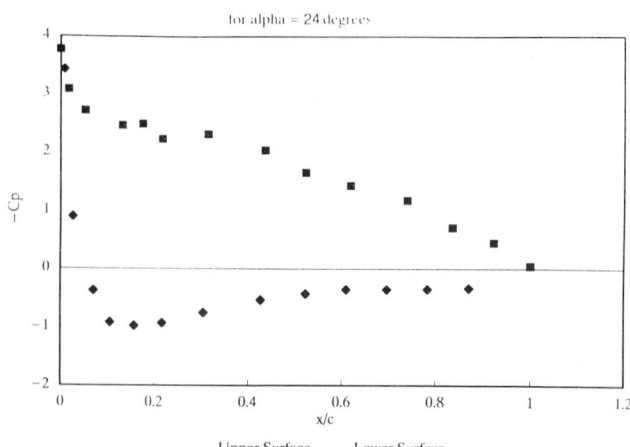

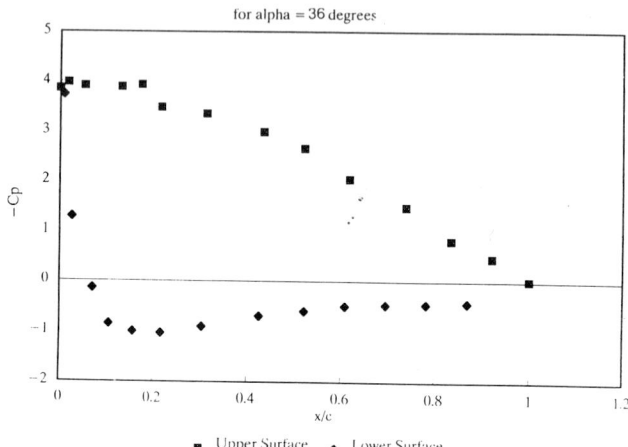

Figure 2.ctd. Pressure coefficient distribution, Section A, r = 0.192m

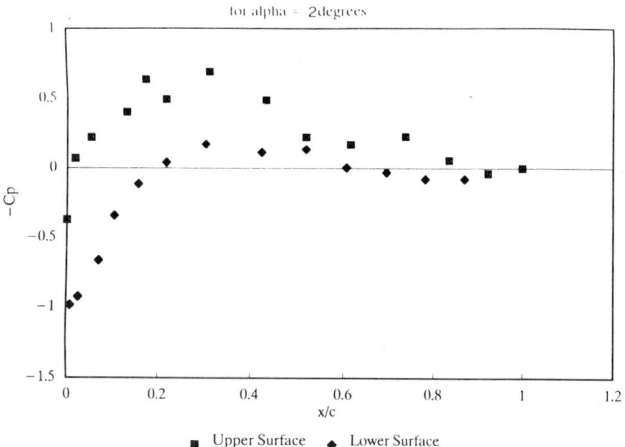

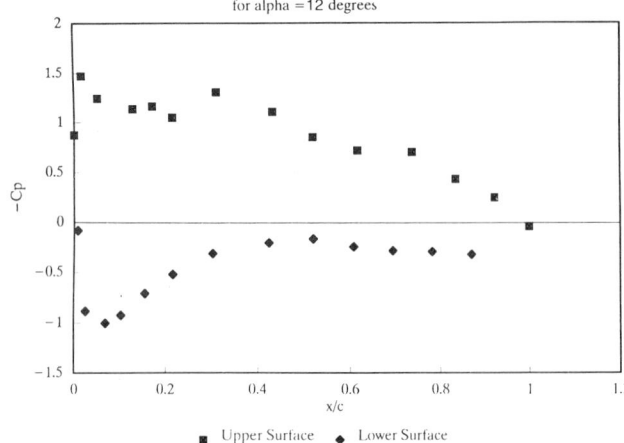

Figure 2. Pressure coefficient distribution, Section A, r = 0.192m

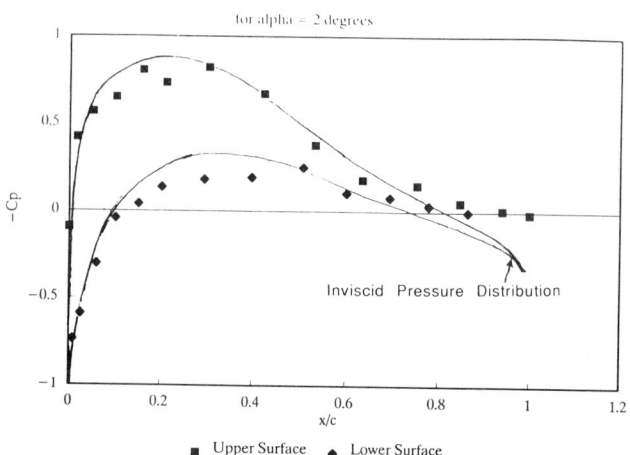

Figure 3. Pressure coefficient distribution, Section B, r = 0.268m

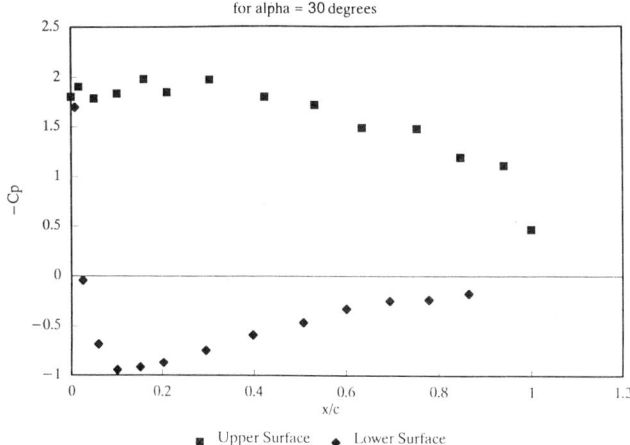

Figure 3.ctd. Pressure coefficient distribution, Section B, r = 0.268m

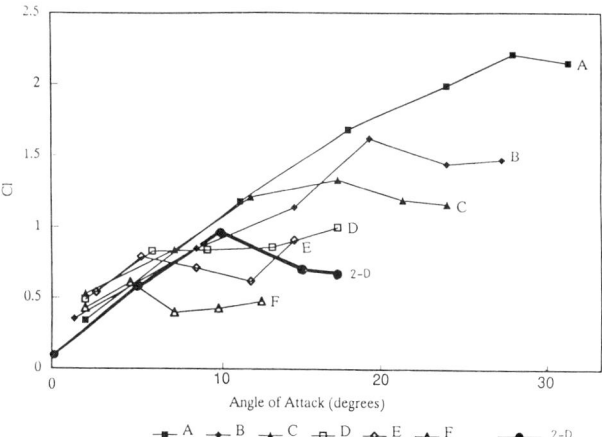

Figure 4. Lift coefficient vs. effective incidence for rotor.

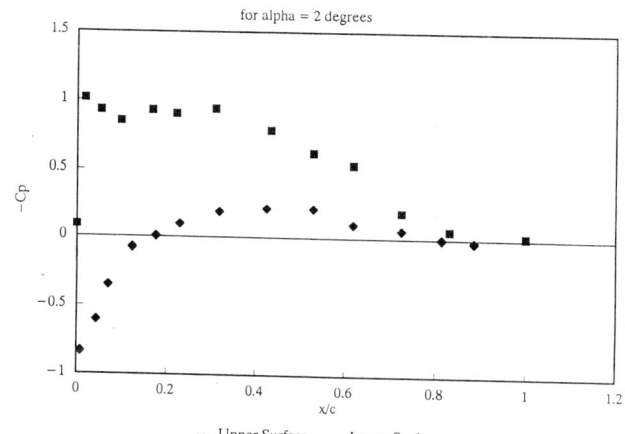

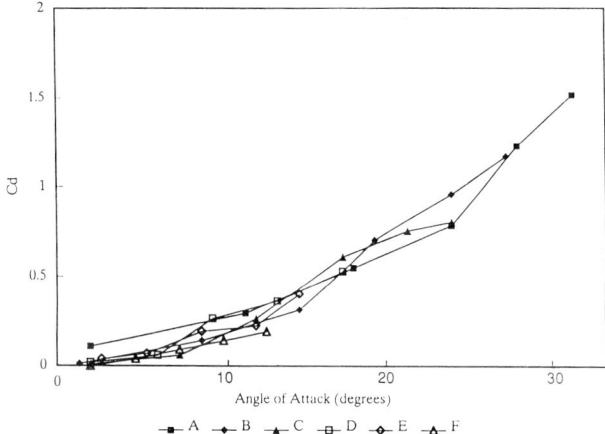

Figure 5. Pressure Drag coefficient vs. effective incidence for rotor.

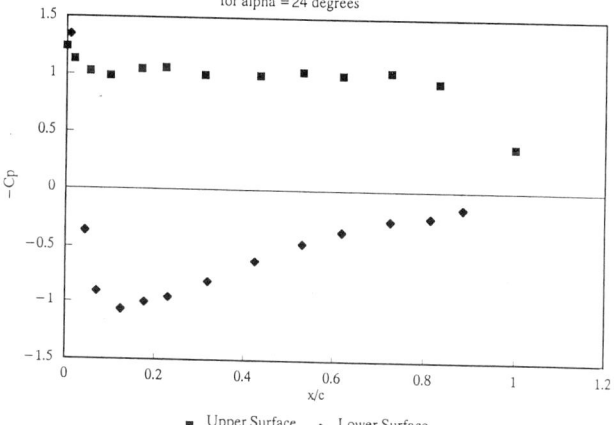

Figure 3.ctd Pressure coefficient distribution, Section C, r = 0.343m

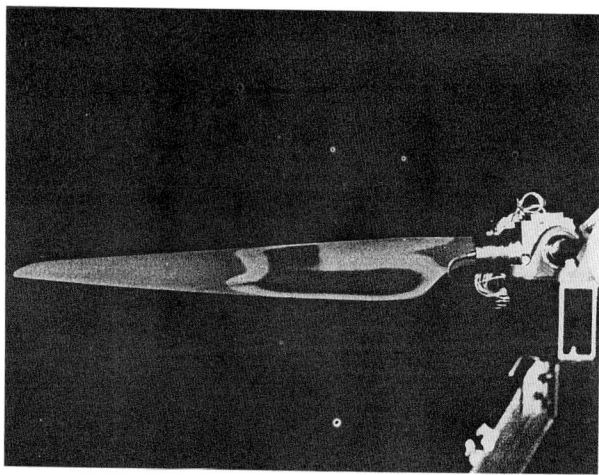

Figure 6. Liquid crystal temperature pattern on rotor blade suction surface taken during rotation. a ~ 25°.

Pre field test development of the 'Fledge' aerodynamic braking and control device

RAYMOND S. HUNTER, W. ALAN. DERRICK, NEIL I. D. ROBERTSON, JIM L. CHAPMAN,
R. ALAN COURT
National Wind Turbine Centre, UK

SYNOPSIS: A prototype 'Fledge' tip device has been designed, constructed and tested in the laboratory. The paper describes the composite structure, the actuation mechanism, and the laboratory test arrangement. A number of practical lessons have been learned which will guide construction of future 'Fledges' for real wind turbines.

1 INTRODUCTION

The 'Fledge' concept of braking and/or control of horizontal axis wind turbines has undergone steady development in the UK over several years. The UK Department of Trade and Industry has recently funded a major programme of work designed to assess the economic and technical feasibility of the idea, and a large number of UK companies comprising consultants, wind turbine manufacturers, and laboratories have participated in the venture.

The 'Fledge' (Flying Leading Edge) device is a lightweight and responsive air brake device which also offers the possibility of very cheap tip control. It has major advantages over conventional pitchable tips particularly in terms of simplicity of actuation and in fatigue duty.

The basic 'Fledge' idea is illustrated in figure 1. Rather than pitching the entire blade tip, only the leading edge and top surface rotates, hence causing development of a high drag arrowhead profile. Initial opening is aided by aerodynamic suction and additional opening force is generated by a radial sliding mass which forces the 'Fledge' open via a link rod. Control or damping can be exercised by a cable which travels back down the blade and thereafter to appropriate assemblies in the hub or nacelle.

Previous papers (refs 1,2,3) have reported the predicted aerodynamic performance and loading characteristics of the 'Fledge' and have also described results of dynamic simulation and control studies. The present paper concentrates on the design and construction of a full scale prototype specimen, which has subsequently been tested in a specially assembled loading simulation rig in the laboratory. The aim has been to demonstrate engineering feasibility and to identify potential problem areas prior to attempting full scale field trials on an actual turbine.

2 COMPOSITE MANUFACTURE

Manufacture of the 'Fledge' structure in GRP has involved conventional plug and mould techniques, but the complex split line geometry and the need for structural reinforcement has required some novel solutions.

Initially a plug was made up of individual sections of high density foam which were cut to the desired profile, bonded together and then shaved down. From this plug a two part (upper and lower surface) GRP female moulding tool was cast. On this were then scribed the future positions of the 'Fledge' split lines. The four main sections of the parent blade stub and the 'Fledge' (figure 2) were then built up by trimming various laminate layers of glass fibre mat to match patterns taken from the scribes. Fabrication of the four main sections then followed a sequence of laying and curing an outer gel-coat skin and then building up on the inner surface alternate layers of mat and resin. Hand rollers were used to ensure fully wetting out and compression. After curing, trimming and grinding were carried out.

Two of the four main sections were fabricated in this way (one upper and one lower surface). They were then used as edge moulds for the other pieces, and during subsequent lay up, polymer sheeting was used to avoid undesired adhesion.

Local bulking out was achieved by including balsa mat in the lay up. The balsa was tapered and surrounded with a fibre mulch to ensure good and smooth load transition.

Structural reinforcement in the stub was obtained by bonding in GRP channels, pre moulded to match the contours of the inner skin surfaces.

Continuity of the GRP construction at the leading edge of the 'Fledge' was obtained by fabricating and attaching a re-entrant, wrap round mould extension to the leeward mould half. Smoothness of joints between opposite blade pieces was obtained by using beeswax to stand off the piece being fabricated from its mould during lay up and then subsequently filling in and finishing off the resultant channel during final assembly.

The 'Fledge' design called for two main pivot bearings. Khaya bearing block housings for these were positioned and bonded in place with the assistance of a mandrel and alignment jig fixed to the main mould housing.

Generally the above processes were found to be technically feasible, but slow and requiring skill and inventiveness. Quicker, more streamlined methods would undoubtedly be required for series production.

3 ACTUATION MECHANISM

After looking at many actuation layout options, the system shown in figure 3 was selected for manufacture.

Several options for the actuating mass were looked at, viz using a linear sliding mass, as eventually selected, or using a rotating arm or folding, elbow joint. The latter options suffer from not having particularly good mechanical advantage at respectively the start and end of their travel and were therefore rejected, however the selected system does have the problem of having exposed elements which would require environmental protection in a real system.

Having established that the mass would follow a linear path, methods for constraining it were then examined. The option of using guide rails (twin or single) were rejected because of space restrictions and the high lateral forces they would have to bear, whilst the idea of a channel guideway was not selected because of uncertainties over bonding, maintenance, and life of the necessary plain bearing strips. The selected solution was to use a proprietary linear bearing slide with recirculating ball bearings, which in the field should help ensure centrifugal effects do not impair lubrication. The rail attachment to the stub was achieved by pre bonding a steel backing plate inside the stub, the skin of which subsequently formed a sandwich.

The design of the ball joint, pivot and link rod assembly do not pose any particular difficulties, other than noting that the ball joint attachment to the 'Fledge' produces a major stress concentration in the composite material.

Figure 3 shows that a travelling pulley was selected to transfer damping, control or closing forces to the 'Fledge' plate. Although this has the disadvantage that the required cable stroke is larger than would be the case were the cable to terminate on the mass, this is offset by the mechanical advantage which can be obtained for holding the 'Fledge' firmly closed at low wind speeds.

A major design requirement of the control cable, which must pass the length of the blade is that it should be stiff. To achieve this a parallel fibre Kevlar cable was selected. Since the required gauge of Kevlar would not be suitable for a tight radius pulley, a PVC coated wire was proposed for the final length of cable.

4 LOADING SIMULATOR

Given that the 'Fledge' is a rather unusual, innovative concept, and given that the actuation system is at an early stage of development, it would be technically and financially undesirable to carry out debugging of the design on an actual wind turbine rotor. For this reason a series of laboratory based experiments were planned, which required the design and construction of a loading rig to simulate the forces which would be experienced on a rotating wind turbine. The major forces are the aerodynamic pitching on the 'Fledge' plate, the centrifugal pull on the 'Fledge', and the centrifugal pull on the actuating mass.

The simulation rig, shown schematically in figures 4 and 5, mimicked the centrifugal forces by application of weights suspended over pulleys, and the aerodynamic force by springs and a pitch loading frame. The latter consisted of a steel arm bonded to the upper surface of the 'Fledge' plate. This arm was hinged via a crank about the same axis as the 'Fledge' and could therefore introduce pure loading moments, the magnitude of which was automatically varied as a function of 'Fledge' position by a twin coil spring and pulley arrangement (fig 5).

Control and/or damping action on the control cable was applied by a computer controlled hydraulic cylinder.

5 STATIC TESTS

Two forms of test were carried out, one simply to confirm the force-deployment relationships and the other to look at skin distortions.

5.1 Deployment

Previous simulation and design work had indicated that for normal rotational speeds, the centrifugal action of the unrestrained actuator mass should have been sufficient to fully overcome 'Fledge' plate aerodynamic closing forces and hence hold the plate fully open.

It was therefore surprising to find on initial loading of the system that the 'Fledge' only partially deployed. It was subsequently hypothesised and demonstrated that by decreasing the length of the link rod, full deployment could be achieved. It is thought however that this was not the main cause of the phenomenon, which was more likely tied in to misplacing of the main 'Fledge' bearings during manufacture. By re-examining the geometrical analysis behind the system's kinematics, it was confirmed that overall behaviour is extremely sensitive to the position of the hinge axis. This could be a major concern for series production, where very high quality jigging and quality control would be necessary.

5.2 Deflection

Because of the complexity of the 'Fledge' structure and shape, there was concern over whether structural distortion under load could give rise to fouling and jamming of the system. To investigate whether this was likely, and if so to pin-point the sources of the potential problems, a series of tests were conducted using the photogrammetric technique of non contact measurement.

This is an extremely powerful method which can measure distortions to 0.001mm. A large number of visual targets were stuck to the 'Fledge' and loading rig surfaces, and thereafter a twin camera photogrammetry system was used to view the specimen from two angles. By taking photographs and subsequently digitising the target positions using a stereographic viewer, the xyz co-ordinates of each point could be derived by an analyser.

This approach was followed for both unloaded and loaded conditions, and this then allowed the change in position of each target to be defined by derived vectors. These were then displayed and visualised using a sophisticated graphics package to allow user interpretation.

Typical results are shown in figures 6 and 7. Figure 6 shows the distortion vectors which were generated when the centrifugal load of 500kg was applied to the 'Fledge' when in the aerodynamically neutral position, ie the deployment angle at which the aerodynamic forces neither wish to close or open the 'Fledge' further. The view is looking at the leading edge, with the main blade on the left and the 'Fledge' on the right. The points higher up are on the trailing edge of the 'Fledge', whilst isolated targets on the periphery of the figure relate to targets on the loading frame. Most vectors have been removed for clarity, the major observation to note being that the distortion on the leading edge of the 'Fledge' is somewhat irregular, with significant downward motion at the outer bearing position. In practice such distortion would give rise to misalignment of the two main bearings which would restrain 'Fledge' motion.

For figure 7, the centrifugal force on the actuating mass has been added to the loading, and the vectors from figure 6 subtracted out so as to focus solely on the distortions developed by the additional loads. The main effect to note is that the 'Fledge' plate has undergone a skewing motion, the pivotal point of which is at the position where the vector length is shortest. This not unexpectedly occurs very close to the link rod attachment point. The source of the skewing is thought to be the couple which results from the compressive force in the link rod and the tensile force in the control cable. The vectors suggest the skewing is not uniform over the entire 'Fledge' implying distortion as well as global rotation is occurring.

6 DYNAMIC TESTS

Dynamic tests were undertaken to simulate transient braking events and also repeated control action.

6.1 Functional Tests

Before carrying out the main dynamic tests the general function of the system was checked. This highlighted that there was a tendency for the control cable to jump the actuator mass's pulley guide. The cause of this problem was that the rotational axis of the pulley support was not coaxial with the cable axis coming from the hub, meaning that the rotational moment being generated by the cable line to the 'Fledge' was being resisted, hence causing the pulley to remain in position and the cable to ride up.

6.2 Braking Simulation

It was not possible to simulate perfectly the deployment schedule of the 'Fledge' for the following reasons: the centrifugal loads were fixed implying changes in rotor speed could not be accommodated; the loads were less distributed than they would be in reality; and more relevantly the aerodynamic loading frame significantly increased the pitching inertia of the assembly.

For these tests the central hydraulic servo actuator was programmed to provide the damping characteristic which previous simulation had suggested would be required. This damping could be varied electronically.

It was found that if the damping was set too high then the 'Fledge' would not fully deploy. This suggests that the bearing misalignment referred to previously, which itself would generate damping, could prevent full deployment in a real system.

6.3 Bearing Temperature

Because the bearings are set in a composite structure which has poor thermal conductivity, there was concern that the heat generated in the bearings by repeated control action could give rise to high temperatures. A series of tests were carried with the 'Fledge' undergoing pitching motions of 5 degrees amplitude at 1Hz for several hours. It was confirmed that the bearings would undergo significant temperature rise, however due to the bearing axis distortion referred to previously, it is not yet known whether this would be a problem were the hinge line to be better restrained.

6.4 Endurance

Although endurance trials were originally intended, ultimately they were not thought particularly appropriate given the problems which had been highlighted with various components in the system.

8 CONCLUSIONS

The feasibility of construction of the 'Fledge' has been demonstrated and a suitable first iteration actuation mechanism designed. Laboratory tests of the prototype have highlighted a number of characteristics and problems, which would have been expensive to put right had they been discovered in a full field test. Photogrammetric techniques have been particularly useful in understanding how the 'Fledge' structure distorts under loading. The main fault in the design as tested was the lack of rigidity in the 'Fledge' hinge line.

ACKNOWLEDGEMENTS

The authors acknowledge the valuable contribution made to the programme by Peter Jamieson of Garrad Hassan and Partners, in inventor of the concept, and to Dave Corbet, latterly of the same company, who provided valuable advice on the composite design.

The work reported in this paper was carried out under contract to the Energy Techniology Support Unit (ref E/5A/6060/2495).

REFERENCES

1 Jamieson, P. and Agius, P. A Comparison of Aerodynamic Devices for Control and Overspeed Protection of HAWTs. BWEA 13, Norwich, March 1990. MEP.

2 Jamieson, P. et al. Innovative Concepts for Aerodynamic Control of Wind Turbine Rotors. EWEC'91, Amsterdam, October 1991. Elsevier.

3 Derrick, W.A. Aerodynamic Characteristics of Novel Tip Brakes and Control Devices for HAWTs. BWEA 14, Nottingham, March 1992. MEP

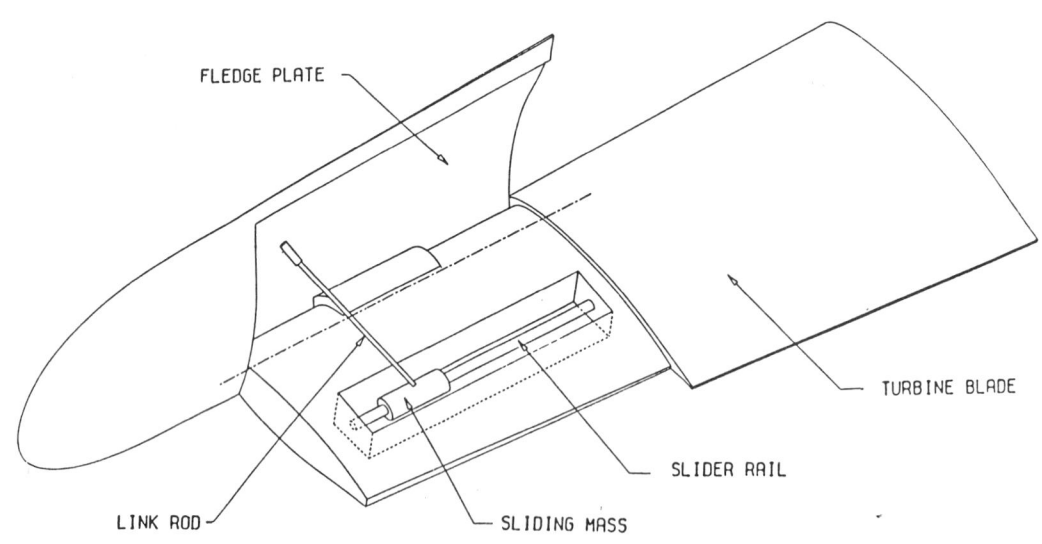

Fig 1   The Basic 'Fledge' Device

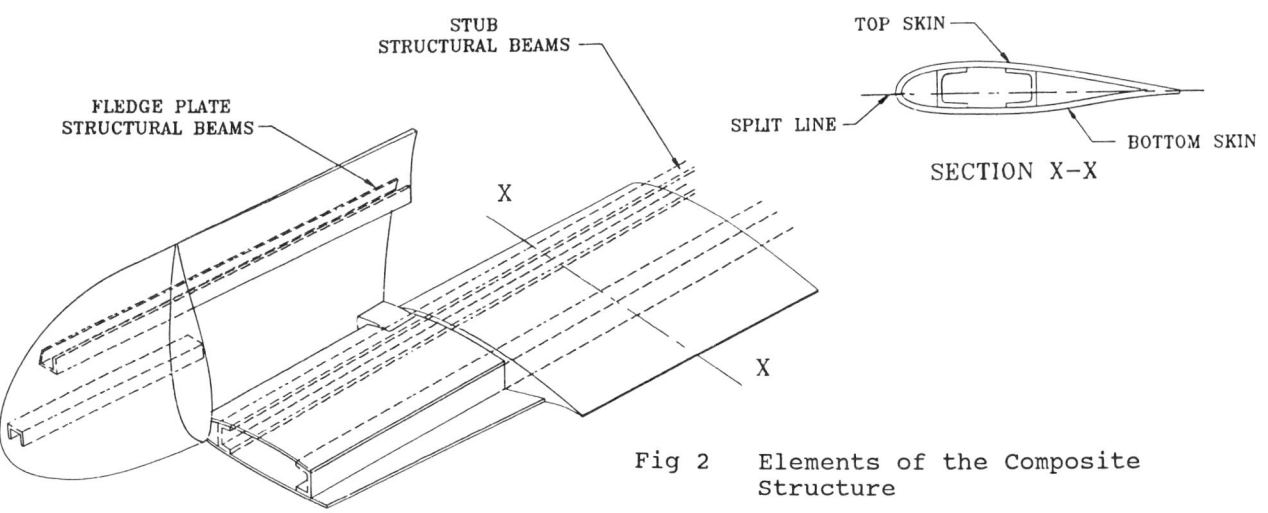

FLEDGE PLATE
STRUCTURAL BEAMS

STUB
STRUCTURAL BEAMS

TOP SKIN

SPLIT LINE

BOTTOM SKIN

SECTION X–X

Fig 2   Elements of the Composite
         Structure

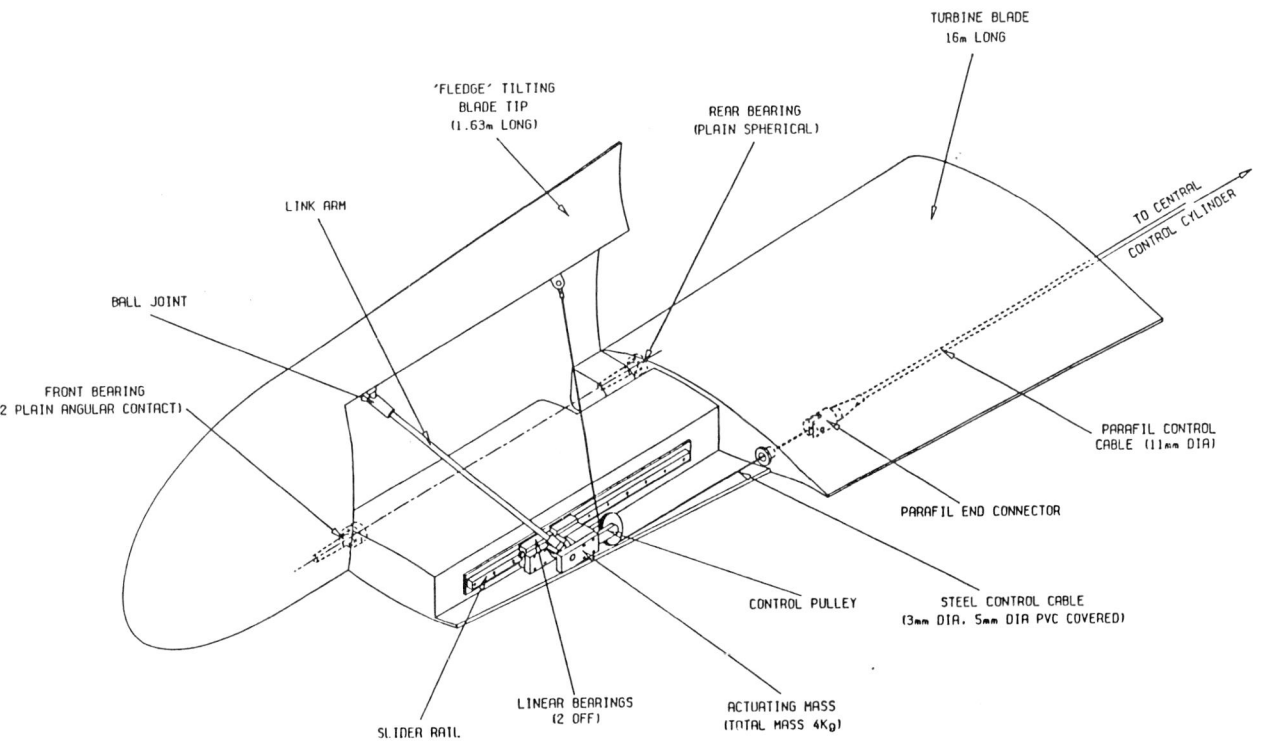

'FLEDGE' TILTING
BLADE TIP
(1.63m LONG)

REAR BEARING
(PLAIN SPHERICAL)

TURBINE BLADE
16m LONG

LINK ARM

TO CENTRAL
CONTROL CYLINDER

BALL JOINT

FRONT BEARING
(2 PLAIN ANGULAR CONTACT)

PARAFIL CONTROL
CABLE (11mm DIA)

PARAFIL END CONNECTOR

CONTROL PULLEY

STEEL CONTROL CABLE
(3mm DIA, 5mm DIA PVC COVERED)

SLIDER RAIL

LINEAR BEARINGS
(2 OFF)

ACTUATING MASS
(TOTAL MASS 4Kg)

Fig 3   Selected 'Fledge' Actuation
         System

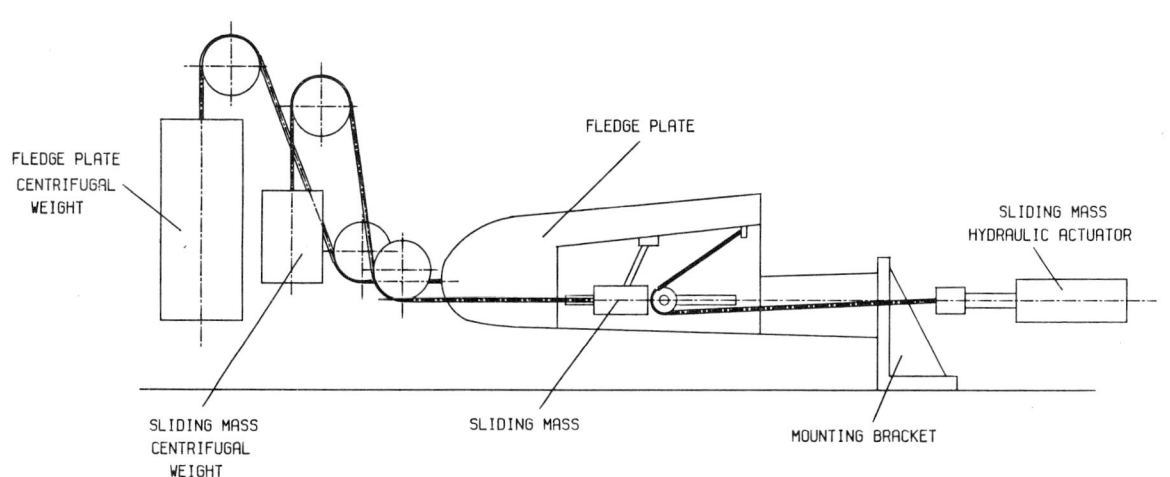

FLEDGE PLATE
CENTRIFUGAL
WEIGHT

FLEDGE PLATE

SLIDING MASS
HYDRAULIC ACTUATOR

SLIDING MASS
CENTRIFUGAL
WEIGHT

SLIDING MASS

MOUNTING BRACKET

Fig 4   Load Simulation Rig – Schematic
         of Centrifugal Force Loading
         Components

303

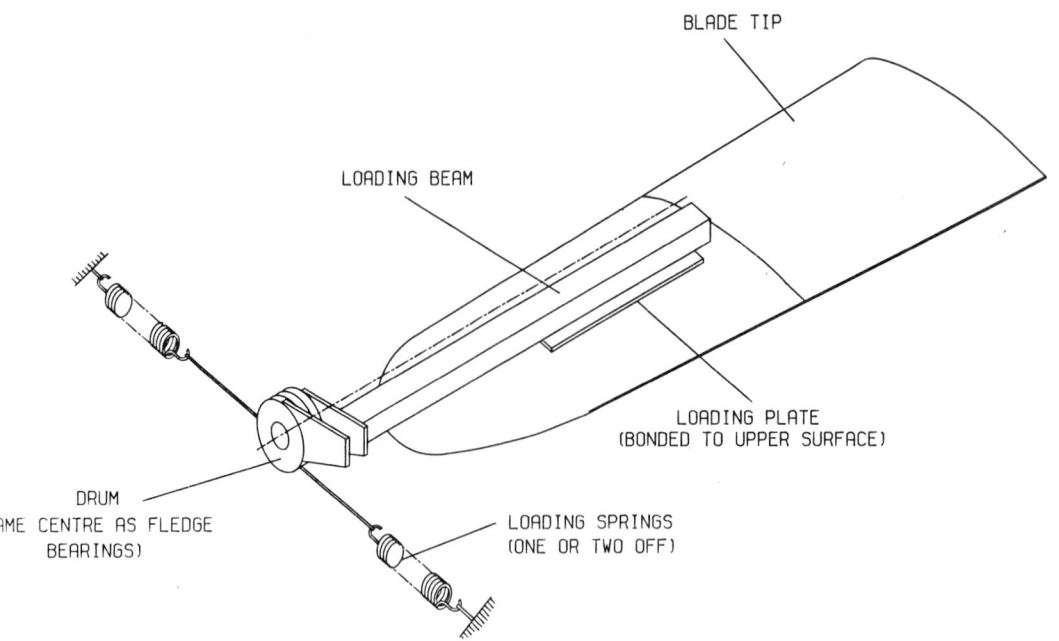

BLADE TIP

LOADING BEAM

LOADING PLATE
(BONDED TO UPPER SURFACE)

DRUM
?ME CENTRE AS FLEDGE
BEARINGS)

LOADING SPRINGS
(ONE OR TWO OFF)

Fig 5   Load Simulation Rig - Principle
of Aerodynamic Force Simulation

Fig 6   Photogrammetric Results -
Distortion of Leading Edge Under
Application of 'Fledge'
Centrifugal and Neutral
Aerodynamic Loading

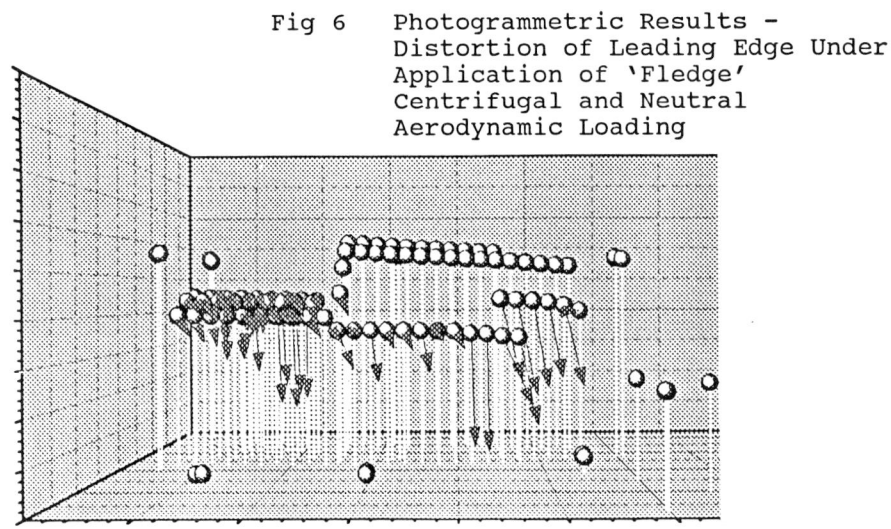

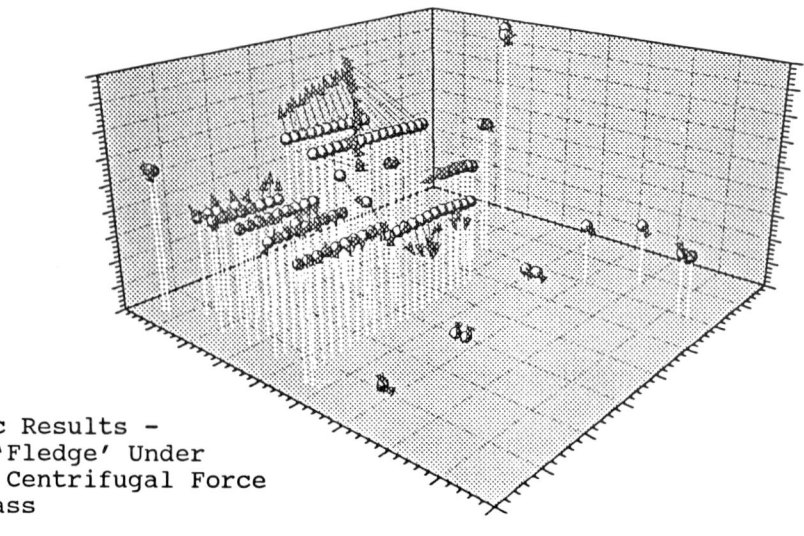

Fig 7   Photogrammetric Results -
Distortion of 'Fledge' Under
Application of Centrifugal Force
to Actuating Mass

An attempt to measure dynamic inflow phenomena on a 1 MW turbine

W. E. LEITHEAD BSc, PhD,
University of Strathclyde, UK
M. C. M. ROGERS BSc,
Industrial Systems and Control Ltd, UK

SYNOPSIS During start-up, the rotor speed of the HWP55/1000 wind turbine is regulated to enable a reasonably smooth synchronisation of the generator with the grid. The rotor speed is made to undergo slow oscillations by varying the pitch angle sinusoidal with a frequency of 0.1 Hz. When the rotational speed matches the grid frequency, the generator goes on-line. Since the frequency at which the tips are oscillating during start-up is very slow, it might be expected that the effect of dynamic inflow would be apparent in measured data. However, the characteristic overshoot associated with dynamic inflow is absent, instead, there is a marked undershoot.

NOTATION

ω_{grid}	grid frequency
p	number of generator pole pairs.
V	wind speed 'seen' by the wind turbine
R	radius of the rotor
$IneLs$	inertia of the low-speed shaft
ρ	air density
C_q	torque coefficient
$OmgLs$	rotation velocity of the low speed shaft
λ	tip speed ratio

1 INTRODUCTION

The Howden HWP 55/1000 1 MW constant speed grid-connected horizontal-axis wind turbine, sited at Richborough Power Station, is a three-bladed machine with an induction generator. The hub height is 45 m and the wood/epoxy blades have a diameter of 55 m with 4 m movable tips. The tips are hydraulically activated and are used to control the power output in above rated wind speed when the machine is on-line (power control). In addition, they are used to control rotational speed when the machine is starting up before connection to the grid (speed control).

* Current address : Industrial Control Centre, University of Strathclyde

Control and simulation models of wind turbines (1), (2) are employed to facilitate the investigation of the control task for pitch regulated machines. Suitable models have been extensively validated for medium scale wind turbines rated at 300 kW. Advantage was taken of an existing monitoring programme being undertaken by James Howden on the HWP 55/1000 at Richborough to obtain data with which to validate the models when extended to cater for large scale machines. In this paper, the validation of these models within the context of speed control is investigated. In particular, an attempt is made to identify and model dynamic inflow during slow oscillation of the pitch angle of the tips. This paper is motivated by the current interest in dynamic inflow (3).

2 MONITORED DATA

The wind turbine itself was fully monitored with additional wind speed measurements taken on two meteorological masts placed 100 m either side of the turbine. Data were collected at various sampling rates with a maximum of 40 Hz. Of particular interest for the purpose of validating the models are the measurements of the wind speed 'seen' by the machine, the speed of the low-speed shaft, and the tip pitch angle.

2.1 Wind speed measurements

The wind speed was measured at a variety of heights on the two meteorological masts and on the wind turbine nacelle. Since air flow at the nacelle is impeded by the movement of the blades, the measurement on the nacelle is not a good indicator of the wind speed 'seen' by the machine; on average it is lower than wind speed measured at hub-height on the masts. As the masts are on either side of the machine and 200 metres apart, it is hardly surprising that time series for wind speed derived from their measurements are very different from each other. However, the power spectra of the two time series are very similar. Therefore it may be assumed that the wind speed 'seen' at the machine also has a similar spectrum to that measured at a height of 45 m on the masts, although the wind speed at any particular time remains unknown. For this reason, when investigating power control, measurement data at hub height from one of the masts is used as the wind speed input for the simulation. The simulation can then be validated against the measured data provided any comparisons are made in the frequency domain. However, for speed control, it is the time series which must be compared. In this case, the average wind speed measured on the masts at hub height is used as the wind speed input for the simulation, since the full time series would be no better a representation of the actual wind speed 'seen' by the turbine.

2.2 The speed of the low-speed shaft

The speed of the low-speed shaft is calculated from a position sensor and the measurement is initially stored in one processor before being transferred to the data-logging processor as a low-priority task. This method of recording the rotor speed involves an unknown time delay.

2.3 Tip angle measurements

Since the three blades are at an angle of 120° to each other, the measurements of tip angle for each blade are averaged to remove the gravitational component, see Figure 1. The calibration of the tip angle measurements assumed that there is a linear relationship between the millivolt signal from the strain gauges and the real tip position. In reality, however, this relationship is only linear for tip angles less than 40°. Unfortunately, during speed control the tip angles recorded by the logging system can reach 62°, which corresponds to an actual position of 80°.

3 SPEED CONTROL

Initially on start-up, the wind turbine starts to rotate with the generator not connected to the grid. The machine's rotational speed must be controlled otherwise it would attain excessive rotational velocity since there is no electrical torque opposing the motion. At the instant of connection to the grid, the high speed shaft velocity is required to be

$$\frac{\omega_{grid}}{p}$$

At grid synchronisation, the rotational velocity of the high-speed shaft for the Richborough turbine should be 1000 rev/min, which corresponds to a rotor speed of 24.04 rev/min. In order to reach this synchronous speed and connect smoothly and efficiently to the electrical grid, a speed control system as in Figure 2 is required. The speed control algorithm used on the Richborough machine is described below.

3.1 Speed control algorithm

The speed control algorithm has three stages (2). First, with the speed set-point fixed, the low-speed shaft brake is released and the rotor speed is allowed to increase in a controlled manner, with discrete movements of the tips, until the speed is 25 percent of synchronous speed. Second, the speed set-point is ramped up to 90 percent of the synchronous speed. Simultaneously, the angle of pitch of the tips are oscillated to superimpose an oscillation on the increasing rotor speed. Third, the speed set-point is kept fixed but the oscillation of the pitch angle and rotor speed are maintained. The tips continue to be oscillated so that the rotor speed oscillates in a range close to synchronous speed. The machine is connected to the grid when the rotor speed is sufficiently close to synchronous speed and the rotor acceleration is zero. The speed control algorithm makes extensive use of the wind speed in terms of three minute average of wind speed.

3.2 Speed control data

The average tip angle during speed control is shown in Figure 1 and the rotor speed is shown in Figure 3. The large slow oscillations in rotor speed caused by varying the tip angle sinusoidally at a frequency of 0.1 Hz, during the second and third stages of the speed control algorithm, is evident. The oscillations would appear to be well-suited for detecting the presence of dynamic inflow.

The tip deployment rate (5°/s) is approximately twice as fast as the retraction rate (10.5°/s) due to being controlled by different valves in the hydraulic system. Although the tip movement is asymmetrical the resulting speed oscillations appear to be symmetrical due to the non-linearity of the aerodynamic torque with respect to tip position. The phase of the oscillations in rotor speed would be expected to lag the oscillations in tip angle by 180°. However, the phase relationship of the two data sets requires clarification, because of the time delay associated with the data logging.

4 THE SIMULATION MODEL

The Richborough wind turbine is simulated during speed and power control by adapting the ACSL simulation model developed by Wilkie and Leithead (1), (2). The only difference between the actual speed control algorithm and that used in the simulation arises from the use of the average wind speed measured at hub height on the masts during speed control rather than the genuine three minute average. Unfortunately the necessary data is not available as the wind speed measurements for the three minutes before start-up were not recorded.

The structure of the simulation is shown in Figure 4. The wind speed input is the measurement data at hub height from one of the masts. It is spatially filtered [4] to account for averaging over the rotor. The spatially-filtered wind speed and the tip angle are used to calculate the corresponding torque coefficient, Cq, from torque tables. The rotor torque is

$$1/2\rho V^2 \pi R^3 C_q$$

The torque coefficients were supplied by James Howden plc. They are tabulated in non-dimensional form as function of tip speed ratio and tip angle. The torque coefficients supplied are well validated for the purpose of describing steady-state loads experienced by a wind turbine operating in attached flow, i.e. not stalled flow. During speed control (see Figure 3) the tips can reach very large pitch angles and hence are at times in stalled flow. For the tips, stall in negative incidence is predicted by the steady-state aerodynamics to occur when the angle of attack is less than -15°. Since the flow angle is approximately 6°, it follows that stall is predicted to occur when the pitch angle is approximately 21°. However, for the main blade, during the third stage of the speed control, the speed oscillations occur when the tip speed ratio is

approximately 8, and the flow over the main blade sections is therefore attached. Hence, the aerodynamics of the main blade should be well-represented by the torque coefficients during the speed oscillations.

Initially, the rotor torque in the simulation is modified by a lead-lag compensator to mimic the dynamic inflow effect (5); that is, the overshoots in torque caused by the delay between a change in wind speed or pitch angle and the consequent adjustment of the downstream wake of the turbine. In response to a step change, there is an instantaneous over-shoot, greater than 50 percent, which then decays slowly (5). The time constant associated with the dynamic inflow effect is related to the mean wind speed. The model employed induces a 50 percent over-shoot in torque with a time constant of 7.5 s. It is a lead-lag filter of the form

$$(1+as)/(1+bs); \text{ where } a = 11.5 \text{ and } b = 7.5$$

However, when validating the simulation for below-rated wind speed on-line operation it is found that data were not of sufficient resolution to validate the requirement for a model of the dynamic inflow effect. Agreement in the frequency domain between the measured and simulated torque was no poorer when this lead-lag compensator is removed.

5 COMPARISON OF MEASURED DATA WITH SIMULATION RESULTS

From Figure 3, it may be observed that the acceleration of the rotor between 0 s and 100 s is almost constant; that is, the aerodynamic torque acting on the turbine rotor is almost constant. It follows that over the time scale of 100 s there is no steady-state error in the aerodynamic torque associated with the dynamic inflow effect, or indeed any lead or lag with a similar time constant. In other words, the simulated rise is speed between 0 s and 100 s must match closely that shown in Figure 3 whether a model of dynamic inflow is included in the simulation or not. By displacing in time the plots of simulated and measured rotor speed, they can be synchronised and the unknown time discrepancy associated with the rotor speed data recording, cf. section 2.2, eliminated.

Because of the lack of evidence for dynamic inflow during power control, cf. section 4, no model of dynamic inflow is included when simulating speed control. Figure 5 shows the measured rotor speed

307

and the rotor speed predicted by the simulation when using the average of the three tip angle measurements and the constant wind speed. The two plots are synchronised. The sensitivity to variations in the wind speed, determined by varying the wind speed input to the simulation, is sufficient to account for the differences between the plots in Figure 5 for time less than 100 s. The simulation appears to reach synchronous speed and level off at about the correct time. However, the magnitude of the wind speed only weakly effects the amplitude of the oscillations. The main discrepancy between the two figures is the amplitude of the simulated oscillations which is twice that measured. In addition, the simulated oscillations lead the measured oscillations by 90° which is, equivalent to a time delay of 2.5 seconds on the measured oscillation. Unfortunately, these discrepancies are not those which would arise from dynamic inflow. The oscillations in tip angle would be expected to induce greater oscillations in speed than those predicted by the simulation. However, in reality the tips appear to have much less effect on the aerodynamic torque than predicted.

To investigate further, the torque coefficient associated with the measured speed is back-calculated using the simulation model. Figure 6 shows, for the first oscillation, the calculated torque coefficient verses the rotor speed for the measured data and the calculated torque coefficient versus the rotor speed for the simulation. The torque coefficient calculated from the measured data follow a hysteresis loop during each speed oscillation, while the simulated torque coefficient does not. The error in torque coefficient is large at small pitch angles. This is surprising because the tip is predicted not to be in stall, (it is below 21°) and is therefore in the region where the torque coefficients have been well validated.

A lag compensator is introduced to the simulation to provide a more sluggish transient response and a phase lag. The compensator is of the form

$$(1+as)/(1+bs); \text{ where } a = 0.01 \text{ and } b = 2.5$$

Note that this is a lag-lead compensator in contrast to that of the dynamic inflow which is a lead-lag compensator.

With the lag-lead compensator, the torque coefficient is again back-calculated using the simulation model. Figure 7 shows, for the first oscillation, the calculated torque coefficient versus rotor speed for the amended simulation. The simulation results now

have the hysteresis loop which agrees well at low pitch angles. At high pitch angles, the measured torque coefficient is greater than the simulated but this is consistent with the tip stalling at the high angles of pitch as expected. It may be observed from Figure 8 that the simulated rotor speed is now in good agreement with that measured. Any differences are minor and can be attributed to actual wind speed being unknown and wind speed being represented by the average measured at hub height on the mast.

Hysteresis loops had previously been observed on the SERI Combined Experimental Wind Turbine (6), Figure 9. In this case, changes in the angle of attack were measured when the turbine was operating in yaw with cyclic stall of the blades. The frequency of the loops is 1-P, the rotational frequency of the rotor. On the Richborough machine, however, the frequency of the tip angle oscillations is considerably slower than the rotational frequency of the rotor and the hysteresis loop cannot be attributed to the same source.

The phenomenon described in this paper is observed on all start-ups, some of which have many oscillations when the turbine fails to synchronise, for which data is available.

6 CONCLUSIONS

Although, the slow oscillations of the tip angles might appear well-suited no dynamic inflow effect could be detected. The characteristic overshoot associated with dynamic inflow is absent and, instead, there is a marked undershoot. With a lag-lead compensator rather than the lead-lag compensator appropriate for dynamic inflow, the simulation is in good agreement with the measured data. The period of oscillation (10 s) is too long for the dynamic effects modelled by the lag-lead compensator to be the result of dynamic stall, stall hysteresis associated with the rotation of the rotor, or to be due to dynamics of the wind turbine. Some aerodynamic explanation is required.

Acknowledgements

The authors wish to thank Peter Jamieson of Garrad Hassan and Partners for very helpful discussions concerning the aerodynamics and the DTI (Department of Energy) for funding the work and for permission to publish this paper.

References

(1) Wilkie J., Leithead W.E., *Dynamic modelling of wind turbine systems using ACSL*, report prepared for the Dept of Energy, Industrial Systems and Control Ltd, 1988.

(2) Leithead, W.E., de la Salle, S.A., Reardon, D., Grimble, M.J., *Wind turbine control systems modelling and design phase I and II*, report prepared for Dept of Energy University of Strathclyde, 1991.

(3) Snel, H., Schepers, J.G., Investigation and modelling of dynamic inflow effects, *Proc. European Community Wind Energy Conference*, Lubeck-Travemunde, 1993, 371-375.

(4) Madsen P.H., Frandsen S., Pitch angle control for power limitation, *Proc. European Wind Energy Conference*, Hamburg, 1984, 612-619.

(5) Stig ∅ye, A.F.M., *Unsteady wake effects caused by pitch angle changes*, Report Tech. University of Denmark, 1986.

(6) Hansen, A.C., *The recent progress in the prediction of yaw dynamics*, 8th A.S.M.E. Wind Energy Symposium in Houston, Texas, 1989.

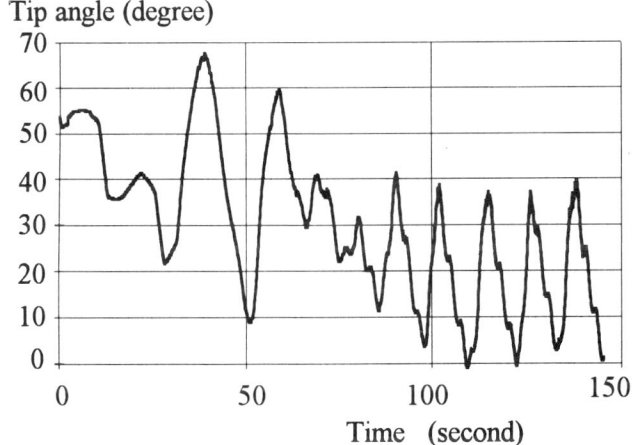

Figure 1 a) Tip angle for blade 1

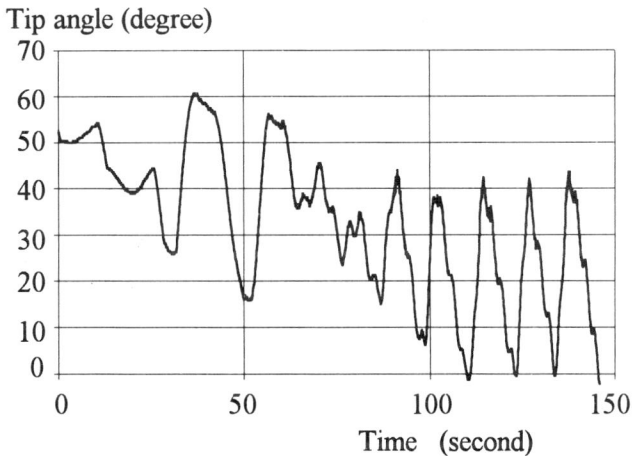

Figure 1 b) Tip angle for blade 2

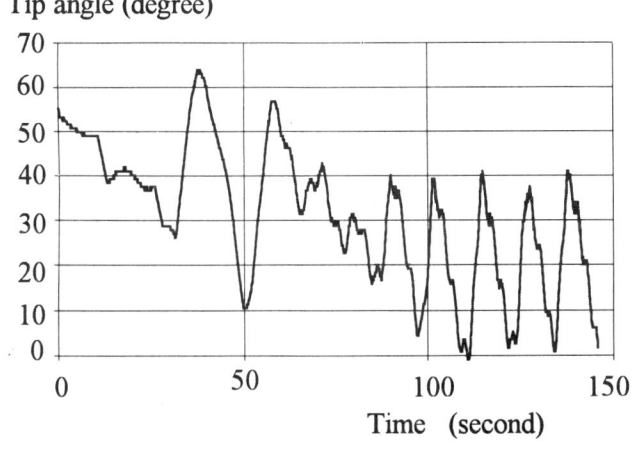

Figure 1 c) Tip angle for blade 3

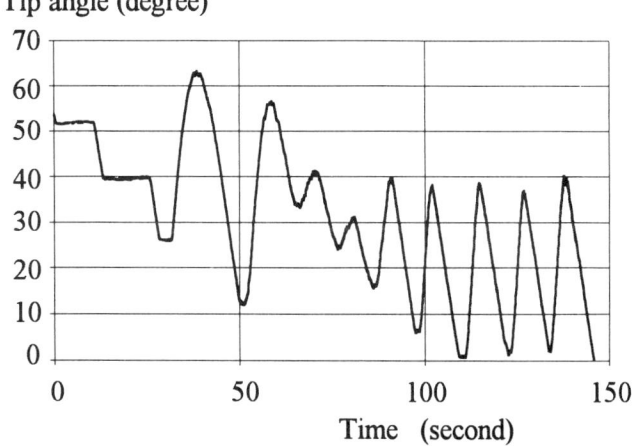

Figure 1 d) The average tip angle

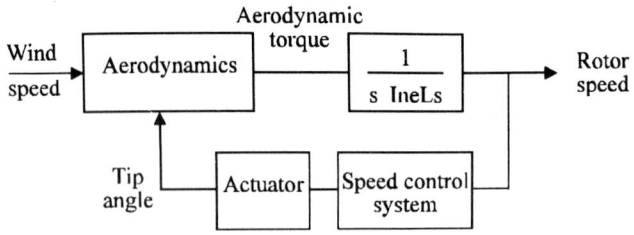

Figure 2 A block diagram of a wind turbine with speed control

Speed of low speed shaft (rev/min)

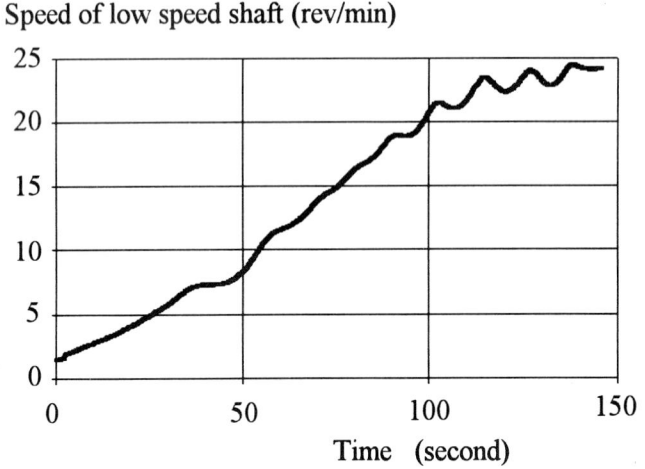

Time   (second)

Figure 3 A typical time trace of measured rotor speed during speed control

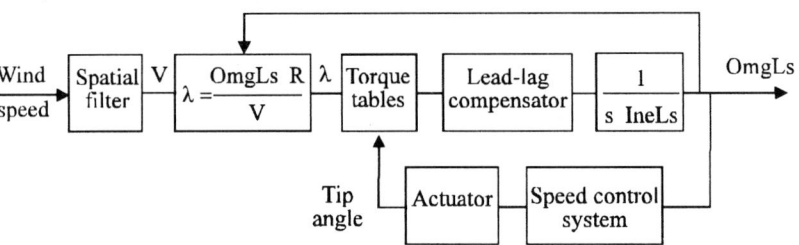

Figure 4 A block diagram of the simulation with speed control

Low speed shaft speed (rev/min)

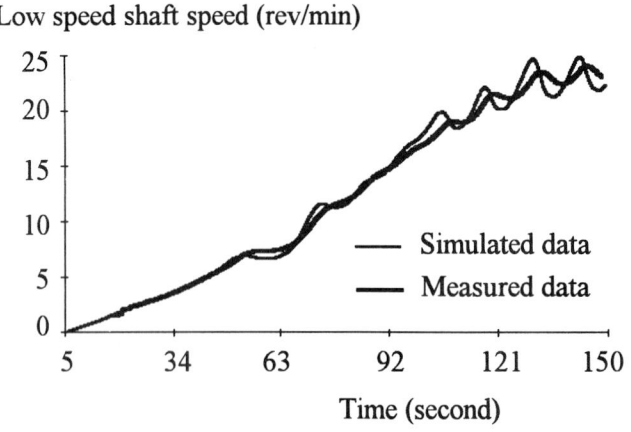

Time (second)

Figure 5 The measured and simulated rotor speed

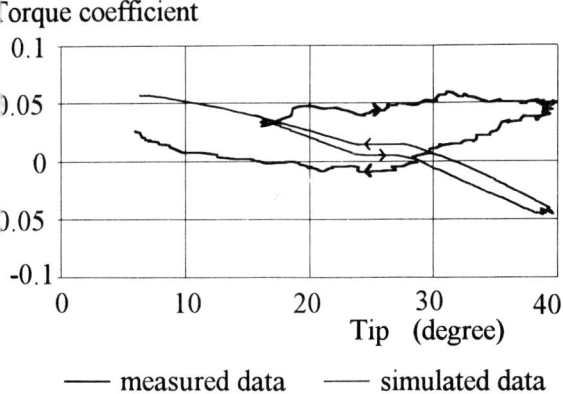

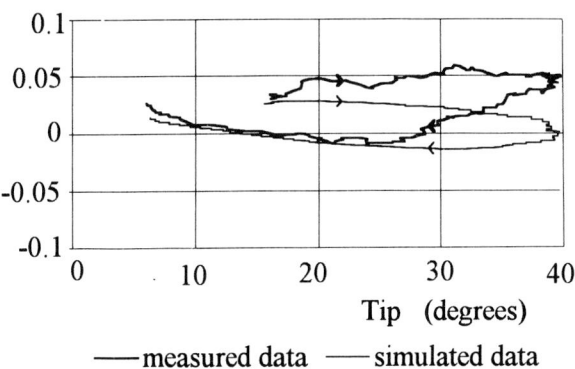

Figure 6 The calculated torque coefficient from measured data with the torque coefficient from the simulation

Figure 7 The calculated torque coefficient from measured data with the torque coefficient from the simulation when the lag-lead compensator is included.

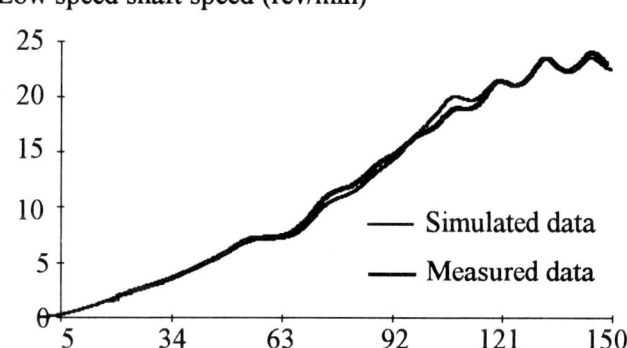

Figure 8 The measured and simulated rotor speed. The simulation includes the lag-lead compensator.

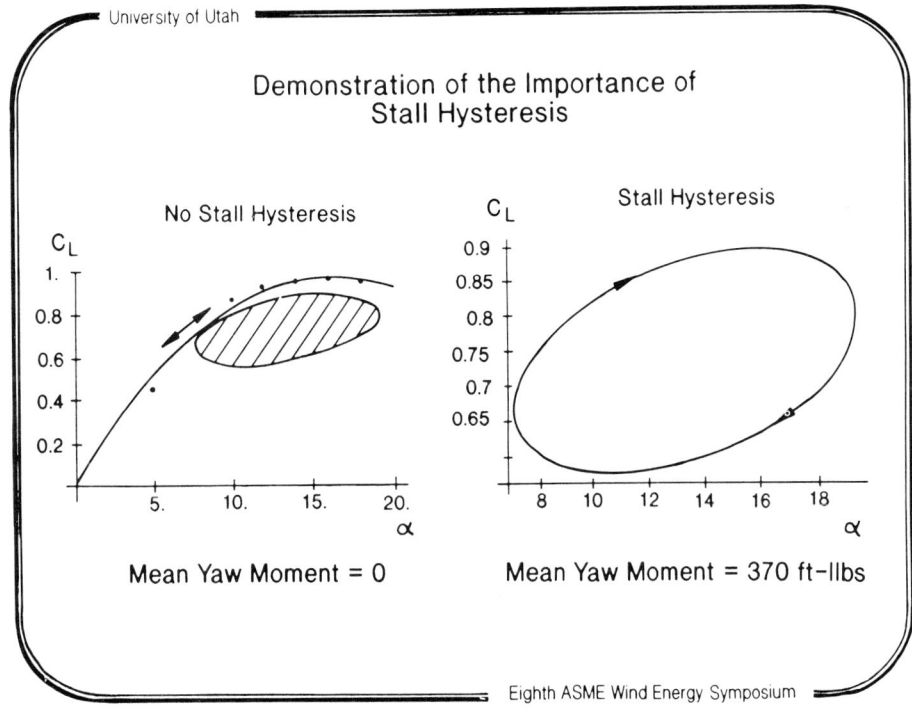

Figure 9 Hysteresis loops observed on the SERI Combined Experimental Wind Turbine

Investigation of a fundamental trade-off in tracking the $C_{p\ max}$ curve of a variable speed wind turbine

B.CONNOR BSc, PhD, and **W. E. LEITHHEAD** BSc, PhD
University of Strathclyde, UK

SYNOPSIS One of the advantages claimed for variable speed wind turbines as compared to constant speed machines is that variable speed wind turbines have the potential for a greater energy capture than constant speed machines. Energy capture is maximised below rated wind speed by a controller causing the wind turbine to track the $C_{p\ max}$ curve. The purpose of this paper is to investigate the choice of error for the control system. The paper shows that there is a trade-off between energy capture and the extent of the transient loads in the drive-train.

NOTATION

C_p	power coefficient
$C_{p\ max}$	maximum power coefficient
P	power from the wind
λ	tip-speed ratio
A	area swept by rotor
V_w	wind speed
V	rotor speed
ω_r	rotor rotational speed
R	rotor radius
T_A	aerodynamic torque
T_e	generator reaction torque
J	rotor inertia
$\ddot{\theta}$	rotor acceleration
α	rectifier firing angle

1 INTRODUCTION

The most common type of horizontal-axis grid-connected wind turbine is the constant speed wind turbine. A comprehensive investigation of the control of constant speed wind turbines can be found in (1). In a constant speed wind turbine, the rotor is connected, by way of the gearbox and an induction generator directly to the electrical grid. As a result, the rotor speed is constant for the whole range of operational wind speeds apart from during start-up and shut-down and a slight variation caused by the generator slip.

The major alternative to the constant speed wind turbine is the variable speed wind turbine. In a variable speed wind turbine, the electrical system frequency and generator speed are made independent of each other by inserting a variable-to-fixed frequency converter between the wind turbine and the electrical system.

There are several reasons that make variable speed attractive (2). First, the energy conversion efficiency of wind turbines depends on the ratio of rotor speed to wind speed, i.e., the tip-speed ratio, λ. If the speed of the rotor can be controlled as wind speed changes, the tip-speed ratio can be adjusted to optimise the power coefficient, C_p, which results in an increase in energy capture. The ability to vary the rotor speed with wind speed is normally only exploited below rated wind speed. Second, the initial effect of any increase or decrease in wind speed is to vary the rotor speed thereby reducing the impact of the wind speed variations on the drive-train loads. This feature is particularly important above rated wind speed.

The purpose of this paper is to investigate some aspects of the control of variable speed wind turbines when operating below rated wind speed. In particular to investigate the effect on performance of different choices of control strategy for below rated wind speed operation.

2 CONTROL STRATEGY OPTIONS

Below rated wind speed the main objective for a variable speed wind turbine is to maximise energy capture. In this paper, the task of designing a controller to achieve this is considered for a wind turbine, the parameters of which are typical of medium scale commercial wind turbines:

Rotor type: Horizontal Axis
Blades: 3
Diameter: 33 m
Rotor speed: Variable
Generator: Synchronous
Electrical
System: AC-DC-AC grid commutated
Rated wind speed: 11.74 m/s
Rated power: 330 kW

The power available in the wind, P, and C_p and λ are related by the following equation

$$P = C_p(\lambda)KAV_w^3 \qquad (1)$$

where $\lambda = \dfrac{R\omega_r}{V_w}$ and k is a constant.

Figure 1 shows the relationship between C_p and λ for a typical commercial horizontal-axis wind turbine. It can be observed that maximum available power from the wind occurs at the maximum power coefficient, C_p max. In this particular case, C_p has a broad flat peak. Variable speed operation might be more effectively exploited with a sharper C_p curve.

A control system is required such that the correct relationship is maintained between rotor speed and wind speed. In control terms, the state of the wind turbine must be caused to track the $C_{p\ max}$ curve, where $C_{p\ max}$ represents the maximum power coefficient. Figure 2 shows the torque-speed curves at various wind speeds.

The $C_{p\ max}$ curve can be approximated by the relation, $T = kV^2$, where T is the torque, V is the rotor speed and k is a constant. It is represented by curve bc in Figure 2. Suppose the wind turbine is operating with rotor speed, V_1 and torque T_1 which is point a in Figure 2. The control system must act to bring the operating point back onto the maximum C_p curve in order to maximise energy capture. A frequently used strategy is to minimise the error $T-kV^2$, which is the

vertical line ab. The large rotor inertia stops the torque from tracking the $C_{p\ max}$ curve instantaneously and results in the torque spiralling around $C_{p\ max}$.

There are two possible choices of the torque in the relation $T=kV^2$, namely the aerodynamic torque, or the high-speed shaft torque, which is a representative measure of the drive-train loads. The former can only be inferred from measurements of the high-speed shaft torque which is related to the aerodynamic torque by the relation $T_A \approx T_e + J\ddot{\theta} + \gamma\dot{\theta}$ where T_A is the aerodynamic torque, T_e the generator reaction torque, J the rotor inertia, $\ddot{\theta}$ and $\dot{\theta}$ the rotor acceleration and velocity and γ the viscous damping coefficient. A suitable estimate of T_A is $\hat{T}_A = T_e + \dfrac{10Js + 10\gamma}{(s+10)}\dot{\theta}$.

Which is the better choice for T as regards energy capture and adequate transient levels is investigated below. For each choice, controllers are designed and their performance assessed by simulation.

3 DYNAMICS OF VARIABLE SPEED SYSTEM

As discussed in Section 1, in variable speed wind turbines the rotor speed and grid frequency are made independent by the insertion of an AC/DC/AC frequency converter. Control action is achieved by varying the rectifier firing angle, α, to induce variations in the generator reaction torque. Consequently, the coupling between the wind turbine rotor and the grid is weakened and the damping of the drive-train dynamics, which is normally supplied by the generator, is greatly reduced. The variable speed wind turbine power generation unit consists of a synchronous generator and frequency converter. The dynamics for the power electronics were identified from experimental data obtained from a test-rig (3).

To maximise energy capture below rated wind speed, the control problem is one of regulating α such that the tracking error $T - kV^2$ is small for a suitable choice of torque T. Choosing T to be the aerodynamic torque acting on the rotor results in an open-loop plant with the Bode plot shown in Figure 3. Choosing T to be the high-speed shaft torque, the open-loop plant has the Bode plot shown in Figure 4. The presence of the expected lightly damped mode, associated with the reduction in damping, is clearly seen at 32 rad/s. Classical Nyquist/Bode loop shaping is employed to achieve appropriate disturbance rejection curves. The controllers must be designed such that the open-loop plant rolls off fast enough to counteract the effect of the mechanical resonance at 32 rad/s. The controllers designed for each of the two choices of torque T have the performance characteristics given in Table 1.

4 PERFORMANCE COMPARISON

The controllers described in Section 3 are designed to minimise the tracking error $T - kV^2$. In the first T is chosen to be the aerodynamic torque. In the second T is chosen to be the high-speed shaft torque. The performance attained by the two controllers are compared. One measure of performance is the efficiency of energy extraction from the wind. It can be estimated as the ratio of the mean value of C_p to $C_{p\ max}$. When so doing, the following should be borne in mind. First, a constant speed wind turbine with the same characteristics as the variable speed wind turbine investigated here attains an overall efficiency below rated wind speed of approximately 95 %. Second, even if the tracking error $T - kV^2$ is reduced to zero the efficiency would not be 100% since reducing $T - kV^2$ to zero is not exactly equivalent to tracking the $C_{p\ max}$ curve. The efficiency quoted here and in the rest of this paper is the percentage of available energy extracted from the wind by the rotor. No internal drive-train losses are included.

The wind speed used to drive the wind turbine for each choice of T has a mean wind speed of 8 m/s and is shown in Figure 6. For the tracking error with aerodynamic torque, the performance is shown in Figure 7 (a) and 7 (b). Figure 7 (b) shows that the aerodynamic torque tracks the curve tightly resulting in high efficiency. However, the drive-train is subjected to excessive torque transients such as those on the low-speed shaft torque in Figure 7 (a). The statistics obtained when tracking the $C_{p\ max}$ curve using the aerodynamic torque are presented in Table 2. These results reveal that the average power coefficient is 0.4506 which is equivalent to an efficiency of 98.6% below rated wind speed.

For the tracking error with the high-speed shaft torque, the performance is shown in Figure 8 (a) and 8 (b). The results show that the low-speed shaft torque tracks the $C_{p\ max}$ curve tightly keeping the torque below its rated value. The aerodynamic torque is shown in Figure 8 (b) confirming that it is not as tight as in Figure 7 (b), as might be expected. The corresponding statistics obtained with this controller are presented in Table 3. The statistics reveal that the energy efficiency for the strategy of tracking the $C_{p\ max}$ curve using the high-speed shaft torque is 97.3%.

The above results suggest that increased energy capture is accompanied by increased drive-train loads. Maintaining the drive-train torques at moderate levels, is incompatible with the aerodynamic torque tightly tracking the $C_{p\ max}$ curve and hence maximising energy capture. A variable speed machine with a high efficiency factor is desirable but the large torque levels required to achieve this efficiency suggests that maximising energy capture is counter-productive. Using a weaker control action results in smaller torque levels but the energy efficiency is considerably reduced. The two controller performances compared in the preceding, are extreme cases. Because of the broad flat shape of the C_p curve, Figure 1, the rotor efficiency is high even for constant speed operation of the wind turbine. In addition, because the rotor inertia is high, 192,000 kgm^2, the transient drive-train loads are great when attempting to control with aerodynamic torque. In order to improve the cost-effectiveness, it might be anticipated that, in future, rotor inertias will become less and the C_p curve will become more sharply peaked. A more sharply peaked C_p curve would make the efficiency of the variable speed wind turbine more sensitive to the choice of controller. Reducing the rotor inertia would reduce the magnitude of the drive-train torque transients thereby increasing the feasibility of control acting on aerodynamic torque. For these reasons, it is worthwhile attempting to determine a control action which is a compromise of the two already investigated, that is, a control action which achieves greater, although not optimal, rotor efficiency whilst not increasing unduly the drive-train torque transients. In order to explore such a possibility, the inertia of the rotor is reduced to 95,060 kgm^2, i.e. half its previous value.

With the rotor inertia halved to 95,060 kgm^2, the open-loop Bode plot is as shown in Figure 5. A weaker controller is designed for this plant with a crossover frequency of 0.42 rad/s. The controller performance characteristics are shown in Table 1. With the tracking error based on aerodynamic torque and the amended rotor inertia, the performance is shown in Figure 9(a) and 9(b). The aerodynamic torque tracks the $C_{p\ max}$ curve tightly, giving increased energy capture, with the low-speed shaft torque considerably reduced from that in Figure 7 (a). The statistics obtained with this slower controller are presented in Table 4. They reveal that, with the rotor inertia halved, the energy efficiency for the strategy of tracking the $C_{p\ max}$ curve with the aerodynamic torque is 98.1%. The high efficiency factor is achieved with the maximum low-speed shaft torque being 90 kNm less than the previously examined case with the original rotor inertia, see Table 2.

5 CONCLUSIONS

In variable speed wind turbines, energy capture below rated wind speed is maximised by tracking the $C_{p\ max}$

curve by a suitable choice of torque, T. When T is chosen to be the aerodynamic torque, the aerodynamic torque can be made to closely track the $C_{p\ max}$ curve resulting in high energy capture. However, the corresponding drive-train torque transients are excessive. Low drive-train torque transients are achieved by choosing T to be the drive-train torque but the tracking of the $C_{p\ max}$ curve by the aerodynamic torque is not as tight resulting in reduced energy capture. There is, thus, a fundamental conflict in the control requirements: to maintain the drive-train torques at moderate levels implies the aerodynamic torque cannot tightly track the $C_{p\ max}$ curve. This conflict is enhanced when the C_p curve becomes more sharply peaked and the rotor inertia decreases. The results and conclusions presented here agree with previous theoretical estimates of the performance of variable speed wind turbines (4).

6 ACKNOWLEDGEMENTS

The support of the UK Science and Engineering Research Council and the Energy Technology Support Unit is gratefully acknowledged.

7 REFERENCES

(1) Leithead, W.E., De La Salle, S.A., Reardon, D., *Classical control of active pitch regulation of constant speed horizontal axis wind turbines*, INT. J. CONTROL, Vol. 55, No. 4, pp. 845-876, 1992

(2) Hinrichsen, E.N., *Variable rotor speed for wind turbines: objectives and issues*, EPRI AP-4261, 1985

(3) Leithead, W.E., Rogers, M.C.M., Connor, B., *Design of a Controller for a Test-Rig for a Variable Speed Wind Turbine*, Submitted to IMACS Mathmod Conference, Vienna, February 2-4, 1994

(4) Leithead, W.E., *Dependence of performance of variable speed wind turbines on the turbulence, dynamics and control*, IEE PROC., VOL., 137, PT., C, NO., 6, pp. 403-413, Nov., 1990

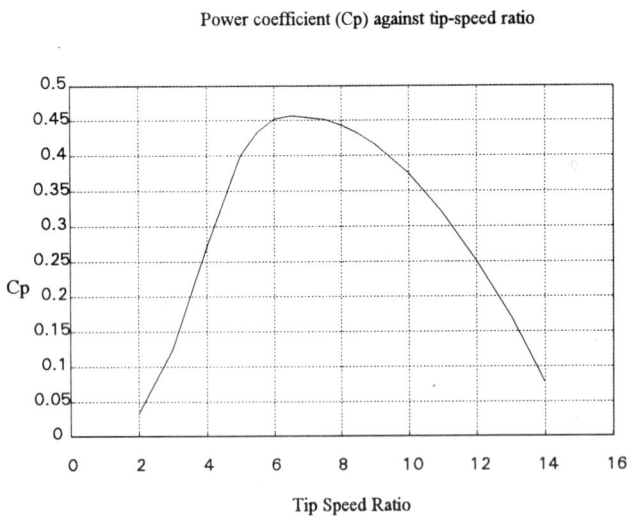

Figure 1 Power coefficient (C_p) against λ for a HAWT

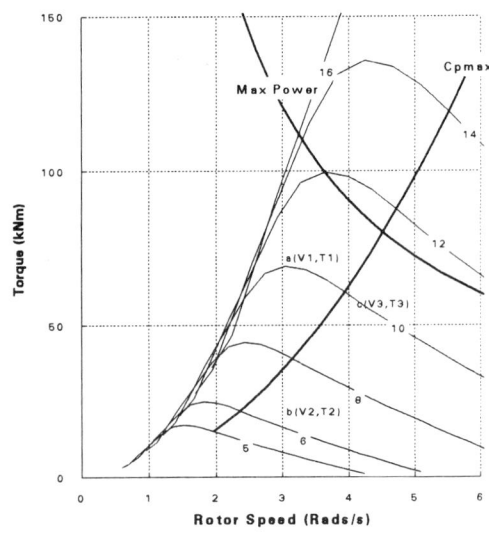

Figure 2 Torque speed curves at various wind speeds

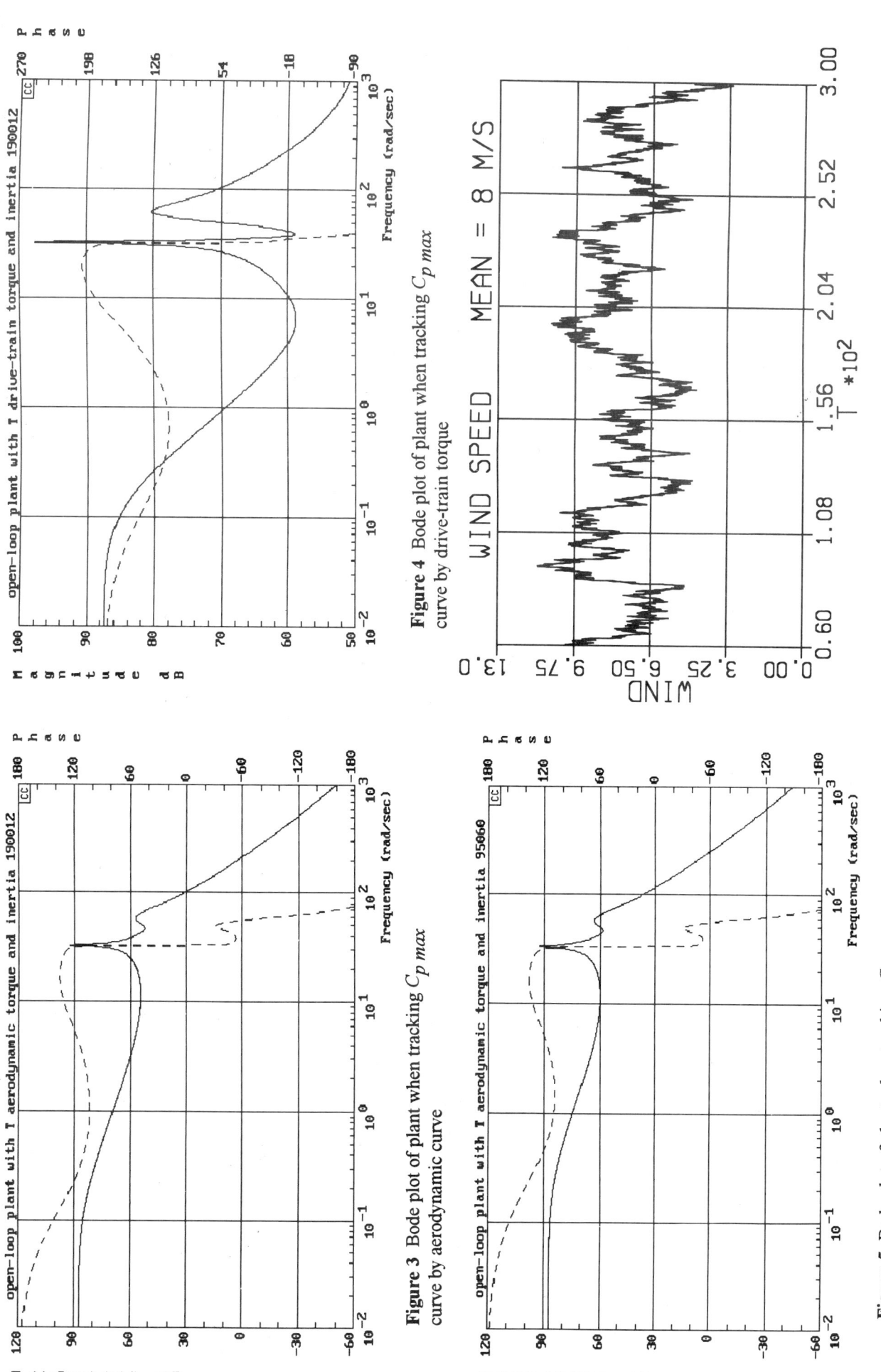

Figure 3 Bode plot of plant when tracking $C_{p\,max}$ curve by aerodynamic curve

Figure 4 Bode plot of plant when tracking $C_{p\,max}$ curve by drive-train torque

Figure 5 Bode plot of plant when tracking $C_{p\,max}$ curve by aerodynamic torque. Rotor inertia: 95 060 kgm^2

Figure 6 Plot of wind speed (m/s) against time (secs)

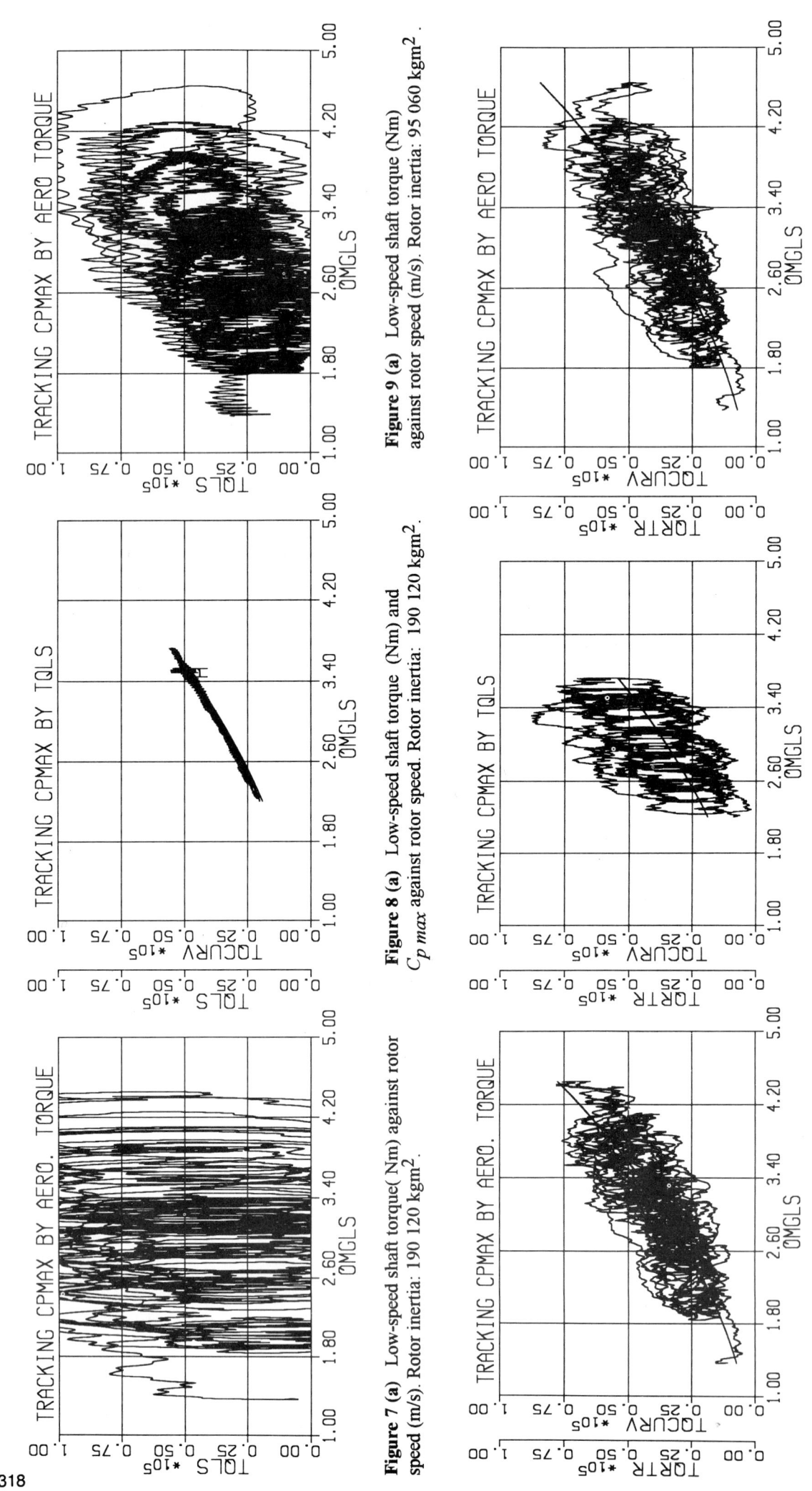

Figure 7 (a) Low-speed shaft torque(Nm) against rotor speed (m/s). Rotor inertia: 190 120 kgm².

Figure 8 (a) Low-speed shaft torque (Nm) and $C_p max$ against rotor speed. Rotor inertia: 190 120 kgm².

Figure 9 (a) Low-speed shaft torque (Nm) against rotor speed (m/s). Rotor inertia: 95 060 kgm² .

Figure 7 (b) Aerodynamic torque (Nm) and $C_p max$ against rotor speed (m/s). Rotor inertia: 190 120 kgm² .

Figure 8 (b) Aerodynamic torque (Nm) and $C_p max$, against rotor speed (m/s). Rotor inertia: 190 120 kgm².

Figure 9 (b) Aerodynamic torque (Nm) and $C_p max$ against rotor speed (m/s). Rotor inertia: 95 060 kgm².

Table 1 Performance characteristics for two choices of T in the tracking error $T - kV^2$.

Torque (T)	Crossover frequency (rad/s)	Phase margin (degs)	Gain margin (dB)	Rotor Inertia (kgm^2)
aerodynamic	1.0	65.0	10.33	190120.0
high-speed shaft	1.0	65.0	15.51	190120.0
aerodynamic	0.42	72.0	10.0	95060.0

Table 2 Statistics obtained when tracking $C_{p\ max}$ by aerodynamic torque. Controller crossover frequency: 1 rad/s, inertia: 190120.0 kgm^2.

signal	units	min	max	mean	standard deviation
wind	m/s	3.117	11.316	7.663	1.397
tip-speed ratio		4.414	8.748	6.487	0.620
C_p		0.3355	0.4570	0.4506	0.0121
low speed shaft torque	kNm	-139.16	195.53	35.402	57.936

Table 3 Statistics obtained when tracking $C_{p\ max}$ by high-speed shaft torque. Controller crossover frequency: 1 rad/s, inertia: 190120 kgm^2.

signal	units	min	max	mean	standard deviation
wind	m/s	3.117	11.316	7.663	1.397
tip-speed ratio		4.339	12.064	6.461	0.930
C_p		0.2459	0.45701	0.44454	0.01818
low speed shaft torque	kNm	20.034	55.947	35.903	8.545

Table 4 Statistics obtained when tracking $C_{p\ max}$ by aerodynamic torque. Controller crossover frequency: 0.42 rad/s, inertia: 95060 kgm^2.

signal	units	min	max	mean	standard deviation
wind	m/s	3.117	11.316	7.663	1.397
tip-speed ratio		4.0892	8.744	6.459	0.7014
C_p		0.2861	0.4570	0.4484	0.0165
low speed shaft torque	kNm	-28.475	104.18	34.64	20.57

Some control aspects of a small isolated wind turbine

M. T. IQBAL, A. H. COONICK and **L. L. FRERIS**
Imperial College, London, UK

Synopsis The modelling and simulation results of an isolated 5 KW wind turbine consisting of a 6.3m diameter, two bladed teetered rotor, with controllable pitch angle, yaw angle and generator field current are presented. For the dynamic control of the system choice and design considerations of a PID type controller and twenty five rules fuzzy controller is discussed. Several dynamic control strategies for this system with PID type and fuzzy controllers are investigated. A comparison of different control strategies in terms of energy capture is also presented.

Introduction A well instrumented and very flexible to control Wind Energy Conversion System owned by Imperial College London is installed at Rutherford Appleton Laboratory (RAL) Wind Test Site. Some major changes in the system have been proposed and are currently under way. In the new system blades pitch angle, turbine yaw angle and its DC generator field current can be actively controlled. A number of fixed and variable speed control strategies can be implemented on this system. Each control strategy requires at least two controllers. One controller for the below rated wind speed operation to optimize the energy capture and another controller for the above rated wind speed operation to keep power or torque or speed constant.

The correct model of such stochastic and highly nonlinear system is difficult to determine, therefore both model based and model independent controllers are considered. The controllers design procedure and comparison in terms of response and implementation considerations is presented. Some results of this full nonlinear simulation based exercise are also presented.

System Description and Dynamical Modelling The isolated wind energy conversion system considered here has rotor power of about 5KW at the rated windspeed of 8m/sec. It consists of two 3.15 m radius blades, teetered rotor, a variable pitch mechanism, a gear box, a separately excited dc generator, a fixed resistance dump load, a dc yaw drive and associated measurement transducers. It is supported on a 10 meter high steel lattice tower. An IBM 386 PC acts as an controller and data acquisition system.

Drive Train Dynamics The turbine rotor is coupled to the generator by a gearbox and rigid shaft. The difference between the wind generated torque Tw and generator torque Te is responsible for the rotation of system. A first order moment of inertia J and friction based dynamic model is considered here. Assuming a stiff shaft (stiffness coefficient $K=10^6$ Nm/rad) and gear ratio N=wg/wm=14 system equation may be written as

$$Jeq \; dwm/dt + Beq \; wm = Tw - N \; Te$$

The Aerpac 6.3 WPX 3.15 radius two bladed rotor's moment of inertia is Jeq=85 Kg m^2. Typical value of damping coefficient for gear box of such system is Beq=.5 Nm sec/rad. Aerodynamic induced torque for the wind turbine is given as Tw = Kq u^2 Cq($\lrcorner$,ß) cos^2(Θ) [Ref 1] where the tip speed ratio $\lrcorner$ = R wm cos(Θ)/ue. Θ is yaw angle, R is rotor radius while ue is the effective wind speed. For this system Kq = .5 ϵ π R^3 =60.143

Aerodynamic Performance of Rotor For the simulation of wind turbine control, system's aerodynamic performance characteristics (i.e. Cp=f1($\lrcorner$,ß) and Cq=f2($\lrcorner$,ß)) of the turbine rotor are required. Imperial College Wind Turbine Analysis Code (ICWTAC) [Ref.1] is used to determine performance characteristics of the rotor. The rotor blade aerofoil changes radially (NACA 4424-4412) The lift and drag coefficients for the blades were supplied to the ICWTAC along with blade geometry and rotor Cp, Cq coefficient were obtained for different tip speed ratios (1 to 15) and blade pitch angle (-1 to 30°). A 16 terms bivariate polynomial was fitted to the ICWTAC results in a least

squares manner using MATLAB. Figure 1.0 shows this steady state rotor operating envelop.

$$Cq = \Sigma_{i=0,j=0 \; 3,3} \; a_{ij} \; \beta_i{}^{\lrcorner}{}_j$$

The wind turbine in consideration is going to rotate over the range 150-200 rpm. Therefore small variations in the output power are expected at about 40 rad/s (2P). This high frequency power component will lead to no servo action because of the slow pitch actuator, that will act as a low pass filter. Therefore, the effects of tower shadow and wind shear are neglected. The effects of other physical phenomena as variation in renoyld number, contamination of rotor blades, influence of rain, influence of icing, influence of air density variations etc are also neglected.

Effective Wind Speed Model The point wind speed u (anemometer readings) is obtained from a file of 2 second sampled data in ASCII format. The spatial variations in wind speed are averaged over the swept area of the rotor. The forces and torque induced on the rotor result from an effective wind speed rather than a point wind speed. The effective wind speed may be obtained from the point wind speed using the spatial filter [Ref. 2] (for R=3.15m ,δ=1.3 and a=.55) With the introduction of state variables u1, u2 and for point wind speed ui, we can express spatial filter in state space form as

$$du1/dt = -0.6467 \; ui \; u1 \; -0.084335 \; ui^2 \; u2 + ui$$

$$du2/dt = u1$$

and effective wind speed ue can be determined by

$$ue = 0.2442 \; ui \; u1 + 0.084335 \; ui^2 \; u2$$

Unsteady Aerodynamics (Induction Lag) The dynamic effect due to adjustment of wake in response to change in the wind or blade pitch angle has been reported. [Ref. 2,3] This induction lag can be modelled by a lead lag filter. [Ref. 2]

$$G2(s) = (K \; s + 1/\tau)/(s + 1/\tau)$$

where K=1.37 and τ = 4R/u.[Ref 3] With the introduction of a state T1 above relationship can be written in state space form as [Ref. 4]

$$dT1/dt = -7.9365e-2 \; uc \; T1 + Tw$$

and the effective rotor shaft torque

$$Twe = -2.93651e-2 \; ue \; T1 + 1.37 \; Tw$$

Model for DC Generator A forced cooled, seperatly excited DC generator is used in the system. It generate 460VDC when rotating at 1500rpm. Its armature is connected to a bank of fan cooled resistors. This set of resistors with the effective value of 28.8 Ω act as the dump load for the system. A three phase full bridge rectifier and a pulse width modulated chopper provide the field current to the generator. The control voltage Vfi (0-10VDC) of the field controller set the voltage (0-560VDC) across the generator field winding leading to a field current of 0-.75A The PC 8-bit digital output is converted to 0-10VDC as a command for the field controller. The field winding of the generator is a simple L-R circuit described by

$$d\ i/dt = (56\ Vfi - 754\ i)/260$$

with a gearbox ratio of 14, the generator saturation characteristics (air gap voltage Ea) can be described as

$$Ea = Kg\ wm\ (87.1\ i^3 - 1001.3\ i^2 + 1282.3\ i + 9.9)$$

where Kg = .08912, wm in rad/sec and field current in A. For the armature loop we can write

$$d\ ia/dt = (\ Ea - (28.8 + 4.062)\ ia)/.0401$$

the generator reaction torque (N-m)

$$Te = Ea\ ia/\ wm\ /\ N$$

and the output electrical power in kW as

$$Pe = 28.8 * ia^2 /1000$$

The armature loop time constant (.00122 sec) is the smallest in system therefore, the effect of armature inductance was neglected during the simulation to save time.

Model for Pitch & Yaw Angle Position Servo The drives to change the blade pitch angle and absolute yaw angle are based on bidirectional 12VDC motors. Power MOSFET based H bridge pulse width modulated convertors act as power amplifiers for these drives. Potentiometer based pitch angle and yaw angle feedback with controllers make these DC position servos. The absolute yaw angle with respect to tower is measured by a servo potentiometer. Reference pitch angle and yaw angle (0-5VDC signal) for these position servos are supplied by the PC through a DACs.

The response of pitch angle position servo can be described by a 2nd order transfer function or

$$dß/dt = ß1\ ;\quad dß1/dt = 28\ ßi - 5\ ß1 - 28\ ß$$

where ß is actual pitch angle and ßi is desired pitch angle.

A windvane installed on a separate tower provides reference yaw position to the yaw angle position servo through the PC. When yaw angle is not being actively controlled the wind turbine simply follow the windvane. The response of the yaw angle position servo can be described by a 2nd order transfer function or

$$d\Theta/dt = \Theta1\ ;\quad d\Theta1/dt = 6.25\ \Theta i - 4\ \Theta1 - 6.25\ \Theta$$

where Θ is actual yaw angle and Θi is the desired yaw angle.

Models for Measurement Transducers The electrical power is computed from current and voltage measurements across the dump load. These are measured using transformer based sensors. The whole power transducer may be considered as a 1st order low pass filter with a time constant of about .1sec Therefore, the measured power Pem may be written as

$$dPem/dt = 10\ (Pe - Pem)$$

The rotor torque is measured by a strain gauge bridge followed by an instrumentation amplifier and a filter. Assuming it to be a 1st order system we may write the measured torque Twm as

$$dTwm/dt = 15\ (Twe - Twm)$$

Rotor speed is sensed by an optical shaft encoder that is followed by a frequency to voltage convertor with a time constant RC=3.3 msec so,

we may neglect the dynamics of the speed transducer. The anemometer is installed on a separate tower with windvane.

The above described system model is presented in block diagram in Fig 2.0

Control Strategies The blade pitch angle, yaw angle or generator field current may be controlled to achieve constant speed, variable speed, constant power or constant torque etc. Below a certain wind speed called the "rated wind speed", turbine speed is regulated by the field current and above the rated wind speed blade pitch angle or yawing can be used to shed the excess power or generator field current can be used to bring the turbine speed down hence, stalling the machine. The decision to operate below or above rated wind speed (switching) may be based on wind speed, electrical power, rotor torque or the turbine speed measurement.

The power output of the wind turbine may have some transient component due to sudden changes in the field current. Such power transients are not indicators of an increase in wind speed. We cannot base switching decisions on the turbine speed because in some cases the turbine will be running in the variable speed mode. The generator itself is capable of stalling the machine and in this mode the machine reaches its rated power at a wind speed about 13m/s Switching to pitch regulation or yaw regulation at such high wind speed is dangerous for the machine. It was decided for the safety of machine to base the switching decision on the basis of the effective wind speed. A hysteresis band of $\approx \pm .5$ m/s of the wind speed was selected to reduce the number of switchings. Simple decay laws are used for the determination of the control variable that is not being actively controlled.

$$ß(k) = ß(k-1)/2\ ;\ \Theta(k) = \Theta(k-1)/1.3\ ;\ Vfi(k) = Vfi(k-1)/5 + 2.04952$$

The decay laws described above were obtained heuristically with the number of switchings as the main criterion. The following control strategies are selected for the simulation and the comparison will be made on the basis of energy captured.

1 BRW: ⌐ optimum by generator field control, ARW: Constant electrical power by pitch control. (LOKP)

2 BRW: ⌐ optimum by generator field control, ARW: Constant electrical power by yaw control. (LOKPYAW)

3 BRW:Hill climbing algorithm based on electrical power measurements, ARW: Constant power by pitch control (HCKP)

Changing the field current to have an increasing power output or in other words 'hill climbing' is a turbine parameter independent way of extracting power from wind. [Ref.5] A simple algorithm based on power measurement is; if Pe(k) > Pe(k-1) then increase the field current. This method require large variations in the field current corresponding to large changes in power. Assuming a proportional relation dVfi = -K2 dPe or in its recursive form Vfi(k)=Vfi(k1)+K2 (Pe(k)-Pe(k-1))[Ref. 5]. Most suitable value of K2 was found to be 0.2

4 BRW: Generator field control by power feedback, ARW: Constant electrical power by pitch control. (PBKP)

It is known that in variable speed range aerodynamic power is proportional to the cube of the rotor speed wm [Ref. 1] Therefore, below rated wind speed the field current can be varied based on the difference of K1 wm^3 and the measured power Pem. A suitable value for K1 was determined to be .496

5 BRW: ⌐ optimum by Generator field control, ARW: Constant mechanical torque by pitch control. (LOKT)

6 BRW: Fixed speed by generator field control, ARW:Fixed speed by pitch control. (FSVP)

7 BRW: Fixed speed by generator field control, ARW: Fixed speed by generator field control (Stall Regulation) (FSSR)

8 BRW: ⌐ optimum by generator field control, ARW: Constant mechanical power by generator field control (Stall Regulation) (LOKPSR)

9 BRW: ⌐ optimum by generator field control, ARW: Constant torque by yaw control. (LOKTYAW)

A block diagram for a typical control strategy is provided in Fig. 3.0 In general controller#1 is used for below rated wind (BRW) speed operation and controller#2 is for the above rated wind (ARW) speed operation. Rated electrical power is 4.16 KW, rated rotor torque 245 Nm and the rated rotational speed is 20 rad/sec. For the stall regulated fixed speed operation rated rotational speed is 13 rad/sec.

Digital PID Controller An IBM 386 20 MHz PC is used as the system controller. It was observed that a sampling time Ts less than 1sec does not improve the control performance therefore, Ts = 1sec was selected. A general transfer function of a 2nd order digital controller may be written as [Ref 6]

$$Gr(s) = (q0 + q1\ z^{-1} + q2\ z^{-2})/(1 - z^{-1})$$

with $q0 = Kp\ (1+Cd)$; $q2 = Kp\ Cd$; $q1 = Kp\ (Ci - 2\ Cd - 1)$ and $Cd = Td/Ts$, $Ci = Ts/Ti$. Ziegler- Nichols tuning rules are used for the determination of controller parameters. [Ref. 6] Since, the wind turbine is highly nonlinear system the gain of controllers must be varied to cope with the time varying system. [Ref 2] Controller gain as a function of effective wind speed was found by tuning controller at different wind speed. Table 1.0 give the gain scheduling scheme for the controllers of different control strategies. BRW controller cover the small range of operation from cut in wind speed of 4m/s to rated wind speed of 8m/s therefore, no gain scheduling is used.

Table 1.0 Controllers for different Control Strategies

Control Strategy	ARW Controller Kp	Ti	Td
LOKP	-.02 ue^3 +.978 ue^2 -15.728 ue +86.54	1	.24
LOKPYAW	-.088 ue^3 +4.127 ue^2 -63.94 ue+334.8	3	.5
HCKP	-.02 ue^3 +.978 ue^2 -15.728 ue +86.54	1	.24
PBKP	-.02 ue^3 +.978 ue^2 -15.728 ue +86.54	1	.24
LOKT	.03 exp(-.3 ue + 2.4)	1	.24
FSVP	-.0117 ue^3 +.566 ue^2 - 9.17 ue+51.13	1	.12
FSSR	1.5	2	.2
LOKPSR	3	1.5	.3
LOKTYAW	if ue<11 then -.1 ue +1.15 else .05	1	.24

Control Strategy	BRW Controller Kp	Ti	Td
LOKP	1	2	.2
LOKPYAW	1	2	.2
HCKP	Vfi(k)=Vfi(k-1)+.2(dPem)		
PBKP	3	1	.24
LOKT	1	2	.2
FSVP	.5	2	.2
FSSR	1.5	2	.2
LOKPSR	1	2	.24
LOKTYAW	1	2	.2

Fuzzy Logic Controllers The methodology of fuzzy logic controllers appears very useful when the processes are too complex for analysis by conventional quantitative techniques or when the available sources of information are interpreted qualitatively, inexactly or uncertainly. A simple fuzzy controller (Fig. 4.0) consist of four major parts 1)fuzzification 2) fuzzy control rules 3) computation unit and 4)defuzzification. Detailed mathematical description may be found in Ref. 7

Fuzzification: It is defined as a mapping from an observed input space to fuzzy sets in certain universe of discourse. Fuzzification interprets input x as a fuzzy set A with membership function $\mu_A(x)$. General principles are a) number of fuzzy sets associated with a variable should generally be an odd number between five and nine. b) each fuzzy set should overlap (10-50%) somewhat with the neighbours. c) the density of fuzzy sets should be highest around the optimal control point of the system and should thin out as the distance from that point increases. d) histogram of the measured data may be used to estimate the membership function. We have selected five fuzzy sets to represent each control variable and their membership functions are shown in Fig. 5.0 These fuzzy sets are positive large (PL), positive small (PS), zero (ZE), negative small (NS) and negative large (NL).

Fuzzy Control Rules: Fuzzy rules ties the inputs values to the output model properties. Set of fuzzy rules is often called fuzzy associative memory (FAM). Rules are defined considering all possible combinations of the fuzzy sets of inputs. Fuzzy rules are to be formed on the basis of a) expert experience and his control engineering knowledge b) human operator control action c)fuzzy model of the process iv) learning. A set of 25 rules was developed and it is shown in Fig 6.0

A typical fuzzy rule can be stated as

IF error = PS AND change in error = NLB THEN change in output = -1

Computation Unit: This perform the necessary inference operation on fuzzy rules. The computational unit of the fuzzy logic controller first compares the fuzzified values of error and change in error, take minimum of those and then calculate degree of fulfilment (DOF) for the use in defuzzification process.

Defuzzification: This transform the fuzzy results of the computation unit to the real number. Therefore are mainly three defuzzification methods 1) Maximum criterion 2) mean of maximum 3) centre of area (COA). It has been shown that COA method yields superior results [Ref 8]. A modified form of standard COA method is more popular in control applications due its computational and analytical simplicity. This defuzzification method is

$$\text{Change in output } c(k) = \frac{\sum_{i=1}^{n} (DOF)_i\ B_i^d}{\sum_{i=1}^{n} (DOF)_i}$$

here DOF is 'degree of fulfilment', B^d is the defuzzified value of the membership function B and n is the number of control rules. B is the single value that best represent the linguistic description. A fuzzy logic controller can be made adaptive but we used non-adaptive fuzzy controllers. The final crisp output of the controller is calculated as

$$c(k) = c(k-1) + GAIN * \text{change in } c(k)$$

here GAIN is selected by trial and error method.

A general method of tuning of fuzzy controller cannot be found. Because any optimum values always depend on specific models of the process and control objectives. So, tuning of fuzzy controllers must be based on expert knowledge of the controlled plant, and not by computation. A good general guide line may be found in Ref. 9. Usually first membership functions are specified then their overlap is decided and finally rules are decided. Rules in the central area of fuzzy associative memory (FAM) (see Fig 6.0) are responsible for the stability of the system while the rules on edges of the FAM are responsible for the responsiveness of the system. By looking at Fig 6.0 it can be realized that a fuzzy controller unlike PID controller is a non-linear controller.

For a typical control strategy of the wind turbine we need one fuzzy controller for BRW operation and one fuzzy controller for the ARW operation. Switching between these controllers is done as described previously. For each control strategy we have to define a new set of membership functions, overlap, rules and the gain of controller. Table 2.0 provide the necessary details of fuzzy controllers membership functions used for different control strategies. where ER mean error and ERB stand for change in error.

Table 2.0 Specifications of the Membership functions of Fuzzy controllers for the wind turbine

Control Strategy	Above Rated Wind Speed Base Width ER	ERB	overlap ER	ERB	GAIN	Below Rated Wind Speed base width ER	ERB	overlap ER	ERB	GAIN
FLOKP	.5	.25	.15	.075	.6	2	2	.6	.6	.5
FLOKPYAW	1	1	.3	.3	3	2	2	.6	.6	.5
FHCKP	.5	.25	.15	.075	.6	Hill climbing algorithem				
FPBKP	.5	.25	.15	.075	.6	1	1	.3	.3	.4
FLOKT	200	200	60	60	.7	2	2	.6	.6	.5
FFSVP	2	2	.6	.6	1.5	2	2	.6	.6	1
FFSSR	2	2	.6	.6	1.1	2	2	.6	.6	1.1
FLOKPSR	1	1	.3	.3	1	2	2	.6	.6	.5
FLOKTYAW	200	200	60	60	5	2	2	.6	.6	.5

Simulation and Results All simulations were done using the simulation language SIMNON from Lund Institute of Technology Sweden. SIMNON is designed for solving ordinary differential and difference equations and for simulating dynamical systems. In all the simulations described a Runge Kutta integration algorithem of order 4/5 due to Dormand and Prince was used. The error was selected as .000001 and integration step in auto mode. Each simulation was started with the same initial conditions. For each simulation seven one hour, 2sec sampled wind speed data files were used. Some characteristics of these recorded wind speeds data are provided in Table 3.0

TABLE 3.0 Some Characteristics of Wind Speed Data

Data File	Mean Wind Speed (m/sec)	Std (m/sec)
ws1.dat	6.31	1.25
ws2.dat	8.27	1.35
ws3.dat	10.0	1.629
ws4.dat	5.13	0.757
ws5.dat	3.97	0.897
ws6.dat	7.98	1.403
ws7.dat	7.87	1.321

The simulation results of a typical control strategy LOKP with gain scheduled PID controller is presented in Fig 7.0 while, simulation result of same strategy with fuzzy controllers is presented in Fig 8.0 Less variation in the power can be noted for simulation with fuzzy controllers. This is because of the robustness of fuzzy controllers. Table 4.0 give the energy capture comparison of the above discussed control strategies with the gain scheduled controllers. Looking at the Table 4.0 we can say that when mean wind speed is below the rated wind speed as in the case of ws1,ws4 and ws5 then a simple control strategy as 'hill climbing' or 'power feedback' can give more or comparable energy capture than fixed tip speed ratio control. At a particular time in the wind data ws5 wind speed drop to a very low level and turbine is not able to maintain the same rotational speed that is the reason of missing results. This also indicate that BRW an variable speed control strategy will lead to higher energy capture than fixed speed control strategy. As for as ARW speed operation is concerned it can be noted from the Table 4.0 that a constant torque control strategy lead to higher energy capture. Reason for this is energy stored in the rotor because of its speed variation. The results of LOKPSR (stall regulated variable speed control strategy) shows that energy capture by this mean may be higher than the pitch or yaw regulated wind turbine. Again reason is energy storage and extraction from the rotor due to its variable speed mode.

Table 4.0 Comparison of energy captured in KW-hr with gain scheduled controllers.

Control Strategy	ws1	ws2	ws3	ws4	ws5	ws6	ws7
LOKP	2.101	3.549	4.055	1.152	0.556	3.379	3.344
LOKPYAW	1.966	3.451	4.034	1.152	0.566	3.268	3.220
HCKP	2.151	3.632	4.085	1.152	0.551	3.455	3.417
PBKP	2.159	3.636	4.088	1.156	0.568	3.460	3.421
LOKT	2.156	3.821	4.303	1.152	0.556	3.622	3.579
FSVP	1.890	3.501	3.942	0.724	low wind	3.320	3.283
FSSR	1.901	3.184	4.274	1.103	low wind	3.001	2.927
LOKPSR	2.134	3.686	4.229	1.152	0.556	3.484	3.440
LOKTYAW	2.030	3.634	4.179	1.152	0.556	3.455	3.400

Table 5.0 gives a comparison of energy captured in different control strategies with fuzzy controllers. Results are almost similar to the results in the Table 4.0 These fixed gain fuzzy controllers are much easy to design which means that they are cheaper to develop. These fuzzy controllers are cutomizable or in other words their rules can easily be modified at any later stage based on the experience of the operator. A particular set of rules lead to a non-linear fuzzy controller able to cope with time non-linear system with out any gain scheduling. If all the possible conditions of the machine are considered while deciding the controller rules then this will lead to a very robust controller covering a much wider range of operation than a PID controller. Determination of a PID controller gain scheduling scheme is model dependent while fuzzy logic controller are completely model independent hence, they bear more chances of success in the real situation.

Table 5.0 Comparison of energy captured in KW-hr with Fuzzy Logic Controllers.

Control Strategy	ws1	ws2	ws3	ws4	ws5	ws6	ws7
FLOKP	2.112	3.617	4.074	1.145	0.548	3.430	3.397
FLOKPYAW	2.072	3.644	4.084	1.145	0.548	3.435	3.435
FHCKP	2.160	3.635	4.068	1.152	0.551	3.463	3.426
FPBKP	2.160	3.640	4.069	1.142	0.540	3.467	3.430
FLOKT	2.175	4.115	4.464	1.145	0.548	3.861	3.836
FFSVP	1.919	3.514	3.937	0.725	low wind	3.334	3.300
FFSSR	1.897	3.179	4.278	1.102	low wind	2.998	2.922
FLOKPSR	2.119	3.676	4.242	1.145	0.548	3.487	3.433
FLOKTYAW	2.101	3.903	4.170	1.145	0.548	3.693	3.678

Conclusions The simulation of number of control strategies for a particular wind energy conversion system has been carried out. Design of PID controllers and fuzzy logic controllers for a particular wind turbine is presented. A Steady state aerodynamics model with some unsteady effects has been used for each simulation. This time domain, non linear simulation leads to the following conclusions:

Fuzzy controllers are simple, model independent, robust, customizable and give better dynamic response. Hence, they are the best choice for the wind turbines.

With the use of fuzzy controllers responsiveness and the stability of the system can be tuned seperatly.

Determination of a suitable gain scheduling scheme for a wind turbine PID type controller is not easy and require a complete system model.

Constant torque pitch regulated control strategy will lead to higher energy capture than constant power because of energy storage in the rotor.

Hill climbing or power feedback method is a better and simple option of control for below rated wind speed operation a wind turbine than fixed tip speed ratio control.

Fixed speed operation below rated wind speed will lead to less energy capture than variable speed.

For a variable speed stall regulated control strategy of wind turbine energy capture might be more or comparable to the pitch regulated machine.

Acknowledgement The authors wish to thank Professor Ryuichi Yokoyama of Tokyo Metropoliton University for his help during the design of fuzzy controllers.

References

1.   White S.K.(1991) Ph D thesis "A state of art cretan water pumping wind turbine using electrical transmission" EE ICSTM
2.   W. E. Leithead, S.A. De La Selle and D Reardon (1992),"Classical Control of active pitch regulation of constant speed horizontal axis wind turbine", Int. J. Control Vol-55, No.4.
3.   Bierboom W A A M (1990), "A dynamic model of a flexible rotor including unsteady aerodynamics", Proceeding of 3rd IEA Symposium on aerodynamics in wind turbine, ETSU-N-117.
4.   Kuo B C (1991) 'Automatic Control Systems', Prentice Hall NJ, USA.
5.   Shi K C & Herapath R (1983), "Performancd optimization of variable pitch HAWT system in random enviornment" Journal of Institute of Energy, Dec.
6.   Isermann Rolf (1988) 'Digital Control systems' ,Springer verlag, Berlin.
7.   Bart Kosko, (1992),"Neural Networks and Fuzzy Systems: A dynamical Approach to Machine Intelligence, Prentice Hall
8.   Lee Chuen Chien, (1990)," Fuzzy Logic in Control Systems Part-I,II", IEE Transaction on Systems, man cybernetics, Vol. 20 No.2
9.   Zheng Li, (1992), "A practical guide to tune of proportional and integral like fuzzy controllers" Proceedings of 2nd IEEE conference on fuzzy control. pp 633-640

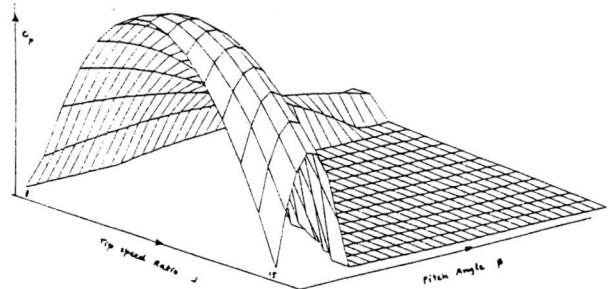

Figure 1.0 Rotor aerodynamic characteristics Cp($\lrcorner$, β)

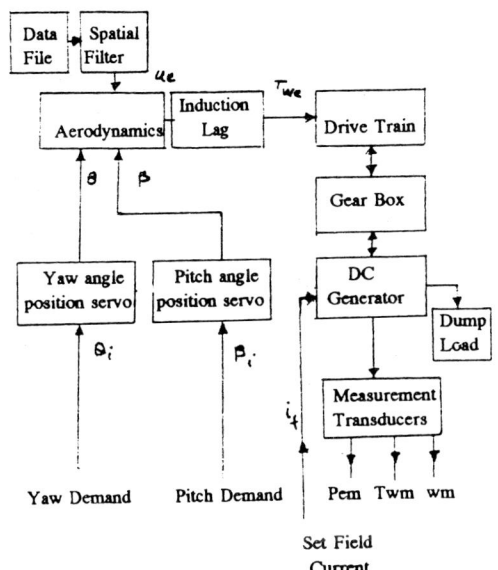

Figure 2.0 The model of wind energy conversion system (WECS).

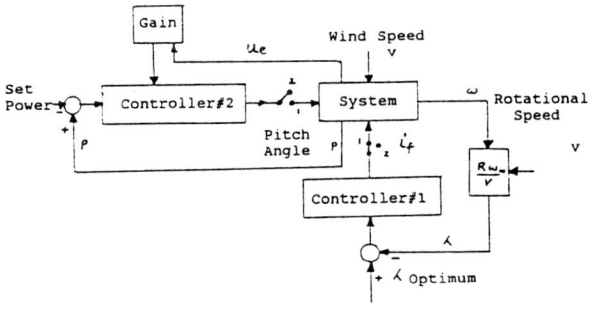

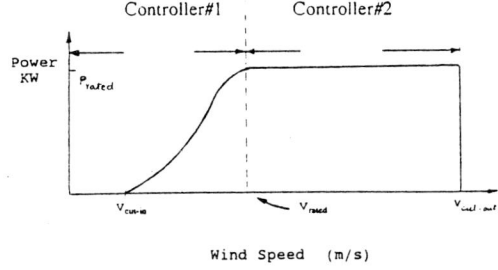

Figure 3.0 The block diagram of WECS in a typical control strategy.

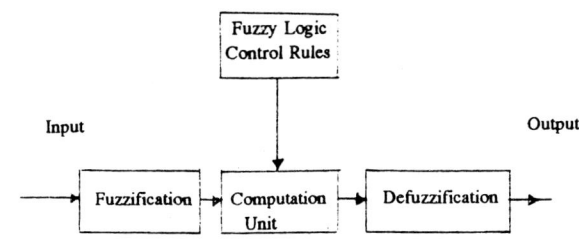

Figure 4.0 Major parts of a fuzzy logic controller.

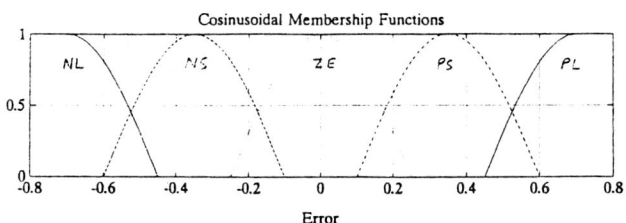

Figure 5.0 The membership functions for the wind turbine fuzzy controller.

Change in Error / Error	NLB	NSB	ZEB	PSB	PLB
NL	-8	-4	-2	-1	0
NS	-4	-2	-1	0	+1
ZE	-2	-1	0	+1	+2
PS	-1	0	+1	+2	+4
PL	0	+1	+2	+4	+8

Figure 6.0 A typical set of fuzzy control rules.

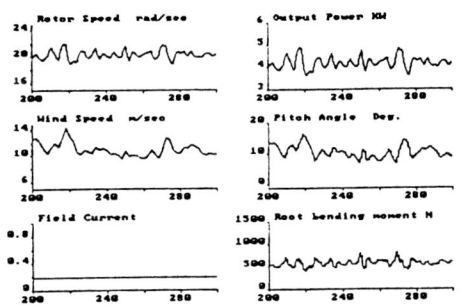

Figure 7.0 A typical result of the wind turbine simulation with gain scheduled PID type controllers.

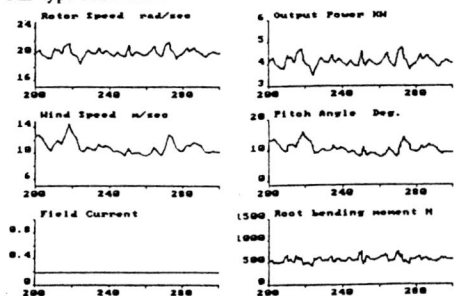

Figure 8.0 A typical result of the wind turbine simulation with fuzzy logic controllers.

Prescribed wake modelling of vertical and horizontal axis wind turbines at the University of Glasgow

F. N. COTON BSc, PhD, **R. A. McD. GALBRAITH** BSc, PhD.(Cantab), CEng, and **D. JIANG** BSc, PhD
University of Glasgow, UK

SYNOPSIS This paper describes the development of the Prescribed Wake method and, in particular, the subsequent coupling of the technique to an unsteady aerodynamic model. The manner in which the method has been used to identify design guide-lines for straight bladed vertical axis wind turbines is also highlighted. Additionally, recent work on the extension of the method for horizontal axis turbines is discussed.

1. INTRODUCTION

For many years, the wind energy community within Britain have had an active interest in the design and development of large-scale vertical axis wind turbines. The prime motivation for this was an anticipation that, for multi-megawatt machines, the economics of the vertical axis wind turbine would be more favourable than that of its horizontal axis counterpart. In addition, the apparent simplicity of their mechanical design and their omnidirectional capability augured well for their placement in large offshore wind farms.

Although, as mentioned above, there was an apparent simplicity of mechanical design, the aerodynamic environment and operation of the blades is indeed complex, for the turbine operation relies on the apparent unsteady flow past the aerofoil (1). As a consequence of this and a lack of detailed knowledge, little was known as to the manner in which the choice of aerofoil section or blade geometry should be made. Several research programmes which went some way to addressing this problem included detailed pressure measurements on wind turbines, two-dimensional dynamic tall of typical turbine sections, and aerofoil design procedures. Albeit much was gained from the research, designers were still without codes for the detailed aerodynamic design of the blades. An aerodynamic performance model which could produce useful information in minutes rather than many hours was required. The main aim of the work at Glasgow was to provide such a scheme.

In the past, Glasgow University, under Department of Energy funding, developed an aerofoil design code (2) and has the skills to produce codes for assessing the unsteady aerodynamic performance of aerofoils. At this time, they were also at the early stages of developing a Prescribed Wake model (3) for vertical axis wind turbines. The technique had the potential of a greatly reduced execution time when compared to that of a Free-Wake model. Early indications were that a target of under ten minutes per run, with the then available mini-computers, was possible. It soon became apparent, however, that the basic technique required detailed extension and the inclusion of an unsteady aerodynamic coefficient assessment (6). This was provided by the Leishman-Beddoes model (4,5).

The combination of these two codes (i.e., the Prescribed Wake and the unsteady aerofoil model) was an interesting and successful activity resulting in a high-performance design tool (7) which elevated the level of aerodynamic analysis of such machines to that of their horizontal axis equivalents. The final version of the model gave the designer freedom to consider the effects of pitch angle, blade taper, blade twist and the use of different aerofoil sections along the span. The code is also useful to structural designers because the detailed output may be the input to an aeroelastic code.

A similar method is currently under development for horizontal axis wind turbines to bridge the gap which exists between simple momentum schemes and Free Wake codes. In this case, the code is being developed in a phased manner with a view to utlimately addressing yaw, dynamic inflow and unsteady effects. It is envisaged that, on completion, the code will be one of the most comprehensive aerodynamic models available but will be considerably more efficient than existing schemes which are less versatile. Currently, a core model has been developed for the head-on steady flow condition and is in the process of gross validation against field data.

2. VAWT PERFORMANCE PREDICTION METHODS

It is well known that a vertical axis wind turbine (VAWT), which generally has a simple geometric configuration, experiences a very complicated flow field environment as shown in Figure 1. As a consequence, when calculating the flow field around a VAWT many parameters have to be considered.

Historically, the performance of vertical axis wind turbines has been satisfactorily assessed by momentum methods (8) in the form of multiple stream tubes with either single or double actuation discs. Such methods are both adequate and fast for the purpose intended but, when time-dependent details of the local flow states are required, other methods must be employed. Currently, detailed aerodynamics can be provided by vortex wake methods and they represent the most sophisticated tools available for aerodynamic design. The purest application of the vortex method is the so-called free wake method (9) in which the turbine is represented by filaments of shed and trailing vorticity that are free to convect at will. The associated computation time, unfortunately, renders the technique inappropriate for day to day design practice and, consequently, hybrid methods which blend elements of momentum theory and vortex theory are generally preferred.

One such method is the Prescribed Wake technique of Basuno et al. The Prescribed Wake model uses a combination of vortex and momentum theory to obtain a solution for the flow around a vertical axis wind turbine. In the technique, vortex elements, corresponding to the spanwise and azimuthal blade loading variations, are shed from the turbine and follow a path prescribed by consideration of momentum theory. The induced effect which the wake has on the blade loadings is then calculated according to vortex theory. This technique can be extremely accurate if the wake is correctly prescribed.

A particular feature of vortex methods, however, is that the detailed information relating to the aerodynamic coefficients of the blade aerofoil section are normally taken to be those from steady experiments. In other words, it is the standard aerodynamic data which can be obtained from most aerofoil catalogues. Thus, whilst the prescribed or indeed the free-wake methods provide unsteady aerodynamic information from the wake, the aerodynamic data for the aerofoil is derived from static data. It is because of this that such methods may be classified as quasi-steady.

Where a vertical axis wind turbine operates at low tip speed ratios, the aerofoil environment is truly unsteady and, indeed, may exhibit excursions in incidence far beyond the static stall incidence and an associated reduced frequency which places it firmly in the full dynamic stall domain. The use of static data under these conditions could lead to significant predictive inaccuracies. This obvious deficiency, however, can be alleviated by the inclusion of an aerofoil unsteady-performance code.

Whilst there are several such predictive codes available, that chosen for the present study is attributable to Leishman and Beddoes and was developed for helicopter rotor performance assessments. The vertical axis wind turbine is not a helicopter rotor and so, as anticipated, a few modifications were incorporated to render it more appropriate to the current application.

3.   THE PRESCRIBED WAKE TECHNIQUE

The Prescribed Wake method follows the philosophy of a fixed wake approach in that the influence which the wake has on itself is not directly considered during a calculation. Instead, it is considered that accurate prescription of the wake shape should adequately account for this effect. This form of wake modelling also removes the requirement to build up the wake in a step-by-step manner and so it is possible to prescribe a wake consisting of many cycles at the outset of the calculation. In addition, the existence of unrepresentative vortex filament strengths at the rear of the wake has been

removed by updating all filament strengths in the wake as the calculation progresses. By constantly updating wake elements in this way, it has been possible to accelerate convergence and thus significantly reduce the computation time associated with the model.

The vortex method which forms the core of the prescribed wake model is illustrated in Fig. 2. In it, the spanwise blade loading distribution is approximated by a series of bound vortex segments of constant strength. In this way, a vorticity imbalance is created between adjoining bound vortex segments. This is resolved by the creation of trailing vortex filaments whose strengths are defined by the difference in vorticity from one bound vortex element to the next. The strengths of the vortex filaments trailing from the blade tips are simply equivalent to those of the corresponding bound vorticity segments.

The change of turbine blade incidence with time necessitates inclusion of the influence of the associated variation of blade bound vorticity. This is achieved by considering the blade incidence variation in discrete time steps and producing shed vortices, equivalent to the resulting change in circulation, from each blade segment at every time step. In this way, a lattice of shed and trailing vortex elements is generated behind the turbine blade.

As indicated above, the vortex systems which trail from the turbine blade, are constrained to follow a pre-determined path derived from momentum theory. Thus, in the initial stages of the calculation procedure, although the wake shape is known, the strengths of the vortex elements in the wake and on the blades must be calculated. It is, therefore, necessary to adopt an iterative procedure in which the wake shape is fixed and the values of shed and trailing vorticity in the wake are systematically adjusted to correspond to the variations in circulation and spanwise bound vorticity on the turbine blades. The starting values for the iteration process are determined from the loadings associated with the variation of blade geometric incidence. The particular scheme adopted involves considering the induced effect of the wake at a given blade position, via the Biot-Savart relationship, and then updating the corresponding shed and trailing vorticity terms before moving to the next blade position. By only changing the specific vorticity values which correspond to the azimuthal position under consideration, the remainder of the vortex wake effectively damps the iteration process and so enhances the convergence characteristics of the system.

The purpose of the momentum model in the prescribed wake method is to provide initial estimates of convection velocities at each azimuthal position on the turbine. These velocities are then used to construct a basic wake shape which, in turn, is used to provide more accurate estimates of blade loadings. A further application of momentum theory, is then employed to provide a final wake shape. It is, therefore, essential that the momentum model used is sufficiently accurate to permit a realistic estimate of the actual wake shape to be generated. A double-multiple streamtube model was ultimately incorporated into the prescribed wake scheme. The main advantage which this type of momentum model has over the simpler methods is its ability to differentiate between the induced velocities on the upwind and downwind passes of the turbine blade. This feature was found to be crucial when vortex convection in the near-wake region is being considered.

As indicated in the introduction, the influence of unsteady aerodynamics on wind turbines can be profound especially at low tip-speed ratios. Under these conditions, predictions of both power coefficients and instantaneous force are often in error when quasi-steady aerodynamics are used for the blade section characteristics; usually values are under estimated. Since dynamic-stall effects can, in the extreme case, produce a difference from static results of up to 100% in air loads, the development of prediction techniques which incorporate unsteady aerodynamics has become an important area in the advancement of wind-turbine technology.

The unsteady aerodynamic response of an airfoil to a specific time history of forcing can now be determined in considerable detail and accuracy using numerical solutions of the Navier-stokes equations. Unfortunately, the required computational resources for this kind of solution method are so extreme that it will not be suitable as a design tool for some time. For this type of application, semi-empirical dynamic stall models are currently more appropriate. One such technique is that due to Leishman and Beddoes.

The main features of this technique may be summarised as follows:

(1)   Unsteady effects during attached flow conditions are represented by an indicial formulation.

(2)   Nonlinearities in the aerofoil behaviour, related to small amounts of trailing edge separation, are represented using a Kirchhoff flow model.

(3)   The onset of vortex shedding during dynamic stall is identified using a criterion for leading edge or shock induced separation based on the attainment of a critical leading edge pressure.

(4)   The induced vortex lift and the associated pitching moment are represented empirically in a time dependent manner during dynamic stall.

In this unsteady aerodynamic model, the effect of the shed vorticity, due to changes in angle of attack, is automatically considered. Likewise, the Prescribed Wake method also includes the influence of the shed wake as an inherent part of the calculation procedure. To avoid a duplicative accounting of this effect, part of the influence of the shed vorticity, on the induced velocity, was removed from the Prescribed Wake scheme , i.e. the contribution from a given blade's shed vortex system to the induced velocity at a control point on that blade was not considered.

When applying the dynamic model, it is necessary to input some data, such as the reduced pitching rates, the local relative velocity, the instantaneous angle of attack and so on. These data can be initialized by running the Prescribed Wake program using static aerodynamic characteristics. This stage of the calculation procedure also produces the wake geometry for the subsequent dynamic calculation. The significance of the dynamic model at low tip speed ratios is highlighted in Fig. 3. where the convergence characteristics of the calculation scheme are presented for a tip speed ratio of two.

4.   THE EFFECT OF BLADE GEOMETRY ON THE PERFORMANCE OF AN H-CONFIGURED VERTICAL AXIS WIND TURBINE

The quality of prediction obtained from the fully dynamic Prescribed Wake model detailed above presented a unique opportunity to study the influence of blade geometric characteristics on the instantaneous loading patterns experienced by a vertical axis wind turbine. A study was, therefore, initiated to identify the possible benefits, in terms of power output, reduced structural loadings and smoother torque distribution, which could accrue from a 'tailored' blade. Substantial modifications were made to the original computer code to provide the flexibility necessary for this work and the resulting scheme has the capability to include the effects of blade pitch, twist, taper and aerofoil section.

In the parametric study, the effect of varying each one of the above geometric features was examined in turn by comparison with a baseline configuration. This configuration was taken to be a vertical axis wind turbine with straight, untapered blades. The profile used for the blade cross section was the NACA 0015 aerofoil. The low tip speed ratio range is associated with the largest unsteady aerodynamic loads and, thus, was the obvious region in which to target this study. For this reason, most of the calculations were made for the tip speed ratio 2 case although some results were computed over the entire tip speed ratio range. Analysis of the results obtained indicated that there was scope to improve the aerodynamic performance of VAWT blades and that, as a consequence, significant structural benefits could be obtained. The implications for power generation can be summarised as follows

4.1.   The Effect of Pitch

The performance of an H-configured vertical axis wind turbine was studied for various degrees of blade pitch. It was clear from the outset that large pitch angles would be detrimental to the overall performance of the machine and would result in a severe loading imbalance between the two blades. For this reason, the range of pitch angle considered in the study was limited to six degrees on either side of the circumferential direction.

In Fig. 4., power and power coefficient curves corresponding to moderate positive and negative blade pitch are compared with the baseline condition over a range of tip speed ratio. Examination of these power curves indicates that a small amount of positive pitch, whilst reducing the power output at low tip speed ratios, can be beneficial to performance at high tip speed ratios. Larger amounts of positive pitch, however, have been found to degrade the performance over the full tip speed ratio range and so there is little scope for power regulation using pitch.

4.2   The effect of taper

The influence of blade taper was examined using a fixed blade area but varying the taper ratio between the blade tip and the cross-arm junction. Thus, any reduction in the tip chord length required a corresponding increase in the chord length at the centre span. In Fig. 5. the variation in power coefficient with taper ratio is plotted for the tip speed ratio 2 case. Unlike the previous case, it is clear that tapering the blade can lead to slightly increased performance. In this case, a power coefficient increase of around 2% is achieved with 50% taper. Lower taper ratios than 0.5 result in very high solidity in the region of the cross-arm junction and performance is adversely affected.

4.3 The effect of twist

Twist may be considered as a prescribed span-wise distribution of pitch and so it would not be unreasonable to expect the characteristics exhibited earlier by the pitch cases to be present when twist is applied. In this study, twist was applied linearly to the blades with -3° of twist indicating that the blade tip chord was offset by -3° from the chord at the cross-arm junction.

The overall effect of blade twist on power and power coefficient is assessed, for the full tip speed ratio range, in Fig. 6. As expected, similar characteristics to the pitching cases are observed with twist of 3° being almost equivalent to the 2° pitch case.

4.4 The effect of aerofoil section.

To examine the influence of blade aerofoil section on turbine performance, two aerofoil profiles were used. The baseline configuration of the rectangular blade with a NACA 0015 aerofoil profile was again taken as the reference. The second configuration consisted of a rectangular blade with the NACA 0015 profile at the tip and an earlier stalling GUAV 10 profile at the cross-arm with the aerofoil performance characteristics varying linearly between the two sections.

It was hoped that the second configuration would exhibit some degree of power regulation as a consequence of the earlier stall on the inboard blade sections. Indeed, calculations indicated that earlier stall and later recovery are, in fact, achieved by the second configuration. The consequences of this are apparent in that the reduction in torque over such a significant azimuthal range inevitably affects the power produced by the turbine. This is illustrated in Fig. 7 where the power and power coefficient variations with tip speed ratio are presented for the two configurations. It is clear that the second configuration does display some degree of power regulation without significant effect at the higher tip speed ratios.

5. PRESCRIBED WAKE MODELLING OF HORIZONTAL AXIS WIND TURBINES

A "fast" Prescribed Wake model for horizontal axis wind turbines is currently under development. The technique is being constructed in such a way that yaw, unsteady aerofoil performance and dynamic inflow can all be considered. Currently, a core model has been developed for the head-on steady flow condition and is in the process of gross validation against field data. A typical wake shape produced by the technique is shown in Fig. 8.

6 CONCLUSIONS

A comprehensive performance prediction scheme for vertical axis wind turbines has been developed. This technique, the unsteady Prescribed Wake model, is extremely accurate and at least two orders of magnitude faster that comparable techniques. Its value as a design tool has been demonstrated with a parametric study to assess the influence of blade pitch, taper, twist and aerofoil section.

The performance prediction techniques developed in this study are now being extended to consider the modelling of horizontal axis wind turbines. Additionally, the new scheme will have the flexibility to calculate yaw and dynamic inflow cases.

Acknowledgements

The authors are indebted to Dr. B. Basuno of Bandung Institute of Technology, Indonesia, for his work on the original Prescribed Wake method. The authors are also grateful to A Harris of Renewable Energy Systems Ltd. for many useful discussions.

REFERENCES

(1) Galbraith, R.A.McD., Niven, A.J., and Coton,F.N.Unsteady aerodynamics and its relevance to wind turbines.Wind Engineering, 1991, Vol, 14 No.5

(2) Coton, F.N., and Galbraith, R.A.McD, 'An aerofoi design methodology for low speed aerofoils', Tenth British Wind Energy Conference, London, 1988

(3) Basuno, B., Coton, F.N., Galbraith, R.A.McD., 'A Prescribed Wake aerodynamic model for vertical axis wind turbines', Journal of Power and Energy, Vol 206, 1992

(4) Leishman, J. G., & Beddoes, T. S. A Semi Empirical Model for Dynamic Stall Journal of The American Helicopter Society, July 1989.

(5) Leishman, J.G. Practical Modeling of Unsteady Airfoil Behaviour in Nominal Attached Two Dimensional Compressible Flow. University of Maryland, 1987, Report UMAERO - 87 - 6

(6) Jiang, D., Coton, F. N., and Galbraith, R.A.McD A Fixed Wake Vortex Model for Vertical Axi Wind Turbines Including Unsteady Aerodynamics Wind Engineering, Vol. 15, No. 6, 1991

(7) Jiang, D., Coton, F. N., and Galbraith, R.A.McD 'The inclusion of unsteady effects in an aerodynami model for vertical axis wind turbines', Fourteenth British Wind Energy Conference, 1992

(8) Paraschiviou, I. & Delclaux, F. Double Multiple Streamtube Model with Recent Improvement Journal ofEnergy, 1982, Vol. 7, No. 3.

(9)   Strickland, J. H. et al.   A Vortex Model of the
        Darrieus Turbine: An Analytical and Experimental
        Study. Journal of Fluids Engineering, Dec, 1979,
        Vol.101

Power Output (KW)

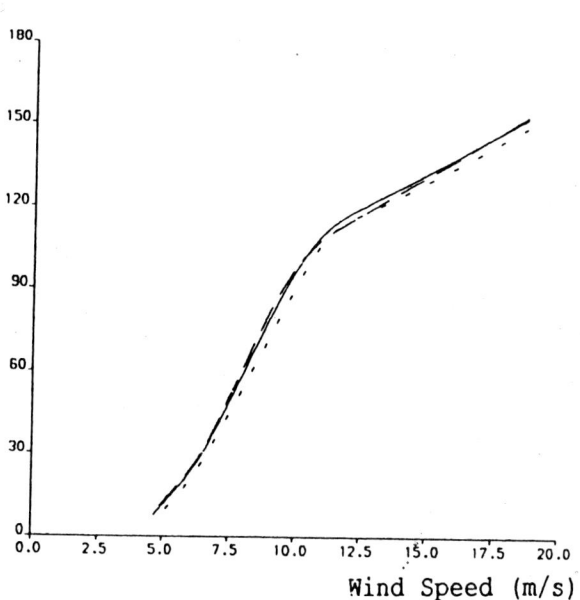

Wind Speed (m/s)

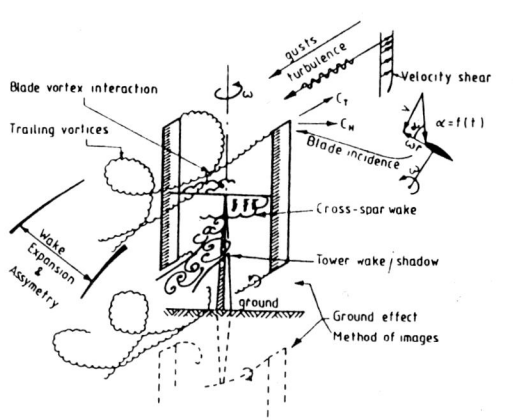

Fig.1. VAWT Flowfield

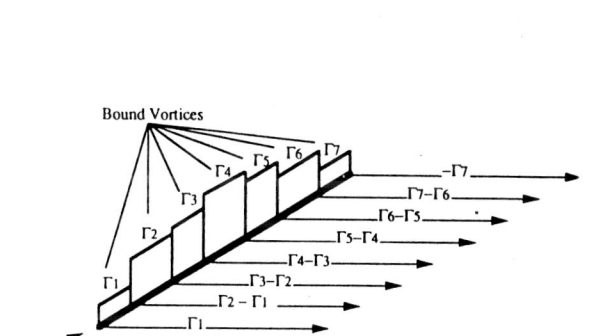

Fig.2. Basic vortex model

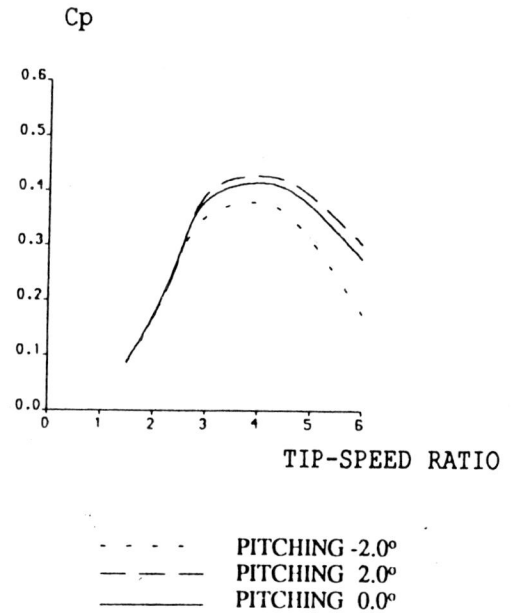

TIP-SPEED RATIO

- - - -   PITCHING -2.0°
— — —   PITCHING 2.0°
————   PITCHING 0.0°

**Fig.4. VAWT power and power coefficient curves
for three blade pitch settings**

Cp

**Fig.3. Convergence   history   of   unsteady
Prescribed Wake model for tip speed ratio
2**

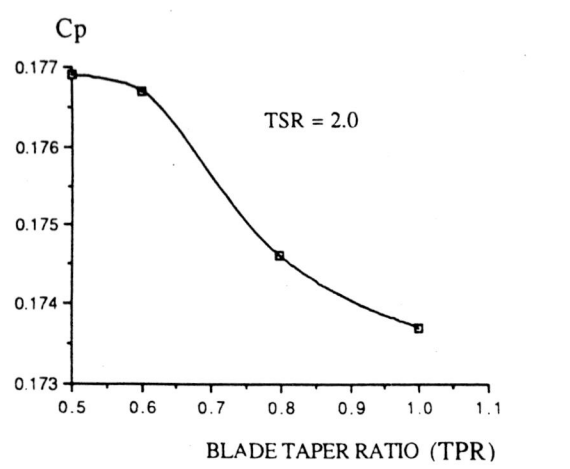

Fig.5. Variation of power coefficient with taper ratio at tip speed ratio 2

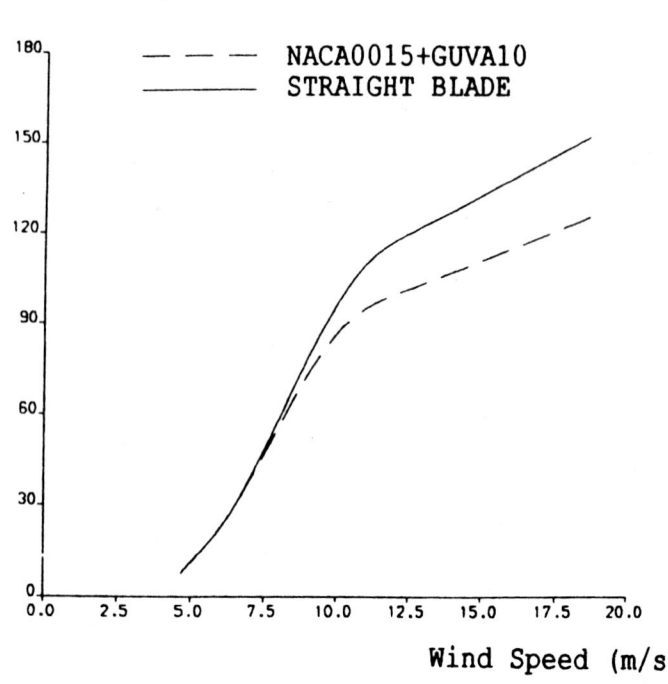

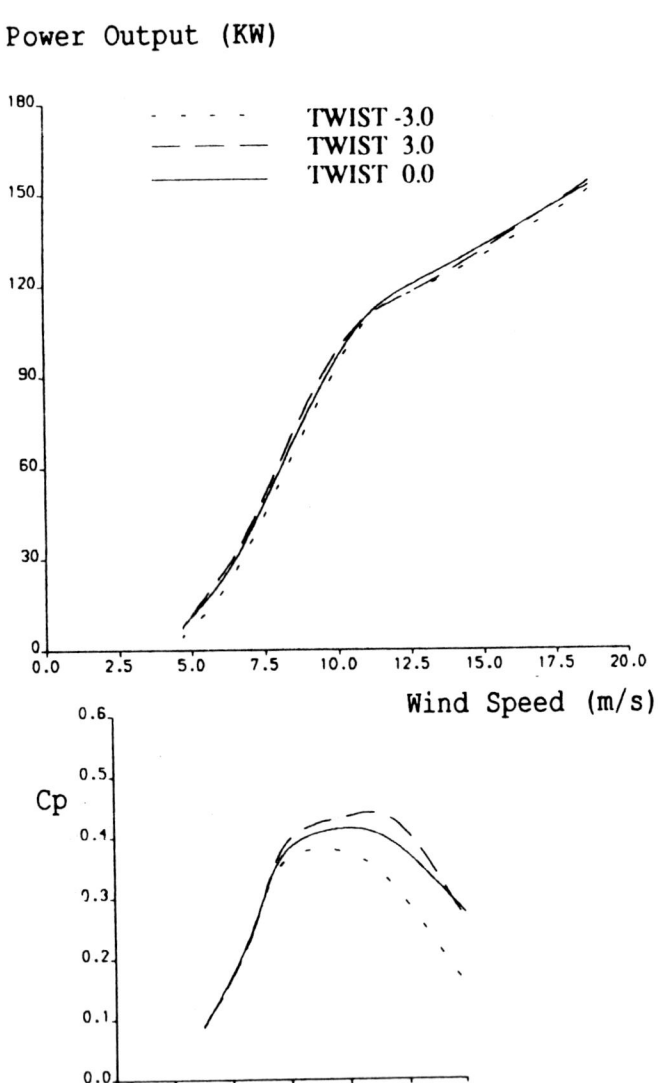

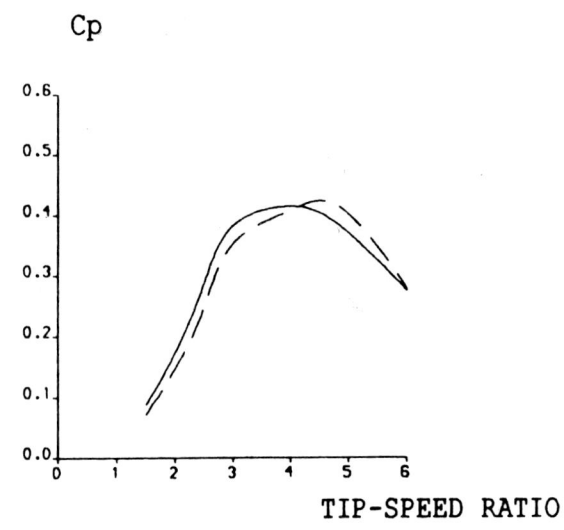

Fig.7. VAWT power and power coefficient curves for two blade profile geometries

Fig.6. VAWT power and power coefficient curves for three blade twist settings

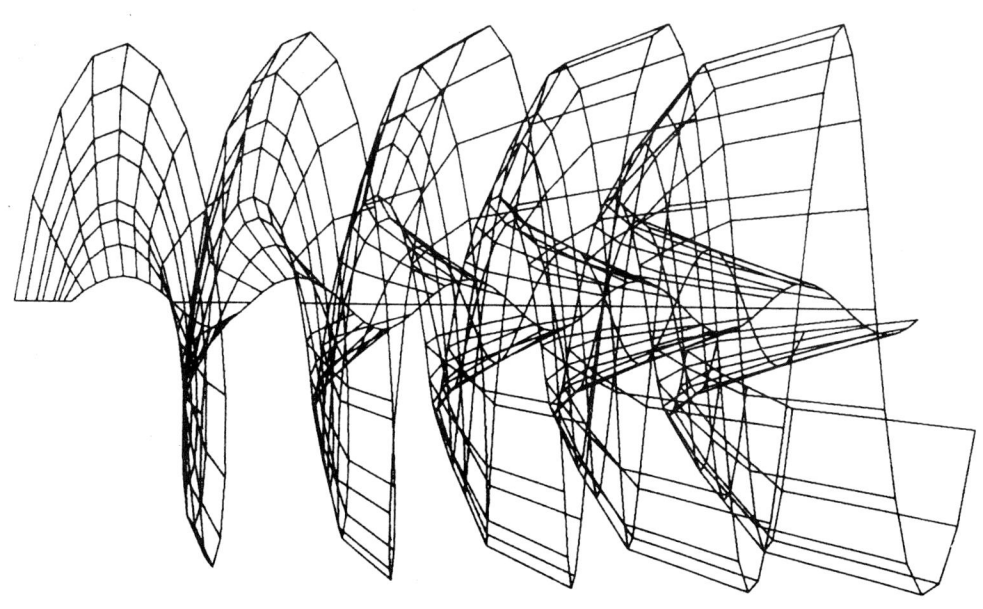

Fig.8. Typical HAWT wake lattice produced by the Prescribed Wake model

Noise control on the BONUS 300 kW wind turbine

H. STIESDAL and **E. KRISTENSEN**
BONUS Energy A/S, Denmark

SYNOPSIS: In 1991-92 BONUS Energy A/S carried out a project for noise control on the BONUS 300 kW, a commercial wind turbine of the Danish concept. The project was successful, with a reduction in the sound power level of 3-4 dB, and with a tonal content in the noise significantly below the objective criterion of the Joint Nordic Method.

The methods for reduction and control of both the overall sound power level and the tonal content of the turbine noise are presented.

1. INTRODUCTION

Noise is one of the most important limiting factors in the dissemination of wind power in the populated areas of Europe. To reach even moderately ambitious goals for installed power it is necessary to reduce the noise output of commercially attractive wind turbines to the largest extent possible.

The overall sound power level of a modern wind turbine is usually dominated by the aerodynamic rotor noise. Several different mechanisms contribute to the aerodynamic noise generation. For each mechanism reduction and control must be considered.

The tonal content of the noise is usually dominated by harmonics of the gear mesh frequencies of the gearbox. The source noise from the gearbox must be reduced as much as possible, and the noise transfer from the gearbox to the surroundings must be controlled.

2. THE TURBINE

The BONUS 300 kW is a typical Danish wind turbine. It has an upwind, three-bladed, fixed pitch rotor with stall regulation, a grid-connected asynchronous generator, and an active yaw system. The 300 kW prototype was erected in 1989, and the production version has been marketed since 1991.

At the time of writing approximately 70 BONUS 300 kW turbines have been installed. BONUS' first windfarm in Britain, the Rhyd-Y-Groes project on Angelsey, consists of 24 300 kW turbines.

The turbine has the following main data:

Rated power .... 300 kW
Rotor diameter .... 31 m
Rotor speed .... 31 rpm
Hub height .... 30 m
Blade type .... LM 14.2 m

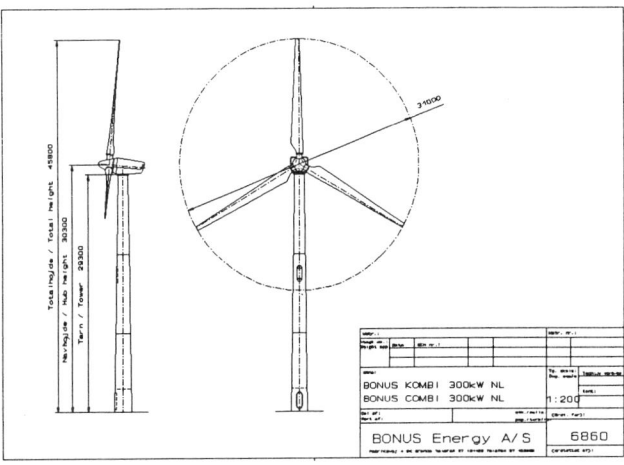

Fig.1. BONUS 300 kW turbine

3.   AERODYNAMIC NOISE

The BONUS 300 kW turbine is fitted with LM 14.2 m blades. The LM 14.2 m blade has a high-efficiency planform with convex leading and trailing edges. At the inboard part of the blades the aerodynamic profiles belong to the NACA 634 series, and at the outboard part they belong to the NACA 632 series.

In the present project five different sources of the aerodynamic noise have been considered. They are

1.   Tip noise
2.   Trailing edge noise
3.   Stall noise
4.   Turbulence noise
5.   Noise from surface imperfections

3.1   Tip noise

The outboard end of the LM 14.2 m blade tip is "triangular". The trailing edge is longer than the leading edge, stretching the tip into a sharp point at the trailing edge. This tip planform generates a rather concentrated tip vortex. A concentrated tip vortex at the trailing edge is attractive because it is assumed to result in rather moderate tip losses and a large efficient radius of the blade. However, the high air speed of the concentrated tip vortex tends to generate more tip noise. The tip noise from this design is also more sensitive to minor imperfections of the geometry than the noise from an ordinary, rounded tip planform.

As soon as the first set of LM 14.2 m blades were tested in operation on a BONUS 300 kW turbine it was realized that the tip was an important noise source on this blade type. The tip vortex was clearly audible when the rotor was operating in moderate wind speeds.

Flow studies with ticklers and a rotating video camera indicated that the tip vortex would separate from the airfoil and re-attach again at the trailing edge a few centimetres from the tip itself. It was thought that this separation and re-attachment could contribute significantly to the tip noise.

BONUS carried out a tip modification project, part of which was undertaken in cooperation with LM Glasfiber. Since an earlier noise project [1] had indicated that the prediction of tip noise behaviour is very difficult, the modifications were not developed through any exact or scientific reasoning. They

evolved through "trial and error", based on the noise engineers' intuitive feelings of the problem as it was indicated by the video studies. Some of the proposed modifications aimed at dissolving the concentrated tip vortex by changing the tip planform itself, and others aimed at methods for prevention of the separation and re-attachment of the tip vortex on the original tip planform. The blade tip modifications were implemented on two different rotors with LM 14.2 m blades on BONUS 300 kW turbines, and after each change the overall sound power level was measured by BONUS according to the Joint Nordic Method.

The uncertainty of sound power level measurements on wind turbines is typically in the range of 1-2 dB. Most of the tip modifications did not result in any tip noise reduction at all, or the reduction was smaller than the measurement uncertainty and therefore not significant. Two of the tip modifications did, however, lead to significant improvements. One was a change in the tip planform to an elliptical, rounded shape, resulting in a more diffuse tip vortex, and the other was the attachment of a small torpedo-shaped body to the outboard end of the trailing edge of the original tip. In both cases the change in overall sound power level for the turbine was in the range of 2-3 dB. Power curve measurements were performed to determine any associated changes in turbine performance. It was found that the annual energy output would be reduced by approximately 5 percent with the elliptical tip planform, while it was unchanged with the tip torpedo. The tip torpedo solution was selected, and the optimum size and location was determined by further experiments.

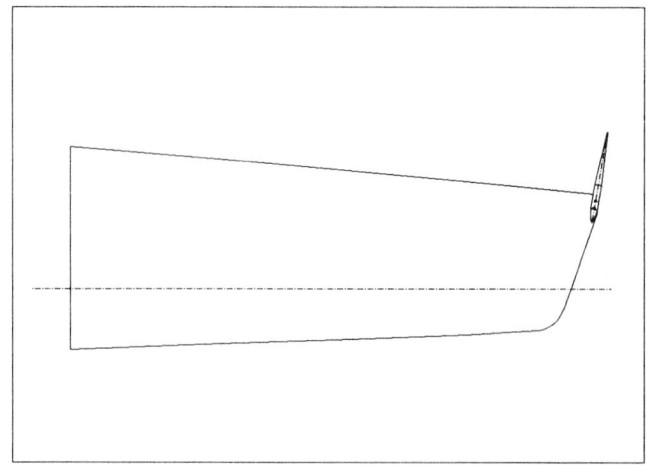

Fig.2.   Tip torpedo

The tip torpedo is now a standard feature of blades with "triangular" tip planform on BONUS turbines.

Development and demonstration of an advanced two-bladed HAWT

R. S. HAINES, R. H. SAUVEN and **P. B. SIMPSON**
Wind Energy Group Ltd, UK

The Wind Energy Group has introduced a new design of horizontal axis wind turbine generator for wind farm application. This is the WEG 400, a 36.7 m teetered rotor machine, available at 400 and 450 kW ratings. It retains successful features from the company's earlier commercial machines but is further optimised to give a marked improvement in cost effectiveness.

A development prototype was tested over a twelve month period at the National Wind Turbine Centre in Scotland. The production version is being demonstrated at a commercial wind farm in south-west England.

This paper describes the development and testing of the design.

1 INTRODUCTION

A newcomer to the medium scale wind turbine market this year is the WEG 400, a two-bladed machine with 36.7 m diameter rotor. Although somewhat novel in its drive train layout, this machine is essentially an evolutionary development of previous designs from the same manufacturer. This paper outlines the course and rationale of this evolution, and concludes by describing the testing and performance monitoring of the production prototype.

2 DEVELOPMENT BACKGROUND

The machine in question is designed and supplied by Wind Energy Group Ltd (WEG), a wholly owned subsidiary of Taylor Woodrow Construction Holdings plc. Over the last eight years WEG has pursued a line of commercial development of horizontal axis wind turbine generators, whilst for over fifteen years it has been a major participant in the programme of investigation of wind energy technology sponsored by the UK Department of Trade and Industry (DTI - formerly as the Department of Energy).

The first commercial wind turbine produced by WEG was the three-bladed MS2, of which a windfarm of 20 was installed in Altamont Pass, California in 1986 [1] following successful operation of a prototype at Ilfracombe. However, WEG had already built two two-bladed, teetering rotor machines

on Orkney, the 250 kW MS1 [2] and the 3 MW LS1 [3]. A major programme of experimental monitoring was carried out for the Department of Energy on both these machines.

Confident in this experience, and recognising the potential of the two-bladed teetered hub concept for achieving economies in machine design, in 1987 WEG adapted the design of the MS2 to incorporate the teetered hub principle. Taking advantage of advances in the laminated wood blade construction, the lighter rotor arrangement and built in reserves in the drive train, the diameter of the turbine and the rating were increased to 33m and 300 kW respectively. A prototype MS3, as this turbine was named, was installed in Altamont Pass in late 1988, followed by a second machine at a test centre at Carmarthen Bay in South Wales [4].

In 1990 MS3's were sold to Aeritalia for ENEL and to Northern Ireland Electricity. Two machines were also built for use in the UK Department of Energy R & D programme, for installation at the National Wind Turbine Test Centre (NWTC), on Myres Hill, near East Kilbride in Scotland [5].

In 1992 WEG Ltd sold 65 MS3 turbines to a UK windfarm developer, First Windfarm Holdings Ltd., for installation in three windfarms at sites near Cemmaes and Llangwyryfon in mid Wales and Cold Northcott in Cornwall [6]. The first example of the WEG 400 was also sold to the same developer.

3 DESIGN CONCEPT

The adoption of a teetered hub with a two-bladed rotor provides the basis for taking maximum advantage of the economy in use of materials offered by the two-bladed configuration, in comparison with a three-bladed configuration. Uncoupling the out of plane blade moments from the rotor shaft relieves both the blade root sections and the pitch bearings (in the case of full span pitch control) of a considerable proportion of the loading they would otherwise have to carry, permitting lighter blade and bearing construction. Although the additional complexity in hub and pitch control mechanism design somewhat erodes the advantage, arguably it does not do so entirely and certainly not when the further benefits accruing to drive train, nacelle bed plate (or equivalent), yaw mechanism and drive are taken into account. In particular, when the low speed shaft bearings are integrated with the gearbox and, also, when the gearbox casing is integrated with the nacelle support structure to achieve further economies, the advantage offered by the teetered hub concept increases. As described below, the WEG 400 incorporates both these principles of integrated arrangement, and therefore derives maximum benefit from the teetered hub concept.

The WEG 400 design was conceived from the outset as a utility-standard, windfarm-oriented machine with due attention paid to features such as safety interlocks, user interface, weather-protected working environment, and noise isolation in addition to efficiency of material use. Moreover, the design is supported by comprehensive calculations based extensively on field-monitored data, and the machine will be sold with certification by Germanischer Lloyd.

4 DESIGN DEVELOPMENT

4.1 Drive Train Arrangement and Support

The general arrangements of the MS3 nacelle, showing the drive train arrangement and bed plate structure, is shown in Fig. 1. (The principles of the MS2 arrangement were similar.) The integration of the low speed shaft bearings with the gearbox is an important feature, contributing to the relatively low total tower head mass of around 11 tonnes.

The central feature of the development of the WEG 400 general arrangement, shown in Fig. 2, is the use of the gearbox casing to carry load horizontally back towards the tower head, not just vertically as before. This eliminates the duplication of load paths formerly provided by the gearbox casing and nacelle bed plate, resulting in integration of the gearbox casing with the nacelle structure, with

corresponding reduction in use of materials, weight and cost in relation to energy capture, yet with structural safety margins maintained. Likewise, the increased performance has not affected the nacelle overall dimensions, which remain suitable for transport of the complete nacelle in standard freight containers.

In practice, the concept referred to above is implemented in the WEG 400 by mounting the gearbox at its rear on a space frame comprising a fabricated box-section framework and a tie-bar to complete triangulation. The generator is also mounted on this frame.

Some form of compliance is required in the drive train of a wind turbine of this class in order to avoid resonance at blade passing frequency. In the WEG MS3 this was achieved by means of an induction generator with high slip. With increased rating, this method incurs an increased cost in terms of the greater generator frame size required, particularly with the move in the industry to totally enclosed generators. The cost of the energy lost in rotor slip must also be taken into account. In view of these factors it was decided to investigate alternative forms of compliance for the WEG 400 in conjunction with a less costly, low slip generator.

Well-known devices such as fluid and elastic couplings were considered, but it was concluded that mounting the entire drive train flexibly was a superior solution in this application. The elements of this solution comprise a relatively light framework connecting the gearbox and generator, and a system of elastomeric mountings, with very high flexibility about the roll axis, connecting this framework to the nacelle bed. The major advantage of this approach is that the flexible mountings work as anti-vibration mountings as well as effectively reducing the drive train stiffness, so that structural transmission of noise is strongly reduced. Moreover, the mountings concerned provide an economic form of flexibility that is well-tested in other industries. With compliance provided in this way, by virtue of the reduced requirement for slip the WEG 400 uses a generator with the same frame size as that for the MS3.

The principle was tested by modifying the existing fabricated pallet to form the bed member and fabricating a new framework to support the gearbox and generator. The dynamic behaviour of the system was satisfactorily tested during both transient and normal running conditions above and below rated. A production version of the system was then developed, using the same concept and torsional characteristics but with components optimised for the purpose. In this version the gearbox/generator mounting frame is itself mounted, by way of a pair of elastomeric hinges and an elastomeric compression spring, on a cast bed; this also provides a stiff support for the rolling element yaw bearing. A linear damper is normally mounted between the frame and bed to limit resonant response, although data collected so far has related to performance with the damper omitted.

Like the generator, the gearbox internals are also unchanged in most respects from those in current production for the MS3, the initial design having intentionally included a margin for uprating. Two specially modified gearboxes have been successfully endurance tested at the new torque level in a back-to-back configuration. Carried out with DTI support, this test used a special set-up incorporating simulation of all the predicted rotor loads including thrust and teeter impact. One change since the initial gearbox design, an increase in the size and span of the main bearings to cater for the larger rotor, has already been incorporated in the current MS3, so as to increase the degree of commonality.

4.2 Blades

Over the last two years, WEG has undertaken a DTI-supported programme of blade development. As a result the rotor design has been improved in terms of weight, energy capture and quietness.

Part of the programme has included material testing at Bath University and fatigue testing at City University, London, which have together provided a very sound data base on which to assess the strength of the blade in service. As was anticipated, the strength reserves of the existing root connection and inner part of the original blade have been found to be well able to withstand the loads from a tip extended radially by over 10%.

Another part of the programme has focussed on the improvement of production methods. The blades tested at City University have progressively incorporated the new production features, which have not only greatly reduced cycle times but have also resulted in further increase in strength.

The wood used in the WEG 400 blades, which is of European origin, was selected after comparative trials (in the same test programme) with a number of candidate woods.

4.3 Noise Reduction

Noise reduction has been a major objective in the design of the WEG 400. The incorporation of vibration isolation in the drive train mounting has already been described. At a more detailed level, the design of the gearbox housing has been modified, without affecting the moving parts, to avoid features likely to serve as efficient noise radiators. The design has a number of other features tending to reduce noise transmission, such as the wrap-around, flexibly mounted cladding and the use of SG cast iron for the nacelle bed.

4.4 User interface

The WEG 400 is linked to a windfarm SCADA (Supervisory Control & Data Acquisition) system. This system logs ten-minute statistical data and WTG event (change of state) data. This data can be processed to provide alarm summary information, power versus windspeed curves, and so on.

The SCADA system also enables the WTG to be monitored at any distance over the public telephone network.

For local control, the WEG 400 controller is provided with an operator keypad and integral backlit LCD display. This operator terminal provides password-protected, menu-driven access for WTG controls such as turbine, start, stop, reset; "manual" control functions; controller set points; limited logging of data. Although normally used at the tower base, the terminal is a plug-in, pocket-sized device that may equally be used for test purposes in the nacelle.

Another aspect of the user interface is access for maintenance. This has been improved by grouping parts requiring routine maintenance on the same side of the drive line as the entrance from the tower, and by remodelling the nacelle cladding front so that most hub tasks (such as pitch bearing greasing, teeter damper inspection) can be carried out while standing inside the cladding. The cladding itself provides ample headroom for standing.

5 EXPERIMENTAL PROTOTYPE TESTS

As mentioned above, the principle of the flexible drive train was tested by modifying an existing fabricated pallet and adding a new framework to support both the gearbox and generator. This was undertaken on one of the two MS3's installed at the NWTC on Myres Hill. Longer, modified blades, with the improved tip geometry described earlier, were also fitted to this machine, increasing rotor diameter to 35.4m. At the rotational speed of 45 rpm chosen for the prototype gearbox, the rated output of the turbine was increased to 450 kW and operational trials were conducted throughout 1992. The performance of the machine was found to be very satisfactory. In particular, the dynamic behaviour was as predicted and the improvement in aerodynamic sound power level achieved with the modified blade tip geometry was confirmed.

6 PRODUCTION PROTOTYPE TESTS

6.1 Evolution to production prototype

The production-version WEG 400 retains the major concepts of the experimental prototype, with the further advances described in Section 4, while incorporating many of the newer details developed to suit wind farm application on the MS3. Adjustments were also made to the rotor diameter and tip speed to reflect changing market needs. The first example of this design was commissioned in the summer of this year at Cold Northcott windfarm in Cornwall.

6.2 Component tests

Ultimate load and/or endurance tests have been carried out on several of the components of the WEG 400. The tests on the blades have already been referred to. Endurance tests on the pitch linkage hinges have been used as a basis for bearing material selection, resulting in a major improvement from the material first selected. The tie bar and the elastomeric spring have also passed appropriate tests.

6.3 Trial build

A trial nacelle using a mixture of real and mock-up components was assembled in advance of the production prototype nacelle, in order to identify and counter any difficulties in assembly or maintenance access. This proved invaluable as a tool for formal Design Reviews, and remains available to trial fit any mechanical modifications suggested by field experience. As a rule, such modifications are then (with the owners' agreement) incorporated in the production prototype so that this remains representative of the current state of the design.

6.4 Monitoring

Monitoring transducers installed on the production prototype include strain gauges on the fabricated frame, a displacement transducer between the frame and bed casting, and pressure transducers on the hydraulic power pack. The data acquisition system used in conjunction with these transducers is also able to record signals from the control system. The control system also has a built-in data capture capability. The monitored data enabled correct behaviour during commissioning tests to be confirmed, and was also used for system identification purposes as described below. Early post-commissioning data collection has concentrated on behaviour during transient conditions, as well as validation of the power curve.

6.5 System Identification

As part of the commissioning tests on the production prototype WEG 400, the response of various parts of the total system was monitored while suitable synthetic signals were added to the pitch control feedback loop. From the data collected, transfer functions were constructed, and matched to functions derived from linear models of the system, so that the model parameters closely represent the true behaviour of the system. By this means, for example, the gain parameters of the closed-loop control algorithm can be optimised rationally. An example of a measured transfer function and fitted model is shown in Figure 3.

6.6 Braking system sequence verification

An example of monitored data used to verify the operation of the hydraulic brake control system during ground tests is shown in Fig. 4. About two seconds from the start of the test, a brake release signal is issued, and the brake line pressure rises accordingly: this operation also fills a hydraulic accumulator. About three seconds later in the test, the power supply is cut off to simulate a grid fault. The pressure promptly falls to an intermediate level which results in a braking torque approximately equal to rated torque. When the speed has been reduced enough by blade pitching to ensure that the rotor is under control, the brake is released by fluid from the accumulator. The final stage of the trace shows the application of full braking at twice rated torque, which is achieved by reducing the hydraulic pressure to zero when the rotor reaches parking speed and a horizontal position. As can be seen, the system approaches this condition progressively, to avoid unnecessarily severe transient torques.

6.7 Noise emission

At the time of writing, a program of noise data collection is under way. Initial results confirm a significant reduction of mean noise power level relative to the MS3.

7 POWER PERFORMANCE

The predicted power versus windspeed curve is shown Fig. 5, together with monitored data. The latter is averaged in bins of power, excluding data

where the pitch is actively controlling. The high speed mode data (squares) is shown separately from the quieter, low speed mode that is used in lighter winds. The isolated point below the curve represents brief periods when the wind has risen but low speed is still selected. Otherwise, the trend clearly confirms the expectation.

A thirty-second trace of the power variation in moderate winds, while running in high speed mode (40 revs/min.) is shown in Figure 6. The expected fluctuation at blade-passing frequency (40 peaks in the figure) is seen to be reduced to a low level by the compliant drive-train support. In fact, the dominant frequency is once per revolution (20 peaks), which is approximately the natural frequency of the system: even this is not particularly severe, despite the omission of the damper at the time of the recording.

ACKNOWLEDGEMENTS

The authors are grateful for the support of the DTI for many of the developments described in this paper. They also wish to acknowledge the contribution made to the project by many of their colleagues.

REFERENCES

[1] P M ELLIOT, C R GAMBLE, D LINDLEY, WEG's California Windfarm, Proceedings Twelfth BWEA Wind Energy Conference, 1990.

[2] J R C ARMSTRONG, G R KETLEY, B J COOPER, The 20m diameter wind turbine for Orkney, Proceedings Third BWEA Wind Energy Conference, 1981, pp 54-62.

[3] P B SIMPSON, A L BURTON, D LINDLEY, The LS1 3 MW horizontal Axis Wind Turbine, Fifth International Conference on energy options - The Role of Alternatives in the World Energy Scene, 1987.

[4] Dr D LINDLEY, Dr C R GAMBLE, Dr J G WARREN, Performance of the WEG MS3 300 kW Wind Turbine, Conference Publication EWEC Part Two, 1989, pp 623-630.

[5] C R GAMBLE, N JENKINS, R H SAUVEN, A L BURTON, Installation and operation of the MS3 wind turbines in Ireland, Scotland and Sardinina, Proceedings Thirteenth BWEA Wind Energy Conference, 1991, pp 33-38.

[6] D LINDLEY, P MUSGROVE, J R C ARMSTRONG, M L HITNER, Early experience in UK Windfarms, European Community Wind Energy Conference, 1993.

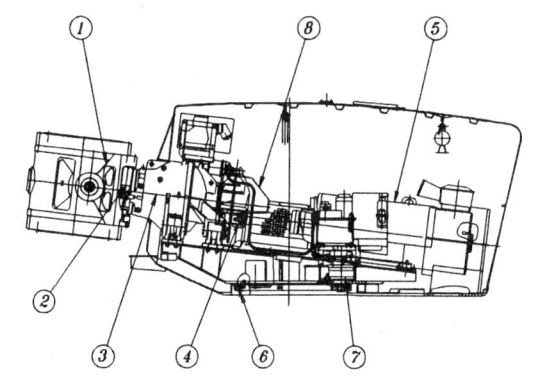

1. Rotor Hub.
2. Teeter Damper.
3. Gearbox.
4. Pitch Servo Actuator.
5. Generator.
6. Yaw Bearing.
7. Yaw Drive.
8. Pitch Sensor.

Figure 1 MS3: General Arrangement of Nacelle

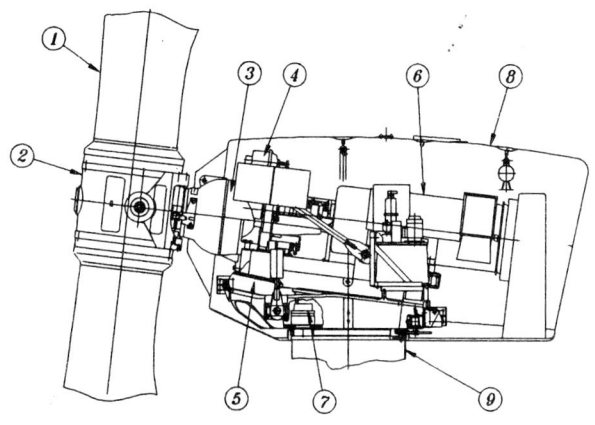

1. Blades.
2. Hub.
3. Gearbox.
4. Mechanical Brake.
5. Compliant Mounted Transmission.
6. Generator.
7. Yaw Drive.
8. Nacelle.
9. Tower.

Figure 2 WEG 400: General Arrangement of Nacelle

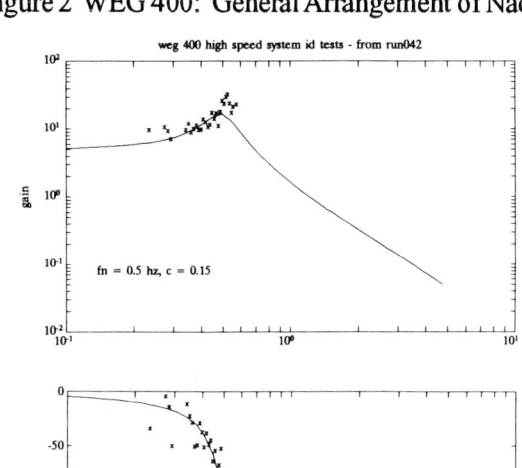
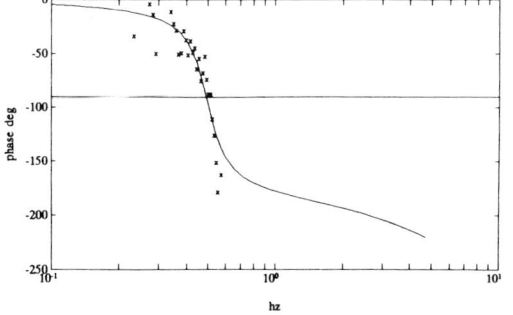

Figure 3 System identification: transfer functions

345

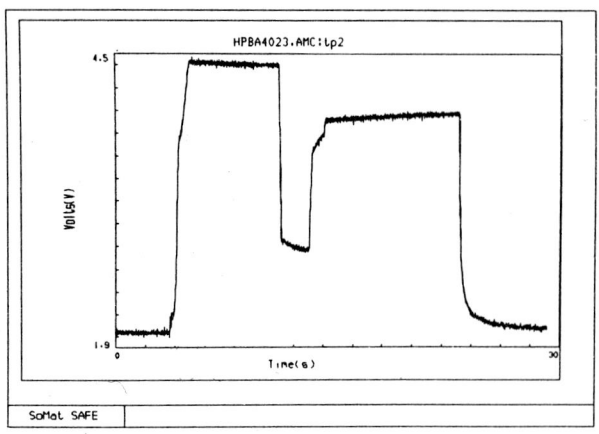

Figure 4 Brake line pressure: grid loss simulation

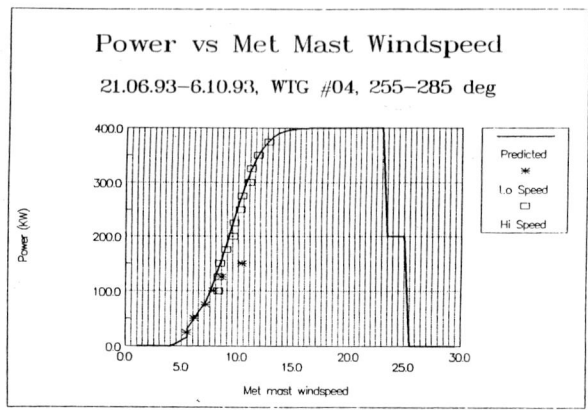

Figure 5 Power versus windspeed data

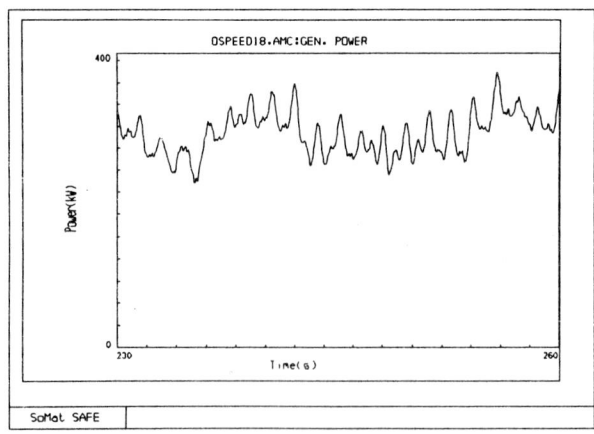

Figure 6 Power fluctuations

Test results from the AOC 15-50 wind turbine development programme: tip brake life test

P. HUGHES, S. CHILDS and **A. FACCHETTI**
Atlantic Orient Corporation, USA

SUMMARY

The AOC 15-50 advanced wind turbine has undergone an extensive test program designed to prove its capability and ensure reliabilty in service. As part of this test program a tip brake qualification test has been completed by simulating, on a test rotor, the centrifugal forces at the blade tips of the AOC 15-50 wind turbine, while subjecting the tip brakes to a 30 year lifetime of expected service demands. This included normal shutdown and overspeed braking of the test rotor. Static testing was done to monitor decrease in damping. Results showed a break-in period of approximately 3,000 cycles. The decrease in damping had a minor effect on the tip brake performance.

1 INTRODUCTION

The AOC 15-50 advanced wind turbine has undergone an extensive series of tests both on laboratory test beds and in the field. These tests have included loads, performance and modal testing of the blades, drive train tower and control system (1),(2). Field testing is continuing at the US Department of Agriculture test station in Bushland, Texas and will begin shortly at three other tests sites in Canada and the USA.

As part of this ongoing test program a Tip Brake Qualification Test has been carried out at the AOC Norwich Test Facility. The objective of the Tip Brake Test program was to qualify components in the laboratory in conjunction with full atmospheric testing of the pre-production prototypes. Through this test program the technical risks associated with design innovations have been minimized.

The life testing carried out at AOC has been used to validate the performance and reliability of the tip brake mechanism. This has been accomplished by cycling the mechanism through an expected lifetime of normal shutdown and overspeed emergency braking conditions.

The testing of the AOC 15-50 tip brake mechanism in Norwich, Vt.has provided an opportunity to detect any crucial problems associated with the operation of the mechanism before field testing. In addition, the tip brake testing apparatus has been a valuable asset in improving the design of the tip brake and the associated rotary transformer as designs mature.

2 METHOD

Two tip brake mechanism were tested by rotating them on a shortened test rotor 2.4 m (8 ft) in diameter. The test apparatus simulated centrifugal forces produced at the tips of the actual 15 meter rotor on the AOC 15-50 machine. Centrifugal force simulation was accomplished by operating the test rotor at 16.96 rad/s (160 rpm), a higher rotational speed than 6.69 rad/s (64 rpm) of the AOC 15-50. Dependability and longevity of the working mechanism was determined by cycling tip brake deployment from the simulated normal operating speed, 160 rpm.

The maximum number of test cycles was determined based on an average of 2 cycles per day and a machine life expectancy of 30 years. This was equivalent to approximately 22,000 cycles per tip brake mechanism. For the laboratory test, the mechanism was cycled every 30 seconds.

Dependability of the tip brake mechanism includes automatic emergency deployment. In the event of a control failure, loss of grid, or main shaft failure the tip brakes will deploy in a fail safe manner by overcoming the electromagnetic retaining force with no controlled outside intervention. The designed emergency deployment speed for the AOC 15-50 is approximately 125% of normal operating speed. The centrifugal forces acting on the tip brake mechanism at 125% of normal speed were simulated in the laboratory. Overspeed testing involved both gradual (slow overspeed) and sudden (fast overspeed) acceleration from normal operating speed in order to quantify any inertial variations acting on the operation of the tip brake mechanism. This portion of the test verified that the mechanisms do work in a fail-safe manner.

The speed at which the tip brakes reset was determined by allowing the test rotor to slow down by means of deployed tip brakes both with and without dynamic braking assistance. The tip brakes close by means of the retraction spring when the centrifugal forces are sufficiently low.

The following parameters were measured and recorded:

a) Magnet Voltage: The voltage supplied to the primary side of the rotary transformer was adjusted such that the voltage to the magnets was 12 volts.

b) Rotor Speed: The rotor speed was calculated directly from a frequency output generated by the AC-Tech variable speed 20 Hp motor drive.

c) Mechanism motion detection for Tips #1 and #2, respectively: The motion of the tip brake plate either opening or completing closure was detected using a micro switch activated when the tip brake plate was within 1° of the fully open or closed position.

d) Magnet De-energization: The moment when magnet de-energization occurred was recorded and time stamped.

The centrifugal force was the governing force simulated by the test apparatus. One consequence of testing the tip brake mechanism on a shortened rotor was that the tip speeds were not equal to those experienced on a turbine in the field. This reduced the aerodynamic effects experienced by the mechanism from 16% to only 3% of the total force.

3 TEST SETUP

The tip brake mechanism is shown in Figure 1. In addition to the mechanism, the rotary transformer and blade tip ends served as functional components in the test.

The test components were as follows:

a) Tip Brake Mechanism: The mechanism consists of four major parts. The hinge block is a 12.7 mm (0.5 inch) thick aluminum plate which is secured to the tip end of the wind turbine blade. One end of the hinge block serves as a hinge point for the brake plate. The hinge block also houses the damper bracket. A retraction spring and damper are fastened to the inside of the damper bracket. An electromagnet is used to provide a retaining force which holds the brake plate closed during normal wind turbine operation

b) Brake Plate: The brake Plate is a 6.4 mm (0.25 inch) thick aluminum plate. The 914 mm (36 inch) long plate is bent approximately 18° from straight. Both sides are equally tapered so that the width at one end is 47.6 cm (18.75 inches) and 22.9 cm (9.0 inch) at the opposite end .

c) Rotary Transformer: The rotary transformer is an air-gap transformer having two separate halves for the primary and secondary windings. Each half contains windings and magnetic laminations in an aluminum shell. The primary half is stationary while the secondary half rotates thus eliminating the need for slip rings to deliver power to energize the electromagnets in the tip brakes. The primary (stationary) side of the transformer was rigidly attached to the test stand. The secondary side was rigidly clamped to the rotating hub of the test rotor. A rectifier on the output of the secondary winding provided 12 V dc supply to the electromagnets. Control of the electromagnets is obtained by switching the primary power supply to the the rotary transformer. The testing of the tip brake mechanism provided an operational test of the rotary transformer.

d) Blade Tip: The AOC 15-50 blade is constructed of a wood - epoxy composite. Two, 91 cm (36 inch) tip sections of the AOC 15 meter blade were modified and used as part of the test rotor. Modifications included installing studs in the 'root end' and adding extra wire for the microswitches used in the test. The 'tip end' was identical to that on the full 15 meter blade.

e) Test bed: The test bed used for the tip brake testing consisted of a 2.4 m (96 inch) diameter rotor, powered by a 20 Hp variable-speed drive. The test rotor consisted of a 51 mm (2 inch) by 152 mm (6 inch) by 6.4 mm (0.25 inch) thick rectangular steel tube 70 cm (24 inch) diameter which joined two, 91 cm (36 inch) long blade tip sections The 1.5° angle of attack at the blade tip and the 6° cone angle associated with the AOC 15-50 rotor design were reproduced on the test rotor. Tip brake mechanisms were attached to the blade section in the same manner as on the actual AOC 15-50 wind turbine. The digitally controlled variable speed drive could be programmed to ramp speed up or down as necessary. The AC drive motor was coupled with the rotor shaft using a pulley assembly with a 6:1 ratio.

f) A computer controlled data acquisition system utilizing LabTech Notebook Version 7.1.1. was used for automated machine control and data acquisition.

4 RESULTS

The AOC 15-50 tip brakes achieved 22,000 cycles plus with successful operation. On completion of the test the tip brake mechanisms were disassembled and examined. Prior and post test dimensions were measured. All mating surfaces received minimal wear with exception to the right hinge eye of Tip # 2. The inside diameter of the bushing increased from 0.376 inch to 0.380 inch. This relatively extreme wear was attributed to an error in the assembly process.

A black film of oil covered parts of the hinge block assembly and brake plate This oil film was also found

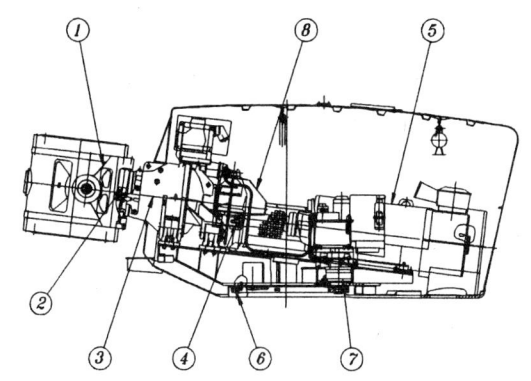

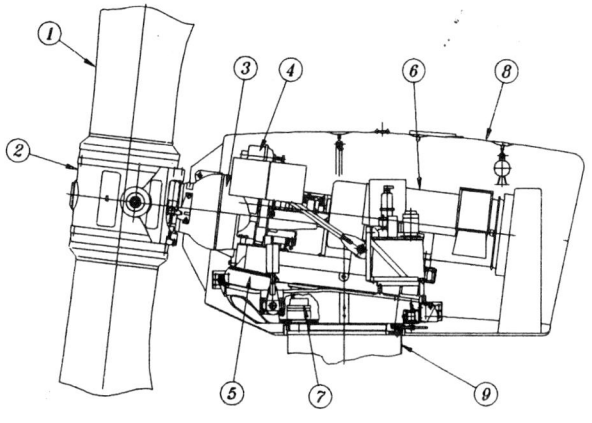

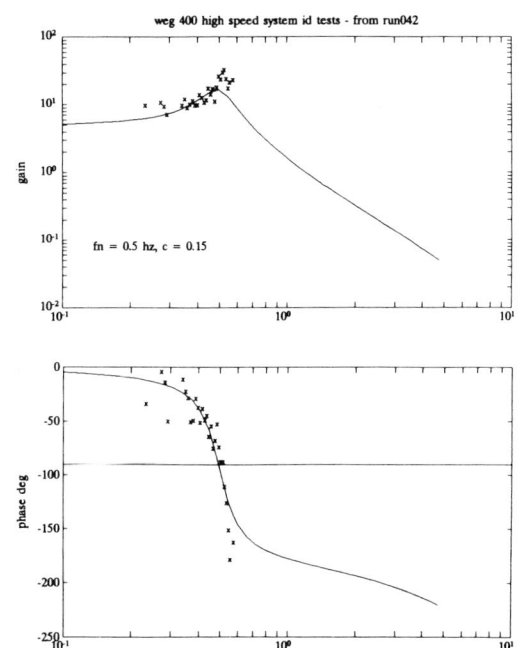

300 kW wind turbines erection in Romania

ANTON GĂRBACEA, PhD(Eng)
Electromontaj S. A. Romania

SYNOPSIS The present paper attempts to asses a part of the activity for erection and mounting of 300 kW wind turbines (W.T.), in Romania. These generators are located in the Banat mountains on an alpine plateau situated between the Semenic, Gozna and Nedeea peaks, at a height of over 1400 m. The realization program for these W.T. of the EOLTIM family (wind generators Timisoara) is mainly due to the activity of technical scientifical center Timisoara a city in the western part of the country.

1 INTRODUCTION

The present papers deal one aspect of wind energy in Romania. So I shall refer to achievements of construction and erection works for first demonstrative wind farm in western Romania , in the Banat mountains, contain only 300 kW wind turbines with horizontal axis.

The construction and erection program for first wind farm are located on an alpine plateau of 1446 meters altitude, between the Semenic and Gozna peaks, in the vicinity of the tourist resort bearing the same name. I think this location for wind farm, the Semenic area is solitary in Europe.

The construction program for first wind farm in Semenic area, has W.T. with 300 kW. power and horizontal axis. The building of first demonstrative farm provides of 10 units of 300 kW. each, total 3 MW.

In the future, a wind power plant will be built which will sum up 300 horizontal axis units, total 99 MW., contain about 20 units wind farms.

The wind farm of Semenic, as, a result of the fundamental and applicative research for the construction of these units, the EOLTIM family - "TIMISOARA WIND UNITS ". The obtained results are localized in technical - scientific center Timisoara (1).

Timisoara center comprises specialists from staffs, of the TECHNICAL UNIVERSITY (UTT), and Center Research Of Aeroenergetical Energy (CCAA), HIDROTIM - Designing Institute for hydromechanical equipments, IPROTIM - Designing Institute of the Timis District,

ELECTROMONTAJ S.A. Bucharest Group Site Banat - Timisoara.

The shop designs, have been made by HIDROTIM, for the mechanical and electric equipments, by IPROTIM, for the civil works - the supervision and command house and on part of foundation of wind turbine. UTT, designed the metal blade of OSPM 7 type. The construction and erection works complete, the part of fundamental research, for foundation and towers, technical constructive solution were carried out by ELECTROMONTAJ Group Site Banat Timisoara (4), (8), (11). The equipment was fabricated at the Steel Construction Works Bocsa.

2 CONSTRUCTION AND ERECTION WORKS

The achievements up to the present, for the 300 kW. W.T., form the EOLTIM family, mark the obtaining of fourth technical generation of wind units. This technical evolution brought about a series of technical and constructive improvements, reaching the necessary maturity for the application of industrial series production.

The technical generations are the following:

I st   technical generation - EOLTIM 1
II nd   technical generation - EOLTIM 2
III th technical generation - EOLTIM 3
IVth   technical generation - EOLTIM 4.

EOLTIM 1 -financed by Technical University Timisoara - CCAA. A photographic view of it is given in Figure 1. The W.T. is in operation since 1990 and is an experimental model of UTT. The W.T. is connected to the national energy system, delivering electric energy of 20 KV., through underground cables witch connect EOLTIM 1 to the transformation station of the Semenic resort (3).

EOLTIM 2 - financed by RENEL Bucharest - ROMANIAN POWER AUTHORITY through the Electric Network Subsidiary in Resita. The works began in 1989 and the commissioning works will be finished in August 1993. Activity to continue the works for erection and connection EOLTIM 2 to the energy national systems. Fig 2 shows a photo in 1992 year of this wind turbine, and Fig 3, erection the line machine with crane.

EOLTIM 3 - financed by the Technical University Timisoara. The works will be finished in 1993. This wind unit is the first unit of the 3 MW demonstrative wind farm,

witch will be built on the Semenic. Fig 5 shows the tower EOLTIM 3, where one can see the supervision and command house, was excluded. (7), (8). This W.T. are the series prototype.

EOLTIM 4 - financed by HIDROTIM Timisoara. The works began in 1991 and will be finished in 1993. The unit is the second W.T. within 3 MW. first wind demonstrative farm. The unit is fabricated also by Steel Construction Works Bocsa. This W.T. of 300 KW is the first unit for series erection in the wind farm (10).

I want to speak about some of its components and constructive solutions, from the point of view of the technical solutions adopted for this W.T. EOLTIM .

Fig 1 W.T. EOLTIM 1

Fig 2 W.T. EOLTIM 2

Fig 3 Erection of line machine for EOLTIM 2

354

2.1 Foundation

In Table 1, the "Technical characteristics of the foundation" the constructive elements of the fourth generations are presented.

I begin with EOLTIM 2 foundation, financed by RENEL Bucharest based on design of ISPH Bucharest (Institute for Hydroenergetic Designing) from RENEL Bucharest. The excavations totalized 1590 cm, in a circular perimeter with 20 m diameter and 4.5 m deep. The volume of concrete, incorporated in the foundation was 790 cm + 35 tons of OL 52 reinforcing steel.

The solution adopted by RENEL, couldn't be accepted by Electromontaj Timisoara, from point of view of environmental protection, and very high costs. A number of 15 concrete mixer lorries, with a capacity of 3.2 c.m., transported 800 c.m.,concrete for continuous casting, from Resita to Semenic. In case that this solution is to be used for the 90 MW wind power station, an excavation of 477.000 c.m. soil and hard rock (2); (3); (4); (7) will be needed. The extent of ruined alpine pastures is enlightening without comments.

The solution was abandoned, and the one of EOLTIM 3 and 4 foundation, satisfying the technical, environmental and economical exigencies, was applied . One photo with their new foundation it is in Fig 4.

The EOLTIM 4 foundation, result of basic and applicative researches takes advantage of the geological structure, in majority rock, of the alpine plateau. The foundation is anchored with 12 rods, located in borings of $\varnothing$ 115 mm and each stressed at an equivalent of 90 tons. So an anchorage at 1080 tons (5); (6) adequate to the wind generator's stress, is obtained.

The foundation, having the characteristics from Table 1, hasa high technicity level, but economic it is realised with only 48% of the cost for foundation model EOLTIM 2. The incorporated quantities are small for exemple at EOLTIM 3 - 98 c.m. concrete, for an excavation of 82 c.m. The execution speed of foundations is good in the case of serial production of wind generators, (11).

Table 1   Technical characteristics of the foundation

Generator type specification	EOLTIM 1	EOLTIM 2	EOLTIM 3	EOLTIM 4
Wind generator's total weight (tons)	121	92	60.58	56
Wind speed for foundation calculation (m/s)	80	80	65	65
Foundation				
Soil excavated for foundation (c.m)	315	1590	82	60
-Concrete (c.m)	178	790	98	74
-Reinforcing steel (tons)	14	35	6	6
-Metallic structure (tons)	6	6	15.5	15.5
-Type of construction	Four legs disposed at 7.25 m between them	Cylinder $\varnothing$ 16 at basis $\varnothing$ 8 at surface	Foundation with 12 stressed anchor rods 1080 t total stress	Foundation with 12 stressed anchor rods 1080 t total stress

2.2 The Tower

The technical characteristics of Tower are presented in Table 2 "Technical characteristics of tower".

The Semenic site, situated at 1446 m height, rises a series of technical problems, connected to the desighn, fabrication, transport from Resita to Semenic. These impose the execution of tower with a length of 30 m, in sections which could be transported easely and cheap to the site (7).

By the erection of tower, Electromontaj has serious technical difficulties, for model EOLTIM 2 (5); (6). The technical condition welding joint first class weld, was realised in open air at Semenic. A special welding technology was needed and extra expenses resulted (6); (7) in the application of this technology with the exigencies of

355

ultrasonic control of welding seam and contour.

The horizontality of the 4 sections at the welding jointsd was realised with a new device for determination of horizontality, wich uses magnetic liquid; clinometer D.B.D.O. The accuracy of this electronic device is of the order of seconds, indicating simultaneously the deviations from the horizontality on two axes "X" and "Y". The device was realised by Electromontaj UTT and AEM Timisoara (Enterprise for Measure and Control Apparatus). (12)

The constructive solution used now at suggestion of Electromontaj Timisoara, is adoppted for EOLTIM 3,4 and serial production. It is a cylindrical tower with variable sections presented in Fig 5 (the photografy of the tower erected at Semenic).

Table 2   Technical characteristics of the tower

W.T. type -Technical specification	EOLTIM 1	EOLTIM 2	EOLTIM 3	EOLTIM 4
Wind speed for tower calculation (m/s)	80	80	65	65
Total weight of tower (tons)	70	59	37	31
Tower -Number of parts -Construction steel 52 4k -Height (m) -Type of construction	7 30 4 legs of H=+10 m -tubular ∅ 2.2 m joined with bolts and nuts	4 30 Cone trunk H= 12.5 m at base -tubular ∅= 2.2 m joined with double bevel butt weld I class	3 30 Tubular ∅= 3 m ∅= 2.2 m ∅= 1.56 m joined with interior flanges and bolts	3 29 Tubular ∅= 3 m ∅= 2.2 m ∅= 1.56 m joined with interior flanges and bolts

Fig 4 Foundation of EOLTIM 3 and 4

Fig 5 The tower erected at semenic for EOLTIM 3

2.4 Machine Line

The gear-train diagram , used for the three generations of 300 kW horizontal axis W.T. is given in Fig 6.

The first constructive solution is used for 300 kW W.T. of the first WT - EOLTIM 1. This is an experimental model, witch, due to its construction is tested in operation and measurements are made on UTT and Hidrotim.

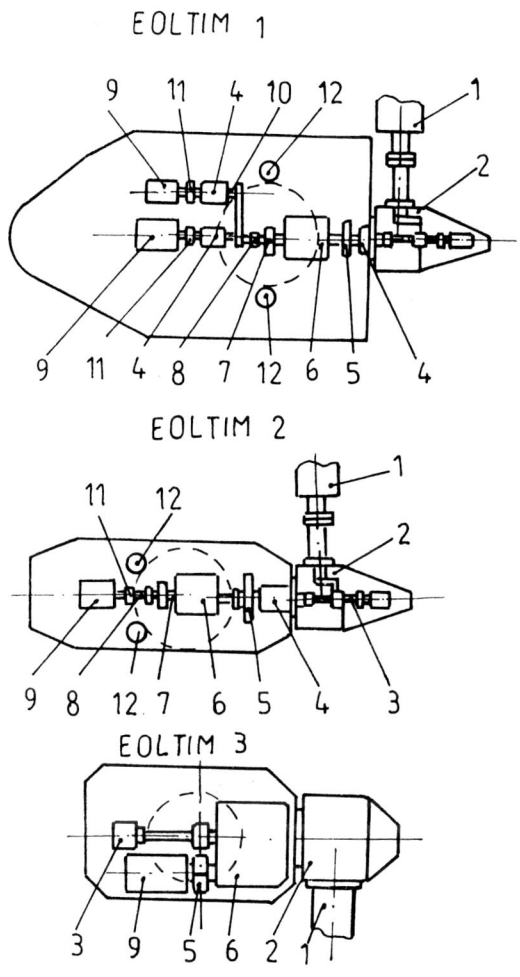

EOLTIM 1

EOLTIM 2

EOLTIM 3

1.Blade;2.Hub;3.Blade control mechanism;4.Bearing;5.Emergency brake;
6.Multiplier;7.Operational brake;8.Torsiometric coupling;9.Generator;10.Belt transmission;
11.Electromagnetic coupling;12.Yawing mechanism

Fig 6   Aggregates line machine components

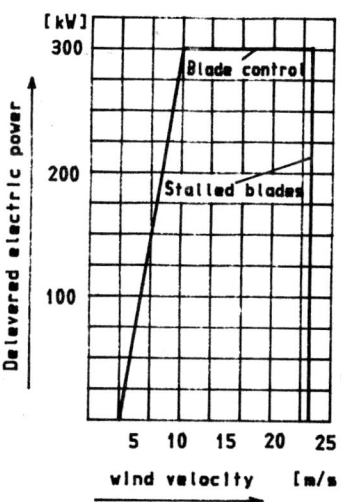

Fig 7   Power diagram wind turbines. 300 kW

The characteristics of this wind generator are: rotor diameter 30 m, speed 50 rpm, cylindric plantary gear box i=31, two synchronous generators (the first of 55 kW at 1000 rpm, the second of 275 kW at 1500 rpm), three blades OSPM 7 model, adjustable with an electromechanical mechanism. The operation range is at wind speeds 3.5 - 32 m/s; controlled by programmable controller.

The W.T. produces electrical energy, which delivered in the national energetical system, through 20 kV cables, from the own conversion station. The generator EOLTIM 3+4, constitutes the third and fourth technical generation and is an industrial unit for the production of electrical energy for the national sytem. The power diagram of this wind generator is presented in Fig 7.

The technical characteristics of EOLTIM 4 are: rotor diameter 30 m, speed 50 rpm, cylindric planetary gear box i=41, asynchronous generators of 315 kW at 1500 rpm, 3 blades adjustable with electromechanical mechanism; the blades are of steel covered with polyester, reinforced with glass fiber. The operation range is at wind speeds 3.5 - 25 m/s. Logic control system with electromagnetic relays, active yow sytem for nacelle having a speed of 0.31 degrees/sec

The improvements of the contructive solution for EOLTIM 4 are remarkable; EOLTIM 4 will be the wind turbine for first demonstrative wind farm of 3 MW on the Semenic plateau (9).

REFERENCES

(1) PROF. DR. ENG. F. GYULAY ; DR.ENG. ANTON GÂRBACEA ,Wind energy research in Romania. *News from Romania - Newsletter of the A.-British and European Wind Energy Association - LONDON,* SUMMER 1992, volume XII ,No.1, pages 26-27.

(2). DR.ENG. ANTON GÂRBACEA , Environment protection by building wind turbines achievements on semenic mountains. *Symposium and demonstration of the use of wind energy in Romania - Timisoara - 6 May 1992 vol. I, pages 33-37.*

(3) DR.ENG. ANTON GÂRBACEA, Aspects of wind energy in Romania. *7 th International "POWER SYSTEMS CONFERENCE" Teheran 7-9 November 1992, pages 505-516.*

(4) DR.ENG. ANTON GÂRBACEA, Construction of 300 kW wind units in Romania. *Symposium and demonstration of the use of wind energy in Romania - Timisoara - 3-6 May 1992 vol. II pages 49-54.*

(5) DR.ENG. ANTON GÂRBACEA, Results of 300 kW wind turbines erection and mounting in the Semenic mountain. Romania Energy Conference - CNE '92, Neptun 15-18 June 1992, Section III, page 59.

(6). DR.ENG. ANTON GÂRBACEA , Erection 300 kW wind turbine in Romania. *WIND POWER '92 - American Wind Energy Association 22- nd National Conference, Seattle - Washington 19-23 October 1992.*

(7). DR.ENG. ANTON GÂRBACEA , Appreciation over constructive solutions for wind turbines of 300 kW in Romania. *WIND POWER' 93 23rd annual conference American Wind Energy Association,* San Francisco12-16 July 1993 .

(8). DR.ENG. ANTON GÂRBACEA, Assessments of constructive solutions for 300 kW wind generator in Semenic area. *Power system conference PSC-93* Teheran,6-8 November 1993

(9). DR.ENG. ANTONIU ANGHEL, ENG. CEZAR GHIȚULESCU, Development of the conception and fabrication of horizontal axis wind generators in Romania. *Symposium and demonstration of the use of wind energy in Romania,* Timisoara - 3-6 May 1992 pages 3-14.

(10). DR.ENG. ANTON GÂRBACEA, Wind farm construction in the Banat Mountains. *HARMONY WITH NATURE - ISES Solar world Congress,* August 23-27. 1993 Budapest

(11). DR.ENG. ANTON GÂRBACEA, New solutions adopted by Electromontaj for erection 300 kW wind turbines in Semenic area. *Romanian Wind Energy Associations,* Buletin No.3 Timisoara 1992, pages 16-18.

(12. DR.ENG. ANTON GÂRBACEA, Device for obtain horinzontality with magnetic liquid clinometer type D.B.D.O.. *Note Book Electromontaj Group Site Banat Timisoara -A.E.M. Timisoara,* Timisoara 1991, pages 1-12.

Harmonisation of certification approaches

J. R. MAGUIRE BSc, PhD, CEng, MICE, MIStruct
Lloyd's Register Industrial Division, UK

SYNOPSIS This paper summarises Lloyd's Register's current thinking in relation to wind turbine certification. Included is a look ahead to how the certification approaches of a number of EC certification bodies could be harmonised, in line with the EC approach to conformity assessment and emerging International Standards.

1 INTRODUCTION

Over the last decade Lloyd's Register (LR) has been involved in a number of wind turbine projects (Ref.1) and over the last 2 years has been developing a set of rules for classification and certification. However, during these last 2 years the wind turbine market has evolved rapidly on both a European and international level, particularly with regard to the development of standards.

The following sections describe how LR's current thinking has developed over the last two years, and how this could fit in with a harmonised approach.

There are a number of benefits to harmonisation of EC certification (conformity assessment) approaches, including:

- Removal of internal barriers to trade
- Reduction of multiple certification
- Reduction of cost of certification
- Improved competitiveness in external market.

2 BASIC ASSUMPTIONS

2.1 Approach

LR have decided to set safety (and quality) standards for different stages of the wind turbine life cycle (Figure 1) and invite a "safety case" (Ref 2) to be put forward, detailing the safety (and quality) approaches adopted. At each stage LR will examine the approach, to ensure it meets the requested standard, and if acceptable will issue an appropriate certificate.

2.2 LR Certification/LR Classification

Both LR Certification and LR Classification are forms of certification, as illustrated in Figure 2. Classification can be seen to be a "cradle to grave" concept, covering all of the wind turbine life cycle, and importantly covering feedback

Figure 1 Idealised Wind Turbine Life Cycle and Relation to LR Certification

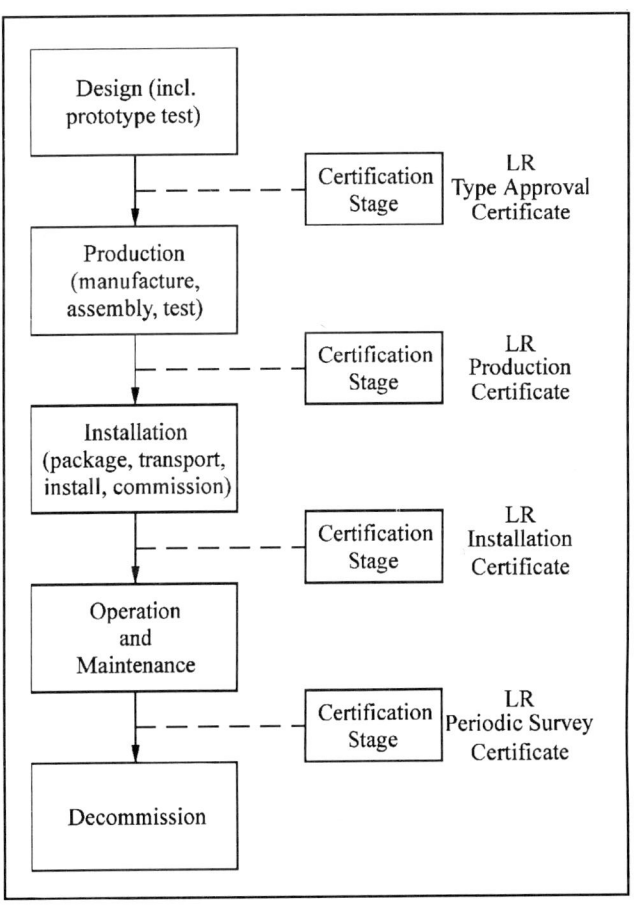

of in-use experience. Classification has traditionally (within LR) been carried out with reference to "in-house" (and national and international) standards, and has traditionally been of interest to all members of an industry (particularly insurers).

Figure 2 Relation of LR Certification to LR Classification for Wind Turbines

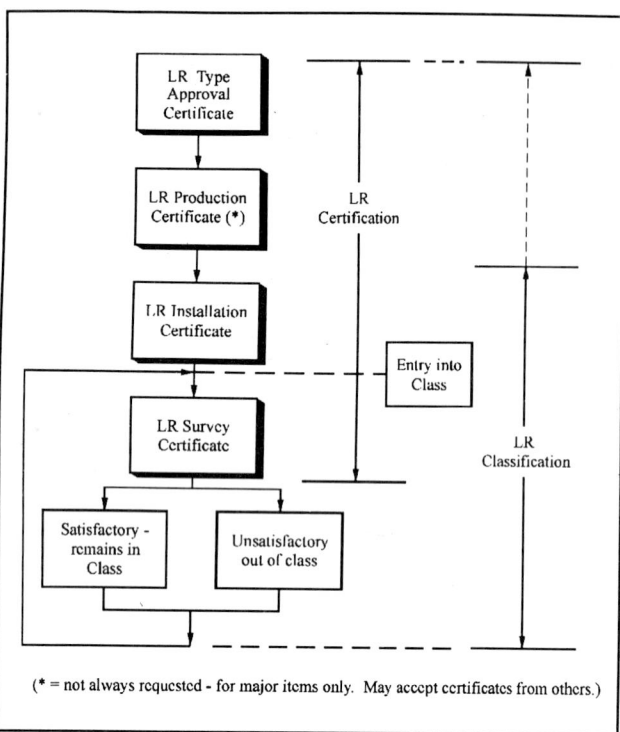

(* = not always requested - for major items only. May accept certificates from others.)

3 TERMINOLOGY

Certain "concepts" come up regularly in relation to wind turbine certification, but these concepts are labelled differently depending on whether one is British, European or international. The sections below detail some of the major concepts and their possible labels.

3.1 The Global Approach

The EC recognises a global approach to conformity assessment (Figure 3) which is modular in nature (Table 1). With respect to wind turbines this needs interpretation in light of the Machinery Safety Directive (Directive 89/392/EEC). This is the subject of much discussion in the UK and Europe at the time of writing (August 1993) and has yet to be finally resolved.

3.2 LR Type Approval

LR Type Approval (Ref 3) is broadly equivalent to EC Type Examination. A manufacturer will submit to LR technical documentation related to a particular type of wind turbine. LR will ascertain conformity with the essential requirements (see later), carrying out (or witnessing) type tests of the example wind turbine. If satisfactory, a LR Type Approval Certificate (EC Type Examination Certificate) will be issued. Figure 4 relates the wind turbine life cycle to conformity assessment.

3.3 LR Production & Installation Certificates

These certificates can be achieved in any of four ways - refer to Table 1. The four ways are:

(1) "Conformity to type" (product checks);
(2) "Production QA" (quality system checks);
(3) "Product QA" (quality system checks);
(4) "Verification of Products" (product checks).

The Machinery Safety Directive does not yet indicate which of these 4 routes should be chosen. Various European/international standards may make the choice. This is the subject of much discussion at the time of writing (August 1993) and has yet to be finally resolved, but in the final event the modules permitted will be restricted for any artefact.

Figure 3 European Conformity Assessment (Source DTI, April 1992)

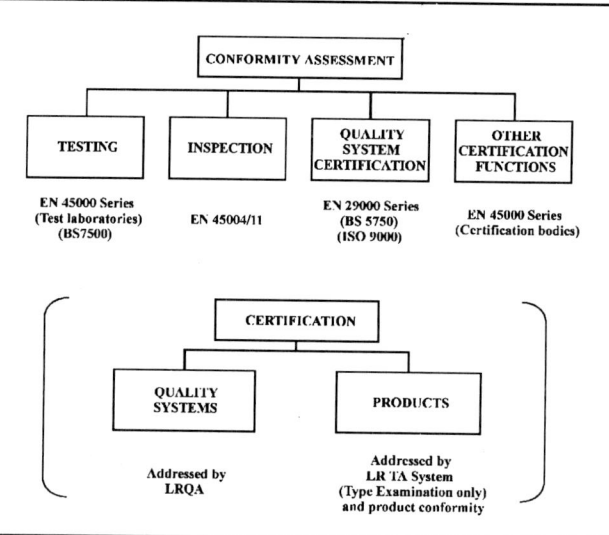

Figure 4 Relation between Life Cycle and Conformity Assessment for Wind Turbines

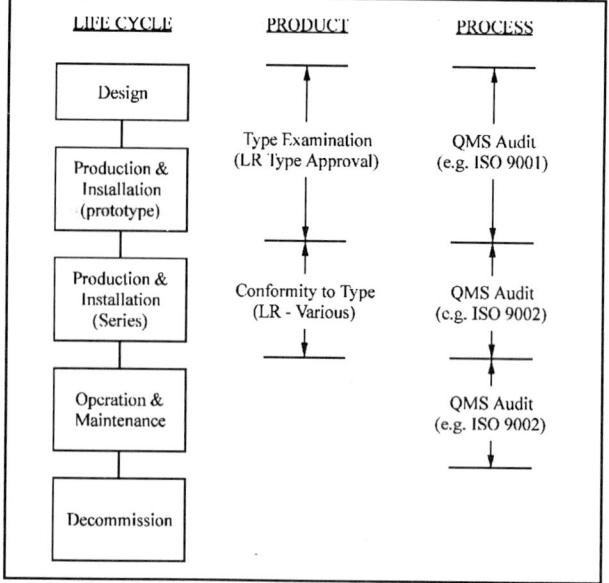

Table 1 - Conformity Assessment Procedure in EC Legislation

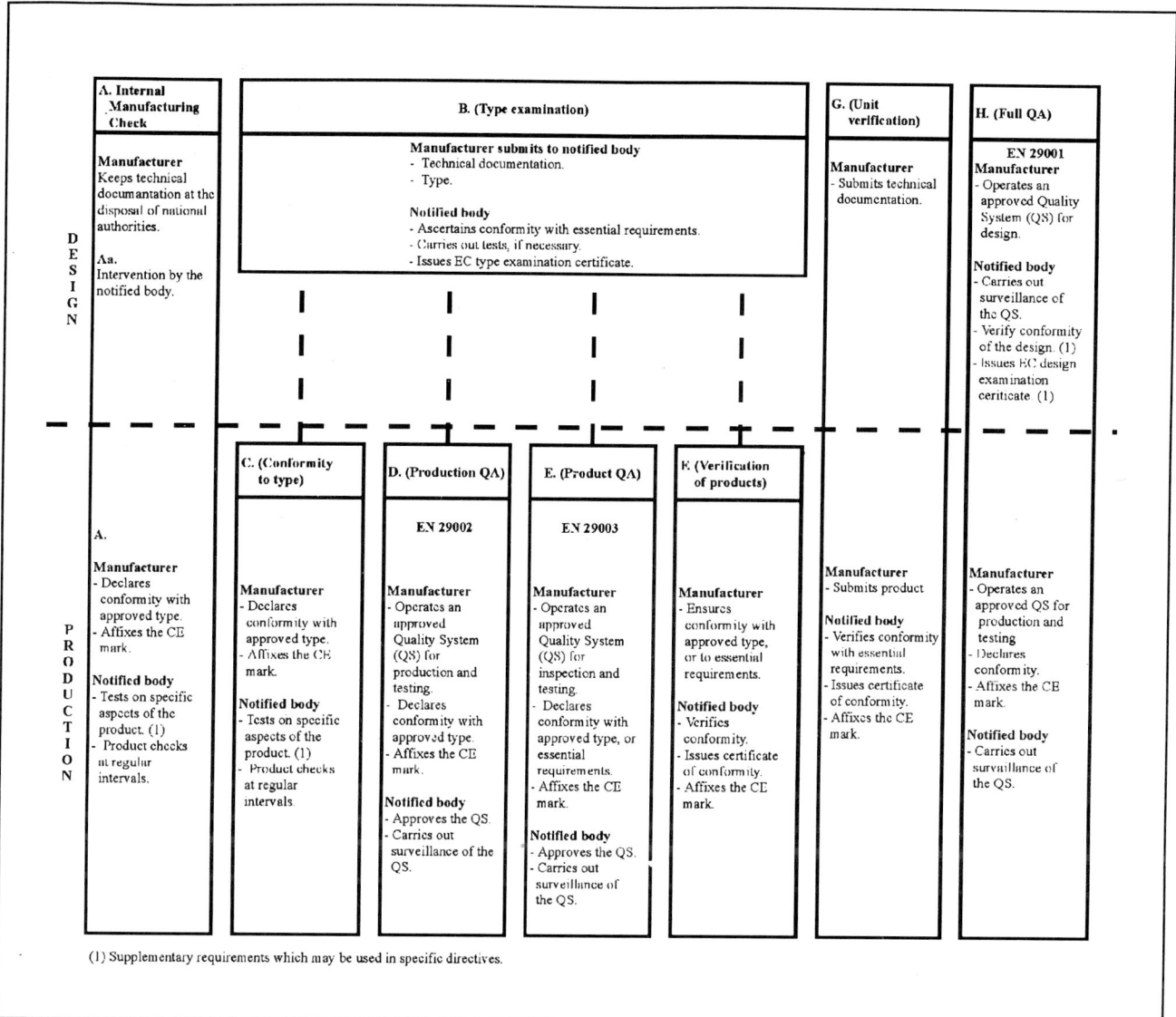

(1) Supplementary requirements which may be used in specific directives.

4 KEY DOCUMENTS

There are a number of standards/certification documents which LR consider to be "key" as regards formulating the essential health and safety requirements (EHSR's) referred to in the Machinery Safety Directive. These are briefly reviewed below.

4.1 International Energy Agency (IEA) Recommended Practices

There are currently 8 practice notes in print, including No. 5 - Structural Safety and No. 8 - Glossary of Terms. A ninth is being written on anemometry. Two (No. 1 - Performance and No. 4 - Noise) are being rewritten. The IEA notes that standards can never be the same from country to country but that it is desirable that the framework for standards be as common as possible. As far as safety philosophy, safety systems, load cases and a safety factors the IEA suggests adopting either Dutch or a mixture

of Dutch/Danish practice. LR's current thinking is to broadly aim to follow Danish (certification) practice at present.

4.2 (EC) Machinery Safety Directive

This entered into force on 31 December 1992. It does not explicitly reference wind turbines yet but the DTI have suggested this is likely to be the most relevant directive. Under the directive most machines must:

(1) satisfy the EHSR's (assume IEA/Danish);

(2) in some cases be type-examined depending on supporting standards;

(3) carry a CE mark and information (not done at present).

The manufacturer must draw up and keep a technical file. The notified body (LR) may need to be involved in the process. In view of the relevance of the EC documents to a large number of machines installed (and planned) in Europe it is thought that LR should take on board the principles of this directive as closely as possible.

4.3 (EC) Proposal for harmonisation - 1991

With a desire to create a common European certification procedure a group of 4 wind turbine test stations (CRES-Greece, ECN-Netherlands, NEL-UK and RISO - Denmark) proposed a harmonisation of procedures and technical regulations in Feb. 1991. The proposal was based around Dutch/Danish certification systems but has never been taken up due to lack of EC funds. It is understood that the Danes, with UK support, put forward their certification system as a EC role model. LR's current thinking is to align broadly with this.

4.4 (Joule II) Proposal for harmonisation - 1993

With a desire similar to that outlined in 4.3, a consortium was formed in 1993 to harmonise the approaches of EC certification bodies involved in the wind energy industry, in line with the EC approach to conformity assessment (3.3 above) and emerging international standards (4.5 below). The consortium comprises Lloyd's Register, Germanischer Lloyd, CIWI and RISO, assisted by a technical specialist (Garrad Hassan). EC funding, under the Joule II project, has been sought and a response is awaited at the time of writing (August 1993).

4.5 International Electrotechnical Commission (IEC)- Safety Standard TC88

The IEC have produced a draft standard on safety of wind turbine systems, which went for third draft discussion in Rotterdam in October 1992 and was technically completed in February 1993. Originally envisaged for completion in 1995 it is now hoped that this draft will be finalised substantially earlier (1993). This standard on safety deals with:

(1) safety philosophy;
(2) QA;
(3) engineering integrity;

The draft covers design, installation, maintenance and operation, with QA being integral at all stages.

There has been a close liaison between the IEC, CENELEC and national (eg. BSI) standard bodies, resulting in very similar technical contents in national, European and international standards. As the IEC draft is the "highest level" standard LR's current thinking is to align with this as closely as possible.

4.6 (Danish) Technical Basis for TA & Certification of Windmills

This document was published by the Danish National Energy Board (DNEB) in May 1991. The DNEB approval system, self-administered, covers type approval, certification of QC systems for production (to EN29002), and certification of QC systems for installation (to EN29002). The technical basis for the system is split into:

(1) principles of approval;
(2) procedures for applicants;
(3) technical requirements.

It assumes the existence of uniform standards, specifications and other technical rules with respect to design, production and operation of wind power installations. Where standards do not exist technical requirements are to be drawn up.

It is interesting to note that the Danish system emulates on the LR "class" philosophy i.e. the Danish system envisages ongoing maintenance of Type Approval by ongoing submissions of information. (This ongoing feedback of information is also a feature of the Dutch certification system).

Comparing the Danish system with LR's system the following broad evolution is under way:

(1) LR has developed a document entitled "Conformity Assessment of Wind Turbines" (Ref 4) which embraces/explains the LR Type Approval/QA/QC system, and interprets it for wind turbines. This emulates the Danish "Technical Basis" document, and includes principles of approval and procedures for applicants.

In turn, the "Conformity Assessment" document refers to another supporting document entitled "Interpretation of LR's Type Approval System for Wind Turbines" (Ref 5). This can be used for "stand-alone" Type Approval.

(2) LR needs to further develop/modify its provisional rules for certification into a document similar to the Danish "Technical Requirements". This should call up all the relevant standards by reference, and only aim to "plug missing gaps" where necessary. It should therefore be a relatively short document compared to the current provisional rules.

(3) The "Conformity Assessment" document has been structured to be country independent (somewhat similar to EC Eurocodes). In this way the LR system can become applicable world-wide, with only local conditions (e.g. local environment) to be inserted as appropriate.

5 CONCLUSIONS

A significant change in LR thinking has taken place over the last 2 years. LR's provisional rules have been changed and restructured in line with current (and emerging) European and international standards. The (proven) LR Type Approval system is now forming the focus for dealing with standards and other technical requirements, as part of an integrated approach to conformity assessment.

It is now clear that LR must more actively support and align itself with co-operative international standards activity. It is also clear that all certification bodies involved with wind turbines should now harmonise their (broadly similar) certification and classification approaches, for the benefit of this industry.

REFERENCES

(1) SURPLUS, D.C., An Introduction to Modern Wind Energy Technology, Lloyd's Register Technical Association, Session 1991-92, Paper No 3.

(2) Professional Brief, Safety-Related Systems, The Institution of Electrical Engineers, Sept. 1992.

(3) Lloyd's Register Type Approval System, 'Procedure', 1990.

(4) Lloyd's Register Industrial Division, Report CIV/9238/R001/jrm, 'A Conformity Assessment System for Wind Turbines'.

(5) Lloyd's Register Industrial Division, Report CIV/9235/R005/jrm, "Interpretation of LR Type Approval System for Wind Turbines".

Infra-red thermography for condition monitoring of composite wind turbine blades: feasibility studies using cyclic loading tests

G. M. SMITH and **B. R. CLAYTON**
University of Nottingham, UK
A. G. DUTTON and **A. D. IRVING**
Rutherford Appleton Laboratory, UK

SYNOPSIS

Infra-red thermography has the potential for providing full-field non-contacting techniques for the inspection of wind turbine blades. One method could exploit the temperature increases produced when blade materials experience dynamic mechanical loads. In order to quantify and understand the thermal behaviour, strip specimens of glass reinforced polyester were subjected to tensile cyclic loading under controlled conditions. Increasing the stress range, the frequency of loading and the specimen thickness all increased the equibrium temperature reached. The development of microstructural damage and its effect on heating effects is also reported: both stress concentrations and regions of damaged material gave rise to hot spots. For application to turbine blades, the sensitivity of the thermal imaging has been shown to be suitable for non-destructive examination during fatigue testing; furthermore, it is thought that for blades *in situ*, the wind loading conditions may be sufficient to create effects detectable by thermal imaging.

1 INTRODUCTION

Infra-red thermography methods have already been shown to have considerable potential for the non-destructive evaluation of wind turbine blades [1-3]. Techniques, described more fully elsewhere [4,5], can involve the external application of heat to the testpiece, whereby disturbances to the heat conduction path beneath the surface cause perturbations in the measured surface temperature field. Alternatively, surface temperature variations may be observed using heat generated within the testpiece material when it is subjected to dynamic loading. Work to date has included thermal imaging of both types and it has been applied to full-scale blades and to test coupons.

In general, thermography methods employ a scanning camera containing a detector which is sensitive to radiation of either 2 to 6 μm or 8 to 13 μm wavelength and thus suitable for measuring temperatures close to ambient. From the heat flux values the temperature of the surface is automatically derived provided the surface emissivity is known. Unlike metals, the polymer composites, gelcoats and paints which commonly form the surface layer of turbine blades all have a similar, very high emissivity making this conversion to temperatures straightforward. Moreover, thermal imaging can be performed without the need for any additional special surface paints.

A choice of lenses made from an appropriate glass can usually be used with the thermal camera so that images may be recorded in close-up or from a distance with corresponding differences in spatial resolution. Thus, thermal imaging provides the desirable possibility of examining testpieces from a distance and giving full-field information.

The great majority of wind turbine blades are manufactured from either glass fibre reinforced polymer or from laminated wood-epoxy and both these families of materials have particularly useful characteristics with regard to thermal imaging. They have relatively low thermal conductivities, typically two orders of magnitude less than ferritic steels, which results in any temperature variations tending to dwell and to remain localised. For example, after the loading was switched off, a hot spot detected in a full scale blade undergoing fatigue testing was found to cool with a time constant of 14 minutes [3]. Furthermore polymer composites and woods both produce pronounced cumulative heating when given dynamic loading. This heating effect is attributed primarily to their viscoelastic behaviour, arising from their long-chain molecular structure. The net absorption of energy manifests itself as hysteresis in the stress-strain loop. Often this heating is considered an inconvenience because it limits the frequency at which the fatigue testing of polymers can be conducted. However the heating effects can be harnessed for non-destructive examination since they will vary in a testpiece where stresses and structures are not completely uniform.

It might be expected that heating effects during cyclic loading might also arise from the creation of microstructural changes which are known to occur as a function of loading level and number of cycles [6]. Additional heating effects might be produced just in the presence of non-developing microstructural damage, arising from stress redistributions on a structural level or friction between newly created surfaces.

In this paper, investigations are described of the behaviour of strip specimens of glass reinforced polyester under tensile cyclic loading using systematically varied loading conditions. The aim was to gain a better understanding of the relationship between applied stresses and frequencies, specimen thickness, and the heating effects produced. Interrelationships between the applied loading, the microstructural changes within the composite and the heating effects were also investigated. The overall findings are related to the conditions of loading likely to be experienced by turbine blades, and the feasibility is then considered of using thermal imaging techniques as tools for the non-destructive evaluation of blades.

2 EXPERIMENTAL METHOD

All the specimens were of glass reinforced polyester, produced in the laboratory in the form of plaques approximately 550 mm square. Rather than imitating the complex layup of production blades which have changes in fibre orientation at different positions in the wall thickness, a simplified and more homogeneous design was chosen. Nominally equal volumes of fibres parallel and transverse to the specimen length were obtained using 0°/90° stitch-bonded glass fabric, laid up with mirror symmetry about the mid-plane to promote fully axial loading. A lightweight 300 g/m^2 fabric was used to keep the inevitable structural variations to a small scale. Hand layup with polyester resin simulated a common method of production for full-scale turbine blades.

Plaques were cured, trimmed and cut into strip specimens of 25 mm x 250 mm. This parallel sided geometry was chosen to simplify subsequent considerations of heat fluxes and load paths. Two plaque thicknesses of approximately 3.2 mm and 10.2 mm were used. Strip thicknesses varied by up to 0.2 mm due to slight surface undulations but the original surfaces were retained so as not to reduce the surface integrity by exposing glass fibres.

For the cyclic loading, a servo-hydraulic Instron system was used in a laboratory with air temperature controlled at 20$\pm$1 °C. The thicker specimens were gripped directly; Tufnol tabs were used on the thinner specimens, some of which were taken to higher stresses. Loading was in tension-tension using a sinusoidal waveform, and an R ratio (minimum/maximum load) of 0.1. Load ranges (maximum - minimum load) of up to 20 kN were employed for the main matrices of test conditions, corresponding to an applied stress range of around 80 MPa for the thicker specimens. This compares with a tensile strength of 250 to 300 MPa for the material and corresponds to a strain range of around 0.5% (5000 microstrain). Loading frequencies of 2, 4 or 8 Hz were applied for most tests with work concentrating on 4 Hz. Combinations of stress and frequency were selected to produce measurable but not excessively high temperatures when material properties would be expected to change markedly and shearing in the grips occurred. In practice bulk temperature rises, presented in Figures 1 to 4, were kept mostly to less than 12 °C.

Thermal images were stored at programmed time intervals during the cyclic loading using an Agema Thermovision 880 system which is capable of distinguishing surface temperature variations of less than 0.1 °C. For each test, a control specimen of similar thickness was fixed adjacent to the specimen under load, to provide a reference temperature during imaging and therefore allow compensation for local changes in laboratory temperature. The system software enables image displays to be subsequently re-scaled and for images or parts of images to be subtracted from one another. Additional software was developed to calculate and plot the average temperature of selected areas as a function of time. The temperature rise of an area of the test specimen compared at any one time with that of the control specimen was also derived using this software.

3 EFFECT OF LOAD RANGE AND FREQUENCY

With the application of cyclic loading, two types of heating effect are observed. One is a temperature fluctuation pulsing in time with the loading frequency, with a temperature range of tenths or hundreths of a degree. This arises from the thermoelastic response found in all materials and its magnitude depends on the sum of the principal strains [7]. The effect is essentially reversible and does not therefore result in the buildup of heat.

The effect of interest in the work reported here, and which dominates, is the heat accumulation occurring in this material by other mechanisms. To smooth out the temperature pulsing, the progress of the accumulated heating was calculated from average temperature values over 10 frames collected with the scanning camera. As time progressed from the start of the cyclic loading, the surface temperature of the specimens gradually rose to approach an equilibrium value after about one hour (3600s) as presented in Figures 1 to 4 for the two thicknesses of specimen.

As the load range and the frequency increase, the heating effect increases. Comparing data in Figures 1 and 2, or 3 and 4, it is seen that doubling the frequency or increasing the stress range by a factor of $\sqrt{2}$ have a similar effect on the temperature reached, as found for example by Dally & Broutman [8] in a similar experiment. In simple terms this is not unexpected since energy absorbed will be proportional to the <u>area</u> inside the stress-strain hysteresis loop, and to the number of times the loop is traced out in unit time. Thus, for the thicker specimens after 30 minutes of loading, a specimen experiencing 14.15 kN (56 MPa) range at 8 Hz gave a similar 11 or 12°C rise to one experiencing 20 kN (79 MPa) at 4 Hz; and one experiencing 14.15 kN (56 MPa) range at 2 Hz gave a similar 2.5°C rise to one loaded with 10 kN (40 MPa) range at 4 Hz.

Changing the mean load whilst maintaining the same load range was found to have little effect on the heat buildup, which would be expected unless the maximum load was raised so much as to alter the microstructure significantly.

4 EFFECT OF SPECIMEN THICKNESS

For the same applied stress range and frequency, the 10.2 mm specimens reached a higher equilibrium temperature than the 3.2 mm specimens. For example, a thick specimen given a stress range of 79 MPa at 4 Hz reached 11°C above ambient after 30 minutes whereas a thin specimen experiencing a similar stress range at the same frequency reached less than 2°C temperature rise.

For a uniform stress distribution throughout the specimen, a uniform rate of heat generation per unit volume would be anticipated. The total heat generated will therefore be approximately three times greater for the thicker specimens. In contrast, the paths for heat loss will be similar since the thicker specimens have only 27% greater surface area. These losses will occur from the exposed surfaces and by conduction through the grips. Higher equilibrium temperatures for the thicker specimens are therefore consistent. Work is in progress to model these thermal fluxes.

5 EFFECT OF MICROSTRUCTURE

In its virgin state, the glass reinforced polyester material is translucent with only the zigzag stitching and occasional voids visible in the microstructure (Figure 5a). The glass fibres cannot be made out because they have similar optical properties to the matrix, and the interface bonding is intact. At low levels of stress, typically half that of the macroscopic yield stress, irreversible damage appears in the structure when it is subjected to either monotonic or cyclic loading. This may be accompanied by audible clicks. The nature of this damage changes with increasing load in a tensile test as charted, for example, by Owen [6]. Initially debonding occurs between fibres lying transverse to the loading direction and the resin matrix. Microcracks then initiate from the debond areas and gradually propagate through the matrix. With higher strains, debonding of fibres parallel to the loading direction occurs. A general intensification of damage is produced with further loading until eventually fibre breakage starts. This development of microstructural damage occurs throughout the stressed volume of the composite. Figure 5b shows the characteristic opacity of a heavily damaged specimen. The debonding reveals the positions of the fibre bundles, and cracks in the resin can be made out.

Only as the ultimate strength is approached does damage development become localised, and for an unconstrained specimen final fracture occurs explosively with the creation of massive fibre breakage and pullout. If the stressed volume is contained, further damage may build up.

Macroscopically the changes in the microstructure produce a curved stress-strain curve, reflecting the reduction in the elastic modulus of the composite. The progression of damage development described above also occurs during fatigue testing with increasing number of cycles [6].

To obtain some idea of the degree of damage achievable in the strip specimens, a fatigue curve for the thinner specimens was derived using a 4 Hz applied frequency. A series of similar specimens was then given fatigue loading to 2×10^4 cycles at different load ranges for further comparison. The opacity of specimens and microcracking occurred as anticipated. To examine the effect of the presence rather than the creation of microstructural damage on the heating produced, thermal imaging of a heavily but uniformly damaged specimen was performed. A less intense loading regime was used compared with that for the original straining so as to minimise any further damage creation. It was also possible to then retain the virgin specimen used for comparison in a relatively undamaged state. Thus, whereas the heavily damaged specimen had previously experienced 4.7×10^4 cycles at 81 MPa stress range, the comparison trial was undertaken using 41 MPa. To maximise any differences in heating effects, the two specimens were taken from the thicker batch and a frequency of 8 Hz was used. Temperature rises of 3.3 and 3.5°C were recorded for the virgin and heavily damaged specimen respectively after the same time: however this 0.2°C difference is within the specimen-to-specimen variability observed.

Specimens containing even heavier damage were therefore examined. In particular, specimens were selected containing some parts more heavily damaged than others. Figure 6 shows the line of compression damage created under the central 25 mm diameter steel roller during a three point bending test. Figures 7a and 7b, with the same temperature scale, show the thermal images produced when the specimen was given 8 Hz tensile cyclic loading with a 41 MPa stress range. A temperature rise along the damaged line appeared from the start of loading, reaching a magnitude of 1°C above the rest of the specimen after just 2.5 minutes. As time progressed and the whole specimen became warmed, the hotter band persisted but developed a more diffuse and circular profile as a result of greater losses towards the edges of the specimen.

Thermal images of specimens containing damage from circular or spherical indentors are reported elsewhere [9]. Different types of damage were created by dynamic impact and static compression loading. Each produced changes in the surface profile with the conical indentor penetrating further than the spherical, and the dynamic impact producing a greater degree of subsurface fracture of the material. Hot spots which appeared during cyclic loading were attributed to both the stress concentrating effects of the changes in profile and the microstructural damage.

6 DISCUSSION

For thermal imaging to be a viable method for examining wind turbine blades the heating effects being measured need to be detectable at the surface, whether the temperature differences are produced by variable loading or by external heaters. Temperature rises as much as 2°C have already been measured at hot spots in wood laminate blades undergoing fatigue testing, and this under the fluctuating ambient conditions of a testing laboratory [3]. The hot spots were associated with joints in the wood. In one blade failure ultimately occurred from this position and there was an attendant temperature increase. Thermal imaging of cyclic loading tests on wood laminate coupons have been reported elsewhere [10]: they give more evidence of the conditions under which temperature rises in blades might be expected.

Likewise, in the tests reported here, the results can be related to the behaviour anticipated in glass reinforced polyester blades. In the glass reinforced polyester specimens with no intentional defects or stress concentrations, a typical result was to produce a 2°C temperature rise after ten minutes of cyclic loading at 2 Hz with a stress range of 56 MPa, equivalent to a strain range of around 0.3% (3000 microstrain). In comparison, glass reinforced polyester turbine blades undergoing fatigue testing may see a loading frequency of 1 Hz or less, but with stress ranges of perhaps 60 MPa. Since the heating effect is related to the square of the stress range, this implies that overall temperature rises of a degree or so might be expected during a blade test but, more importantly, that where there are local stress concentrations detectable hot spots will be produced.

Other blades are tested by applying a spectrum of loads and frequencies to reproduce more exactly the stochastic loading seen in service. Tests are accelerated by omitting low load cycles which are thought to contribute little to the fatigue damage accumulation. These blades too would be experiencing the level of loads and frequencies likely to produce measurable hot spots at stress concentrations and damage sites.

The generation of heat from obvious stress concentrations such as drilled holes has been previously reported [10] and the investigations reported here indicate that heating is also produced at regions of local microstructural damage. The work has also shown that the mechanisms of heat generation are difficult to separate. Regardless of this, it seems clear that when a blade is subjected to cyclic loading, local heating is produced at the very types of region which would be liable to initiate failure if beyond a critical severity. Furthermore, the blades are experiencing largely an overall bending moment, with compression on one external face and tension on the other. Maximum loading stresses are found at the surface, with a shallow stress gradient through each wall thickness. Features nearer the surface, and therefore the more detectable by thermal imaging, would be expected to be the more important.

During the fatigue testing of blades, thermal imaging would provide a very suitable method of inspection since it is carried out whilst the test is in progress and without any prior preparation of the blade being required. In view of the way damage accumulates throughout a fatigue test, sometimes accompanied by noise emission, thermal imaging would provide a means of locating more active areas rather than having to wait for the onset of visible surface features or the final failure itself.

The external heating method of thermal imaging can also play its part in the inspection of blades, providing complementary information. Compared with the applied loading method, the external heating method is more sensitive to planar defects parallel to the surface, such as delaminations. Little heating is produced at these under cyclic loading since the defects lie predominantly parallel to the loading plane. Both thermal methods should detect structural differences. A further feature of the external heating method is that the testpiece is stationary which can be an advantage.

Whichever non-destructive evaluation method is used, a good knowledge of the structural design, materials and manufacturing method of the blade is important. Glass fibre reinforced polymer blades are often made with a load-bearing spar and then a thinner outer shell to provide the aerofoil section. As well as the step changes of wall thickness at the boundaries of the spar, the wall thickness may be slightly greater around the leading edge and will increase towards the blade root. Features such as root fixings and glued joints will also create changes in the load path. Local features such as voids, resin-rich areas and fibre waviness are other possibilities. All may show up as indications during non-destructive examination and may or may not be of concern.

The materials used in blade construction are also various and will provide different thermal responses. Some blades include rigid expanded foams which are extremely poor heat conductors. Carbon fibre is also being explored either as a major material in the blade construction or as part of a mixed fibre layup. Since carbon fibre has very anisotropic thermal properties, with a conductivity along the fibre length which is 20 times that of glass fibre, this would be expected to cause the uneven dispersion of hot spots.

As a tool for the monitoring of wind turbine blades in service, thermal imaging also continues to show potential. Blades on some turbines, particularly in California and Denmark, have now been in service for ten or fifteen years. Weakening of the blades may be occurring through the effects of the weather, and they may have sustained damage from lightning strikes or impact which may not be visible at the surface. The integrity of newer blades also needs to be confirmed. Designs are continually being improved, and are less likely to be overconservative than previously. Problem areas, in terms of integrity prediction, tend to be in the vicinity of rapid changes in section or around features such as root fixings. A practical means of verifying calculated stress distributions would be useful.

The ability of thermal imaging to provide full field information from a distance, the sensitivity of the imaging systems to temperature differences of less than 1°C, and the persistence of hot spots in terms of minutes rather than seconds all play their part in making the techniques look attractive.

For elderly blades which had been removed from their machines, a combination of external heating and applied loading techniques could be used. For blades still *in situ* there may also be possibilities for using the techniques since it is thought that wind loading conditions might be sufficient to create measurable heating effects in the blades. Whether imaging could be carried out with the blade rotating still needs to be ascertained. Otherwise hot spots could be observed after the machine had been braked. Some blades on megawatt machines have inspection cradles which might be used for access. Alternatively, blades could be imaged from ground level.

7 CONCLUSIONS

* The heat generated in uniform strip specimens of glass reinforced polyester increases with increasing frequency of cyclic loading and range of applied stress. Higher equilibrium temperatures are reached in thicker specimens.

* During monotonic or fatigue loading, microstructural damage develops throughout the stressed region in the form of fibre debonding and matrix cracking. Damage localisation only occurs in strip specimens shortly before final fracture.

* Both localised microstructural damage and stress concentrations produce hot spots with cyclic loading. Absolute temperature values reached will depend on testpiece details and on external factors; however, more relevant are surface temperature differences.

* Thermal imaging trials on blades undergoing fatigue testing have shown that hot spots of up to 2°C temperature rise can be produced at locations of structural weakness. It has also been shown that the thermal imaging technique can detect much less severe effects, being capable of resolving hot spots of just 0.1 or 0.2°C warmer than the surrounding material.

* Thermal imaging is suitable for the examination of blades manufactured from both glass reinforced polyester and laminated wood.

* To use thermal imaging for the non-destructive evaluation of turbine blades it is essential to have prior information on the structural design, manufacturing methods and materials of the blade.

* Thermal imaging is suitable for locating damage and its development during the fatigue testing of turbine blades. Most information could be gained by the complementary use of external heating and applied loading methods.

* Thermal imaging could also be used for the assessment of the integrity of blades at any stage in their life. It is thought that wind loading conditions on a blade *in situ* may be sufficient to create measurable heating differences at stress concentrations.

ACKNOWLEDGEMENTS

Thanks are due to the manufacturers and others who have willingly provided information on the design of wind turbine blades and access to fatigue tests.

The authors are pleased to acknowledge that this work was supported by SERC grant GR/H15080.

REFERENCES

[1] CLAYTON B R, DUTTON A G, AFTAB N, BOND L, LIPMAN N H & IRVING A D
Development of structural condition monitoring techniques for composite wind turbine blades.
Proc. European Community Wind Energy Conf., Madrid Sept.1990, H S Stephens & Ass., pp 343-346

[2] AFTAB N, DUTTON A G, IRVING A D, LIPMAN N H, CLAYTON B R & BOND L J
Development of thermal condition monitoring techniques for composite wind turbine blades.
Wind Energy: Technology and Implementation
Proc. European Wind Energy Conference Amsterdam Oct.1991 (Eds. F J L van Hulle *et al.*), Elsevier 1991, ISBN 0-444-89117-X, pp 241-245

[3] DUTTON A G, IRVING A D, LIPMAN N H, CLAYTON B R, AFTAB N & BOND L J
Infra red condition monitoring of wind turbine blade fatigue tests.
Wind Energy Conversion 1992 Proc. 14th British Wind Energy Association conference, Nottingham Mar.1992, (Ed. B R Clayton), MEP 1992, ISBN 0-85298-827-3, pp 221-227

[4] SMITH G M, CLAYTON B R, DUTTON A G & IRVING A D
Thermal inspection of composite structures.
BWEA/DTI Workshop on Safety, Reliability and Condition Monitoring of Wind Turbines Nottingham Sep.1992 (Eds. G M Smith & B R Clayton), ETSU-N-124 Paper 6

[5] PUTTICK K E
Chapter 3 on Thermal NDT Methods, in *Non-destructive testing of fibre-reinforced plastics composites Vol.1* (Ed. J Summerscales), Elsevier 1987, ISBN 1-85166-093-3

[6] OWEN M J
Chapter 18 on Fatigue, in *Glass Reinforced Plastics* (Ed. B Parkyn), Iliffe, 1970

[7] STANLEY P & CHAN W K
The application of thermoelastic stress analysis techniques to composite materials.
J. Strain Analysis,23.3,1988,pp 137-143

[8] DALLY J W & BROUTMAN L J
Frequency effects in the fatigue of glass reinforced plastics.
J.Composite Materials,1,1967,pp 424-442

[9] M D WAKEMAN, SMITH G M & CLAYTON BR
Thermographic evaluation of static and dynamic damage to composite wind turbine blades.
Proc. 15th BWEA conference, York Oct.1993- *this volume*

[10] DUTTON A G, IRVING A D, SMITH G M & CLAYTON B R
Infra red thermography as a tool for identifying flaws and stress concentrations in composite wind turbine blades.
Proc. European Community Wind Energy Conference, Lübeck-Travemünde Mar.1993 (Eds. A D Garrad *et al.*) H S Stephens & Ass./CEC 1993, EUR-15083-EN, ISBN 0-9521452-0-0, pp 168-171

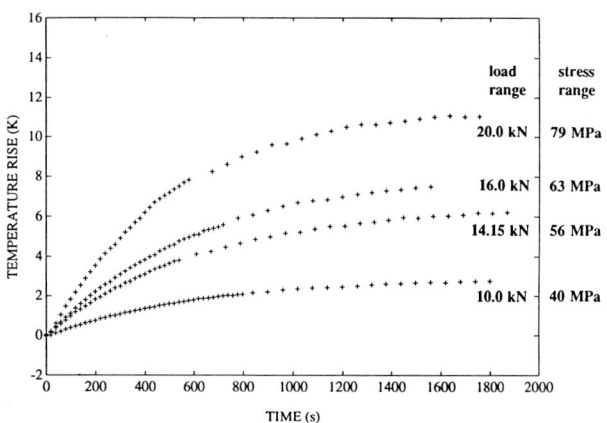

Fig 1 10.2 mm thick specimens, 4 Hz loading frequency, different stress ranges

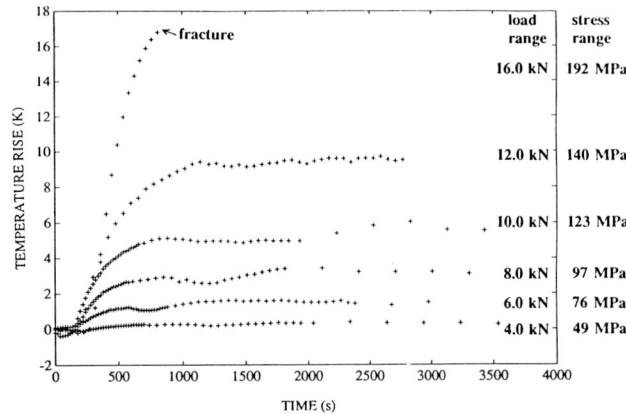

Fig 3 3.2 mm thick specimen, 4 Hz loading frequency, different stress ranges

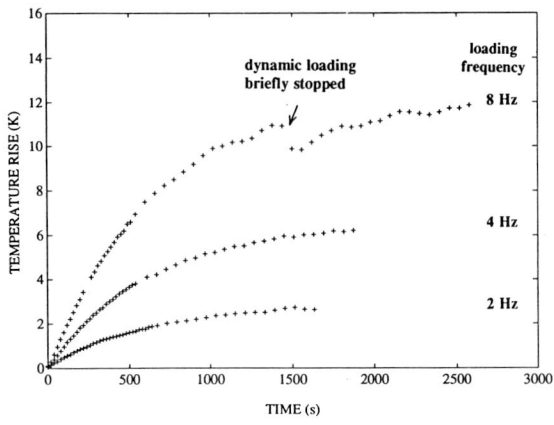

Fig 2 10.2 mm thick specimens, 56 MPa (14.15 kN) stress range, different loading frequencies

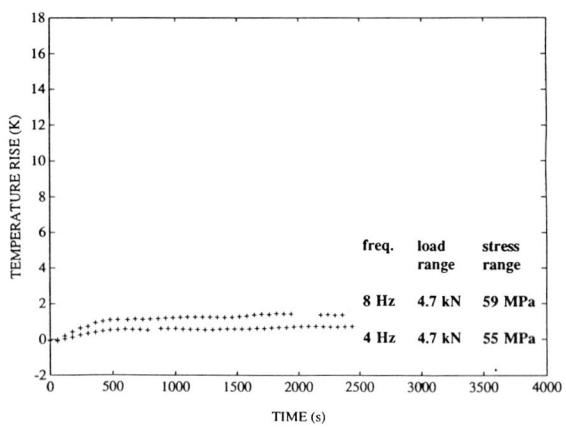

Fig 4 3.2 mm thick specimens, similar stress range, different loading frequencies

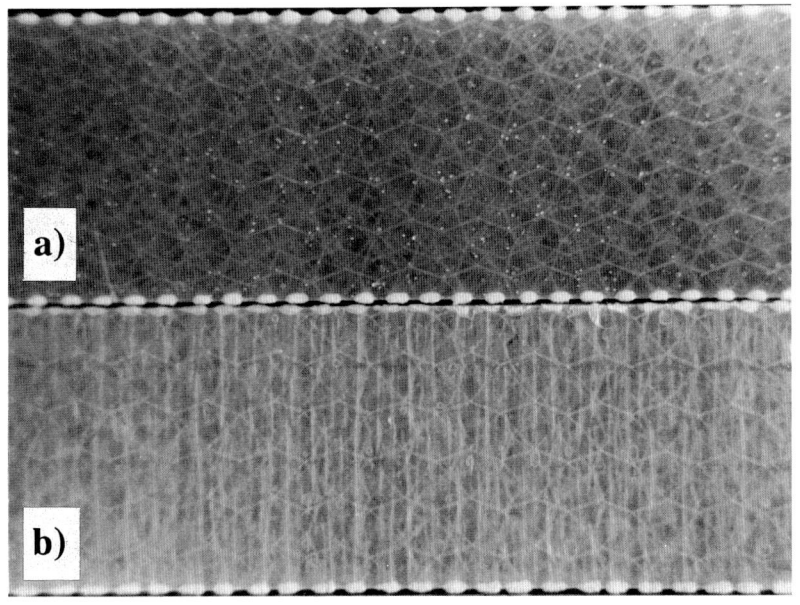

Fig 5   Glass reinforced polyester material
a) in virgin condition;   b) in heavily deformed condition

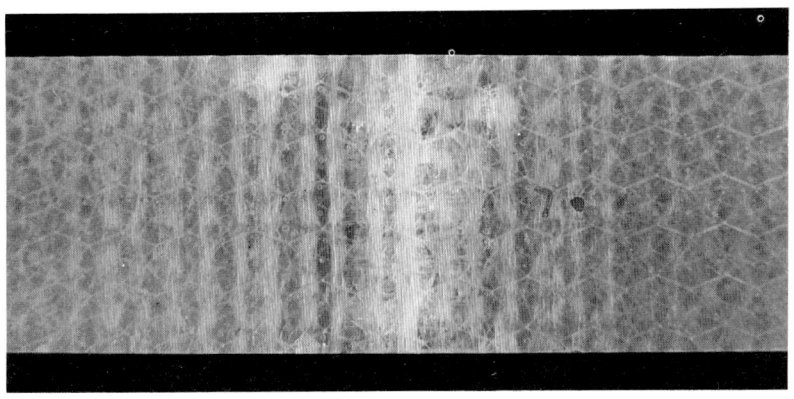

Fig 6   Damage produced under roller during three-point bend loading

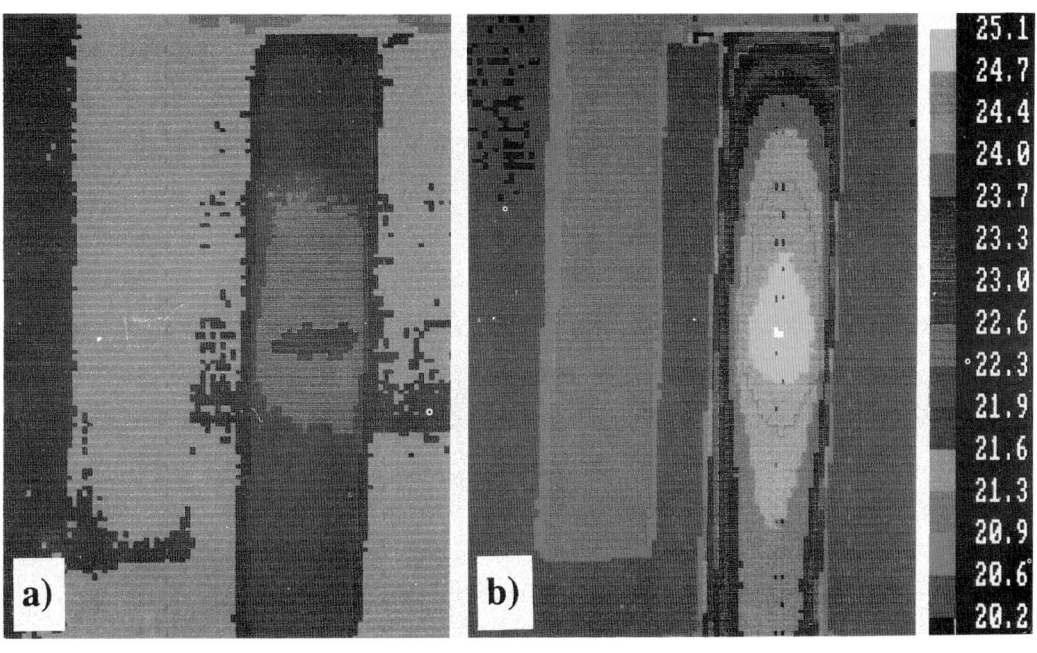

Fig 7   Thermal images of the specimen shown in Figure 6 produced by tensile cyclic loading
at 8 Hz, 40 MPa (10 kN) stress range
a) after 2.5 minutes;   b) after 60 minutes   Scale in °C

Thermographic evaluation of static and dynamic damage to composite wind turbine blades

M. D. WAKEMAN, G. M. SMITH, and **B. R. CLAYTON**
University of Nottingham, UK

SYNOPSIS   A series of tests was performed to examine the damage created in a composite material which simulated that which could conceivably occur in wind turbine blades during construction, transport, assembly or operation. Penetrators with different geometric profiles were used to statically and dynamically damage specimens of 10mm thick 0°/90° balanced stitch-bonded glass reinforced polyester. It was found that damage width increased with load, was of greatest extent for a conical-shaped indenter and decreased in extent when protective rubber sheet was used. Thermographic investigations during cyclic loading of damaged specimens revealed temperature rises at the damage site. These allow analysis of damage characteristics which were also observed by sectioning.

1 INTRODUCTION

However careful the intention may be, it is always possible that some components of a wind turbine could suffer damage even before operation. In particular, the blades are generally large and unwieldy structures which are subject to the vagaries of transport, site assembly to hubs and mounting on towers. Unintentional damage, through applied stress concentrations, can occur from uneven support on lorries in transport which could develop excessive bending stresses. During loading and unloading of the blade, impact stresses may occur from collisions with other hard surfaces. For example, hard contact may be made between the blade and tower or crane during assembly to the hub at the top of the tower. Even the dropping of hand or power tools on the blade surface can cause substantial local damage which can manifest as a progressive weakness during subsequent periods of operation, resulting in a shortened life.

Most wind turbine blades are now constructed from a composite of different materials, for example glass reinforced polymer in several different lay-ups, laminated wood-epoxy and sometimes metallic components. Detection of any damage and its extent now calls for a rather different approach from those for structures in most other engineering applications. Glass reinforced polymers are poor conductors of electricity and heat, they have complex structural and heat conducting properties, and the blade is made up as a layered and inhomogeneous structure. A blade is also likely to have deliberately varying wall thicknesses. Techniques should therefore allow for or take advantage of these characteristics.

A series of major investigations is underway [1-5] to evaluate the benefits of infra-red thermography as a non-destructive evaluation tool for detecting and analysing damage in wind turbine blades. Much of this work has concentrated on crack and flaw development which has initiated at features of the blade construction. Delamination, debonding, failure at glue lines and so on

have been the main sources of detected hot spots in full-size blades that have been considered, and laboratory test pieces have been designed to simulate such defects.

The purpose of the present investigation was to determine the seriousness of applied damage to the surface of a blade and the extent of the subsurface damage. Surface examination and sectioning have been used together with infra-red thermography to provide a picture of damage created under controlled conditions. All tests were conducted in the laboratory using plate specimens made from 0°/90° balanced stitch-bonded glass reinforced polyester. Dynamic tests were performed using a purpose built pendulum arm impact machine which was calibrated to obtain impact energy and absorbed energy. Penetrators with different head profiles were used in both dynamic and static tests. An instrumented loading frame was used to create the static damage when the specimens were subjected to compressive loading.

It is also worth noting that the overall purpose was to create the minimum visible surface damage. Clearly, a readily observed area of damage would alert the operator and remedial action would be expected. The more critical situation arises from apparently insignificant surface damage but unseen subsurface damage which could act as a stress concentrator for crack propagation under variable loading during turbine operation.

2 MANUFACTURE OF SPECIMENS

Two plaques of 500 mm square glass reinforced polyester were used for the work. To simplify the composite structure, a glass fabric was selected which was stitch-bonded rather than woven and with equal volumes of fibre in the 0° and 90° directions. It had a weight of just 300g/m² requiring 32 sheets for the full 10 mm plaque thickness. This was a similar layup to that used for other thermography investigations [5]. It had been chosen to be more uniform and of a finer scale than is found in production blades, with the aim of reducing effects due solely to local material variations. The plaques were laid up by hand with polyester resin to simulate a common method of blade production.

3 PROCEDURE FOR DYNAMIC TESTS

A number of possibilities were explored when seeking a design for the impact machine. Since our purpose was to produce small amounts of visible damage rather than extreme damage leading to structural failure, the sophisticated pressure tank design of Choi *et al.* [6] was not adopted. Similarly a machine of the type used by Broutman & Rotem [7] was also considered inappropriate for the time and resources available. The technique adopted in [7] called for a framework almost 4 m high supporting precision steel guides for the large dropweight. For our purpose the size of the electromagnet required to hold the impactor would have been prohibitive. Consequently, a pendulum-type impactor was designed and built as shown in Figure 1: this offered a relatively simple technique which also allowed impact energy to be adjusted straightforwardly.

The equipment comprises essentially of a rigid double A-frame mounted on a concrete and steel bedplate. Rigidity of the structure was a primary consideration in order to minimise the impact energy absorbed by the frames. The cross-member connecting the tops of the two A-frames also supports the swinging cradle which carries a variable added mass at its lower end. A variable mass device was chosen rather than constant mass and variable swing angle (to vary impact velocity) in order to simplify the swinging arm release mechanism. The test specimen was clamped to a backing plate rigidly fixed to the bedplate at the bottom of the swing.

The velocity of the pendulum arm could be derived either before or after impact using a timer which was triggered by light switches. Hence the incident energy and the energy lost during impact could be calculated.

4 PROCEDURE FOR STATIC TESTS

It was considered important to investigate static loads applied by the indenters in order to simulate steady pressure loading which could occur on blades, for example when stored or awaiting assembly on site. A screw-driven Instron loading frame was used, with load applied to the test specimen over a period of two minutes. This was effectively a short-term creep test. The crosshead could be programmed to travel into the specimen at a constant velocity until the preset maximum load was reached. Under position control, it could then remain stationary for the required time. As a result of the viscoelastic behaviour of the material, creep occurred during the two minute period which decreased the load on the material. The Instron was equipped with a computer interface enabling simple parameter setting and data acquisition. As with the dynamic tester, it was desirable to measure the energy absorbed by the specimen during the test. This was found from the area under the force/displacement curve using numerical integration.

5 TEST PROGRAMME

In order to create damage in the material specimens and thus simulate damage that could occur in wind turbine blades two shapes of indenter/impactor were used: blunt-ended damage was created by a 4.95 mm radius hemisphere and sharp-ended damage by a 45° half-angle cone with a 0.25 mm radius tip.

For the dynamic tests, three tests at each of six energy levels were carried out for each of the two geometries of impactor, achieved using added masses from 0 to 8 kg. For the static tests, four tests at each of six loads from 1 to 12.5 kN were performed, again for each geometry of indenter. One additional test at each of the dynamic energies and static loads was undertaken with a sheet of 1.5 mm thick Neoprene rubber on top of the specimen surface to simulate a possible protective layer. The effect of rubber thickness was examined briefly using sheet up to 5 mm thick with one dynamic testing condition.

Tests were carried out on 150 mm square pieces cut from the glass reinforced polyester plaques, with several damage sites per piece. These were then cut down to individual specimens for further examination of the sites using optical microscopy and thermal imaging.

6 RESULTS OF DYNAMIC TESTS

Using added masses of 0 to 8 kg on the pendulum arm of the test machine, incident impact energies of 5.9 to 11.5 J were achieved reproducibly. By subtracting the pendulum rebound energy for each test, the energy loss was calculated and plotted against incident energy as shown in Figure 2 for the two different impactor geometries. It was assumed that the energy lost by the pendulum was absorbed entirely by the specimen. The size of the damage created was measured in the first instance as the diameter of that visible from the top surface in the translucent material, and is plotted in Figure 3. Both the damage width and the energy absorbed increase as the impact energy increases, and are greater with the conical than the spherical impactor.

Damage sites given approximately 10 J dynamic impact by the two impactors were examined in more detail. The average damage width was 10.7 mm for the spherical impactor and 14.3 mm for the conical. Sectioning revealed subsurface fibre debonding and breakage, delamination, and matrix fracture to a depth of 2 to 3 mm, with the conical impactor creating slightly deeper and more severe damage than the spherical impactor (Figures 4 and 5).

With the addition of the 1.5 mm rubber sheet, the energy losses of the pendulum showed similar values to those without the rubber (Figure 6). The energy this time is being absorbed by both the rubber and the composite material, and the method did not allow the separation of the two components. However, the rubber did reduce the extent of the damage created in the specimen, especially for the spherical impactor (Figure 7).

The effects of increasing rubber thickness with constant 10 J impact conditions and a spherical impactor are presented in Figure 8: damage width decreased and energy absorbed increased until with 5 mm of rubber, no visible damage was present.

7 RESULTS OF STATIC TESTS

Energy absorbed and damage size created for the six different applied loads are presented in Figures 9 and 10. Over the range, energy absorbed was around 50% greater for the conical than for the spherical indenter. Damage depths and widths were also higher.

The appearance of the damage (Figures 11 and 12) was less violent than that observed in the dynamic tests with some delamination and localised fracture occurring but with translucency being retained in a greater volume. The surface profiles are more clearly those of the indenters, with material having flowed to form a lip around the edge. The static tests tend to take on the characteristics of short creep tests.

Figures 13 and 14 show the effects of including rubber sheet on the surface. As with the dynamic tests, the general trend is for the rubber not to give a quantifiable change in the energy absorbed during the test. Again energy was being absorbed by both the rubber and the composite. Damage widths decreased by between 10 and 15%.

8 THERMOGRAPHY

The four specimens selected for thermography had seen nominally identical damage conditions to those in Figures 4, 5, 11 and 12: conical or spherical indenters and with 10 J dynamic impact or 7.5 kN static load. For these the visible widths of surface damage ranged from 10 to 14mm.

The thermal imaging was performed whilst the specimens were subjected to cyclic loading, using the heating effects this produces within the material. The rate of heat generation increases with increasing stress range and frequency [5].

Strip specimens 25 mm wide were used with the damage site centrally located. The applied loading was in tension-tension with an R-ratio (minimum load/maximum load) of 0.1, a load range of 14 kN (56 MPa) and a frequency of 4 Hz. For the thermal imaging, an Agema Thermovision 880 system was employed which can detect surface temperature differences of less than 0.1°C. Thermal images were recorded at one minute intervals from the start of the cyclic loading.

All four samples showed a hot spot which developed over time. After about 30 minutes an equilibrium temperature distribution was reached, when the rate of heat generation was balanced by heat losses through the grips and from the surfaces. To obtain a clearer picture of the heating effects caused just by the cyclic loading, the image taken at the start of the sequence was subtracted from that taken after 30 minutes. These results are presented in Figure 15.

For each specimen the hot spot was located at the damage site. The smallest heating effect in terms of both magnitude (2.7°C) and localisation occurs in the specimen statically loaded with the spherical indenter. This corresponds with the previous observation that, of the four, it appeared to have the least subsurface damage and smallest change in surface profile. In contrast, the well-defined impression created by the statically loaded conical indenter resulted in a local temperature rise of 4.0°C. The dynamically damaged specimens gave hot spots of around 3.5°C. Other work [5] has shown that in an undamaged specimen approximately uniform heating would be expected throughout the gauge length.

9 DISCUSSION

The aim of the tests was to provide a close simulation of damage that could occur in the real environment. It is therefore necessary to relate the energies used in these experiments with those which could occur in practice.

The range of energies used in the dynamic tests was 5.9 to 11.5 J. To put this in perspective, a spanner of mass 0.2kg would have to be dropped through 5 m to have an energy of 9.8 J. A blade of mass 1000 kg, typical of a commercial wind turbine, would need only to be dropped through 1 mm to gain similar energy. This could create damage of the order found in these tests if a suitably sharp object was hit. Likewise, the static loads of up to 12.5 kN may also be encountered in practice.

According to the geometry and velocity of contact, the damage created in glass reinforced polyester has been shown to vary in width, depth and nature. Impact damage in particular causes fibre and matrix fracture in the subsurface layers, sometimes with little change in surface profile. In the specimens used, the translucent nature of the undamaged composite material enabled the lateral extent of subsurface damage to be estimated, but for turbine blades which have an opaque outer skin not even this observation would be possible.

The generation of heat at damage sites, when the composite is subjected to cyclic loading, enables their presence to be demonstrated nondestructively through sensitive thermal imaging. Such a loading situation is encountered during the fatigue testing of full scale blades, and the variable loading experienced by blades in service might also be harnessed for this purpose. The heat generated at damage sites is thought to be related to the combination of changes in surface profile causing stress concentrations and also to the degree of structural damage [5]. It follows that thermal imaging can detect subsurface damage when there is little evidence observable at the surface, and also that even when there are some visible surface indications of damage, thermal imaging can help to give a more complete picture of its extent.

10 CONCLUSIONS

* The contact loads and impact energies sufficient to create damage in glass reinforced polyester specimens are of a magnitude which might inadvertently be experienced by turbine blades.

* The proportions of visible surface damage and subsurface damage varied according to the geometry of the indenter and the velocity of loading. Static loading allows time for the material to creep around the indenter; impact loading creates more fracture and subsurface damage.

* In general more energy was absorbed and more damage was caused by the conical indenter than the spherical indenter under otherwise similar damage conditions.

* The use of rubber sheet decreased the damage width and decreased the subsurface damage for both static and dynamic loading.

* When damaged specimens were subjected to cyclic loading, thermography revealed areas of both surface and subsurface damage.

* Thermography can be used for detecting damage in composites and for assessing its significance.

ACKNOWLEDGEMENTS

The work was undertaken as a final year undergraduate project of MDW under the supervision of GMS and BRC at the University of Nottingham.

REFERENCES

[1] DUTTON A G, IRVING A D, LIPMAN N H, AFTAB N, CLAYTON B R & BOND L J
Infra red condition monitoring techniques for composite wind turbine blades.
Proc. 13th British Wind Energy Association Conference, Swansea 1991, MEP pp 305-310

[2] AFTAB N, DUTTON A G, IRVING A D, LIPMAN N H, CLAYTON B R & BOND L J
Development of thermal condition monitoring techniques for composite wind turbine blades.
Wind Energy: Technology and Implementation Proc. European Wind Energy Conference Amsterdam Oct.1991 (Eds.F J L van Hulle *et al.*) Elsevier 1991 ISBN 0-444-89117-X pp 241-245

[3] DUTTON A G, IRVING A D, LIPMAN N H, CLAYTON B R, AFTAB N & BOND L J
Infra red condition monitoring of wind turbine blade fatigue tests.
Wind Energy Conversion 1992 Proc. 14th British Wind Energy Association Conference, Nottingham Mar.1992 (Ed.B R Clayton) MEP 1992 ISBN 0-85298-827-3 pp 221-227

[4] DUTTON A G, IRVING A D, SMITH G M & CLAYTON B R
Infra red thermography as a tool for identifying flaws and stress concentrations in composite wind turbine blades.
Proc. European Community Wind Energy Conference, Lübeck-Travemünde Mar.1993 (Eds. A D Garrad *et al.*) H S Stephens & Ass./CEC 1993 EUR-15083-EN ISBN 0-9521452-0-0 pp 168-171

[5] SMITH G M, CLAYTON B R, DUTTON A G & IRVING A D
Infra-red thermography for condition monitoring of composite wind turbine blades: feasibility studies using cyclic loading tests.
Proc. 15th British Wind Energy Association Conference, York Oct.1993 - *this volume*

[6] HYUNG YUN CHOI, DOWNS R J & FU-KUO CHANG
A new approach towards understanding damage mechanisms and mechanics of laminated composites due to low-velocity impact: Part 1 -Experiments, *Journal of Composite Materials*, Aug 1991, 25, pp 992-1008

[7] BROUTMAN L J & ROTEM A
Impact strength and toughness of fibre composite materials, *Foreign object damage to composites*, ASTM-STP 568, 1975, pp 114-133

Fig 1   Dynamic impact testing rig showing pendulum arm with weights on left, specimen backing plate on right and timing switches in centre

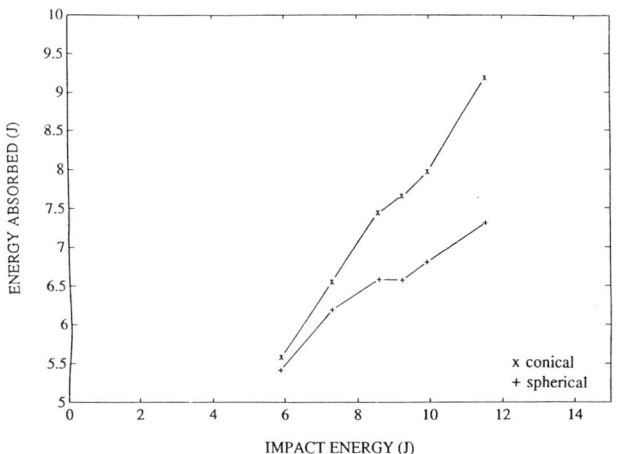

Fig 2   Energy absorbed in dynamic impact

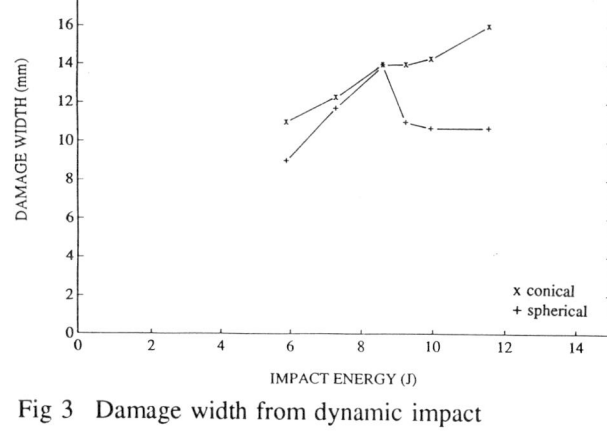

Fig 3   Damage width from dynamic impact

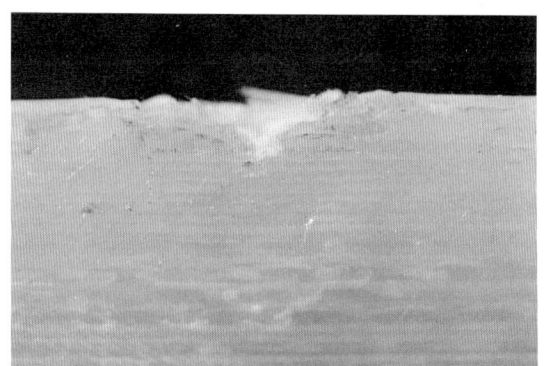

Fig 4   Damage from dynamic impact
(conical impactor, 9.9 J)

Fig 5   Damage from dynamic impact
(spherical impactor, 9.9 J)

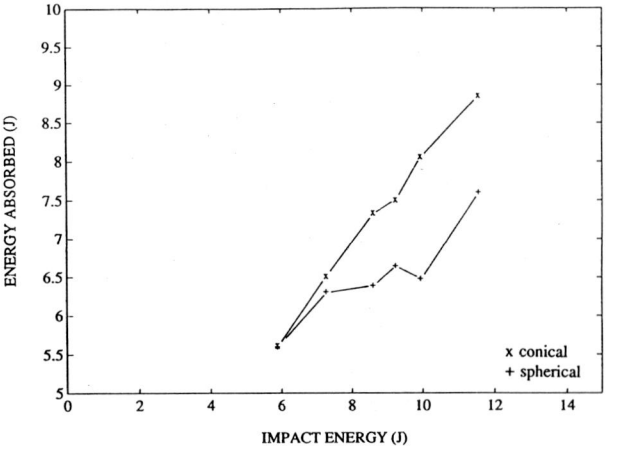

Fig 6 Energy absorbed in dynamic impact -
with 1.5 mm rubber sheet

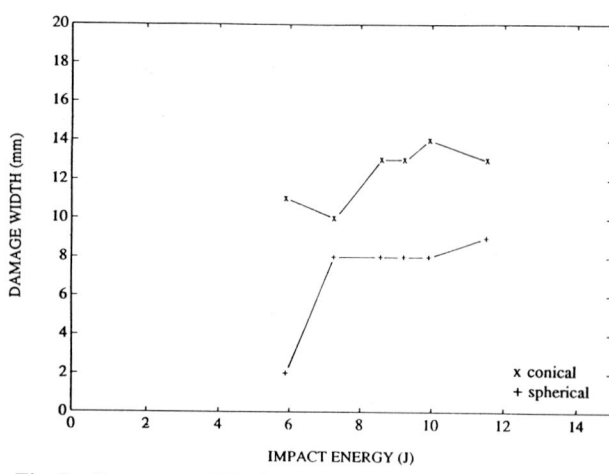

Fig 7 Damage width from dynamic impact -
with 1.5 mm rubber sheet

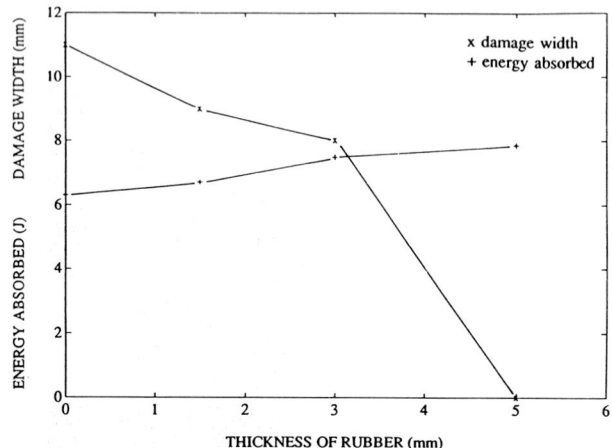

Fig 8 Energy absorbed and damage width from dynamic
impact using different rubber thicknesses

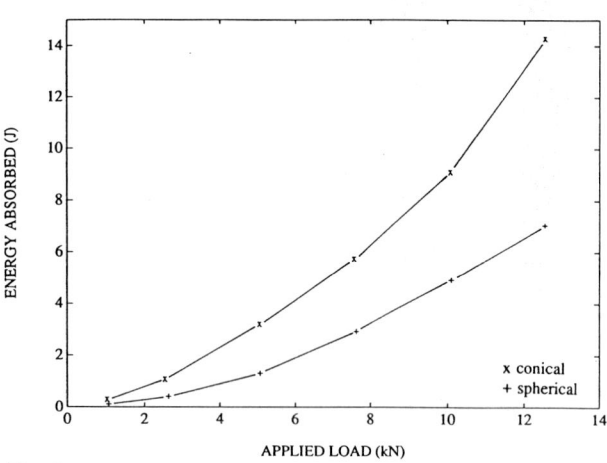

Fig 9 Energy absorbed in static loading

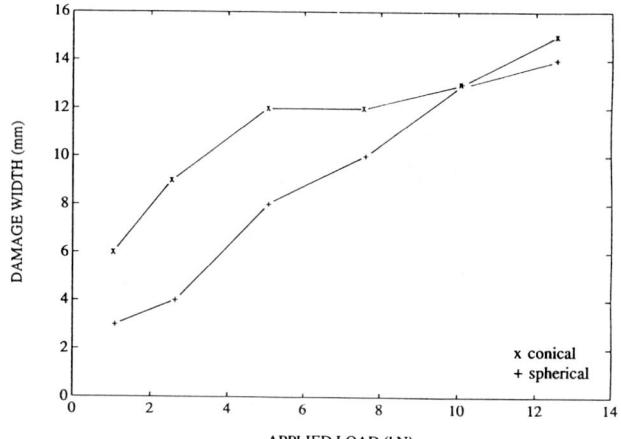

Fig 10 Damage width from static loading

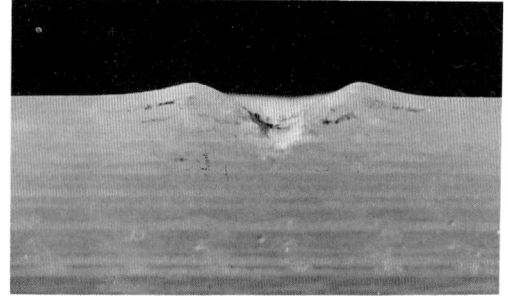

Fig 11 Damage from static loading
(conical indenter, 7.5 kN load)

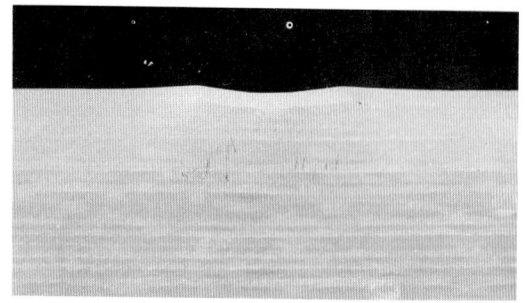

Fig 12 Damage from static loading
(spherical indenter, 7.5 kN load)

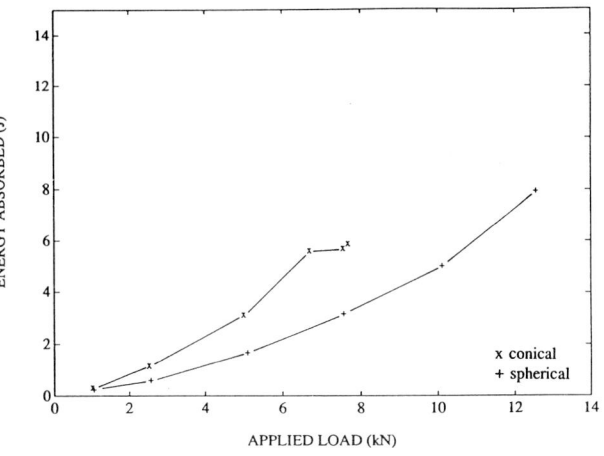

Fig 13 Energy absorbed in static loading -
with 1.5 mm rubber sheet

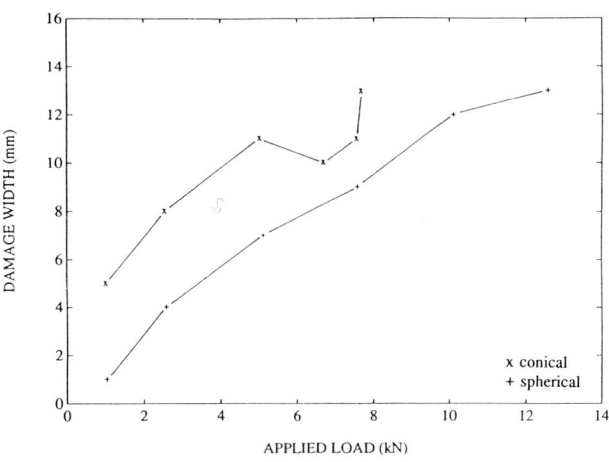

Fig 14 Damage width from static loading -
with 1.5 mm rubber sheet

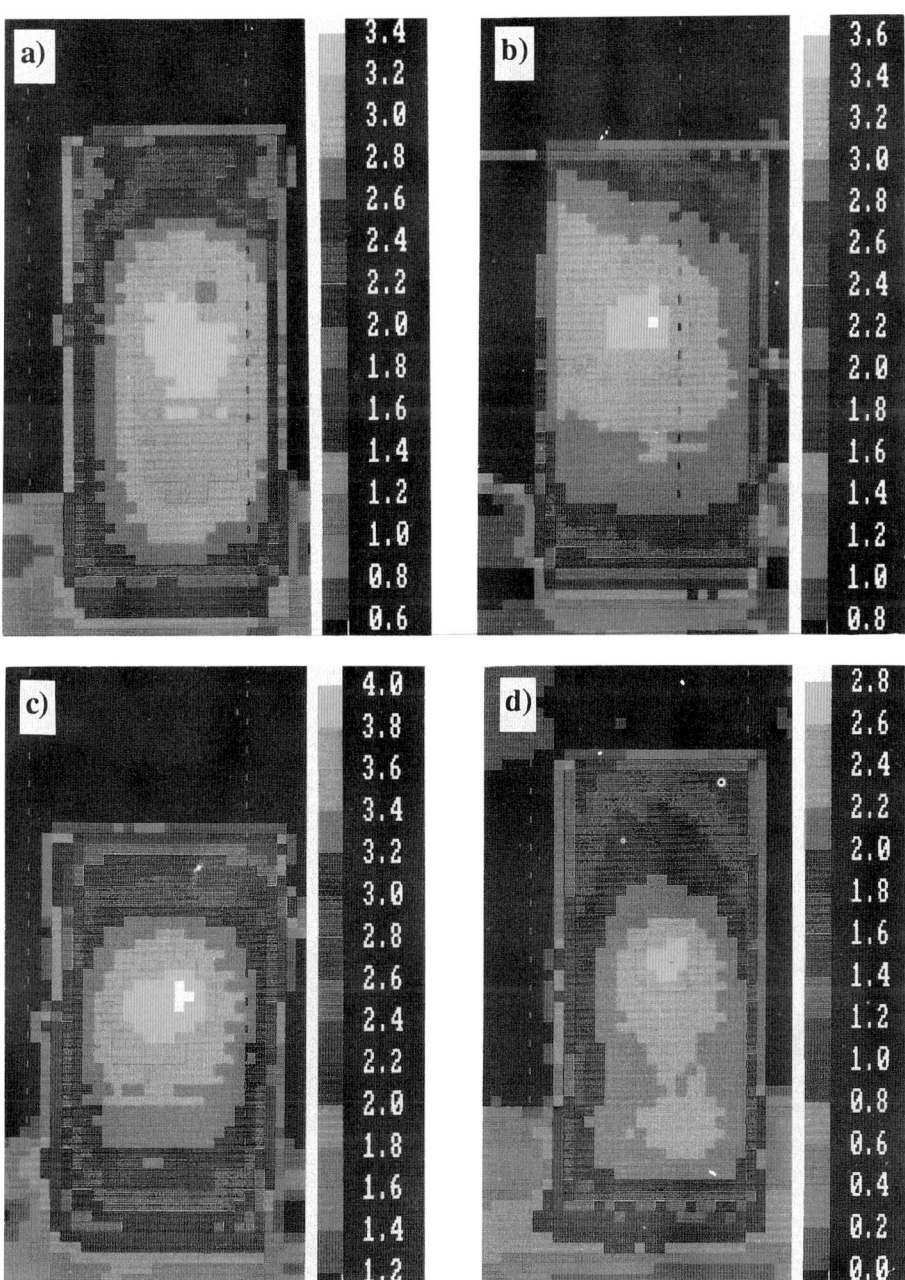

Fig 15 Thermal images of damage sites (temperature rise in °C)
Top: 9.9 J dynamic impact; a) conical, b) spherical impactor
Bottom: 7.5 kN static load; c) conical, d) spherical indenter

Systematic comparison of prediction and experiment for wind turbine aerodynamic noise

M. V. LOWSON and **J. V. LOWSON**
Flow Solutions Ltd, UK

SYNOPSIS   A comparison is made between predictions from a new theory of wind turbine aerodynamic noise and measurements taken from a variety of wind turbines. On average an under prediction of 1.4 dBA overall has been found, with an rms error of 1.9dBA . The error is of the same order as that in the experimental measurements.

1   INTRODUCTION

The widespread desire to use pollution-free energy has seen wind turbine generators erected in increasingly large numbers in recent years. Since the rural areas in which wind turbines are often sited experience background noise levels as low as 20 dBA, a typical wind turbine with a reference sound pressure level of 55dBA can be a major intrusion, and noise issues generally are becoming crucial to planning applications. A new theory of wind turbine noise has been developed (ref. 1), the aim of which was to predict the noise output of a given design of wind turbine using as few and as simple parameters as possible. The theory has now been tested against experimental data from a wide variety of horizontal axis wind turbines, both commercial and prototypes. This paper presents the results of the sound pressure comparisons.

2   THEORETICAL BASIS

There are many sources of noise on a wind turbine, discussed fully in ref. 2, and their relative importance changes with the windspeed, as well as with the design of the machine.

Although it can be a significant noise source, mechanical noise is not taken into consideration by the present theory. In pure decibel terms, the gearbox on a typical wind turbine is likely to be louder than all the other noise sources combined. However, since the gearbox is enclosed within a nacelle, it is possible to control the mechanical noise using conventional techniques, e.g. lining the nacelle with cladding. Such techniques will typically reduce the noise radiated from the gearbox by at least 6 dB, and it will be seen later that mechanical noise is not generally a significant noise source on fully developed machines. Ultimately it is the aerodynamic noise sources that are the most important, as they are essentially fixed by the design of the machine. The current theory enables the wind turbine designer to predict the aerodynamic noise, and thus to work towards low noise from the outset. Work is currently in hand to deduce the exact nature of these aerodynamic noise sources, but at the current stage of the work, only two such sources are considered. They are; Trailing Edge noise and Turbulence Induced noise. Another potentially significant noise source is that due to aerodynamic instabilities, which is discussed in section 2.3.

2.1 TRAILING EDGE NOISE

Trailing edge noise is intrinsic to any aerofoil passing through the air, and, on a wind turbine, produces noise which is concentrated in the middle range of frequencies to which the ear is most sensitive. Trailing edge noise is dependent on the speed of air passing over the blade. Since a typical air speed for the tip of a wind turbine rotor is about 70 m/s, compared to typical wind speeds of 8-10 m/s, the amount of trailing edge noise produced is essentially fixed by the diameter and rpm of the machine. The rated electrical power output of the machine is also largely dependent on the machine's diameter and rpm, which places considerable constraints on the designer. This noise will be produced as long as the rotor is rotating, irrespective of windspeed or power output. Therefore this source of noise is most important at lower windspeeds, near to cut-in. At higher windspeeds the trailing edge noise is essentially unchanged, whilst the noise from another source, Turbulence Induced noise, becomes increasingly loud.

2.2 TURBULENCE INDUCED NOISE

Turbulence induced noise is caused by the interaction of airborne turbulence with the wind turbine. When turbulent eddies interact with the rotor they cause fluctuating pressures, which cause noise to be radiated, predominantly in the low frequency range. Turbulence induced noise increases markedly with windspeed, typically 1 dB per m/s change in windspeed, thus the noise performance of a given machine at higher windspeeds, i.e. at rated power is largely governed by this noise source. Turbulence induced noise is also dependent on the precise size and speed of the turbulent eddies. Air is naturally more turbulent when blowing through built up areas as compared to open plains, and it is necessary to quantify exactly how built up or obstructed the area is. This is done by means of the Ground Roughness constant (Z_0).

The ground roughness constant, Z_0 lies between 0.001m for open water and 0.3m for built up areas, and is 0.03m for typical farmland. A higher Z_0 corresponds to increased air turbulence and thus increased noise. Turbulence induced noise is concentrated in frequencies below the most sensitive range of the ear. However, variations in the ground roughness of a site between ideal open, unobstructed territory and built up areas are predicted to cause an increase of 2dBA in the noise at rated power.

2.3 AERODYNAMIC INSTABILITIES

Aerodynamic Instabilities arise as a result of the intrinsic properties of certain rotor blades (c.f. ref.2). The resulting noise is a characteristic whistle which can be very loud. The presence of aerodynamic instabilities can increase the overall noise level of a given machine by over 6 dBA. The current prediction does not take account of this source of noise. Predicting it is very involved, and in any case it is possible to reduce or eliminate this type of noise during the development stage. The significance of this effect will be illustrated below.

3 PREDICTIONS

Predictions are made by calculating the amount of noise generated in each frequency band by Trailing Edge noise and Turbulence Induced noise. This is described more fully in ref.1. Since only two noise sources are considered, the resulting prediction is a minimum; real wind turbines are expected to be louder. The following details must be known.

```
RPM
Number of Blades
Rotor Diameter      (m)
Hub Height          (m)
Blade Tip Chord     (m)
Blade Root Chord    (m)
```

Table 1:   Parameters required for prediction

These parameters are all easily determined, and are used as input to the equations. The computation produces a third-octave band spectrum. A typical result for a 2.5 MW machine, in open terrain at 9 m/s windspeed is shown in fig. 1. The contributions of both noise sources are illustrated along with the overall predicted noise spectrum. The noise in each frequency band is summed to give a prediction for the overall level. No account is taken of directionality effects. The noise is assumed to be radiated equally in all directions, and to originate from the hub. It can be seen that the trailing edge noise, or self noise of the blades is the dominant noise source in this case.

4   COMPARISON WITH MEASURED DATA

Fig. 2 shows the total prediction of fig. 1 together with the spectrum measured from the machine, the MOD-2. In this example we see excellent agreement between prediction and measurement. Notice the extra measured noise between 40Hz and 250Hz. This is likely to be due to mechanical noise from the gearbox. Although measurement appears to exceed prediction by 5dB in this frequency range, it has little bearing on the dB(A) rating of the machine, since the ear is most sensitive to higher frequencies of about 1 to 2.5 kHz. The difference between measurement and prediction in overall level is 0.2 dBA.

We also have comparable data available for another version of the same wind turbine, in which aerodynamic instabilities are thought to be present (fig.3). The extra noise is clearly visible above 600 Hz. This type of noise is concentrated in the area of the frequency spectrum to which the ear is most sensitive, and thus the wind turbine is 9 dBA louder overall.

Figs. 4 & 5 show prediction and measurement for two medium power Vestas machines, this time in octave band plots. In both cases the prediction is in error by under 1 dBA

overall. The machine in Fig. 5 appears to be exhibiting aerodynamic instabilities, judging by the high level of measured noise in the 2 and 4 kHz bands. However these instabilities do not appear to be contributing much to the overall noise level, unlike the machine of fig. 3, in which the instabilities became the dominant noise source.

5   ANALYSIS

A number of measured spectra, taken from a range of machines from 50kW to 4.2MW at a range of windspeeds from 5 to 15 m/s were compared with the prediction. For the purposes of performing statistical analysis, the predicted overall level (in dBA) has been subtracted from the measured overall dBA level. The results are given in table 2 below.
Four entirely different machines are predicted to within half a decibel.
It should be pointed out that noise measurements, and especially spectral measurements, are often subject to as much as a $\pm$ 3 dB error. Several different measurements (typically by the same author) have been cross checked on each machine in an attempt to determine the validity of the data, and over 70 per cent of all the data thus examined has been internally consistent to within 1dB. In cases where discrepancies between 1 and 3 dB arose, several results were compared and a median value was used. Data from machines exhibiting aerodynamic instabilities, and data from prototype machines with excess mechanical noise has been excluded. Such machines typically radiate 4 to 12 dB more noise than predicted. Mechanical noise can be recognised by local peaks in the measured spectrum between 100Hz and 1kHz, and it appears that much of the excess noise noted in the table above may be due to mechanical noise.

Few machines have exceeded prediction by more than 3dB, which tends to indicate that mechanical noise is of approximately the same significance as aerodynamic noise in practice. The 200kW and 440kW machines listed are both Danish commercial machines which have a great deal of mechanical noise suppression, and it is interesting to note that both these machines exceed prediction by less than 0.5dBA. This tends to support the initial premise of the prediction method, i.e. that aerodynamic noise may present a lower limit to wind turbine noise reduction. One machine listed (75kW) produces slightly less noise than predicted, which is surprising. A 50 kW machine has also been tested, and found to produce at least 2dBA less noise than predicted. Analysis of the spectra for these machines indicates that they may produce less trailing edge noise than anticipated. The theory may need to be modified for machines under 100kW.

Several machines have been predicted more accurately at high windspeeds. This is felt to be due to the increasing relative significance of mechanical noise at lower windspeeds.

In a previous report (ref. 3), prediction and experiment were compared in a robust fashion using a fixed Z_0 value. Throughout the current analysis, Z_0 has been adjusted to correspond to ground roughness details noted in the texts of the original noise measurement reports. Although the prediction errors for certain individual cases in table 2 may thus have changed by more than 1dB compared to ref.3, the average error in the predicted overall level across all the data has remained virtually unchanged, at 1.4dB. It will be recalled that the data itself was typically reliable to within better than 1dB. This, taken with the fact that the two highly developed Danish machines exceeded prediction by less than 0.5dB indicates that the prediction is probably at least as accurate as existing noise measurements. In fact, in ref.3, it was demonstrated that two measurements identical to IEA standard differed by 1.6dB. More accurate measurements will be needed to improve the noise prediction.

6   CONCLUSIONS

The comparison has demonstrated that a prediction method based on Trailing Edge and Turbulence Induced noise sources gives good comparison with measured data.   The most highly

developed machines so far tested have exceeded their overall predicted noise level by approximately 0.5 dBA. Good agreement between predicted and measured spectra is also observed. Many machines appear to have only partially suppressed mechanical noise, and these may exceed their predictions by 3 dBA at lower windspeeds. The self noise prediction becomes unreliable for machines under 100kW, which appear to generate less trailing edge noise than expected. The mean underprediction of 1.4dBA reflects the presence of mechanical and/or aerodynamic instability noise in the data. The rms error of 1.9dBA indicates the potential reliability of the prediction. This is very close to the reliability of experimental data discussed in section 5. Thus it is felt that more accurate measurements are needed in order to further develop the theory.

ACKNOWLEDGEMENTS

This work was supported under contract to Flow Solutions Ltd Bristol from the Department of Trade and Industry, monitored by Dr M.L. Legerton.

REFERENCES

1   Lowson, M.V., "A New Prediction Model for Wind Turbine Noise." Paper at IEE Conference Oct 1993

2   Lowson, M.V., "Application of Aero-acoustic Analysis to Wind Turbine Noise Control" Wind Engineering Vol 16 No 3 1992 pp 129-140

3   Lowson, M.V. & J.V., "Systematic Comparison of Predictions and Experiment for Wind Turbine Aerodynamic Noise." Flow Solutions Rep 93/03 April 1993

Measured (dBA)	Predicted (dBA)	Error (dBA)	Rated Power (kW)	Windspeed (m/s)
66.6	63.14	3.46	4200	10.2
65.8	62.57	3.23	4200	7.2
67	64.28	2.72	300	10
66	63.55	2.45	4200	12.1
68	65.74	2.26	600	9.5
67	64.77	2.23	600	7
69	66.81	2.19	600	12
58.8	56.93	1.87	2500	9
65	63.36	1.64	2000	14
63	61.94	1.06	3000	9
57.8	57.10	0.70	75	5
55	54.66	0.34	200	9
53	52.71	0.29	440	9
64.4	64.28	0.12	4200	15.2
63	63.15	-0.15	75	12
58.8	59.15	-0.35	75	7
59.6	60.16	-0.56	75	8

Mean Error   1.38 dBA
RMS Error   1.87 dBA

Table 2 Error in prediction

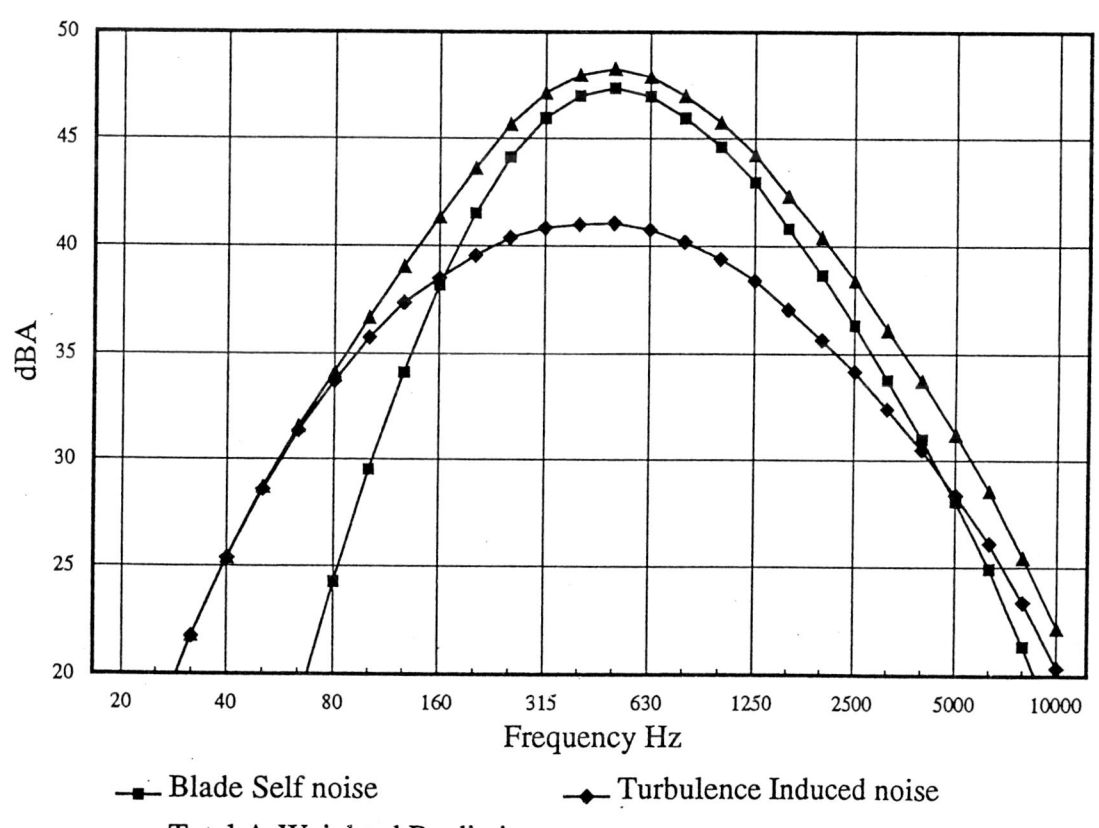

Figure 1   Typical Noise Prediction

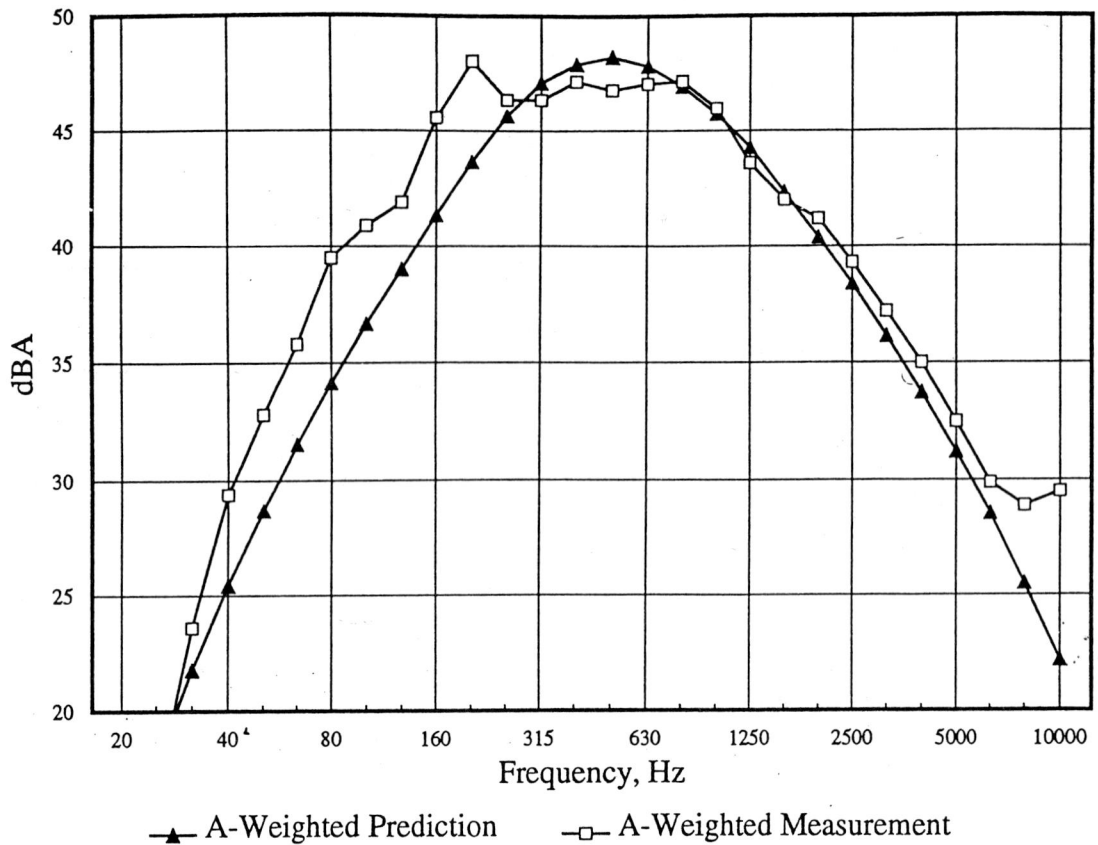

Figure 2   Prediction and Measurement for MOD-2

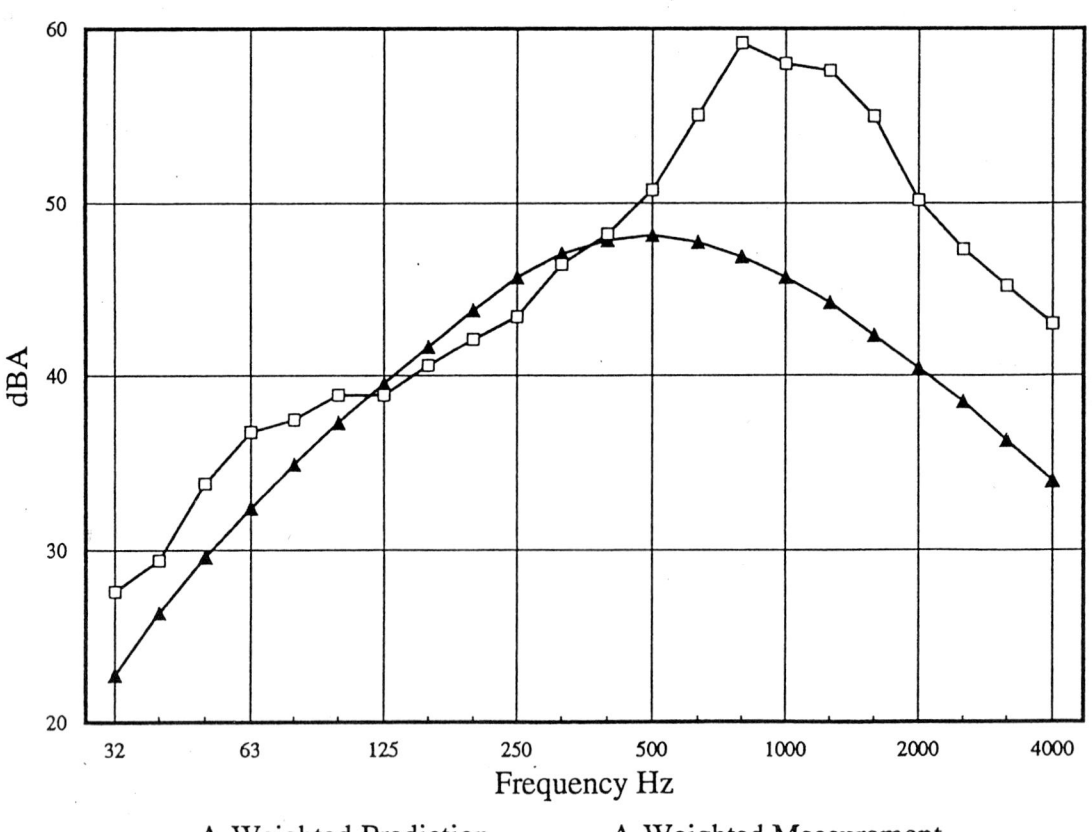

Figure 3   MOD-2 Exhibiting Aerodynamic Instabilities

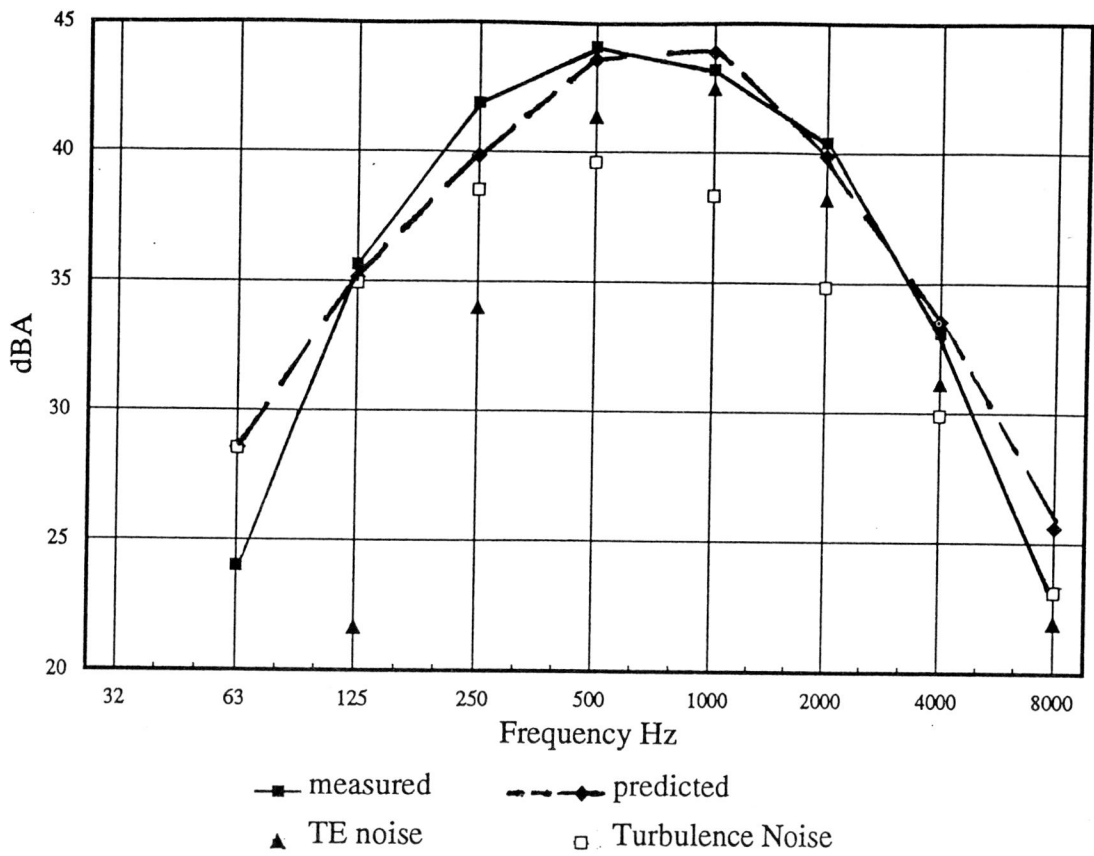

Figure 4   Prediction and Measurement for Vestas V27

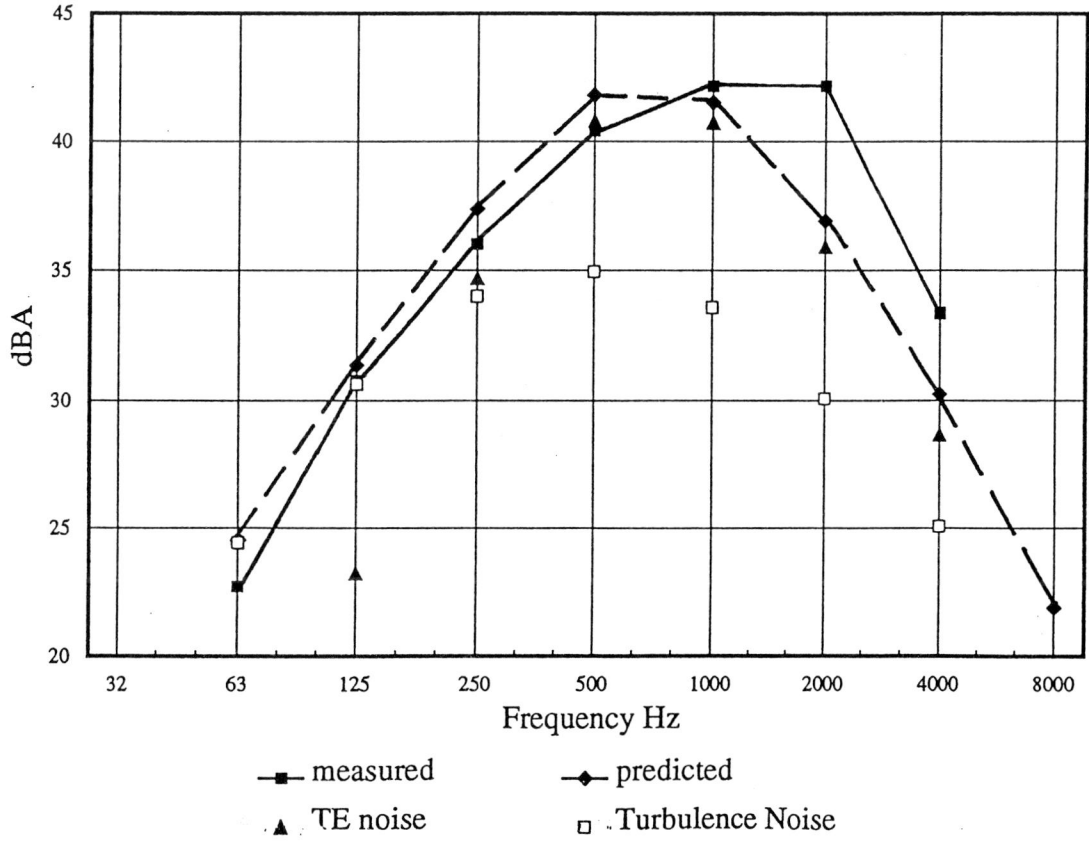

Figure 5   Prediction and Measurement for Windane 34

Machinery noise investigation and reduction on wind turbine generators

P. S. WATKINSON BSc, PhD, MIOA and **A. R. CLARK** BSc, MSc, PhD
GEC-Marconi Naval Systems, UK

SYNOPSIS The support tower of a wind turbine generator has been identified as a significant radiator of machinery generated tonal noise, particularly those components due to gear meshing in the gearbox. Palliatives for this noise include the use of a quieter gearbox, better isolation between the gearbox and the tower and tower damping. This paper concentrates on the analysis of the latter palliative which concludes that there is a need for a more effective treatment than the use of the common free or constrained layer damping treatments. A candidate to fulfil that need is described.

1 INTRODUCTION

Over the last decade many wind farms have been constructed in Europe and the USA. A continuing issue has been that of noise, both from the point of view of direct complainants and the resultant need for a noise specification which is realistic, measurable and relevant. Meanwhile, effort from manufacturers, educational establishments and Government Agencies is directed towards understanding the noise generation, radiation and propagation processes with a view to noise reduction.

The content of this paper is based on the general analysis and conclusions from one such investigation.

2 SOURCE AND PATH IDENTIFICATION

The identification of the sources of machinery generated tonal noise is generally not a difficult matter. Spectral analysis of the noise, spectral analysis of vibration measurements on and around the machinery and a knowledge of the rotation speeds, gear meshing frequencies etc is sufficient to relate the frequency of tones to specific stages in the machinery.

Path identification is rather more difficult. Numerous techniques exist which range from the technique of breaking paths (which can have practical limitations) to techniques of making sufficient measurements to solve sets of simultaneous equations, each one for a different path and degree of freedom, by matrix inversion (which can have credibility limitations).

As an ad-hoc technique the contribution of the tower at a particular tonal was estimated by artificially exciting the tower and measuring the frequency response function between the vibration level on the tower and the far-field noise. This was then used to predict the noise level by multiplication with the measured vibration level with the WTG generating power. The prediction was then compared with the noise level measured with the WTG running. This estimated the proportion of the total noise radiated by the tower, the rest of the noise was attributed to direct radiation from the gearbox as other measurements eliminated the blades and the nacelle walls as significant radiators at this particular frequency. Independent measurements made more recently using a matrix inversion technique confirmed this result.

The radiation from the gearbox was treated by a traditional noise shielding technique but the noise radiation from the tower proved a more complicated problem to treat. The following analysis sets out the approach taken.

3 WAVENUMBER ANALYSIS AND SOUND RADIATION

3.1 Wavenumbers on cylinders

Bending waves on an infinite cylinder can be represented as:

$$\text{Radial Displacement} = C.\exp[in_c\theta].\exp[ik_zz]$$

where $n_c = k_\theta r$ = integer = number of wavelengths around circumference

k_z = axial wavenumber component

k_b = wavenumber of bending wave

and $k_b{}^2 = k_z{}^2 + k_\theta{}^2$

k = wavenumber = $2\pi/\text{wavelength}$

k = f(material properties and geometry)

The characteristics of a particular geometry can be represented by how the wavenumber varies with frequency, for example, on a flat plate the wavenumber varies as the square root of frequency. Whether or not a bending wave radiates sound is dependent on its wavenumber being equal to or less than the wavenumber in air. The wavenumber in air varies linearly with frequency: if graphs of wavenumber versus frequency for sound propagation in air and bending waves in an infinite flat plate are superimposed then at low frequency the wavenumber of the plate vibration is greater than that in air and therefore does not radiate. Above a certain frequency the wavenumber on the plate becomes lower than that of sound in air and therefore radiates. This frequency is known as the coincidence frequency.

Cylinders also follow the above coincidence frequency condition for sound radiation, however, they follow a different characteristic behaviour at low frequency from that of a flat plate. At high frequency where bending wavelengths are short in comparison with the radius of curvature then the cylinder will behave similarly to a flat plate. At low frequency where wavelengths are longer then the curvature of the cylinder plating imposes additional stiffness in bending and therefore the wavelengths are longer than those encountered on a flat plate at the same frequency. The dividing line between these two types of behaviour is the 'ring frequency'. Therefore, at low frequency and small values of n_c then the wavenumbers of bending waves will be low and may well fulfil the radiation condition of being less than the wavenumber of sound in air.

By way of example, a cylinder of 4m circumference (1.27m diameter) and 8mm thick has a ring frequency of about 1.3 kHz.

3.2 Wavenumber analysis

One reason for using wavenumbers rather than, say, wavelength or phase velocity is that space and wavenumber are related through the Fourier transform as are time and frequency. Thus by making measurements of frequency response on the structure (acceleration out due to force in) at a number of points the wavenumbers of the vibration components on the structure can be determined. An example of a frequency/wavenumber spectrum is shown in Figure 1. This shows the predicted circumferential wavenumber components on a cylinder of 1.27m diameter and with a ring frequency of about 5kHz. Note how for $n_c = 0$ there is no strong response below the ring frequency, $n_c = \pm1$ has no strong response below a frequency just below the ring frequency etc. The modes of vibration 'stiffened' by the curvature are not particularly responsive, the marked areas of the wavenumber spectrum are the modes which dominate the vibration field.

3.3 Theoretical predictions

We have the capability to calculate, via an analytical model, the structural response and far field radiation from a cylinder due to a point or line force. This model can be used to calculate the modes (characterised by n_c) which significantly contribute to sound radiation.

Taking our generic WTG case of a 4m circumference 8mm thick cylinder at, say, 250Hz we find that significant acoustic pressures are generated by the $n_c = 1, 2, 3,$ and 4 modes. Figures 2 and 3 illustrate the result for $n_c = 3$ and 4. The lowest wavenumber will be associated with the $n_c = 1$ mode and corresponds to a wavelength of about 3.5m. Referring back to Figure 1, although the low frequency low n_c modes may not respond with as high a level as the higher n_c modes, they are efficient radiators of sound and their contribution dominates the far field sound pressures.

4   FREE AND CONSTRAINED LAYER DAMPING

There are two principal ways to treat the tower to reduce noise radiation. The first is to incorporate structural modifications to the basic design; this can be 'designed' through use of the analytical model and is not the subject of this paper. The second, and most commonly tried, is to damp the tower vibration. Conventional free or constrained layer damping treatments behave approximately as:

$$\frac{13.6 \times \text{loss factor}}{\text{wavelength}} \quad \text{dB/m}$$

A typical loss factor for a good free layer treatment is

about 0.1 and a typical loss factor for a good constrained layer treatment is about 0.2. If we assume a loss factor of 0.2 and use the worst case (ie longest) wavelength from above (3.5m) then an attenuation of a mere .8 dB/m is achieved. Higher attenuation rates will be achieved for shorter wavelengths.

Referring back to Figure 1, the vibration field is dominated by the shorter wavelength components but the sound radiation is dominated by the long wavelength components which have a much lower amplitude. Fitting a damping treatment as per above can produce a significant reduction in the level of vibration measured at individual accelerometer positions yet have little impact on the noise radiation. How many manufacturers and consultants have concluded that the tower is not a significant radiator of noise because reducing the vibration level significantly has little impact on the measured noise level?

Further, a free or constrained layer damping treatment must be <u>continuous</u> over a wavelength scale. This is illustrated in Figure 4. A constrained layer damping treatment was applied to a beam 1.8m long and 8mm thick. The cover plate was the same length as the beam. Loss factor was measured at each natural resonance of the beam. The result is that labelled 'series 1' on the graph. The cover plate was then sawn in half and the result is that labelled 'series 2'. 'Series 3' and 'series 4' correspond the the cover plate divided into 4 and 8 equal segments respectively. Each beam mode corresponds with more wavelengths in the length of the beam with increasing frequency. The conclusion of this experiment is that there is a definite relationship between length of cover plate and wavelength: at low frequency the performance is reduced for a smaller cover plate segment length and each set of results has a definite mode at which the original loss factor is achieved, the smaller the segment length the higher the frequency of this mode. Note that this condition applies both circumferentially and axially; to damp $n_c = 1$ with maximum efficiency then the treatment must be continuous around the circumference.

5 AUXILIARY MASS DAMPING

Auxiliary mass damping tiles are, effectively, a distributed tuned dynamic absorber. However, unlike the usual 'lumped system' tuned absorbers, the spring element has an associated loss factor high enough to give a useful performance over a range of a few tens of Hz. The diagram below illustrates the concept:

Mass per unit area M	
Stiffness K/l	$K = B + 4G/3$ B = Bulk Modulus G = Shear Modulus l = Thickness

The resonant frequency of the tile is $\sqrt{(K/Ml)}$.

Taking the above mentioned worst case of a 3.5m wavelength at 250Hz, about $2m^2$ of these tiles spread over a 1m length of tower are predicted to give about 10dB attenuation. Vibration components with a shorter wavelength will be attenuated by more than this. The treatment does not have to be continuous over a wavelength scale, though it must be spread sufficiently to ensure it is coupled to the vibration (ie not all fitted at a node). It can be fitted as easily handleable sized tiles.

Tiles tuned to a number of frequencies can be mixed to treat more than one tonal component.

Tiles are currently under development with particular emphasis on providing a treatment which will give good performance over a sufficiently wide temperature range. A prototype set of treatment will be fitted to a WTG in the near future to demonstrate effectiveness.

6 CONCLUSIONS

(a) Noise from Wind Turbine Generators is a current issue.

(b) The support tower can be a significant radiator of this noise.

(c) The modes which radiate the noise are particularly difficult to damp using free or constrained layer damping.

(d) Auxiliary mass damping may be an effective treatment which can be applied to the tower.

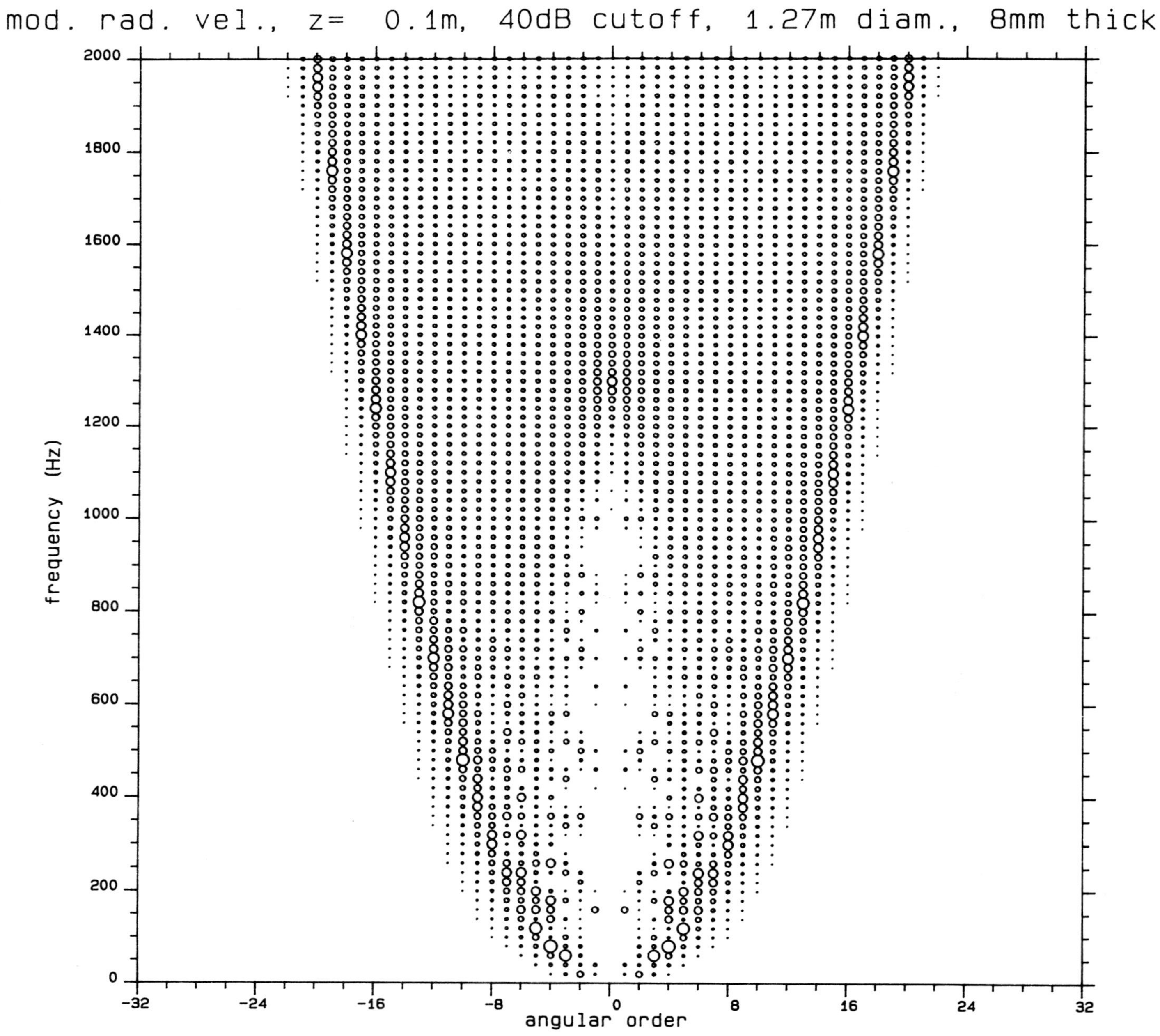

Figure 1 Circumferential wavenumber spectrum for a 1.27m diameter 8mm thick cylinder

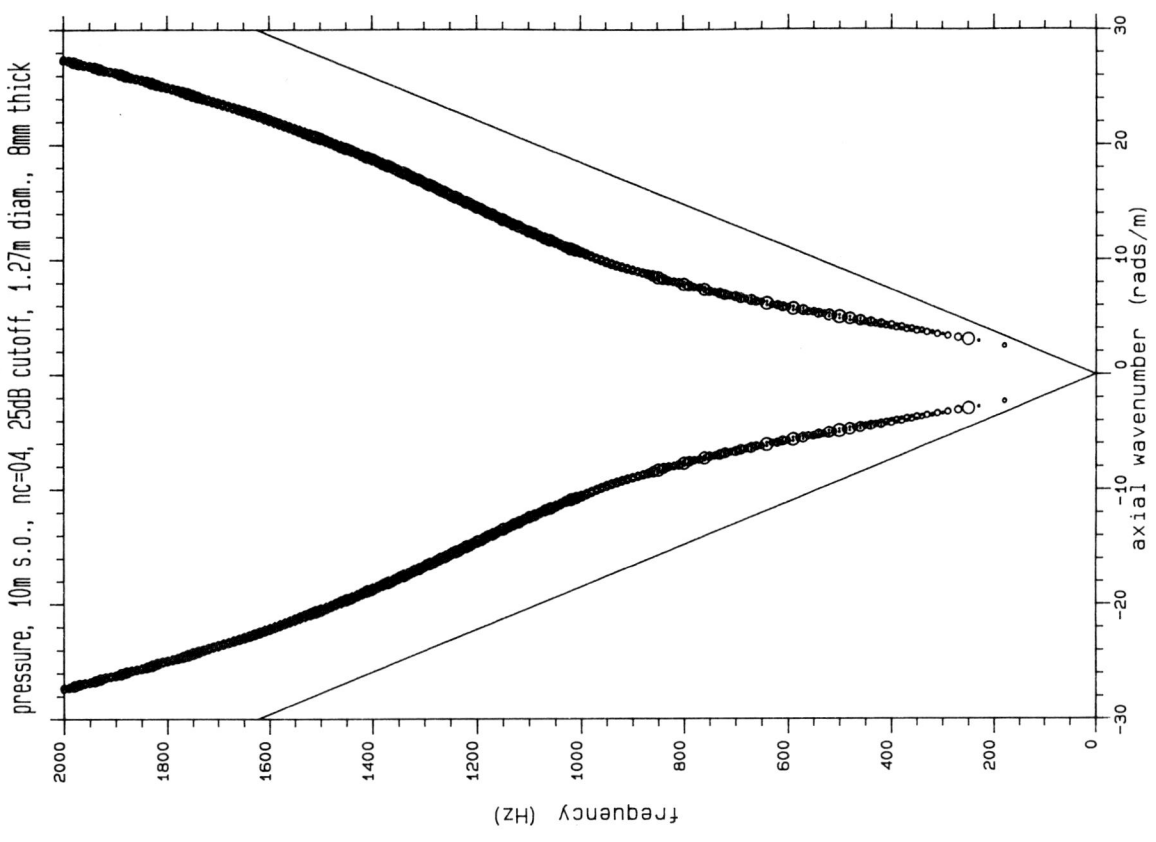

Figure 3 Acoustic pressure due to a $n_C = 4$
component of vibration on cylinder

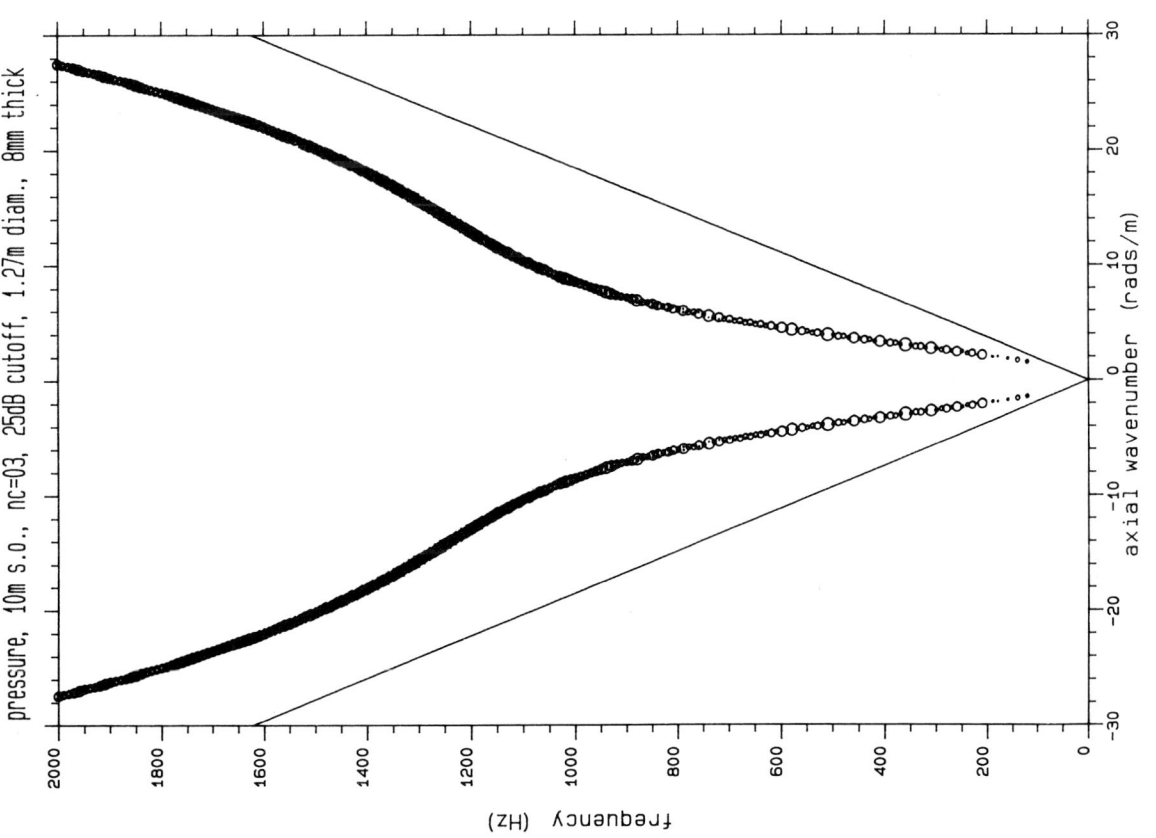

Figure 2 Acoustic pressure due to a $n_C = 3$
component of vibration on cylinder

393

Figure 4 Loss factor for a 1.8m long beam with a constrained layer damping treatment and various segment lengths of cover plate.

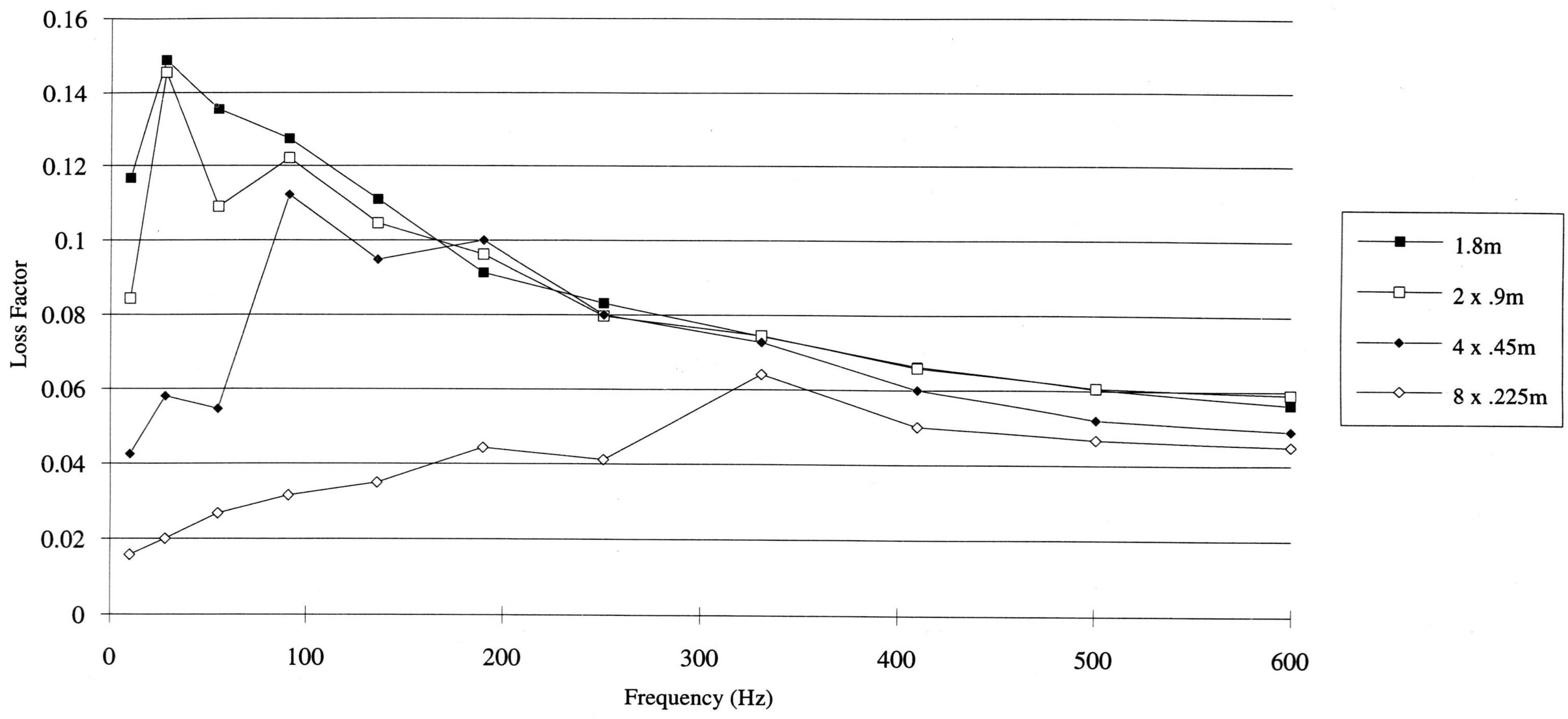

Viability of variable speed

E. A. BOSSANYI MA, PhD
Wind Energy Group Ltd, UK

SYNOPSIS: This paper describes a detailed study of electrical variable speed drives, energy capture, structural implications, control and grid connection issues. Economic assessment shows that the cost of variable speed drives still has to fall before their use becomes economically viable, but this is likely to happen in the near future.

1. INTRODUCTION

This paper summarises and describes the main results of a recently completed study of variable speed operations of wind turbines. The project reviewed the current state of electrical variable speed drive technology for wind turbines, studied a number of aspects of variable speed wind turbine design, estimated cost-effectiveness of variable speed designs, and investigated grid connection and control issues.

The main topics covered were:

- a literature review
- an assessment of commercially available drives
- energy capture and losses
- structural implications
- alternative turbine configurations including pitch and stall regulation
- control
- grid connection issues

Reference 1 is the full report of the project.

2. ELECTRICAL DRIVE OPTIONS

As an integral part of the project, academic and industrial consultants were employed to assess the technology of variable speed drive equipment. The availability, applicability, cost and state of development of various drive and device types were reviewed, and predictions of the likely availability of advanced systems were made.

A detailed specification was prepared for a variable speed drive for use in wind turbines of three different sizes, with ratings from 450 kW up to 1.2 MW. The specification was reviewed by the consultants before being submitted to a number of manufacturers to obtain budget quotations. Detailed discussions were also held with some of these manufacturers. The type of drive was left open to the manufacturers, some of whom quoted for more than one drive type.

Quotations were received for four different types of drive:

- Synchronous generator with DC link
- Induction generator (cage) with current source inverter and DC link
- Synchronous generator with PWM drive using GTOs
- Induction generator with PWM drive using GTOs

Prices varied quite widely. The cheapest quotes were for an induction generator/CSI/DC link system at 450 kW, and for synchronous generator/DC link systems at 750 kW and 1.2 MW. One manufacturer offering all four types of system offered a synchronous/PWM system as their cheapest for 1.2 MW rating.

Figure 1 shows, for each of five suppliers, how the cost per kW for complete drive systems varies (the comparison between manufacturers is approximate since some quotes did not include certain items such as power factor correction). For all except the induction generator/CSI systems (cheapest at 450 kW) the cost per kW for each manufacturer decreases as rating increases, in some cases dramatically (by up to 55%). Taking the cheapest quote for each rating, the cost per kW, relative to 100% at 450 kW, is 85% at 750 kW and 69% at 1.2 MW. This suggests that variable speed is more likely to be cost- effective for larger turbines.

3. ENERGY CAPTURE

In order to compare fixed and variable speed operation, the difference in energy capture must be evaluated rather carefully. The following effects are all taken into account:

- Aerodynamic efficiency: in low winds, the variable speed turbine will stay closer to peak C_p than a fixed speed turbine.

- Transmission losses: variable speed systems incur additional losses in the frequency conversion equipment, and these must be taken into account. In addition, the variation of gearbox and generator losses with both speed and power is taken into account, since this can significantly affect the comparison. Losses in harmonic filters are very small.

- Tower exclusion zone: a variable speed turbine may need a "speed exclusion zone" to avoid excitation of tower resonance. Thus at some wind speeds the rotor will not be at peak C_p. The effect on overall energy capture is however shown to be very small.

- Turbulence: taking into account rapid wind speed fluctuations is shown to have little effect on the results.

Energy capture is calculated for a whole range of different variants of the MS-3 design, all based on a rotor diameter of 33m. This includes both pitch-regulated and stall-regulated designs, since the comparison of fixed against variable speed is quite different in the two cases. For the pitch-regulated case the options include single-speed, two-speed and variable speed. For variable speed there is also the option of increasing the maximum rotor speed above the nominal 48 rpm of the fixed speed variants. If the gearbox design torque is kept constant, the increased speed requires an increase in electrical power rating, but results in a considerable gain in energy capture in high winds. There is an increase in thrust loading, but the structural cost implications of this are taken into account (see below). This increase in top speed is necessary if variable speed is to be cost effective against two-speed.

For the stall-regulated case, no such increase in top speed is possible, as stalling would be lost. However, if aerodynamic tip brakes can be used routinely rather than just for emergency or high-wind shut-down, they can also be used for speed-changing in a two-speed stall-regulated design or for bringing the rotor into stall in the case of a so-called "hybrid" design. In this case, two-speed seems to be the most preferable option.

Figure 2 shows the energy capture for a whole range of fixed and variable speed variants of the MS-3 design, both pitch and stall regulated. The change in energy output compared to a single-speed pitch regulated machine is shown. It illustrates the advantage of two-speed over variable speed unless the rating is increased - this can be done at constant torque by increasing the maximum operating speed. This figure also shows the relatively poorer performance of stall-regulated options. However, with a variable speed stall-regulated turbine there is the option of reoptimising the blade design, to allow higher peak Cp to increase energy capture in low winds without losing the stalling capability in high winds, as the speed can be reduced to maintain stall. Although a full reoptimisation of blade geometry was outside the scope of the project, the study was extended to look at a geometry closer to a pitch-regulated blade. This showed that a considerable energy gain could be achieved, with only a marginal increase in torque rating, and a reduced thrust rating. With full reoptimisation the stall-regulated option may therefore compete against pitch regulated designs.

The study went on to examine the use of a two-speed generator, using tip brakes to achieve speed changes. This proved more favourable than using variable speed. Even a variable speed "hybrid", in which the tip brakes are used to slow the rotor into stall, while achieving similar energy capture to two-speed, suffered much greater thrust loads.

A further option in the stall regulated case is to use the variable speed drive itself to bring the rotor into stall. Although this means that the rotor deceleration torque passes through the whole drive-train, which may be quite severe in around-rated winds, there is an additional energy capture benefit. This option was not included in the original study but is now being studied by WEG as part of another project which will lead to experimental testing of the principle in early 1994.

4. STRUCTURAL AND COMPONENT COSTS

Switching from fixed to variable speed operation has implications right through the design of the wind turbine.

Obviously the cost of the variable speed drive needs to be taken into account, but because operational loads are different there are also changes to the costs of various other components. For some components the changes are relatively straightforward, but for others it was necessary to run some simulations using a detailed dynamic simulation model for the whole wind turbine, driven by a realistic wind input. For the tower and blades, the resulting loadings were fed through a fatigue analysis program to estimate the fatigue life implications. For the gearbox, the results were used to construct a torque exceedance curve, from which the design torque and cost were estimated. The high compliance of a variable speed system (the ability to use the rotor as a flywheel, absorbing aerodynamic torque transients before they reach the gearbox) should allow a reduction in gearbox cost. Initial calculations revealed rather a small reduction. Figure 3 shows the change in gearbox torque exceedance curve in going to variable speed. Although there is a significant reduction in extreme torques, there is also an increase in torque in low winds due to lower operating speeds, and these conditions are much more frequent. A gearbox manufacturer was consulted, and produced a slightly greater cost reduction, but still only of the order of 10%. This figure still contains a large uncertainty over the design factor to be used, which reflects uncertainty over the loads which will be experienced.

Tower costs are dominated by fatigue load cycles largely accumulated at the tower's resonant frequency, which with a soft tower lies in between the 1p and 2p forcing frequencies for the fixed speed machine. Simulations showed the excessive tower movement caused by operation in variable speed when the blade passing frequency is near to the resonant frequency of the tower (figure 4). Further simulations were carried out with the speed restricted to prevent it approaching too close to tower resonance. The results were subjected to fatigue analysis to determine whether the response was acceptable. The results showed that quite a broad speed exclusion zone would be required in order to avoid increased tower fatigue, but the effect of this on energy capture was actually found to be very small. Since an exclusion zone can be placed wherever it is needed, the choice of tower resonant frequency is actually less critical than with a fixed speed machine.

Variable speed in itself was not found to affect blade costs significantly, although increasing the top speed does increase blade costs somewhat due to increased thrust loads.

Other components were examined in less detail, but some cost changes were identified in a number of them, and also in balance of plant costs, in particular electrical system costs if machine ratings are increased along with top speed. Combining these estimated cost changes with the energy capture calculations allowed the cost-effectiveness of different options to be compared.

5. COST EFFECTIVENESS

For the stall-regulated case, it appeared that variable speed (even with the "hybrid" concept) was unlikely to be competitive with two-speed, as it could only match the energy capture of two-speed by sustaining considerably greater thrust loads. However, since the study was completed, this conclusion has been revised for the case in which braking is fully adjustable, as in the case when load torque is used to stall the turbine in high winds.

For the pitch-regulated case, it is clear that variable speed can only compete against two-speed if the top speed is raised, giving a higher rated power. It is then necessary to collect together all the structural and component cost changes and the differences in energy capture to arrive at a comparison of cost-effectiveness.

For a 33-m diameter machine, starting from the 300 kW 2-speed MS-3 production machine, variable speed does not appear to be quite cost-effective even for a 20% increase in top speed and rated power. The results are shown in Table 1.

	2-speed	Variable speed	
	300 kW	330 kW	360 kW
Capital value (£k)[1]	357	387	397
Energy yield (MWh/year)[2]			
6.5 m/s	723	742	758
7.5 m/s	965	1000	1033
8.5 m/s	1183	1236	1289
Unit capital cost (p/kWh)[3]			
6.5 m/s	49.4	52.2	52.4
7.5 m/s	37.0	38.7	38.5
8.5 m/s	30.2	31.3	30.8

[1] Fully installed including all site works plus margins, etc.

[2] Assuming a Rayleigh distribution, 100% availability, no array losses.

[3] Ratio of capital cost to annual energy. A useful comparative measure independent of discount rate, lifetime, etc.

Table 1: Costs and benefits for a 33 m diameter machine

However, with the reduction in unit cost of variable speed equipment with rating it is possible that variable speed may be cost-effective for larger turbines. A comparison at 1 MW rating for the LS-2 design, which currently uses mechanical/ hydraulic variable speed, indicates that electrical variable speed may be more cost-effective in this case.

Also the price of variable speed equipment is expected to fall significantly with time and with volume production, which may make its use cost-effective in smaller turbines.

6. CONTROL ASPECTS

A comprehensive, albeit relatively simple, control algorithm for a variable speed turbine was developed, and tested on a detailed non-linear dynamic simulation model of the turbine driven by a realistic wind input. The control algorithm uses measured rotor speed as input, and outputs a torque demand for the variable speed drive and a pitch demand for the pitch servo system. The torque demand is used below rated wind speed to maximise power by tracking optimum tip speed ratio where possible, but also to give compliant operation at the maximum speed limit, and also to implement the speed exclusion zone to avoid tower resonance. The pitch demand is kept at fine until maximum torque is reached at rated wind speed. Then the torque demand is kept constant and the pitch demand is used to regulate rotor speed. It is important to ensure that there is a clean handover between torque and pitch control. Bivariate control strategies were not considered, as the need for anything as complex has not been demonstrated.

Figure 5 shows an initial comparison of fixed and variable speed operation below rated using the simulation model. This shows smoother (and higher) power available from the variable speed system, with the variations at blade passing frequency (seen in the wind speed signal) eliminated.

The operation of the control algorithm is shown schematically in figure 6, and is divided into four modes: below tower resonance, above tower resonance, up to maximum speed, and above rated. Modes 1, 2 and 3 follow optimum Cp except where restricted by the tower exclusion zone or the maximum speed limit. Pitch is kept at fine and torque demand is varied to maintain the optimum characteristic or using PI control in the constant speed regions, where compliance can be selected by appropriate tuning of the closed loop control. Mode 4 maintains torque demand at maximum and varies the pitch demand to maintain maximum speed.

In fact, the control strategy which was developed demonstrated very good tracking of peak Cp, successful avoidance of tower resonance, good compliance at all times, very good regulation of maximum transmission torque and of rotor speed, and very clean switch-over between control modes. Figure 7 illustrates the small deviations from peak Cp during operation around the tower exclusion zone. Figure 8 shows operation in modes 2, 3 and 4 showing the smooth transition between modes, good speed control with adequate compliance both in torque and in pitch control. Transitions through the tower exclusion zone are shown in figure 9, which also illustrates how limiting the rate of change of torque demand has a beneficial effect on both power smoothness and tower excitation.

This exercise indicated that good control of a variable speed wind turbine with a stiff transmission is actually fairly straightforward. Advanced or multivariable control techniques are not required although they may provide scope for some minor amelioration of structural loadings. The situation may be different in the case of a turbine with a softer drive-train.

7. HARMONICS AND POWER FACTOR

Conventional variable speed drives with a DC link and line-commutated inverter have the disadvantage of producing a poor power factor, and high levels of harmonics due to the non-sinusoidal inverter currents. This part of the project was intended to assess the effect of this on the grid system and the possibility of mitigating the effects using appropriate filters.

The investigation made use of a power systems analysis

program developed at UMIST. This was used to model a typical weak grid situation based on the example of a wind farm located in rural mid-Wales. The effects of firstly a single 300 kW wind turbine and then a wind farm of 25 such turbines were examined. Both 6-pulse and 12-pulse inverters were investigated.

For a single variable speed turbine with a 6-pulse inverter it was shown that the addition of suitable power factor correction capacitors could also bring the harmonics within statutory limits. However, for turbines larger than 300 kW the use of a 12-pulse inverter may be more prudent, since the harmonic distortion is significantly lower.

For a wind farm of 25 300 kW turbines, harmonic distortion is reduced if alternate star and delta connection of 6-pulse inverters is used. Two different designs of filter were investigated, and acceptable power factor and harmonic distortion were found to be achievable. The cost of the filters would be under £4000 per turbine, incurring additional energy losses of 0.3% at full load.

8. OVERALL CONCLUSIONS OF THE PROJECT

Variable speed operation is technically feasible, and no major obstacles are envisaged. In terms of energy capture, two-speed operation may actually be more efficient overall, with higher electrical efficiency compensating for slightly lower aerodynamic efficiency. For stall-regulated machines, this makes variable speed hard to justify economically, even if a "hybrid" concept is used in which the rotor is braked into stall for high wind speed operation, although with variable braking torque, as is possible using load torque control, this may no longer be true. For pitch-regulated machines however, variable speed allows significant additional energy capture in high winds if the maximum rotor speed is increased. Although this incurs higher structural costs, these are justified by the additional energy capture. Variable speed also allows a lower rotor speed in lower winds, which is of significant benefit in reducing aerodynamic noise.

Some uncertainties still surround the reduction in gearbox cost which is possible due to the high compliance of variable speed systems. The estimated cost reduction is not as great as originally expected.

Tower cost is not expected to increase. Tower resonance can be prevented using a "speed exclusion zone" which allows more freedom to select the tower natural frequency and, although it may have to be quite wide, results in very little loss of energy.

Control design is relatively straightforward, allowing excellent tracking of peak Cp despite the effects of turbulence, good compliance at all times, even while operating at a nominally fixed speed, straightforward implementation of the tower exclusion zone, and smooth transitions between torque and pitch control.

The results of this study indicate that it should be possible to overcome grid interconnection problems. Suitable filters can be provided at reasonable cost to allow 6-pulse inverters to be interfaced to a weak grid, giving acceptable power factor and harmonic levels. 12-pulse inverters may be preferable for larger wind turbines. Advanced drives using

sine-wave inverters will eliminate the need for filters. These are now becoming available at suitable rating.

At MS-3 size (33m diameter) variable speed operation is at present not quite cost-effective compared to two-speed operation, although noise considerations may weigh in favour of variable speed. The unit cost of variable speed drives decreases significantly with rating, so for large turbines variable speed may be cost-effective.

However, variable speed drives are unlikely to be widely used on competitively priced commercial wind turbines in the near future unless the price of commercial drives decreases. Consideration of the component cost of advanced drives suggests that large price reductions should be possible, and may be realised with volume production.

9. REFERENCES

1. BOSSANYI, E. A., Electrical aspects of variable speed operation of horizontal axis wind turbine generators, WEG-R040-6202, August 1992.

10. ACKNOWLEDGEMENTS

This work was funded by the Department of Trade and Industry through the Energy Technology Support Unit under contract no. E/5A/6051/2426.

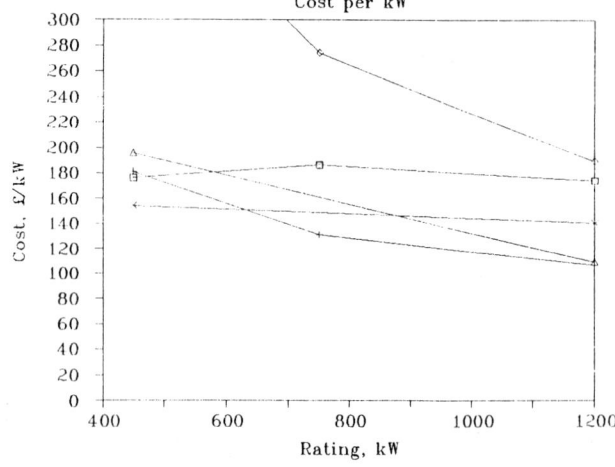

Figure 1: Cost reduction of variable speed drives with rating

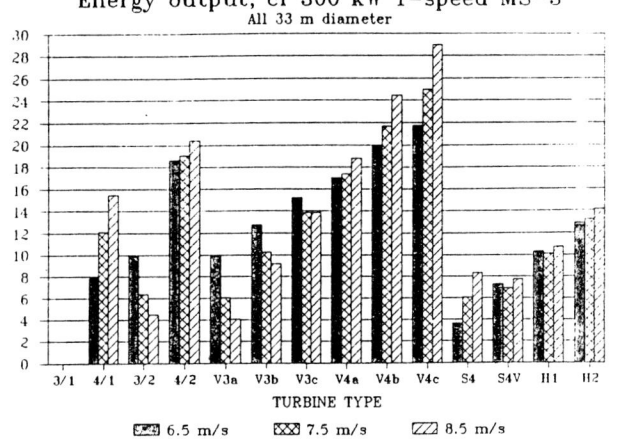

Energy output, cf 300 kW 1-speed MS-3
All 33 m diameter

6.5 m/s · · · 7.5 m/s · · · 8.5 m/s

Key:
3/1: Single speed 300 kW 48 rpm
4/1: Single speed 400 kW 48 rpm
3/2: Two speed 300 kW 32/48 rpm
4/2: Two speed 400 kW 32/48 rpm
V3a: Variable speed 300 kW
V3b: Variable speed 330 kW, top speed +10%
V3c: Variable speed 360 kW, top speed +20%
V4a: Variable speed 400 kW
V4b: Variable speed 440 kW, top speed +10%
V4c: Variable speed 480 kW, top speed +20%
S4: 40 rpm stall-regulated single speed
S4V: 40 rpm max stall-regulated variable speed
H1: Hybrid 48 rpm 400 kW stalling at 40 rpm
H2: Hybrid 48 rpm 450 kW stalling at 40 rpm

Figure 2: Energy yield comparison

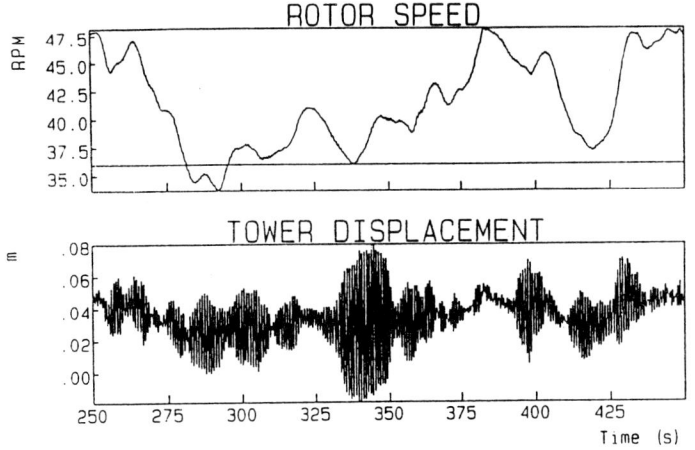

Figure 4: Tower excitation at critical rotor speed

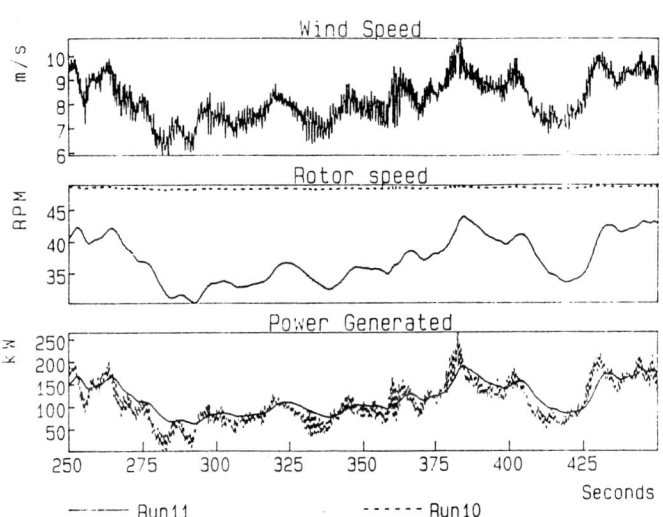

Run11 · · · · · Run10

Figure 5: Simulation of fixed and variable speed running

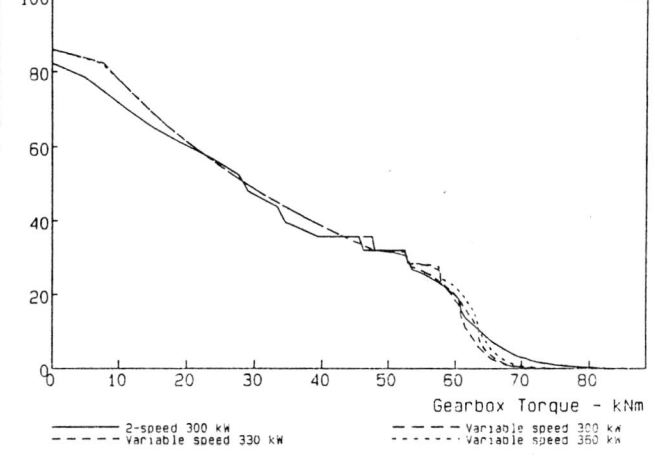

2-speed 300 kW · · · · Variable speed 300 kw
Variable speed 330 kW · · · · Variable speed 360 kw

Figure 3: Gearbox torque exceedance curves

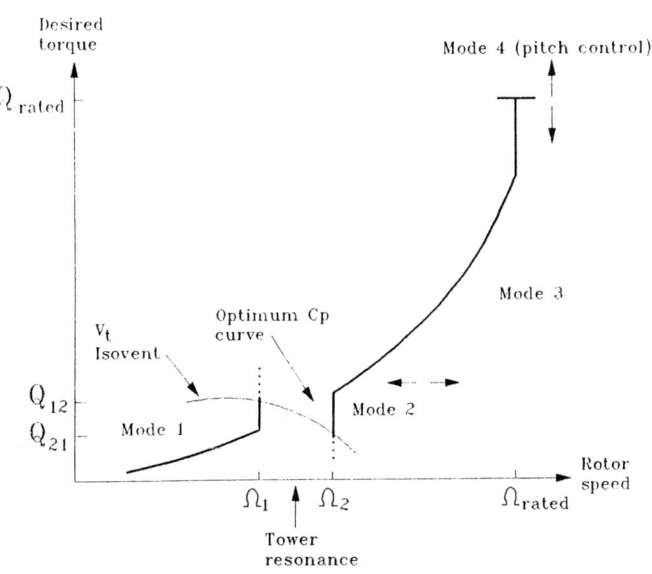

Figure 6: Schematic showing control algorithm operation

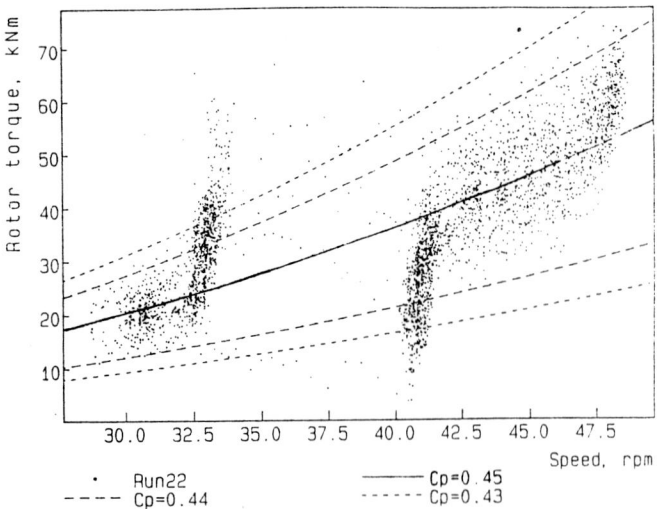

Figure 7: Accuracy of tracking optimum Cp

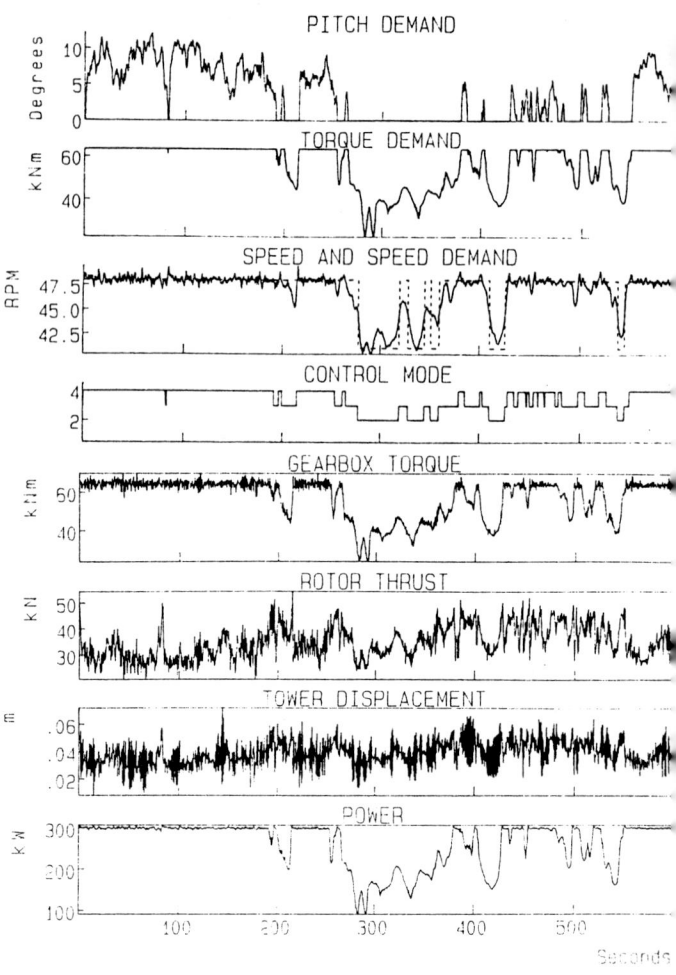

Figure 8: Example of variable speed simulation with full control algorithm in operation

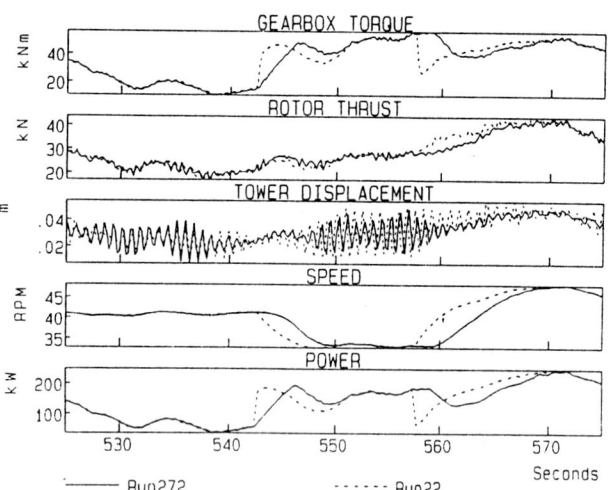

Figure 9: Operation of tower exclusion zone

An assessment of the offshore wind potential in the EC

A. D. GARRAD and **B. M. ADAMS**
Garrad Hassan and Partners Ltd, UK
H.MATTHIES
Germanischer Lloyd, Germany, UK
M. SCHERWEIT and **T. SIEBERS**
Windtest Kaiser Wilhelm Koog, Germany

1 Introduction

The work reported in this paper is one part of a four part project undertaken jointly by Germanischer Lloyd in Germany and Garrad Hassan and Partners in the UK. The project is entitled "Offshore Wind Energy in the EC". It is funded by the Commission of the European Communities, Directorate General for Research and Development, the Bundesministerium fur Forschung und Technologie and the UK Department of Trade and Industry. This part of the project focusses on the wind energy resource in the EC. The other three address the following areas: Review of Existing Offshore Practice, Dynamic Loading of Offshore Wind Turbines and Design Guidelines for Offshore Wind Turbines.

Some countries, notably the UK, Denmark, Sweden, the Netherlands and Italy have already undertaken detailed studies of the offshore resource in their individual countries. Their approaches were not, however, uniform and part of the brief from the EC for this study was to provide a uniform approach so that the scope of the resource in the EC as a whole could be determined in a very broad brush, but nevertheless logical and consistent, fashion.

The approach taken was therefore to look for a way of determining the wind speed over the offshore waters of the Community in a sensible, but manageable, fashion and to include as many physical limits as possible. Political limits were only considered as far as could be gleaned from the marine charts. It is recognised that the political limits are, in fact, more severe than the physical ones. They may, however, change whereas the physical ones will not.

The objectives of this part of the project are listed below:

o To assess the offshore wind energy resource in broad terms
o To determine how that resource is limited by physical constraints

2 Calculation of the resource

In the early stages of the project it was envisaged that the resource should, where possible, be calculated locally using local studies and local knowledge. It soon became apparent that such an approach was likely to be fruitless. The Deutsche Wetter Dienst - Seewetterampt - the German Meteorological Office was therefore commissioned to use the voluntary observer fleet data (VOF) to provide a summary of the offshore wind statistics for the EC. This work has been reported in detail by Schmidt [1]. The authors have noticed a considerable degree of scepticism amongst meteorologists about the validity and hence usefulness, of this body of data. It is, therefore, worth providing a few words of explanation. The VOF consists of ships, above a certain size, who report, routinely, once every three hours on the state of the weather. These observations include, amongst other characteristics, the height and nature of the waves. Such an observation may then be used to make an estimate of the wind speed at the site.

These data are gathered centrally and are archived. Schmidt used data from the 1960's to the present day. He reported the data to this project as a series of mean values for different parts of the EC waters. The data he computed included not only mean wind speed but also air temperature, wave height and Weibull parameters. The areas which he used together with the corresponding spot values of wind speed are shown in Figure 2.1.

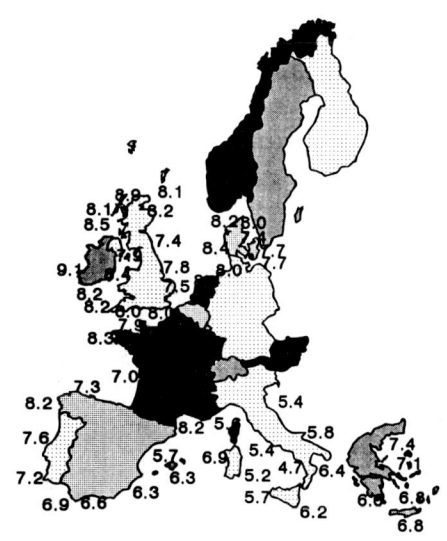

Figure 2.1 Spot values of wind speed from Deutscher Wetterdienst. 25m height

In all of the observer areas there were more than 50 000 observations and in some cases up to 400 000. When the number of observations is taken into account it becomes clear that these data may well provide a reasonable estimate.

Private communication with Erik Lundtang Petersen of Risø has allowed some comparisons to be made between WAsP type calculation performed by him and the DWD VOF data. In general it was concluded, by the authors, that where it was acknowledged that the WAsP approach worked well then the agreement was reasonable and where WAsP did not work agreement was not as good. Lundtang Petersen has also reported some estimates of the resource himself [2]. Schmidt [1] has discussed his procedure in some detail and his arguments will not be repeated here.

There are some data of offshore winds which are available: the Dutch K5 platform and the West Sole (UK) platform. These data were compared with that developed by Schmidt [1] and the general conclusion was that the DWD data was consistent with the platform observations.

One of the objectives of this study was to consider the offshore resource as a function of the distance from the shore. It was therefore important to try and include some estimate of the way in which roughness changes affect the offshore wind. It was decided to model the EC in 12 sectors using the WAsP approach. No topographical data was incorporated but the shore line was used to define the location of the roughness changes. In each of these twelve sectors a selection of up to five WAsP stations was used to initiate calculations. The WAsP stations were eventually selected using those that gave minimum deviations from the DWD data over the area. Each region of the sea was calculated using only one station. The results of these calculations were clearly unrealistic since large discontinuities in wind speed occured. However, within the results the effect of roughness changes and the influence of the wind roses were included. Such influences are not included in the DWD data which provides only a single figure as a spatial average over a very large area and which are derived from observation further offshore.

It was considered that a sensible, if rather pragmatic, approach to improve these results was to take the two sets of data and marry them thereby retaining the correct physical features from each one. To achieve this end the DWD data was interpolated around the coast line so that at each point around the coast the correct, interpolated, value was known. The "free sea" value from each of the WAsP analyses was then found and the pattern of wind speeds determined by WAsP was rescaled to give a "free sea" value equal to that of the DWD data. This process is shown schematically in Figure 2.2.

Figure 2.2 Derivation of wind speed data

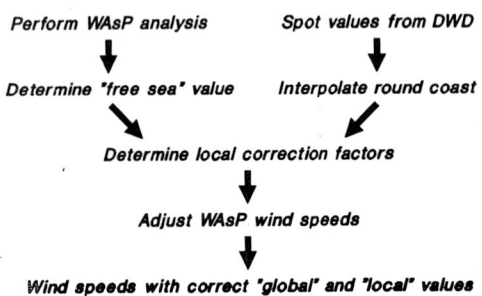

It is clear that such an approach is open to considerable criticism and, indeed, the general subject of offshore wind speed estimates appears to be one about which many people have views but few seem to have practical solutions. The result of the analysis described above was a set of maps and associated figures for all the European waters which exhibited realistic looking patterns, both on the broad scale, in terms of the variation across the Community, and locally, in terms of the variation with distance from the shore.

3 Geographical Information Systems

Early on in the project it was clear that it would be important to be able to handle all the data in an efficient and sensible way. Geographical Information Systems, (GIS's) are intended for just such tasks and a relatively simple system, called IDRISI, was chosen as a suitable tool. It was therefore necessary to obtain all the data in a form which would allow analysis by IDRISI.

4 Physical Data

It proved surprisingly difficult to obtain digital data which described the European waters in sufficient detail for this project. Eventually data, intended for use in computer aided navigation, was found which was apparently a source of suitable information to initiate the project. However the data proved to be lacking in some important ways and much effort was required to provide a uniform set of self-consistent information which could be successfully manipulated by a GIS system. Careful quality control checks of the data were performed and extra data was added from standard Admiralty, or similar, charts. Close inspection of these maps and, indeed, the digital data retrieved from them, demonstrates some of these difficulties. For example, many contours are unclosed which makes the analysis of the data difficult. Before the data could be used in the chosen GIS system it was necessary to close all the contours. Even the coastline data was found to be fragmented and, in some sea areas, the water depth contours were very sparse. Also the only source of digital data which was identified was a literal digitisation of the paper charts and hence dotted lines on the charts appeared as a long series of small lines in the digital image which had to be linked manually. It is important to stress the effort which was required to undertake this task but the important outcome was a set of digital maps which could be used for analysis. The bit-mapped information had to be converted to a vector image which again was time consuming.

5 Constraints

Some of the earlier studies which were performed by the various European countries took great care over the constraints which were applied to offshore development. Given the scope of work which was involved in this project: this subject is only one of four separate items, such detail was impossible. To cover the coastline of a single country might involve consultation with hundreds of consultees. The result of this simplification is that the present project will tend to over estimate the available energy by a large amount. However, it is noted, that in the

Danish study 90% of the Danish waters were removed as a result of the constraints imposed by the military. It is considered that offshore wind energy is a strategic resource and hence, should it be exploited, it will be done so after a strategic decision. It will be necessary at that time to determine the relative importance of energy and defence. A decision which is clearly outside the remit of this study!

The data gathered from the charts and used in this study are listed below:

    Water depth
    Sea bottom slope
    Navigational routes
    Defence areas
    National parks
    Gas pipelines
    Oil pipelines
    Undersea electrical cables
    Oil and gas platforms

In addition IDRISI was able to calculate distance from the nearest shore for every point on the map and hence that parameter could also be used as a constraint.

6 Example calculation

The sections above have described the general approach which was used for the determination of the wind resource and the the constraints. In this section the means of combining them will be described. To provide a vehicle for this explanation one of the sample areas will be used as an example. For convenience of presentation, in addition to using arbitrary sea sectors, the analysis has been performed for each country and hence it is possible for the results to be presented on a national, as well as a regional, basis. The example taken here is the Danish and German coastline - the Jutland Peninsular, but results are presented by country.

Figure 6.1 shows the result of the wind speed calculations described in Section 2. The speeds are presented for a height of 60m.

Figure 6.2 shows the constraints: defence areas, shipping lanes etc together with the Wattenmeer national park.

Figure 6.2 Constraints for Denmark

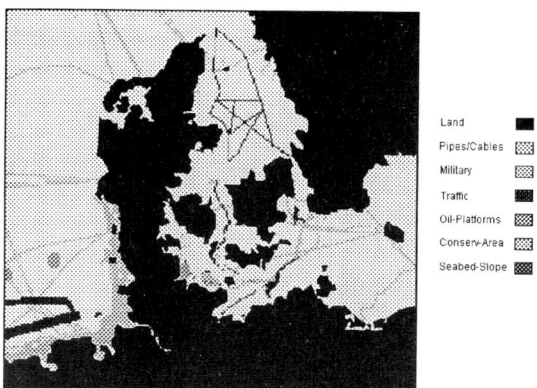

Figure 6.3 shows water depth. Distance offshore and sea bed slope are also included as constraints but are not shown graphically.

Figure 6.3 Maximum water depth for Denmark

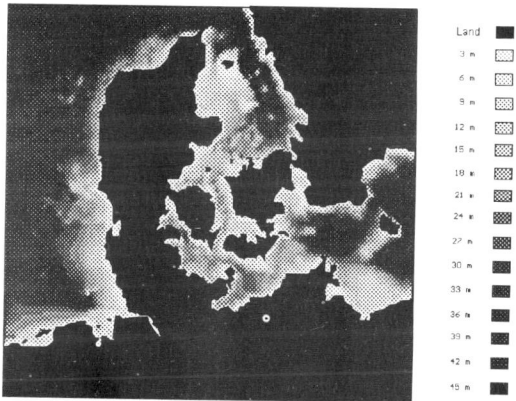

Figure 6.1 Offshore wind speeds for Denmark

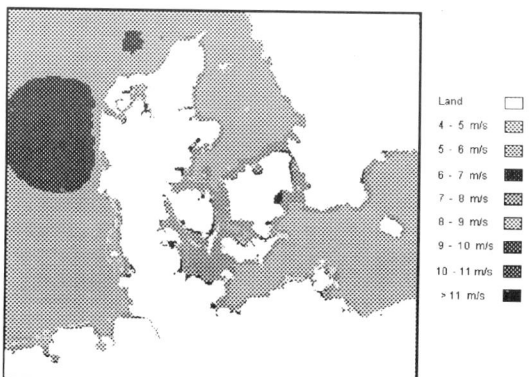

7 Energy Calculations

Now that all the constraints have been assembled it is possible to use them, in conjunction with the wind resource map, to exclude areas which are considered inappropriate, to identify the remaining areas and to use IDRISI to compute the energy which they contain. They are shown, for the Danish example, in Figure 7.1.

Figure 7.1 Available areas & annual wind speeds

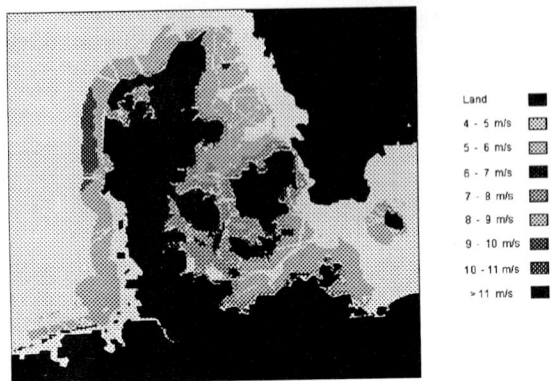

IDRISI allows the calculation of the areas and their attributes to be determined easily. From this combined resource-constraint map it is possible to determine the energy available outside the non-physical constraints. A maximum distance of 30km offshore was set together with a maximum depth of 40m. The results are presented in several different physical categories as defined in the table below:

Distance offshore	0-10km
	10-20
	20-30
Water depth	0-10m
	10-20
	20-30
	30-40

There are various ways in which the energy may be calculated and presented. It was decided that a sensible approach was to use large wind turbines which were well spaced out. The wind turbine chosen was one designed for offshore application in the UK and also used in one of the other parts of this study to illustrate some dynamic loading calculations. It is 100m in diameter and rated at 6MW. The inter-machine spacing used was 1km. It is a fairly easy task to take the data developed for this machine and to recompute the performance using other candidate machines. For convenience the array losses have been assumed to be zero and the machines are deemed to have an availability of 100%. The DWD data contained within it the way in which the Weibull shape parameters vary over the EC and also the mean air temperatures. These two characteristics were therefore included in the energy calculations.

8 Results

Using these data, and the resource-constraint map shown in Figure 7.1, it is found that Denmark's offshore resource is very large. The results are illustrated in Figure 8.1 where the resource is shown as a function of both distance offshore and water depth.

Figure 8.1 Offshore potential in Denmark
Annual consumption 32.2 TWh

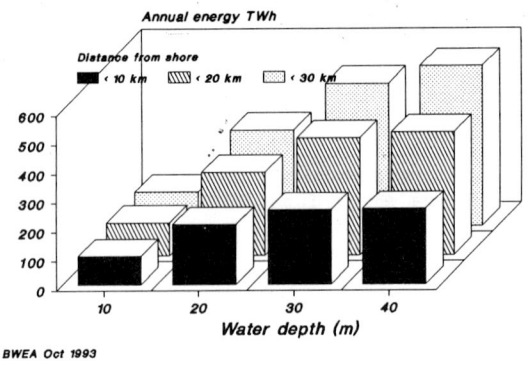

The total national electricity consumption is also shown to put the wind resource figures into perspective. The figure shows that Denmark has three times its electricity demand available from the offshore resource less than 10km offshore and in a water depth of less than 10m.

Denmark has been used as an illustration. Exactly the same procedure has been carried out for all the other countries. Figure 8.2 shows analogous results for the UK and Figure 8.3 those for Germany. Figure 8.4 gives a summary for all the EC countries as a whole. Finally, Figure 8.5 shows the resource for each of the EC countries together with their total consumption.

Figure 8.2 Offshore potential in the UK
Annual consumption 321 TWh

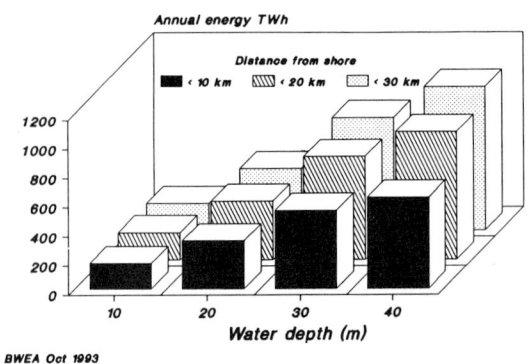

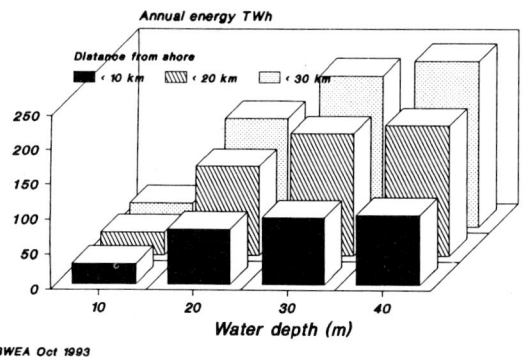

Figure 8.3 Offshore potential in Germany
Annual consumption 431.5 TWh

BWEA Oct 1993

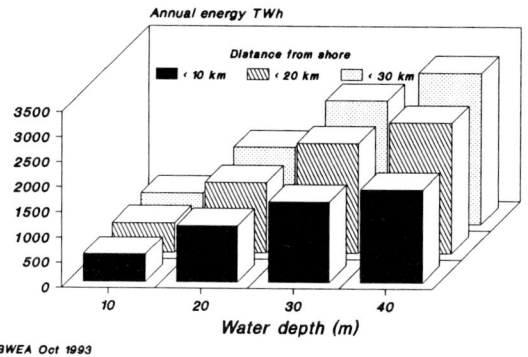

Figure 8.4 Offshore potential in the EC
Annual consumption 1727.3 TWh

BWEA Oct 1993

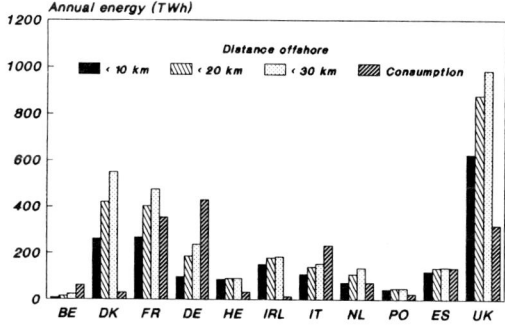

**Figure 8.5 Energy within 40m depth
All EC states**

The results are tabulated in full in the main report [3]. In this report there is some discussion about the way in which the resource is limited by different constraints in the different member countries. For example in Italy and Ireland the major constraints are water depth and sea bed slope whereas in Germany the wind speed and extent of coast line are dominant. It should be stressed that the real constraints will considerably diminish this resource.

The maps presented for each country can be used to help identify likely areas of search and the tables can be used to determine whether it is correct to aim research efforts at deep or shallow sites and also to determine the importance of cable costs.

9 Conclusions

This study is the first to consider the EC offshore resource as a whole. It has not entered into as much detailed evaluation of constraints as some of the local studies. Perhaps the most important result of this study is the fact that a major effort has been devoted to the construction of a well- validated and comprehensive database into which other data can easily be added. Within the GIS system it is now possible to include all the additional constraints as they appear and to continue to re-calculate the resource as better data becomes available.

One thing is sure – the offshore wind resource of the EC is truly enormous.

10 References

1   Schmidt H, "Wind and Wave Conditions in 55 Coastal Sea Areas of the European Community Determined from Weather Observations of Voluntary Ships". Proceedings of ECWEC Conference, Travemünde, Germany, 8-12 March 1993.

2   Pedersen E-L, "Wind Resources Part I, The European Wind Climatology". Proceedings of ECWEC Conference, Travemünde, Germany, 8-12 March 1993.

3   Matthies H, "Study of Offshore Wind Energy in the EC". Report of work performed under CEC JOULE Contract JOUR-0072, Germanischer Lloyd, Hamburg. To be issued December 1993.

FLOAT - a floating offshore wind turbine system

K. C. TONG PhD, CEng, MRINA,
Tecnomare (UK) Ltd, UK
D QUARTON BSc,
Garrad Hassan & Partners Ltd, UK
R. STANDING PhD, BMT
Offshore Ltd, UK

SYNOPSIS Development of the conceptual design for FLOAT - an offshore floating wind turbine - is described in this paper. This design represents a marriage of the wind power and offshore oil and gas technology. The objective of the FLOAT project is to develop a floating wind turbine system enabling the economic generation of electricity from wind power in offshore locations, typically between 100m and 300m water depth.

1 INTRODUCTION

The environmental considerations surrounding electricity generation by conventional means are forcing the renewable alternatives to be seriously considered worldwide. The economic attraction of wind energy places it at the forefront of exploitation. The lack of space however, particularly in Europe, may present an obstacle to high penetration of wind energy in countries with a relatively high population density where wind turbines may be considered visually intrusive. This is currently leading to increased public pressure to site wind turbines somewhere else but 'not in my back yard' !

The logical conclusion of this development is that offshore sites for wind turbines which would allow the generation of large amounts of electricity from wind energy, but without public opposition must be explored. Going offshore enables turbines to be installed in plentiful numbers without the planning and space constraints found onshore. It also means that the turbines can be designed to exploited at highest possible efficiency the generally higher wind speeds found offshore, freed from the environmental constraints onshore.

Interest in offshore wind turbine technology has been increasing. Denmark and Sweden have already constructed fixed structure systems offshore. However, the use of conventional machines on fixed seabed supports limits the useful sites to relatively shallow water. To overcome this limitation, the concept of floating turbines is being developed to extend the flexibility of siting and the number of potential sites.

Offshore windfarm development using floating support structures, being relatively insensitive to water depth and seabed conditions, enables installation in plentiful numbers without the planning and space constraints found onshore. They can be further coupled to offshore sites with higher wind energy potential and turbine design with higher tip speed to increase the overall efficiency and economy. However, the overall wind farm must be considered and integrated as a system to achieve the best overall economy.

The FLOAT project was undertaken by Tecnomare (UK) Ltd., Garrad Hassan & Partners Ltd., and BMT Offshore Ltd. to assess the feasibility of using floating platforms for wind turbines. The project was funded by DTI under the Wealth from the Oceans Programme. The project aimed to provide technical definition of a specifically designed offshore windfarm system together with associated operational and economic considerations. Tecnomare (UK) Ltd. is the offshore engineering and design company responsible for design of the floater, mooring system, fabrication and installation considerations, IMR, and overall project management. Garrad Hassan is the turbine design consultant responsible for the turbine design specifically for the floating windfarm, and electrical aspects. BMT is the offshore consultant responsible for the environment data analysis and model testing.

2 OUTLINE OF DESIGN PROGRAMME

The project was carried out in three phases. Phase 1 was a preliminary design comparative study aiming to select an optimum configuration for more detailed design and analysis in Phase 2. The design basis were also laid down at early stage of this Phase. A range of floater concepts were considered, including the most common conventional forms (such as barge, SPAR buoy) to the more unconventional concepts (such as semisubmersible, Donut, twin turbine semisubmersible and catamaran). Simple approximation methods for design, dynamics, materials, and cost assessment were used. The turbine configuration was selected to suit the offshore floating system but within the reach of proven technology. All aspects of the turbine and generator system were reviewed and considered. A concept was selected for the more detailed design studies in Phase 2 by assigning scores to the system performance, system complexity, dependence on unproven technology, and economic criteria.

A SPAR buoy concept was selected for the detailed design study for Phase 2. A conceptual design for the floater structure, tower, mooring system, turbine and electrical transmission systems were developed. Detailed dynamic analysis of the floater motions and the turbine system under the combined wind and wave action were carried out. Fabrication, installation, IMR (Inspection, Maintenance, Repair), safety and environmental impact were also considered at length. A detailed cost analysis were also carried out assess the economic viability of the wind farm.

Because of the novelty of the FLOAT system, heavily influenced by both wind and wave actions, a comprehensive model test programme was conducted in Phase 3 to validate the design and identify any unexpected behaviour.

3 DESIGN SCENARIOS

Two design scenarios were specified at the beginning of the project to focus the development while allowing reasonable assessment of the sensitivity of the system to various parameters. The selection was based on wind resources, water depth, distance to land, perceived energy market, and relatively benign environment. Two locations around UK and European waters were nominated as possible installation sites : Northern Irish Sea and Central Aegean Sea around the Greek Islands.

BMT undertook detailed assessment [1] of environmental conditions including current, tides, and joint distributions of wind speed, wave height and period. This assessment was based primarily on visually observed wind speed and wave height statistics, enhanced using BMT's NMIMET analysis package, together with published estimates of 50-year wave heights and current data. Parameters in the model are based on the joint distribution of observed wind data distribution only. Joint probability tables of wave height and wind speed, and of significant wave height H, and period T_Z, were constructed for each of the two selected areas. The 50-year design wave height was determined from this study together with the recommendations in the Department of Energy's Guidance Notes [2] for offshore installations.

It was recommended that the design current speed for Area 1 should combine a tidal current of 1.0 m/s with a storm surge component of about 0.4 m/s. Available evidence indicated that the tidal current in the Aegean is negligible, and a storm surge component similar to that around the UK coast was recommended.

In view of the dynamic nature of the FLOAT structure, it was recommended that extreme and fatigue analyses should generally be based on the spectral approach, using the JONSWAP wave spectrum, rather than on a non-linear regular wave model. Breaking waves were considered unlikely to be of significance for overall loads and motions of the structure in water depths between 100 and 300m. Local impact loads on the deck structure might be caused by spilling breakers in extreme wave conditions. If these loads are considered a matter of concern in later detailed design, it was suggested that they should be analysed using a suitable non-linear regular wave theory.

As a result of this in-depth environmental study, the design conditions were specified and summarised as :

	Max. Operating	Max. Survival
Hub Height Wind Speed, m/s	26	40
Significant Wave Height, m	4.0	8.0

and a correlation formula for wind speed with mean significant wave height, and mean wave period with wave height are given as :

$$H_s = \sqrt{(a \cdot U)^2 + h^2}$$

and

$$T_z = \alpha + \beta \cdot \sqrt{H_s}$$

were used as design conditions throughout the operating range.

4 FLOAT WINDFARM SYSTEM DESIGN

The complete wind farm system was considered in the design. Systems integration with regard to aerodynamic interferences, mooring and anchoring arrangements, in-field and export power transmissions were carefully balanced.

The final selected design is shown in Fig. 1. The system consisted of a three bladed turbine rated at 1.4MW supported by a steel tripod spaceframe tower at 45m above SWL. The tower was bolted onto the deck of a concrete cylindrical buoy hull with a wider bottom disk to improve the dynamic behaviour. The buoy was moored onto the seabed by 8 lines.

4.1 Floater System Design

The floater support system for FLOAT includes the tower, the floater hull, and the moorings to the seabed. It was decided to set the hub height at 45m, for the 60m diameter rotor, instead of the conventional land based practice of one diameter. This being a compromise between reducing the lever arm for the weight and wind moment due to the turbine and nacelle and to avoid wetting of the blades. Stability turned out to be one of the driving design parameter for the floater due to the large wind overturning moment. This gave rise to a compelling need to reduce top weight, i.e. weight of the tower and turbine, to maximise overall floater stability.

The final design was a compromise between conflicting requirements : minimising size (cost), maximising stability, and minimising dynamic motion response. Free yawing nacelle was adopted due to the doubt on the ability of the floater o provide sufficient reactive moment for the active yaw control mechanism.

4.1.1 Hull Design

A variety of possible floater configurations were

investigated in the phase 1 study including simple barge form to complicated four column semi-submersible hull form. Twin turbine configurations were also considered. A simple SPAR buoy concept of concrete construction was selected for Phase 2 studies.

After investigations on various materials of construction, concrete was considered most cost effective given the size of the buoy, material unit cost, fabrication cost and the requirement for counter ballast for stability.

The final buoy design has 3570 tonne displacement, featuring a 12.0m diameter circular cylinder of 28.5m height standing on a 19.0m diameter bottom disc of 2.5m thick (see Fig. 1). The buoy was a simple tubular concrete shell construction selected to be easy and cheap to fabricate. The reinforced concrete hull structure consisted mainly of 250mm thick side shell, 400mm thick top and plant room decks, 1800mm thick bottom slab with reinforcement rings and corbels for supporting mooring and tower connections locally.

Approximately 780 tonnes of ballast was required. One option consisted of approximately 660 tonnes of mass concrete and 120 tonnes of synthetic foam. The concrete ballast would occupy the lowest void space to provide stability. The synthetic foam would fill up the vast space between the top of the concrete ballast and bottom of the plant room deck to avoid free flooding if the concrete shell cracked.

4.1.2 Mooring System Design

The primary function of the mooring system was to keep the buoy staying in position under the drifting action of wind and wave. It does not affect the wave induced oscillatory motion of the buoy in general. Adhering to good offshore engineering practice of keeping the buoy in place when one line is broken resulted in 8 mooring lines used per buoy.

The design was adaptable to water depth between 75m and 500m. The mooring lines were catenary chain or taut wire synthetic fibre rope depending on the water depth. The mooring lines were connected to piled anchors at the seabed thus allowing multiple lines sharing a single anchor to reduce the mooring cost which represents a significant contribution to the overall system cost. (see Fig. 2)

Most of the mooring equipment (winches, cable gripper, guide sheave) is located on the main deck. The top section, consists of steel wire rope, of each mooring line passes through the bent shoe fairleaders 4m below waterline and then finally held in position by wire grippers. Line tensioning is provided by a pair of linear winches which move around the deck on a circular rail to access all lines. This arrangement provides for tensioning of pairs of opposite facing lines during installation and occasional tensioning its lifetime. A hydraulic power pack supplying the spooling and linear winches is housed in the control/plant room.

4.1.3 Tower Design

The critical characteristics of the tower were to avoid resonance at the nP (i.e. multiples of the turbine rotation frequency) and its weight, which translates into cost. Phase 1 study concluded that the most suitable tower configuration, in terms of costs and structural efficiency, would be a tripod lattice tower constructed from grade 450N/mm2 steel tubulars.

FE structural analysis were carried out to demonstrate the structural integrity and strength of the tower for the various types of loading condition and to derive a suitable simplified model to incorporate into the integrated turbine dynamics analysis model.

The geometry of the tower at its base and top is governed by the size of the hull and turbine slew ring respectively. With 12m diameter cylindrical support, the tower base plan is an equal 8.8m side triangle. Local thickenings in the concrete shell are provided to facilitate the connection of the tower legs to the hull by holding down bolts and base plates. The height (39m) of the tower is governed by the rotor hub siting at a height 45m above the sea level.

4.2 Turbine Design

The selection of a wind turbine suitable for an offshore floating system raises some basic design issues. It is reasonable to assume that to justify all the balance of plant costs beyond the direct costs of the wind turbine support structures themselves, large units must be used. Evidently a minimum tower height and associated machine size will also be dictated by wave heights and hydrostatic stability considerations.

Rather than simple marinisation of land based versions, the turbine was specifically designed to exploit the benefits of offshore locations - including higher wind speeds, no noise constraints - to aim at a machine with higher overall efficiency. The possibility of complicated interaction with the floating support structure was considered carefully. One of the consequences being the adoption of a free yawing nacelle with downwind rotor.

4.2.1 Design Features

At an early stage of FLOAT project, a rotor diameter of 60m was selected as being as large as within the reach of proven wind turbine technology. Although larger wind turbines have already been built, they have been neither reliable nor economical.

The selection of a downwind, coned, three bladed configuration was driven mainly by concern over the yaw system design. It was assumed early in the project that the floating support structure would be unable to provide adequate reaction in yaw for active powered yaw control. The selected configuration offers reliable yaw stability when the wind turbine is operated in a free yaw mode.

Analysis carried out at later stage indicated that the floater system does, in fact, provide sufficient inertia for active yaw control. However, due to the extremely lightweight and flexible rotor, the downwind configuration is necessary in order to avoid excessive shaft overhang and imbalance of tower top weight, and at the same time provide adequate clearance of the blades from the tower.

Since an offshore wind turbine is not constrained by restrictions on noise emission, it is able to operate at a high rotational speed with consequent benefits in terms of significantly reduced tower head weight and cost. The main limitation on increasing the blade tip seed are marginal reduction in energy yield, the feasibility of blade structural design and the incorporation of a system for power and speed regulation of the rotor. Following consideration of all the issues involved, a blade tip speed of 120m/s was selected for the FLOAT rotor.

The result is a lightweight rotor of GRE/CFRP construction with free yawing nacelle weighing 51 tonne. The machine was rated at 1.4MW at 12.5 m/s hub wind speed, cutting in at 6 m/s and cutting out at 26 m/s. The rotor is three bladed of 60m diameter with low solidity operating downwind with 10° cone angle. Rotating at 38.2 rpm, a very high tip speed of 120m/s is achieved. The power train consists of two stage epicyclic gearbox driving an 4.16 kV induction generator at 1000 rpm.

4.2.2 Dynamic Analysis

The dynamic behaviour of the floating wind turbine has been investigated under operating and extreme environmental conditions. Existing computer programs developed by Garrad Hassan for the calculation of wind turbine loading have been extended to enable analysis of the coupled behaviour of a floating system subject to both wind and wave loading. A schematic of the analytical approach to the dynamic analysis is shown in Fig. 3.

It is evident from the results of the dynamic analysis that the influence of wave loading of the floater on the fatigue of the rotor and drive train is minimal. The fatigue loading of the tower structure is more sensitive to wave loading but this is of little concern since fatigue is unlikely to be design driver for this component. The calculations of extreme loading have indicated that for the rotor, drive train and tower, the extreme loads occur during operation of the wind turbine. This is explained by the very low solidity of the rotor and hence the inability to develop large drag loads when the rotor is parked.

A study of gyroscopic yaw loading of the wind turbine as a result of floater pitch motion has indicated that the loads are rather small and pose no serious problems for free yaw operation.

4.3 Wind Farm Design

A typical windfarm consisting of 3 arrays of 3 machines, is depicted in Fig. 2. The centre to centre distance between adjacent machines exceeded 1 km so that turbine aerodynamic interference was negligible. Most anchors were shared by 2 or 4 mooring lines so reducing the cost of the piles and installation.

Power generated at 3.3 kV will be carried by a dynamic cable through the bottom of the buoy to a submerged buoy connector near the seabed. Power will be gathered to a submerged transformer by in farm static cables connected to each connector buoy. Transmission to shore is via 33kV static cable after stepping up by the underwater transformer.

4.4 Fabrication and Installation

The overriding criterion for any fabrication site is the water depth for float out and subsequent tow to the open sea. The present FLOAT configuration has a relatively high lightship float out draft in excess of 25 metres. This will have to be reduced to suit the selected fabrication facility and can be achieved by a number or means including the placing of ballast and equipment after float-out, the use of air bags or the partial fabrication in the dock with afloat completion in deep water.

The completed FLOAT structures can be transported from the fabrication facility to the wind farm location by either "wet" or "dry" tow. "Wet" having the buoys floating and being towed by a suitable vessel and "dry" having the buoys on a suitable transportation vessel that could off load the buoys at site.

The installation of the FLOAT wind farm can be split into two distinct phases. Phase 1 is the laying of the pre-installed mooring lines consisting of piled anchor, and lower chain or synthetic rope section. It can be completed by a suitably equipped semi-submersible multi service vessel or a small crane vessel. Phase 2 is the arrival on site of the FLOAT buoys and their subsequent hook-up of the pre-installed mooring lines with the upper wire section of the mooring line and the buoy. This will require 2 to 3 vessels.

Following the initial placing of the buoy in the mooring spread and the low powered tensioning of the lines the linear winches will be used to obtain the final working tension. Diagonally opposite lines will be tensioned together such that the buoy does not move significantly off station. Once a pair of lines has been tensioned to the working condition the lines will be locked off and the winches moved on rails to the next pair of diagonal lines.

4.5 Safety and Environmental Review

BMT undertook a marine safety review addressing navigational and operational hazards to personnel and other marine traffic, and the legislative framework within which they must be managed. It also identify the likely effects of safety provision on the system's economics and performance, in order that these aspects might be addressed more fully in further phases of the development.

BMT also undertook a review [4] of the environmental impact during the development of the FLOAT project. The study reviewed project design, relevant authorities and advisory bodies that should be consulted during site selection, and baseline environmental conditions in each area.

The main environmental concerns for wind farms are noise and visual impact which would be avoided by selection of suitable offshore site. Effects on human activities would include limited access for vessels in the area for safety reasons. The site should avoid area for fishing, recreation, and main navigational channel.

The main physical environmental effects would occur during the installation phase. Localised areas of the seabed would be damaged due to piling of anchors and subsea trenching operations for laying cables to the shore. Most areas of the seabed can be expected to recover from these operations in a relatively short time span, and might eventually benefit from the presence of the buoys due to additional colonisation on the 'artificial reefs'. More sensitive seabed areas, such as coral formations, which support a wide range of sea life, might be irreparably damaged and should be avoided.

5 MODEL TESTING

BMT undertook a systematic programme of model tests, at 1:48 scale, on the FLOAT structure in collinear waves and wind. These tests took place in BMT's No. 2 Towing Tank, and were intended to establish resonant periods and damping values, together with responses in regular and irregular waves. The objective of the model test programme was to validate the design and theoretical models, to find any unexpected features of the structure's behaviour, and to provide high quality photographic and video records.

The tests included both regular and irregular wave conditions, the latter including three operational sea states and three survival sea states, together with so-called 'pink noise' wave spectra.

Motions in surge, sway and heave were measured by means of a non-contacting optical system, with duplicate measurements to provide confidence in the data. A heave accelerometer was also installed. Roll and pitch motions were measured by means of a gyro, and wave heights were measured by capacitance wave gauges. Mooring forces or tensions, wind speed or applied wind load were also measured.

Damping values and natural periods were obtained from systematic series of free decay tests in still water. These test are essential for validation of the predicted motion responses and mooring loads. The heave damping was confirmed by a synthetic decrement analysis of motions in the 'pink noise' wave spectrum. This same technique showed that the heave damping in survival waves was approximately double that in still water and 'pink noise' conditions.

These tests were technically demanding, because of the mechanical complexity of the model, and because of the need to simulate both wave and wind loads simultaneously. The results therefore had to be examined carefully, with particular regard to consistency between different measurements and between runs, in order to gain confidence in the results.

It proved difficult to simulate wind conditions satisfactorily. It was not possible to model real wind conditions realistically, because of scaling difficulties, and so tests in real wind were expected to be of a qualitative nature, for validation purposes only. The majority of tests were undertaken with the in-line wind force simulated by means of a servo-controlled winch and line arrangement. Large motions of the turbine rotor at the point of attachment of the line made servo-control difficult. Inaccuracies in the wind loading seemed, however, to have relatively little effect on the measured motions.

Consistent sets of response amplitude operators (RAOs) were obtained from regular wave tests, operational and survival irregular wave tests, and from the pink noise tests, for the same five degrees of freedom. These RAOs generally proved to be insensitive to most variables: to changes in the mooring system, to whether the turbine was on or off, and to whether the nacelle was fixed or free. The heave RAO was reduced, as expected, in survival wave conditions, and when a heave damping plate was fitted to the bottom of the buoy.

6 COST ESTIMATE

Cost analyses were carried out to assess the economy of the FLOAT system. The analysis followed the fundamental principles of land based windfarm development cost analysis, and referenced to the recent RES and WEG studies, albeit for nearshore windfarms.

Costs have been estimated for both capital, CAPEX, and operating, OPEX. CAPEX have been determined for all major items required for design, fabrication and installation. OPEX, on an annual basis, have been determined for the operation, inspection, repair and maintenance.

The cost breakdown for a 9 buoy windfarm in Northern Irish Sea is shown in Fig. 4. The site was assumed 10 km from shore at 100m water depth with a nominal farm output of 12.6 MW. The overall capital cost for the farm was approximately £30 million. Mooring system was found to be the biggest cost element (28%) followed by buoy fabrication (19%) and turbines (18%). Power cables cost was also significant (15%).

For the alternative Aegean Sea site, assuming to be 5 km from shore at 300m water depth, the overall development cost was found to be almost the same. Slight variations of proportions were found.

Fig. 5 shows the cost of power in pence per kWh for the Northern Irish Sea site. The fundamental assumptions were : 20 year field life, 8% discount rate, income inflation of 5%, operating cost inflation of 1%, overall capacity factor of 41.1%, plus 10% profit. The selling price was found to be around 6 p/kWh for 5% rate of return, rising to around 11.5 p/kWh for 15% rate of return. However, the component of cost attributed to operating cost remained virtually constant at around 2 p/kWh while the capital cost component accounted for the selling price increase with rate of return.

REFERENCES

(1) Final Design Report, FLOAT Report FL3-330-001-TEC, 1993.

(2) Offshore Installation : Guidance on Design, Construction and Certification, Department of Energy, HMSO, 1990.

FIGURES

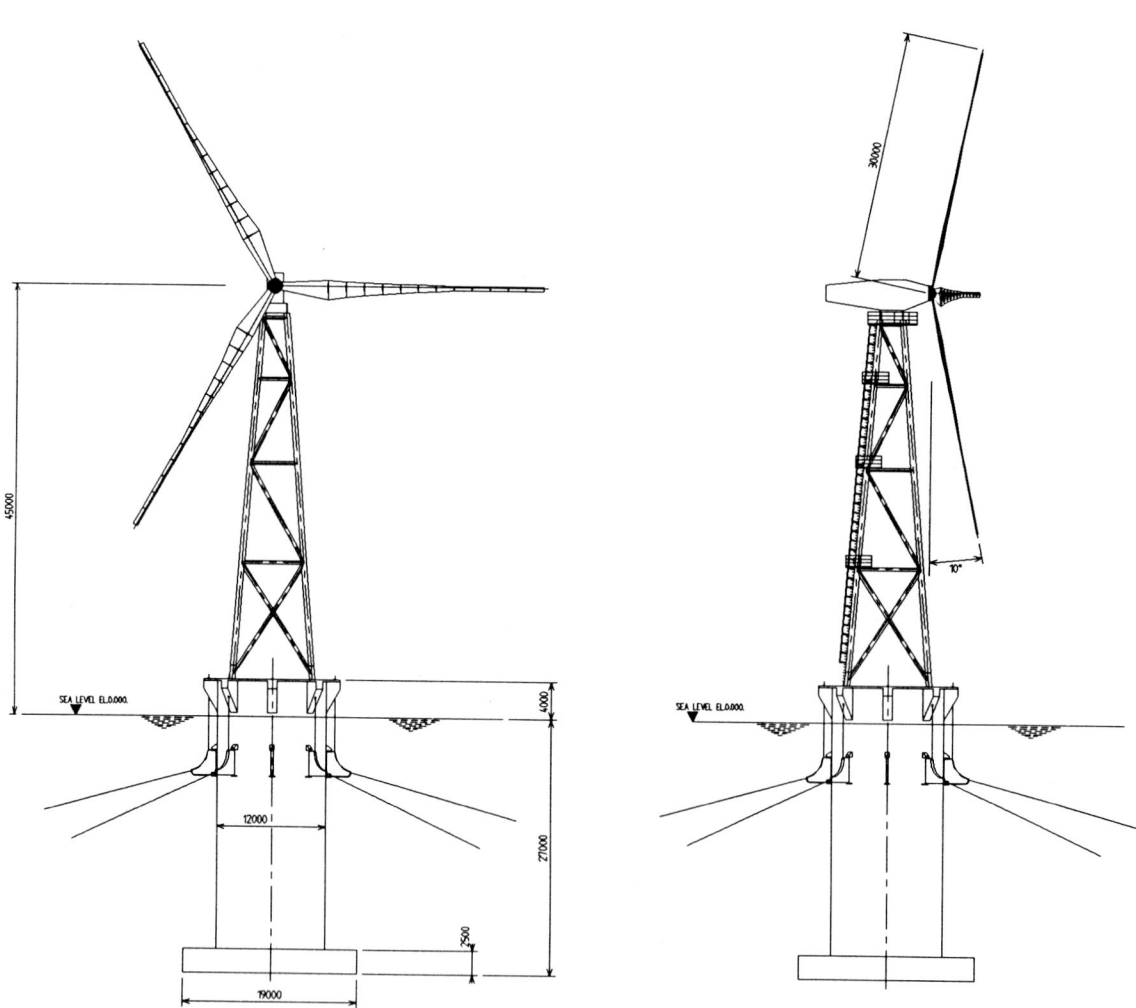

Figure 1 FLOAT design - general arrangement elevations

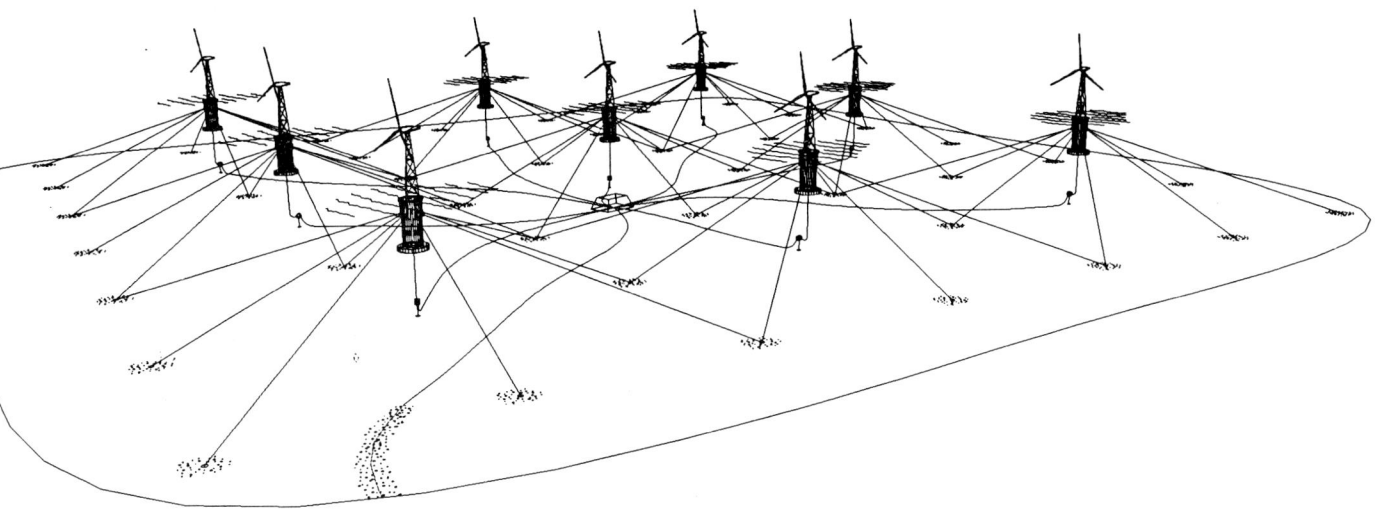

Figure 2 FLOAT design - perspective view of windfarm

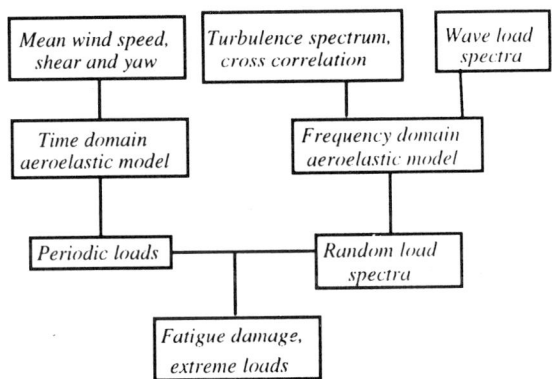

Figure 3 FLOAT turbine dynamic analysis

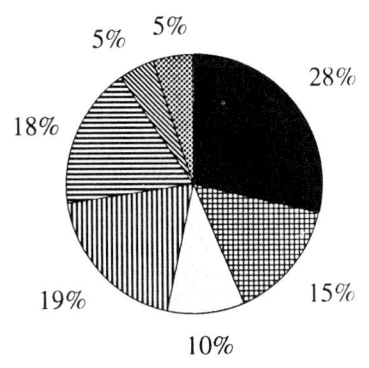

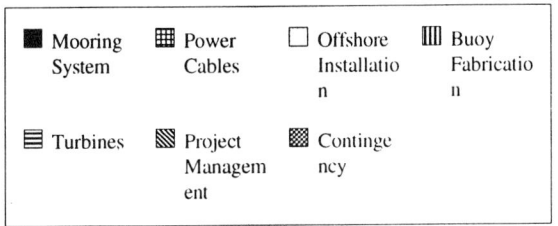

Figure 4 FLOAT capital cost - Total £30.1 M

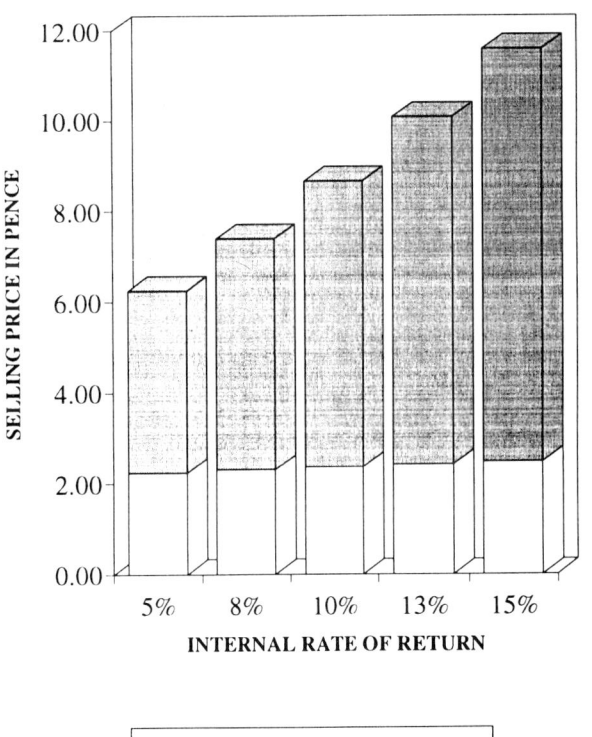

Figure 5 FLOAT energy cost, UK location

Delegates to the Conference

Mr Ben Adams
Garrad Hassan & Partners Ltd
The Coach House
Folleigh Lane
Long Ashton, Bristol
BS18 9JB
UK

Dr M B Anderson
Renewable Energy Systems Ltd
Eaton Court
Maylands Avenue
Hemel Hempstead, Herts
HP2 7TR
UK

Dr M P Ansell
School of Materials Science
University of Bath
Claverton Down
BATH
BA2 7AY
UK

Mr W Armstrong
17 Deepdale Avenue
Scarborough
Yorkshire
YO11 2VO
UK

Mr Nigel Barltrop
WS Atkins
Woodcote Grove
Ashley Road
Epsom, Surrey
KT18 5BW
UK

Mr Michael Birks
ETSU B154
AERE Harwell
Didcot
Oxon
OX11 0RA
UK

Mr Rod Blunden
National Wind Power Ltd
Riverside House, Meadowbank
Furlong Road, Bourne End
Bucks.
SL8 5AJ
UK

Dr Colin Anderson
Dept of Physics, R2107 JCM
King's Buildings
Mayfield Rd
Edinburgh
EH9 3JZ
UK

Mr Leif Anderson
Guttervej 60
9990 Skagen
DENMARK

Dr J R Armstrong
Wind Energy Group Ltd
Greenford House
Ruislip Road East
Greenford, Middx.
UB6 9BL
UK

Mr Brian Armstrong
EcoGen Ltd (Sea-West)
13 Pentrerhedyn Street
Machynlleth
Powys
SY20 8DJ
UK

Dr R J Barthelmie
Riso National Laboratory
Meteorology and Wind Eng Dept
P.O. Box 49
DK-4000 Roskilde
DENMARK

Dr J A M Bleijs
Dept of Engineering
University of Leicester
University Road
Leicester
LE1 7RH
UK

Mr I P Bond
School of Materials Science
University of Bath
Claverton Down
Bath
BA2 7AY
UK

Dr Ervin Bossanyi
Wind Energy Group Ltd
Greenford House
309 Ruislip Road East
Greenford, Middlesex
UB6 9BL
UK

Mr P C Botha
National Wind Power Ltd
Riverside House
Meadowbank, Furlong Road
Bourne End, Bucks
SL8 5AJ
UK

Mr Steven Brown
East Midlands Electricity
Generation Division
Caythorpe Road, Caythorpe
Nottingham
NG14 7EB
UK

Mr A L Burton
Tynyreithin
Carno
Caersws
Powys
SY17 5JS
UK

Mr Greame P Carney
Waltons & Morse
Plantation House
31-35 Fenchurch Street
London
EC3M 3NN
UK

Mr Jay Carter
Carter Wind Turbines Ltd
Beaufort Suite
Lockington Hall, Lockington
Derby
DE74 2RH
UK

Dr Susan Childs
European Orient Ltd
P O Box 1097
Norwich
VT 05055
USA

Dr Allan Clark
GEC Marconi Underwater System
Wilkinthroop House
Templecombe
Somerset
BA8 0DH
UK

Mr Andy Boston
PowerGen plc
Power Technology Centre
Ratcliffe-on-Soar
Nottingham
NG11 0EE
UK

Mr Nick Bristow
Markham & Co
Boad Oaks Works
Chesterfield
Derbyshire
S41 0DS
UK

Mr Andrew J Bullmore
Hoare Lee & Partners
140 Aztec West Business Park
Almondsbury
Bristol
BS12 4TX
UK

Mrs Wendy Capstick
ETSU
Bldg 149, Harwell Laboratory
Harwell
Oxon
OX11 0RA
UK

Herr Uwe Carstersen
c/o Henning Holst
WINKRA GmbH
Zingel 5 D-25813
Husum
GERMANY

Dr Ing Franco H Cavallini
44 Polhill Ave
Bedford
MK41 9DU
UK

Mr A J Chivers
Hojgaard & Schultz (UK) Ltd
Eastleigh House
Upper Market Street
Eastleigh, Hants.
SO5 1TN
UK

Dr Tim Claypole
Dept of Mech Engineering
University College
Singleton Park
Swansea
SA2 8PP
UK

Prof B R Clayton
Dept of Mech Engineering
University of Nottingham
University Park
NOTTINGHAM
NG7 2RD
UK

Mr John Coneybeare
SWEB
800 Park Avenue
Aztec West
Almondsbury, Bristol
BS12 4SE
UK

Mr D C Corbet
Carter Wind Turbines
Beaufort Suite
Lockington Hall, Lockington
Derby
DE74 2RH
UK

Miss Katy Cox
GEC Marconi Sonar Systems
Wilkinthroop House
Templecombe
Somerset
BA8 0DH
UK

Mr Peter J Crone
Farm Energy
Parkhills Industrial Estate
Rectory Road, Combe Martin
Ilfracombe
EX34 0LP
UK

Mr John D'Ardenne
New World Power Co. Ltd
179 Great Portland Street
London
W1N 5FD
UK

Mr Bill Davies
Knill Farm
Presteigne
Powys
LD8 2PR
UK

Mr M Davies
AEA Technology
Dounreay
Thurso
Caithness
KW14 7TZ
UK

Mr Aidan Cleary
Hydro Group HQ
ESB Generating Station
Ardnacrusha
Nr. Limerick
IRELAND

Dr Barry Connor
Industrial Control Centre
50 George Street
University of Strathclyde
Glasgow
G1 1QE
UK

Mr Frank Coton
Dept of Aerospace Engineering
University of Glasgow
Glasgow
G12 8QQ
UK

Mr Alan Creighton
Yorkshire Windpower Ltd
Wetherby Road
Scarcroft
Leeds
LS14 3HS
UK

Eng Pablo Martinez Cutillas
Aries Complex SA
Avda de la Industria
19 Tres Cantos
Madrid
28760
SPAIN

Mr Magnus Davidson
9 Little Boltons
Station Road
Marlow
Bucks.
SL7 1NR
UK

Mr Clive L Davies
Woolmer Forest Composites
44 Hilland Rise
Headley
Bordon, Hants.
GU35 8LZ
UK

Mrs Maureen De Pietro
Colham Energy Ltd
3 Cavendish Road
Henleaze
Bristol
BS9 4DZ
UK

Prof He Dexin
P O Box 211
Mianyang
Sichuan
People's Republic of
CHINA

Mr John Dodds
Mearsdon Manor Trading
Ballbrake Cottage
Moreton Hampstead
Devon
TQ13 8NL
UK

Mr Peter D Edwards
Windelectric Ltd
Deli
Delabole
Cornwall
PL33 9BZ
UK

Ms Patrina Eiffert
The Lodge
Hill Farm Lane
Duns Tew
Oxon
OX5 4JH
UK

Mr George Elliot
National Wind Turbine Centre
National Engineering Lab
East Kilbride
Glasgow
G75 0QU
UK

Mr John F Fawkes
Marlec Ltd
Rutland House
Trevithick Road, Corby
Northants
NN17 5XY
UK

Mr Ian Fletcher
ETSU B156
Harwell
Didcot
Oxon
OX11 0RA
UK

Dr Leon L Freris
Dept of Elect & Electronic
Engineering, Imperial College
Prince Consort Road
LONDON
SW7 2BT
UK

Mr J Diplock
Sen Projects Engineer (Gen)
South Wales Electricity plc
St Mellons
Cardiff
CF3 9XW
UK

Dr A G Dutton
Energy Research Unit
R63, Rutherford Appleton Lab
Chilton
Didcot, Oxon
OX11 0QX
UK

Mr Martin J Edwards
Windelectric Ltd
Deli
Delabole
Cornwall
PL33 9BZ
UK

Dr Nihad M El Chazly
National Research Centre
Tahrir Street
Dokki
Cairo
EGYPT

Ms Ana Estanqueiro
Az Lameriros
Lisbon
PORTUGAL

Dr Andrew Fellows
Garrad Hassan & Partners Ltd
6.05 Kelvin
West of Scotland Science Park
Maryhill Road, Glasgow
G20 0SP
UK

Mr Peter Fraenkel
I T Power Ltd
The Warren
Bramshill Road
Eversley, HANTS
UK

Ms Juanita Fromme
Enercon
Export Department
Schwachhauser
Heerstr 299 D-28211, Bremen
GERMANY

Mr Tim Fulford
Garrad Hassan & Partners Ltd
The Coach House
Folleigh Lane
Long Ashton, Bristol
BS18 9JB
UK

Dr Roderick Galbraith
Dept of Aerospace
University of Glasgow
GLASGOW
G12 8QQ
UK

Dr Andrew D Garrad
Garrad Hassan & Partners
Coach House, Folleigh Lane
Long Ashton
Bristol
BS18 9JB
UK

Mr Andrew Gibbs
Carter Wind Turbines
Beaufort Suite
Lockington Hall, Lockington
Derby
DE74 2RH
UK

Mr Christopher Glen
New World Power Co. Ltd
179 Great Portland Street
London
W1N 5FD
UK

Dr William Grainger
Border Wind Ltd
6 Station Cottages
Hexham
Northumberland
NE46 1EX
UK

Mr Marc Groves-Raines
EcoGen Ltd
Papyrus
Pentrerhedyn Street
Machynlleth, Powys
SY20 8DJ
UK

Dr Roger Haines
Wind Energy Group Ltd
Greenford House
309 Ruislip Road East
Greenford, Middlesex
UB6 9BL
UK

Dr Norman N Fulton
Switched Reluctance Drives Ltd
Springfield House
Hyde Terrace
Leeds
LS2 9LN
UK

Dr Eng Anton Garbacea
18A Take Ionescu Avenue
1900 Timisoara
ROMANIA

Mr Guang Geng
University of Durham
SECS Science Laboratories
South Road
Durham
DH1 3LE
UK

Mr Paul Gipe
606 Hillcrest Drive
Bakersfield
California
G3305
USA

Prof J M R Graham
Dept of Aeronautics
Imperial College
Prince Consort Road
LONDON
SW7 2BY
UK

Dr Andrew D Grant
Dept of Mech Engineering
University of Strathclyde
75 Montrose St
Glasgow
UK

Mr Rodney J Hacker
Halcrow Gilbert Associates Lt
Burderop Park
SWINDON
WILTS
SN4 0QD
UK

Mr Richard Hales
School of Mech Engineering
Cranfield Inst of Technology
Cranfield
Bedford
MK43 OAL
UK

Mr Stuart Hall
5 Dalmahoy Road
Ratho
Edinburgh
EH28 8RE
UK

Mr John F Hall-Craggs
The Lawn
Brightwalton
Newbury
Berks
RG16 0BP
UK

Mr Nick Hall-Stride
ETSU
Bldg 156, Harwell Laboratory
Harwell
Oxon
OX11 0RA
UK

Dr Jim A Halliday
Room 1-12, R63
Rutherford Appleton Lab
Chilton
Didcot, Oxon
OX11 0QX
UK

Mr Phil Hamilton
ICL
West Avenue
Kidsgrove
Stoke-on-Trent
ST7 1TL
UK

Mr Mark Hancock
Wind Energy Group Ltd
Greenford House
309 Ruislip Road East
Greenford, Middx
UB6 9BL
UK

Dr Paul Hannah
National Wind Power Ltd
Riverside House, Meadowbank
Furlong Road, Bourne End
Bucks.
SL8 5AJ
UK

Mr Michael Harper
54 Coity Road
Kentish Road
London
NW5 4RY
UK

Mr Kevin Hartwell
PowerGen plc
Power Technology
Ratcliffe-on-Soar
Nottingham
NG11 0EE
UK

Mr John Haynes
SWEB
800 Park Avenue
Aztec West
Almondsbury, Bristol
BS12 4SE
UK

Prof Constantine Helmis
33 Ippcratous Street
University of Athens
Athens
106 80
GREECE

Mr Till Hermjakob
Hanssenweg 14
20303 Hamburg
GERMANY

Mr Nicholas Hill
Acoustic & Engineering Consultants Ltd
6 Vernon Crescent
Ravenshead, Notts.
NG15 9BH
UK

Dr Jorgen Hojstrup
Riso National Laboratory
PO Box 49
DK-4000 Roskilde
DENMARK

Mr Dan Hollis
The Met Office
Johnson House
London Road
Bracknell, Berks.
RG12 2SY
UK

Miss Sarah C Holmes
Bond Pearce Solicitors
1 The Crescent
Plymouth
PL1 3AE
UK

Dip Ing Henning Holst
c/o WINKRA GmbH
Zingel 5 D-25813
Husum
GERMANY

Mr Raymond S Hunter
National Wind Turbine Centre
National Engineering Lab
East Kilbride
GLASGOW
G75 0QU
UK

Mr Peter Jamieson
Garrad Hassan & Partners Ltd
6.05 Kelvin
West of Scotland Science Park
Maryhill Road, Glasgow
G20 0SP
UK

Mr Michael Jefferson
British Energy Association
34 St James Street
London
SW1A 1HD
UK

Dr Garry Jenkins
Renewable Energy Centre
University of Sunderland
Chester Road
SUNDERLAND
SR1 3SD
UK

Mr William Brian Jenkins
22 Burfield Road
Old Windsor
Berkshire
SL4 2RD
UK

Mr Jonathan Johns
Ernst & Young
Broadwalk House
Southernhay West
Exeter
EX1 1LF
UK

Mr Arthur Jones
Dept of Mech Engineering
University of Nottingham
14 Shakespeare Street
Nottingham
NG1 4FJ
UK

Mr Brian Honeyben
c/o Markham & Co
Broad Oaks Works
Chesterfield
Derbyshire
S41 0DS
UK

Dr David G Infield
Crest, Dept of Electronic &
Electrical Engineering
Loughborough Univ of Tech
Loughborough
LE11 3TU
UK

Mr Frans Janse
Wind Power Monthly
Burg Beststraat 2
1647 BC Berkhout
NETHERLAND

Mr Kevin Jenden
New World Power Co. Ltd
179 Great Portland Street
London
W1N 5FD
UK

Dr Nicholas Jenkins
Dept Electrical Engineering
UMIST (Ferranti Building)
PO Box 88
Manchester
M60 1QD
UK

Mr John Jeremy
10 The Lorne
Great Bookham
Surrey
KT23 4JZ
UK

Mr Charles Johnston
Scottish Hydro-Electric
Blackfriars
Perth
PH1 5LT
UK

Mr Gareth Jones
PowerGen plc
Rheidol Power Station
Cwm Rheidol
Aberystwyth, Dyfed
SY23 3NF
UK

Ir Peter Joosse
Stork Product Engineering
P O Box 379
Amsterdam
NETHERLANDS 1000 AJ

Mr Peter Thoft Knudsen
Kraftvaerksvej
53 DK-Fredericia
DENMARK DK-7000

Mr John Kuhns
New World Power Co. Ltd
179 Great Portland Street
London
W1n 5FD
Uk

Mr Lars Landberg
Meteorology & Wind Energy
Risoe National Laboratory
DK 4000 Roskilde
DENMARK

Mr Simon Lawrence
Quay House
Paterson Street
Lochgilphead
Argyll
PA31 8JP
UK

Mrs Jackie Leach
c/o SEEBOARD plc
Grand Avenue
Hove
East Sussex
BN3 2LS
UK

Mr B Clarke Lees
New World Power Co. Ltd
179 Great Portland Street
London
W1N 5FD
Uk

Dr W E Leithead
Industrial Control Unit
University of Strathclyde
Marland House, 50 George St
GLASGOW
G11 1QE
UK

Mr Timothy Kirby
EcoGen Ltd
Papyrus, Pentrerhedyn Street
Machynlleth
Powys
SY20 8DJ
UK

Mr Jan Kristiansen
Kraftvaeksvej 53
DK-Fredericia
DENMARK DK-7000

Mr Philip Ashton Lacey
General Manager
Shoreham Port Authority
84/86 Albion Street
Southwick, Brighton
BN42 4ED
UK

Ms Charmian D Larke
Atlantic Energy Ltd
1 Riverside House
Heron Way, Newham
Truro
TR1 3XN
UK

Sir Henry Lawson-Tancred
Aldborough Manor
Aldborough
BOROUGHBRIDGE
North Yorkshire
YO5 9EP
UK

Mr Edward Martin Leeming
73 Goonown
St Agnes
Cornwall
TR5 0XG
UK

Dr Mark L Legerton
ETSU
Harwell Laboratory
Oxfordshire
OX11 0RA
UK

Mr David J Leivesley
Wind Energy Group Ltd
Greenford House
309 Ruislip Road East
Greenford, Middlesex
UB6 9BL
UK

Dr David Lindley
National Wind Power Ltd
Riverside House
Meadowbank, Furlong Road
Bourne End, Bucks
SL8 5AJ
UK

Ms Carolyn Lord
Hammond Suddards Solicitors
2 Park Lane
Leeds
LS3 1ES
UK

Prof M V Lowson
Alpenfels
North Road
Leighwoods
Bristol
BS8 3PJ
UK

Mr Sam Macdonald
Gorstain
Taynuilt
Argyll
PA35 1JS
UK

Mrs Sheena Mackenzie
NWTC
NEL
East Kilbride
GLASGOW
G75 0QU
UK

Dr John R Maguire
Lloyd's Register
29 Wellesley Rd
Croydon
Surrey
CR0 2AJ
UK

Mr A Marmont
West Beacon Farm
Dean's Lane
Woodhouse Eaves
Loughborough
LE12 8TE
UK

Ms Janice Massey
Wind Power Monthly
Lyndale
Court Lane
Hadlow, Kent
TN11 0DS
UK

Prof Norman H Lipman
Energy Research Unit (R63)
Rutherford Appleton Lab
Chilton, Didcot
Oxon
OX11 0QX
UK

Mr Peter Lowry
TNI Tyne Brewery
Gallowgate
Newcastle-upon-Tyne
NE99 1RA
UK

Ms Morag MacCormick
Dulas Engineering
The Old School
Eglwysfach, Machynlleth
Powys
SY20 8SX
UK

Mrs Evelyn Macdonald
Gorstain
Taynuilt
Argyll
PA35 1JS
UK

Mr Birger Madsen
Association of Danish Windmill
    Manufacturers
Lykkesvej 18
DK-7400 Herning
DENMARK

Mr Christopher Mansfield
PowerGen plc
Electricity Enterprises
Haslucks Green Road, Shirley
Solihull, West Midlands
B90 4PD
UK

Mr I M Martin
Fairey Hydraulics Ltd
Claverham
Avon
BS19 4NF
UK

Mr Richard Masterman
Windelectric Ltd
Deli
Delabole
Cornwall
PL33 9BZ
UK

Dr Rayner M Mayer
Sciotec
9 Heathwood Close
Yateley
Camberley, Surrey
UK

Prof Jon G McGowan
Mechanical Engineering Dept
Univ of Massachusetts
Amherst
MASS 01003
USA

Dr Conor McMahon
ESBI
18/22 St Stephens Green
Dublin 2
EIRE

Ms Sara Metcalfe
Nicholas Pearson Associates Ltd
22 Gay Street
Bath
BA1 2PD
UK

Mr David J Milborrow
21 Church Road
Broadbridge Heath
Horsham
West Sussex
RH12 3LD
UK

Mr Vincent Moeyersoms
2 Newton Executive Park
Newton Laver Falls
MA 02162
USA

Mr Chris Morgan
Renewable Energy Systems Ltd
Eaton Court
Maylands Avenue
Hemel Hempstead, Herts
HP2 7TR
UK

Mr Niels Mortensen
Riso National Laboratory
P O Box 49
DK-4000 Roskilde
DENMARK

Dr I D Mays
Renewable Energy Systems Ltd
Eaton Court
Maylands Avenue
Hemel Hempstead, Herts
HP2 7TR
UK

Mr Andrew McKenzie
83 Ash Tree Road
Bitterne Park
Southampton
SO2 4NA
UK

Mr Robert Meir
Energy Technology Division
Department of Energy
1 Palace Street
LONDON
SW1E 5HE
UK

Dr Victor Middleton
Dept of Mech Engineering
University of Nottingham
14 Shakespeare Street
Nottingham
NG1 4FJ
UK

Ms Catherine Mitchell
Brook Farm
Launde
Leicestershire
LE7 9DF
UK

Mr Blake Moore
396 Kings Road
London
SW10 0LL
UK

Dr W J Morris
Dept. of Civil Engineering
City University
Northampton Square
London
EC1V 0HB
UK

Mr Jocelyn Mortimer
Ainderby Hall
Ainderby Steeple
Northallerton
DL7 9QJ
UK

Mr Alan Mortimer
Scottish Power Technology
45-47 Hawbank Road
College Milton North
East Kilbride
G74 5EG
UK

Dr Pat Murphy
ESBI
18/22 St Stephens Green
Dublin 2
EIRE

Dr P J Musgrove
National Wind Power Ltd
Riverside House
Meadowbank, Furlong Road
Bourne End, Bucks
SL8 5AJ
UK

Mr Tsuneo Nakano
Mitsubishi Heavy Ind. Europe
Bow Bells House
Bread St
London
EC4M 9BQ
UK

Mr Guy Nicholson
EcoGen Ltd
Hamilton House
Main Street, Acomb
Northumberland
NE46 4PT
UK

Mr Per Norgaard
Riso National Laboratory
P O Box 49
DK-4000 Roskilde
DENMARK

Mr Hideo Ogata
NEDO
Sunshine 60 29F
1-1 Higashi-Ikebukuro
3-chome, Toshima-ku, Tokyo
170
JAPAN

Dr D I Page
ETSU , Building 156.7
Harwell Laboratory
Didcot
Oxon
OX11 ORA
UK

Dr Edwin Mowforth
Dept of Mech Engineering
University of Surrey
Guildford
Surrey
GU2 5XH
UK

Mr Ross Murray
Newbridge Construction Ltd
23a Gold Tops
Newport
Gwent
NP9 4UL
UK

Mr Yoichi Nakamura
Mitsubishi Research Institute
Time & Life Building
3-6 Otemachi 2-chome
Chiyoda-ku, Tokyo
100
JAPAN

Mr Paul Newman
Semikron Ltd
4 Marshgate Drive
Hertford
Herts
SG13 7BO
UK

Mr Jack Noakes
Renewable Energy Systems Ltd
Eaton Court
Maylands Avenue
Hemel Hempstead, Herts.
HP2 7TR
UK

Mr Patrick J O'Neill
East Midlands Electricity
Generation Division
Caythorpe Road, Caythorpe
Nottingham
NG14 7EB
UK

Dr David A Olivieri
Department of Engineering
Simon Building
University of Manchester
Manchester
M13 9PL
UK

Mr Ian Parker
East Midlands Electricity
Generation Division
Caythorpe Road, Caythorpe
Nottingham
NG14 7EB
UK

Mr Bob Paul
New World Power Co. Ltd
179 Great Portland Street
London
W1N 5FD
UK

Miss Catherine Peasley
Rose Cottage
Machynlleth
Powys
SY20 8TN
UK

Mr Keith Pitcher
3 Greenhow Park
Burley-in-Wharfedale
West Yorkshire
LS29 7LZ
UK

Dr S J R Powles
3 St Helena
The Graig
Burry Port
DYFED
SA16 0ED
UK

Mr Peter Quilleash
Renewable Energy Systems Ltd
Eaton Court
Maylands Avenue
Hemel Hempstead, Herts
HP2 7TR
UK

Mr Peter J Radmall
Environmental Resources Ltd
106 Gloucester Place
London
W1H 3DB
UK

Mr James Richards
Seawest (UK) Ltd
Sixth Floor, Tomen House
13 Charles II Street
London
SW1Y 4QT
UK

Mr Bill Richmond
The Lawn
Brightwalton
Newbury
Berks
RG16 0BP
UK

Mr Robert J H Paynter
R63, Rutherford Appleton Lab
Chilton
Didcot
Oxon
OX11 0QX
UK

Mr Nigel J Phillips
Newbridge Construction Ltd
23a Gold Tops
Newport
Gwent
NP9 4UL
Uk

Mr Rowland Pound
HGA Consulting Engineers
Halcrow Gilbert Associates
Byderop Park
Swindon, Wilts
SN4 0QD
UK

Mr Pep Prats
Demostenes 6
08028 Barcelona
SPAIN

Mr Chris P Quine
Forestry Commission
Northern Research Station
Roslin
Midlothian
EH25 9SY
UK

Mr Martin A Read
East Midlands Electricity
Generation Division
Caythorpe Road, Caythorpe
Nottingham
NG14 7EB
UK

Mrs Barbara Richardson
The Met Office
Sutton House
London Road
Bracknell, Berks.
RG12 2SY
UK

Mr Jean-Baptiste Richon
Fluids Group, Dept of Physics
R2701, JCMB, Univ Edinburgh
Mayfield Road
Edinburgh
EH9 3JZ
UK

Mr J C Riddell
Occaney Grange
Copgrove
Harrogate
North Yorkshire
HG3 3TD
UK

Mr Glenn Robinson
Carter Wind Turbines Ltd
Beaufort Suite
Lockington Hall
Lockington, Derby
DE74 2RH
UK

Mr Barry Rodwell
Trinity House Lighthouse Serv
East Cowes
Isle of Wight
PO32 6RE
UK

Mrs Mary Rogers
Industrial Control Centre
50 George Street
Strathclyde University
Glasgow
G1 1QE
UK

Dr Alan Ruddell
Rutherford Appleton Lab
Energy Research Unit
Chilton, Didcot
Oxon
OX11 0QX
UK

Mr Zouhir Saad-Saoud
Ferranti Building
UMIST
P O Box 88
Manchester
M60 1QD
UK

Mr N H Scott-Barrett
Hambros Bank Ltd
41 Tower Hill
London
EC3N 4HA
UK

Dr Peter Sharpe
Bristol Transputer Centre
UWE
Coldharbour Lane
Frenchay, Bristol
BS16 1QY
UK

Ms Maria Rivera
23 Avenue de Lamballe
75016 Paris
FRANCE

Dr A J Robotham
MEMS, Coventry Polytechnic
Coventry University
Priory Street
Coventry
CV1 5FB
UK

Mr Peter Rogers
Faculty of Design & Technology
Luton College of Higher Education
Park Square
Luton
LU1 3JU
UK

Mr Torben Ronnow
Bonus Energy A/S
Fabriksvej 4
DK-7330 Brande
DENMARK

Mr Ian Rutherford
Hawkswick Lodge
Childwickbury
St Albans
AL3 6JG
UK

Mr Haydn Scholes
CSM Associates Ltd
Rose Manowes
Herniss, Penryn
Cornwall
TR10 9DU
UK

Dr Ezio Sesto
ENEL S.P.A./CRE
Via Volta 1
Cologno
Monzese
20093
ITALY

Dr Giora Shatil
24 Sturdon Rd
Ashton
Bristol
BS3 2BA
UK

Mr P B Simpson
Wind Energy Group Ltd
Greenford House
309 Ruislip Road East
Greenford, Middlesex
UB6 9BL
UK

Mr Huw P F Smallwood
'Bron Craig'
Llangwm
Corwen
Clwyd
LL21 0RL
UK

Prof Edward Spooner
School of Eng & Comp Science
University of Durham
South Road
Durham
DH1 3LE
UK

Mr David Still
Border Wind Ltd
6 Station Cottages
Hexham
Northumberland
NE46 1EX
UK

Mr Ewan Stott
Generation Wholesale Division
Cathcart House
Stean Street
Glasgow
G44 4BE
UK

Dr Philip L Surman
3 Queen's Road
Harrogate
N Yorks
HG2 0HE
UK

Dr D T Swift-Hook
Swift-Hook Associates
Bourne Place, Horsell Common
Horsell
Surrey
GU21 4XX
UK

Mr C C Tan
Renewable Energy Systems Ltd
Eaton Court
Maylands Avenue
Hemel Hempstead, Herts.
HP2 7TR
UK

Dr David Skyner
Dept of Physics
University of Edinburgh
Mayfield Road
Edinburgh
EH9 3JZ
UK

Dr Gillian M Smith
Dept Mechanical Engineering
University of Nottingham
University Park
Nottingham
NG7 2RD
UK

Mr Henrik Stiesdal
Bonus Energy a/s
Fabriksvej 4
DK-7330 Brande
DENMARK

Mr Chris H Stockton
Yorkshire Electricity
  Generation Division
Wetherby Road, Scarcroft
Leeds
LS14 3HS
UK

Mr Shigeru Sudo
Jetro
Leconfield House
Curzon Street
London
W1Y 7FB
UK

Mr Gerald J Swarbrick
SWEB
800 Park Avenue
Aztec West
Almondsbury, Bristol
BS8 3RZ
UK

Mr Sohail Syed
Fairey Hydraulics Ltd
Claverham
Bristol
BS19 4NF
UK

Dr Derek A Taylor
Energy & Environment Research
Open University
Walton Hall
MILTON KEYNES
MK7 6AA
UK

Mr G J Taylor
5 Thornton Close
Guildford
Surrey
GU2 6KE
UK

Mr Bill Thomas
Maesymilwr
1 Penllwyn Park
Carmarthen
Dyfed
SA31 3BU
UK

Ms Emily Tomalin
EcoGen Ltd
Papyrus
Pentrerhedyn St
Machynlleth, Powys
SY20 8DJ
UK

Mr Keichi Tsuchiya
Tohoku Electric Power Co.
2-1 Nakayama 7-chome
Aoba-ku
Sendai
981
JAPAN

Mr Yukinobu Uchida
c/o Tomen Power Corp (UK) Ltd
13 Charles II Street
London
SW1Y 4QT
UK

Mr John VandenBosche
Beechwood Bungalow
Cwmynysminton Road
Llwydcoed
Aberdare
CF44 0UP
UK

Mr Dale Vince
Partfield Farm
Tinckley Lane
Nympsfield
Stonehouse, Glos
GL10 3UJ
UK

Mr Steve Wade
Wind & Sun
The Howe
WATLINGTON
OX9 5EX
UK

Mr Jeffrey Thomas
Westwind Generators Ltd
75 Westbourne Road
Penarth
South Glamorgan
CF64 3HD
UK

Mr Jozef Timmermans
Eickhoff (GB) Ltd
Mining Machinery
10 Carver Street
Sheffield
S1 4FS
UK

Mr Marcus Trinick
Bond Pearce Solicitors
1 The Crescent
Plymouth
PL1 3AE
UK

Prof. John W Twidell
AMSET Centre
Sch of Engineering & Manufacture
De Monfort University
Leicester
LE1 9BH
UK

Dr William Unkel
442 Marrett Road
Suite 9
Lexington
MA 02173
USA

Mr Andrew Vaudin
National Wind Power Ltd
Riverside House, Meadowbank
Furlong Road, Bourne End
Buckinghamshire
SL8 5AJ
UK

Dr Spyridon Voutsinas
NTUA-FS
P O Box 64070
15710 Zografou
Athens
GREECE

Mr Richard Wade-Smith
Hammond Suddards Solicitors
2 Park Lane
Leeds
LS3 1ES
UK

Ms Sue M Walker
Llwyn Melyn
Llandinam
Powys
Wales
SY17 5AZ
UK

Mr Ian Ward
W S Atkins
Woodcote Grove
Ashley Rd
Epsom, Surrey
KT18 5BW
UK

Dr Peter Watkinson
GEC Marconi Underwater System
Wilkinthroop House
Templecombe
Somerset
BA8 0DH
UK

Dr Goeff Watson
Manxwind Energy Services
1 Church Road
Port Erin
ISLE OF MAN
UK

Dr Simon J Watson
R63, Rutherford Appleton Lab
Chilton
Didcot
Oxon
OX11 0QX
UK

Mr Nicholas Wedgwood
43 Netherford Road
London
SW4 6AF
UK

Mr Harvey West
Welsh Water Enterprises
Beacon House
Llantarnam Park
Cwmbran, Gwent
NP44 3AB
UK

Mr W J Wilkinson
Yorkshire Electricity
  Generation Division
Wetherby Road, Scarcroft
Leeds
LS14 3HS
UK

Dr J M Ward
ETSU
Harwell Laboratory
Oxon
OX11 0RA
UK

Dr John G Warren
National Wind Power Ltd
Riverside House, Meadowbank
Furlong Road, Bourne End
Buckinghamshire
SL8 5AJ
UK

Dr Lewis Watson
Royal Military Col of Science
CISM Group
Shrivenham
Swindon
SN6 8LA
UK

Mr James Watson
Moat Farmhouse
High Street
Swineshead
Bedfordshire
MK44 2AA
UK

Mr Arthur Waud
Yorkshire Metal Fabrications
110 Hunslet Lane
Leeds
W Yorkshire
LS10 1ES
UK

Ms Fiona Weightman
Friends of the Earth
26/28 Underwood Street
London
N1 7JQ
UK

Mr Patrick H Whitworth
The Mount
Glenferness
Nairn
UK

Mr Brian Williams
Technical Services Dept
Humberside County Council
County Hall
Beverley
HU17 9XA
UK

Miss Angela Willis
Yorkshire Water
P O Box 500, Western House
Halifax Road
Bradford
BD6 2LZ
UK

Dr Donald M A Wilson
Cavendish Laboratory
Madingley Road
CAMBRIDGE
CB3 OHE
UK

Mr Michael Wolff
The Dower House
Lethen
Nairn
IV12 5PR
UK

Mr W Wreglesworth
Dunlop Ltd
Evington Valley Road
Leicester
LE5 5LY
UK

Ir Don van Delft
Delft Univ. of Technology
Faculty of Civil Engineering
Stevinweg 1
2628 CN Delft
NETHERLANDS

Dr John A B Wills
BMT Fluid Mechanics
Orlando House
1 Waldegrave House
Teddington, Middx
TW11 8LZ
UK

Mr Aloys Wobben
Enercon
Export Department
Schwachhauser Heerstr
299 D-28211 Bremen
GERMAY

Mr David Bruce Woodnorth
21 Mount Nod Road
Streatham Hill
London
SW16 2LQ
UK

Mr Sayoshi Yamada
Tohoku Electric Co.
2-1 Nakayama 7-chome
Aoba-ku, Sendai 981
JAPAN